SURFACE ENGINEERING

SURFACE ENGINEERING

— *Editors* —
D. Srinivasa Rao
Shrikant V. Joshi

CENTRE FOR SCIENCE & TECHNOLOGY OF THE
NON-ALIGNED AND OTHER DEVELOPING COUNTRIES
(NAM S&T CENTRE)

2010
DAYA PUBLISHING HOUSE
Delhi - 110 035

ISBN 81-7035-628-8
ISBN 978-81-7035-628-8

Centre for Science and Technology of the Non-Aligned and Other Developing Countries (NAM S&T Centre)
Core-6A, 2nd Floor, India Habitat Centre, Lodhi Road,
New Delhi-110 003 (India)
Phone: +91-11-24644974, 24645134, Fax: +91-11-24644973
E-mail: namstct@gmail.com
Website: www.namstct.org

Published by : **Daya Publishing House**
1123/74, Deva Ram Park, Tri Nagar, Delhi - 110 035
Phone: 27383999, Fax: (011) 23260116
e-mail: dayabooks@vsnl.com
website: www.dayabooks.com
Showroom : 4760-61/23, Ansari Road, Darya Ganj
New Delhi - 110 002
Phone: 23245578, 23244987

Laser Typesetting : **Classic Computer Services**
Delhi - 110 035

Printed at : **Chawla Offset Printers**
Delhi - 110 052

Printed in India

Foreword

The concept of incorporating engineered surfaces, capable of combating the often encountered degradation phenomena like wear, corrosion, oxidation, high heat load etc., to improve component performance, reliability and life has gained increasing acceptance in recent years. The recognition that a vast majority of engineering components fail catastrophically in service through surface related phenomena such as above has further fueled interest in this approach and led to the development of the broad interdisciplinary subject of Surface Engineering. A protective coating deposited to act as a barrier between the surface of the component and the aggressive environment that it is exposed to during routine operation is now globally acknowledged to be an attractive means to significantly reduce/suppress damage to the actual component by acting as the first line of defence.

The increasing utility and industrial adoption of surface engineering is a consequence of the significant recent advances in the field. Very rapid strides have been made on many diverse fronts–science, processing, controls, modelling, application development etc.–and this has been primarily responsible for surface engineering today being valued as an invaluable tool now increasingly considered as being an integral part of component design. The newer surfacing techniques, along with the traditional ones, are eminently suited to modify a wide range of engineering properties. The properties that can be modified by adopting the surface engineering approach include tribological, mechanical, thermomechanical, electrochemical, optical, electrical, electronic, magnetic/acoustic and biocompatible properties. By virtue of the versatility of the approach, we now routinely encounter products whose surfaces have been modified for functional or decorative reasons and there is also growing realization that, if appropriately harnessed, surface engineering can not only solve component degradation problems but also provide added value.

Notwithstanding the fact that surface engineering is a discipline of science and technology which is being increasingly relied upon to meet most modern-day technological requirements of materials savings, improved efficiencies, environmental friendliness etc., the subject is yet to find a place in the academic curriculum in most countries. The awareness of the wide ranging benefits that can accrue through use of suitably engineered coatings is also lacking in most industry segments. The lack of knowledge of the field to promote its use and dearth of human resources to drive its implementation are particularly acute in the developing world and this needs to be urgently addressed.

Although the effort to compile the contents of this book began with the limited intent of providing course material to the participants of a two-week school on surface engineering, I truly believe that this publication can benefit a much wider cross-section of readers. In particular, it will provide a valuable reference resource for students and researchers in the field of surface engineering who wish to be better informed about the vast portfolio of surface modification technologies available today to the designer to enhance performance and augment durability of parts performing in aggressive environments. It will also be a useful guide to practising engineers responsible for designing engineering components.

G. Sundararajan

Director,

International Adavnced Research Centre for Powder Metallurgy and New Materials,

Hyderabad, India

Preface

This publication is an outcome of an *International Training Course and Business Opportunities Workshop on Surface Engineering* organized by the International Advanced Research Centre for Powder Metallurgy and New Materials (ARCI) in collaboration with the NAM S&T Centre, with financial assistance from the Science and Engineering Research Council (SERC), under the Government of India's Department of Science and Technology (DST). SERC has been regularly funding schools targeting frontier areas of technology, with the primary purpose of creating awareness among young researchers and encouraging them to pursue purposeful R&D activities in an emergent area of science and technology. Human resource development through generation of specialized manpower is another important goal achieved by organizing schools of this kind. Surface Engineering being a field of growing relevance in the modern industrial world, it was proposed by the NAM S&T Centre that it would be meaningful to organize a thematic workshop on this rapidly evolving topic for the benefit of young researchers both in India and the NAM countries. The enthusiastic response to the event clearly demonstrated that the idea was most timely as well as apt.

Notwithstanding its rapid industrial acceptance globally, the surface modification approach has not been utilized by the industries in the developing countries to its full potential and R&D efforts in this area are also far short of what this important field demands. This provided the main motivation for conducting the above event with the following broad objectives:

- Provide an overview of the concept of Surface Engineering and its all-expansive domain of applicability
- Review the wide spectrum of surface modification techniques available to the designer and the individual advantages and limitations of each technology
- Present procedures and techniques adopted for characterizing coatings and evaluating their performance

- Outline current state-of-the-art for each technology and identify gaps in technology, which could constitute meaningful research efforts
- Familiarize participants with coating procedures and characterization and performance evaluation techniques through laboratory sessions
- Acquaint participants with industrial scale applications by organizing industry visits

While the initial intent was to request all the faculty members, carefully drawn from the academia, research institutions and industry, to provide comprehensive course-notes for the benefit of the Workshop participants, it soon became apparent that a special volume addressing the current state-of-the-art in the growing field of surface modification technologies would have considerable archival value. As Editors of this compilation, we would like to particularly thank all the faculty members who readily contributed technical chapters and country reports to make this book possible and the NAM S&T Centre for the initiative taken to publish it. We sincerely hope that this book would serve as a useful resource for the surface engineering community.

D. Srinivasa Rao

Shrikant V. Joshi

Introduction

Surface Engineering is a generic term applied to a variety of diverse technologies that can be gainfully harnessed to achieve increased reliability and enhanced performance of degradation-prone industrial components. Typically, surface engineering encompasses coating and surface treatment processes that are capable of modifying surface properties of critical components to provide improved resistance against deterioration due to mechanisms such as corrosion, oxidation, wear or failure under an excessive heat load. The popular surface technologies include electro-deposited coatings, diffusion coatings, thermal sprayed coatings, sol-gel coatings, plasma and laser assisted surface modification etc. Several new variants of these technologies are also becoming increasingly available. Considerable advances in the field of coatings technology in recent years have made it possible to realize the immense potential of surface engineering in diverse industry segments. Although Surface Engineering is of growing relevance in the modern industrial world, the industry in the developing countries is yet to exploit the state-of-the-art surface modification technologies to their full potential and ongoing R&D efforts in this area are also far short of what this important field demands.

Motivated by the above, the Centre for Science and Technology of the Non-Aligned and other Developing Countries (NAM S&T Centre) organised an 8-day International Training Course–cum–Business Opportunities Workshop on Surface Engineering at Hyderabad, India during 19-26 July 2005 jointly with the International Advanced Research Centre for Powder Metallurgy and New Materials (ARCI), Hyderabad. Prof. P. Rama Rao, ISRO Dr. Brahm Prakash Distinguished Professor and former Science and Technology Secretary to the Government of India inaugurated this event, which was attended by 42 specialists from 6 countries to have an overview of the basic concept of Surface Engineering and the all-expansive domain of the applicability of the same and get acquainted with a broad spectrum of current state-of-the-art surface modification technologies that are available for use by the industry. 19 lectures were delivered by the resource persons drawn from premier institutions of excellence, which enabled the participants to have a general understanding on the principles and usefulness of different types of surface processing technologies. Visits of the participants to the industries in the adjacent Technology Park as well as to GE-BHEL, a joint venture company between General Electric and Bharat Heavy Electricals Limited (BHEL) and BHEL R&D Centre, to acquaint them with the industrial scale applications of surface modification technologies were also organized.

The present publication is a follow up on the above Hyderabad Training Course-cum-Workshop and comprises a compilation of the presentations made during this scientific event as also the scientific articles of some eminent scientists actively working in this field. I am particularly indebted to Dr. Srikant Joshi and Mr. D. Srinivasa Rao for undertaking to edit this publication and acknowledge with thanks the kind words from Prof. G. Sundararajan, Director, ARCI, Hyderabad, about this publication. Last, but not the least, I acknowledge the dynamic involvement of the entire team of the NAM S&T Centre, in particular, Mr. M. Bandyopadhyay, Dr. V.P. Kharbanda, Mr. Gaurav Gaur and Mr. Pankaj Buttan in giving a shape to the volume.

The publication establishes a distinct relation of surface engineering advancement with industrial development and resultant economic gains and prosperity of the masses. I sincerely trust that the information present herein will catalyze the awareness on this subject in the minds of policy makers, technology managers and industries in the developing countries.

Arun P. Kulshreshtha

Director, NAM S&T Centre

Contents

Country Reports

Chapter 1
Diffusion Coatings

D.K. Das
Defence Metallurgical Research Laboratory,
Kanchanbagh, Hyderabad – 500 058, India
E-mail: dkdas@dmrl.drdo.in

ABSTRACT

The present paper describes the basic features of diffusion coatings. Subsequently, several industrially important diffusion coatings such as carburized and aluminized coating have been discussed in some detail encompassing the aspects such as fundamentals of coating formation, processing methods, coating microstructure and application of these coatings.

Keywords: Engineering component surface, Surface engineering, Surface modification, Diffusion coatings.

Introduction

A major challenge in technological development has been the need to continuously meet the stringent materials requirements for use in progressively demanding conditions. Exposure to increasingly aggressive environments has made modern engineering components more susceptible to rapid degradation due to interactions between the component surface and the environment. In the material-environment configuration, the surface of a component serves as the first line of defence against attack by an adverse environment. Therefore, the component surface is always of vital importance in determining its performance and durability. This has formed the basis for improving materials properties by adopting the surface modification approach. This approach usually involves the formation of an appropriate protective coating on the component surface to impart the desired property to combat premature degradation of the component. Generally, the surface modification approach, which is also referred to as "surface engineering" enables achievement of properties that neither the bulk material nor the coating is capable of imparting on its own.

With the present day availability of a wide spectrum of surface modification technologies, the surface of a component can be virtually "tailored" by selecting a proper surfacing method in order to meet the required properties without compromising the mechanical characteristics of the bulk material. Surface engineering, however, is not confined merely to improvement in component properties against surface-related degradation processes such as fatigue, corrosion, wear etc. This approach can also be applied for cases where specialized surface properties that are not necessarily related to surface protection and/or performance enhancement of the component, are desired as exemplified by the category of abradable coatings for aero-engine applications which are sacrificial in nature. Surface modification methods also prove invaluable for saving/substituting valuable and scarce bulk materials, and for reclaiming damaged components. Furthermore, this approach provides the designers considerable flexibility in use of materials. All the above mentioned benefits of surface modification methods translate into significant cost effectiveness. Table 1.1 provides a generalized list of various materials properties that can be modified by the application of coatings along with some typical examples. The enormous application potential of the surface engineering is amply clear from this table. Numerous surface modification techniques have been developed over the years and the designer today has the advantage of choosing from a broad array of surfacing technologies that offer a wide range of quality and cost.

Table 1.1: Range of Surface Properties Modifiable by Application of Coatings

Property	*Details*	*Examples*
Tribological	Sliding wear, abrasive wear, erosive wear, fretting wear, friction	Al_2O_3, Ni-Cr self-fluxing alloys, WC-Co, stellite, Cr_2C_3-Ni-Cr alloys, high carbon iron, TiO_2, Cr_2O_3, Cr_2O_3-TiO_2 blend, Al_2O_3-TiO_2 composite, carburized coatings, nitrided coatings
Thermal	Oxidation, thermal insulation	Aluminide coatings, MCrAlY coatings, stabilized zirconia coatings
Electrochemical	Aqueous corrosion, hot corrosion	Chrome plating, phosphatic coatings, MCrAlY coatings, diffusion aluminide coatings
Optical	Refection, absorption, refraction	Gold coating on laser mirrors, antiglare coatings on spectacles
Electrical	Conductivity, breakdown potential	Gold plating on printed circuit board
Electronic	Electron mobility	Gold plating
Bio-compatibility	Tissue growth propensity	Hydroxyapatite coatings

Coatings are generally categorized into two major groups based on the nature of their interaction/association with the substrate. They are: (*a*) diffusion coatings and (*b*) overlay coatings. Formation of diffusion coatings involves chemical interaction by diffusion of the coating-forming element(s) with the substrate. In the case of overlay coatings, however, coating remains as a distinctly separate entity from the substrate in the as-coated condition. Carburized, nitrided, boronized and aluminized coatings obtained on steel/Ni-base alloys are some of the examples of diffusion coatings. In the overlay category, spray-deposited, electroplated, brush-coated and physical/chemical vapour deposited coatings can be cited as examples. Both these categories of coatings have their inherent advantages and disadvantages, and also their exclusive application areas. Often these coatings also compete with each other for similar applications. In the present paper, which is introductory in nature, describes the basic features of diffusion coatings. Subsequently, several industrially important diffusion coatings such as carburized and aluminized coating have been discussed in some detail encompassing the aspects such as fundamentals of coating formation, processing methods, coating microstructure and application of these coatings.

Salient Features of Diffusion Coatings

Differences between Diffusion and Overlay Coatings

As stated above, coatings in general fall into two major categories based on the nature of their interaction/association with the substrate. They are: (*a*) diffusion coatings and (*b*) overlay coatings. The formation of diffusion coatings involves chemical interaction by diffusion of the coating-forming element(s) with the substrate. However, virtually no such interaction occurs in the case of overlay coatings. Consequently, while an overlay coating remains as a distinctly separate entity from the substrate in the as-coated condition, a diffusion coating becomes an integral part of the substrate. The adherence of overlay coating with the substrate is primarily mechanical in nature. Because of the this reason, the coating adherence to the substrate is considerably inferior in case of overlay coatings as compared to the diffusion coatings. The two categories of coatings have been schematically represented in Figure 1.1. Carburized, nitrided, boronized and aluminized coatings obtained on steel/Ni-base alloys by pack cementation methods are some of the examples of diffusion coatings. In the overlay category, spray-deposited, electroplated, brush-coated and physical/chemical vapour deposited coatings can be cited as examples.

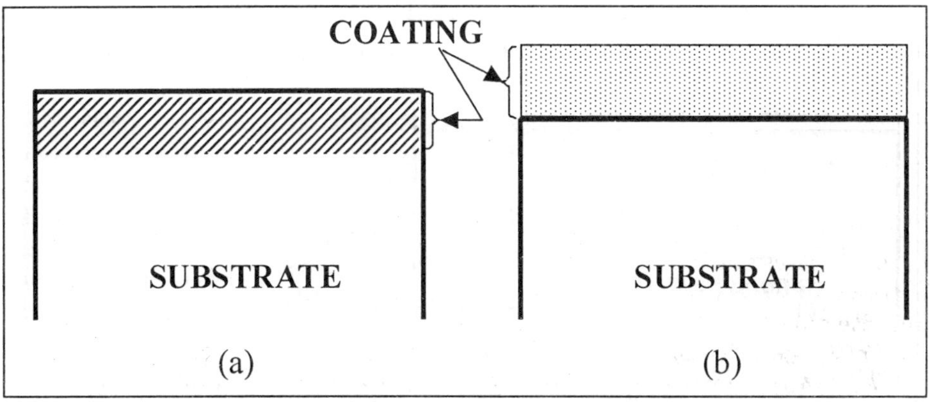

Figure 1.1: Schematic Representation of (a) A diffusion coating and (b) An overlay coating

Coating Forming Elements

Diffusion coatings are usually employed in two broad application areas: (*a*) tribological and (*b*) environment protection. Improvement in tribological properties such as wear requires enhancement of surface hardness which can be achieved by application of carburized, nitrided and borided coatings. These three diffusion coatings are produced by introducing carbon, nitrogen and boron, respectively. Primarily, steels are surface-hardened by application of the carburized, nitrided and boronized coatings. Some of these coatings also improve other important properties of steel like fatigue, anti-galling, hot hardness etc. For improvement of environment properties such as oxidation and corrosion resistance, Al, Cr and Si are added to substrate surface. The resultant coatings are referred to as aluminized (or aluminide), chromized and siliconized coatings, respectively. These coatings are usually applied on steels, Ni and Co-base super alloys, and Ti-base alloys. Often, more than one element are added for the formation of diffusion coatings of improved performance. For instance, carbon and nitrogen can be added together to steel surface to form nitro-carburized or carbonitrided coatings. Whether the coating is to be termed as nitro-carburized or carbonitrided coating depends on which of the two is the primary alloying element in the coating. Similar to the above coatings, chrom-aluminized, silicon-aluminized type of oxidation/corrosion resistant coatings can also be formed.

Apart from coating forming elements, certain other elements called as "modifiers" are also added to diffusion coatings for enhancing their properties. The resultant coatings are called modified coatings. Some of the major modifying elements are Pt, Hf, Y, Rh, Ta. In recent times, elements such as Ru and Pd are also being evaluated as modifiers for aluminide coatings.

Basic Mechanism of Coating Formation

Diffusion coatings are formed by externally adding one more coating forming elements onto the surface of the job to be coated. The added elements diffuse into the substrate and forms a coating by chemically interacting with the substrate atoms. Coating formation is carried out in a medium rich in coating forming elements. These elements are transported from the medium onto the surface of the specimen as active atoms and get absorbed. Although the medium can be solid, liquid or gaseous, the actual transport of atoms always occurs in a gaseous state. The above mentioned mechanism of formation of diffusion coatings can be schematically described as shown in Figure 1.2. The substrate is completely immersed in the medium containing the elements needed for coating formation. On increasing the temperature of the medium, suitable vapour species containing the these elements form. When the vapour species come in contact with the substrate surface, they dissociate releasing the coating forming elements in active (atomic) form. These elements readily diffuse into the substrate where they form the coating after chemically reacting with the substrate atoms. Thus, the coating becomes a part of the substrate as mentioned previously [see Figure 1.1(a)]. It may be noted here diffusion of coating forming elements into the substrate is involved in case of formation of diffusion coatings. Therefore, these coatings almost always are formed at considerably high temperatures, typically several hundreds of degree centigrade.

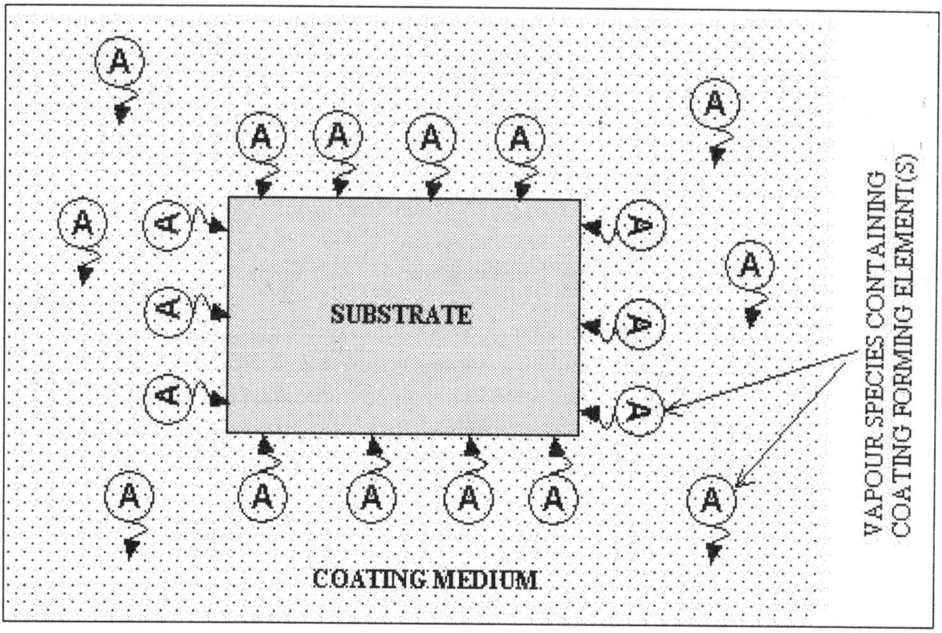

Figure 1.2: Schematic Representation of Formation of a Diffusion Coating

Coating Medium

As mentioned earlier, solid, liquid and gaseous media can be used for the formation of diffusion coatings. When a solid medium is used, the process is called pack cementation method. The pack, or 'cement' as it sometimes called, consists of three primary constituents: (*a*) the element source which can be in pure form or in form of an alloy, (*b*) an activator and (*c*) an inert filler. While the element source supplies the required element for coating formation, the activator facilitates the transport of coating forming element from the pack to the surface of the substrate. The activator usually reacts with the coating forming element to form vapour species which move to the surface of the substrate. The vapour species decompose on the surface of the work releasing nascent atoms of the coating forming element. These atoms readily diffuse into substrate to form coating. For example, during pack aluminization, the activator ammonium chloride (NH_4Cl) reacts with Al of the Al-source to form chlorides of Al such as $AlCl_3$ which are vapour at aluminizing temperature. On coming in contact with substrate surface, these vapour species decompose releasing nascent Al atoms which react with the substrate to form aluminide coating. The presence of inert filler like alumina in the pack serves two purposes. Firstly, it prevents sintering of the pack at the aluminizing temperature. Secondly, it acts as a dilutant and, thereby, controls the rates of various reactions that occur in the pack. In certain cases of pack cementation such as pack carburizing, inert filler is not used.

Diffusion coating formation can also be carried out in liquid and gaseous media. In these two media, the compounds containing the coating forming elements remain as a liquid or in gaseous state. At the processing temperature, various chemical reactions occur among the compounds leading to release of nascent element on substrate surface and the subsequent coating formation. There are certain advantages in adopting liquid/gaseous phase diffusion coating formation. The heat transfer between the medium and the substrate is much superior in liquid or gaseous medium as compared to that in a powder pack. Further, it is much easier to carry out post-coating treatment of the work such as quenching from liquid/gaseous medium than from a powder pack mixture. In the latter case, the removal of the job from the pack for post-coating treatment can be very cumbersome.

Advantages and Disadvantages

The biggest advantage of diffusion coatings is the cost effectiveness of the process. Usually, high levels of sophistication are not involved in the coating processes. Further, relatively inexpensive raw materials are used for coating formation and also large batches of components can be coated simultaneously. As a result, good process economy is achieved in case of diffusion coatings. Pack coating is suited to treat even large substrates, either singly or in bulk, and can handle intricate shapes as it is not a line-of-sight process.

One of the major drawbacks of diffusion coatings is the contribution of the associated processes to environmental pollution. Since hazardous gases such as ammonia and liquids such as cyanides are used for formation of coatings, always there remain chances of these gases/liquids to escape to the environment and cause pollution. Further, disposal of toxic and hazardous wastes also pose a big problem. In recent times, however, eco-friendly processes have been developed which are being increasingly adopted in order to lessen the impact of the coating processes on the environment. Relatively poor process control is another disadvantages of conventional diffusion coating processes. This aspect has been considerably improved over the by incorporating automation and sophisticated instrumentation to the processes.

Major Diffusion Coatings

In this section, major diffusion coatings have been discussed individually in terms of their process details, mechanism of coating formation, coating microstructure and typical applications. The coatings that have been considered include carburized, nitrided, boronized (borided), aluminized and other similar coatings. The general process by which the first three coatings, *viz.* carburized, nitrided and nitrided, are formed is referred to as case hardening since all these coatings lead to a hardened surface. These three coatings almost always are applied to steel surface. Aluminized and other similar coatings such as chromized, siliconized coatings are applied on both steels and superalloys and provide resistance against oxidation and corrosion damage. The various aspects of the diffusion coatings that have been discussed here are widely reported. Further knowledge on these coatings details can be obtained from refs. (Thelning, 1984; Leslie, 1981; Metals Handbook, 1981; Lakhtin, 1983).

Carburized Coatings

Carburized coating is formed by a process called case carburizing. This treatment leads to the formation of a hard surface layer or 'case', as it is often called, on low carbon steels which enhances their wear resistance significantly. Since this treatment does not affect the core which remains comparatively soft and tough, the component as a whole displays high impact strength. Due to the case hardening treatment, compressive stresses also often develop in the surface layers which result in increased fatigue strength of the steel. For attaining the hardened case, the steel is first carburized during which the carbon concentration of the surface layer is increased significantly. Subsequently the steel is hardened by quenching from the carburizing temperature.

Basic Principles of Hardening

The most important mechanism by which steels are hardened is by martensitic transformation. Hardening by this method involves first autenitizing the steel followed by quenching it. This process can be better appreciated from Figure 1.3 which shows the equilibrium Fe-C diagram (Thelning, 1984). For martensitic hardening, the steel is taken into the austenite (γ) region by heating it to the appropriate temperature above A_3 and holding it there for the requisite time period. Subsequently, the component is

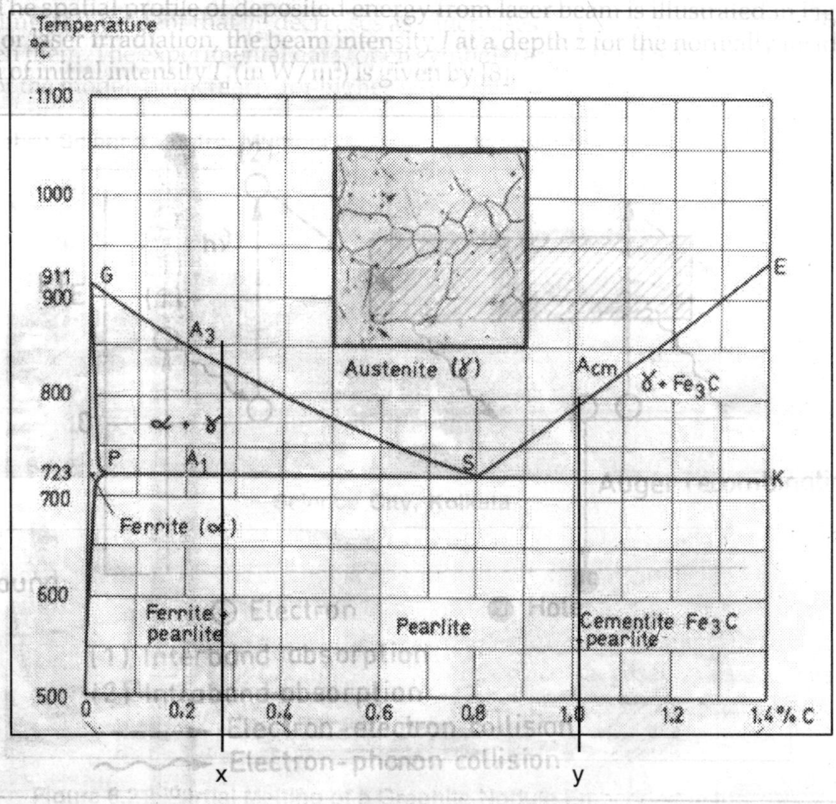

Figure 1.3: The Lower Left-hand Portion of Iron-carbon Equilibrium Diagram (1)

rapidly cooled by quenching in oil/water which leads to the transformation of γ phase to hard martensite phase. The hardness of the martensite phase is directly dependent on the C concentration of the parent γ phase which means that higher the C concentration of the austenite, harder would be the resultant martensite. This aspect can clearly be seen from Figure 1.4. (Leslie, 1981). This direct dependence of C concentration of parent austenite on the hardness of the resultant martensite is taken advantage of in case hardening of steels. It is evident from the above figure that martensite hardness decreases above approximately 1 wt. per cent C. Carbon being an austenite stabilizer, its content beyond the above value leads to retention of significant quantity of untransformed austenite (also called retained austenite) which is much softer than martensite. Consequently, the overall hardness decreases as indicated in Figure 1.4. In case hardening, a carbon concentration in austenite beyond about 0.9 wt. per cent is usually not aimed for during carburization.

Low carbon steels, typically having a hypo-eutectoid compositions, possess good toughness and machinability. However, being low in C concentration, they cannot be hardened to high hardness values by martensitic transformation. Therefore, carburizing these steels increases the C concentration of the surface layers alone while the bulk of the steel remains unaffected. For instance, a steel having x per cent C which lies in the hypo-eutectoid region (see Figure 1.3) can have its surface C concentration increased to y per cent which is a hyper-eutectoid composition. This local increase in C composition is affected during carburization when the steel is in

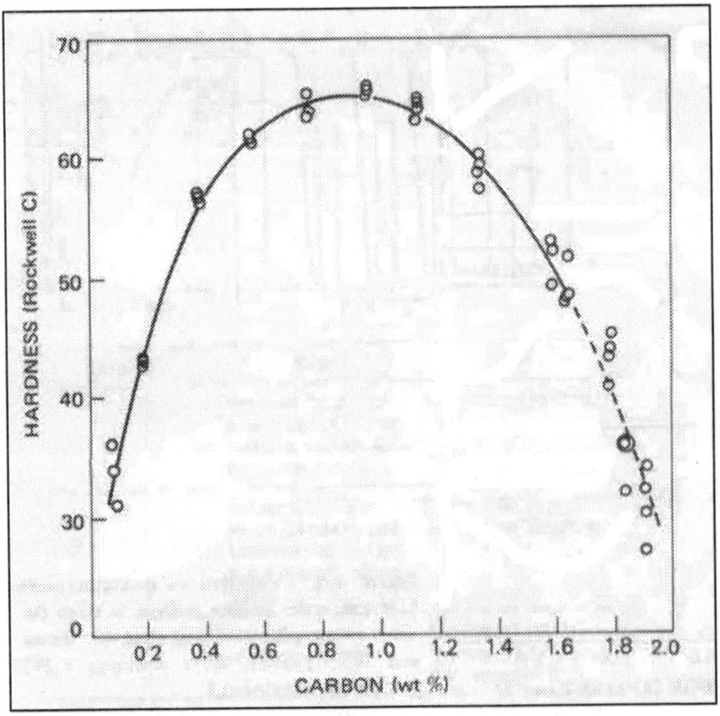

Figure 1.4: Hardness of Fe-C Alloys, Brine Quenched from Austenite (2)

austenitic condition. On subsequent quenching of the carburized steel, the surface layer transforms to martensite having significantly higher hardness by virtue of its enhanced C content. The core, on the other hand, would transform to softer martensite or a mixture of pearlite and soft martensite depending on the dimensions of the steel component. Thus, the component would have a hard surface layer (case) and a tough core which is a desirable combination from application point of view.

The hardness variation of a case hardened steel from the surface to the centre can be schematically represented as shown in Figure 1.5 (Thelning, 1984). The hardness profile is expectedly a function of the variation C concentration of the steel from surface to the interior. The depth of case hardening, or simply the depth of hardening, is defined as the distance from the surface to a plane at which the hardness is 550 HV. This depth is usually determined by hardness measurement across the cross-section of the hardened case. As mentioned previously, carburized coatings are used on steels requiring high hardness and wear resistance. Table 1.2 lists some of the applications of carburized steels.

Table 1.2: Typical Applications of Carburized Coating on Steels

Part	*Steel*	*Case Depth to 50 HRC (mm)*
Mine-Loader bevel gear	2317	0.6
Flying-shear timing gear	2317	0.9
Crane-cable drum	1020	1.2
High-misalignment coupling gear	4617	1.2
Continuous-miner drive pinion	2317	1.8
Heavy-duty industrial gear	1022	1.8
Motor-brake wheel	1020	3
High-performance crane wheel	1035	3.8
Blooming mill screw	3115	5
Die block	1020	1.3
Bushing	CR	1.5

Methods of Carburizing

Carburization is effected in solid media (pack carburizing) or in salt baths (liquid carburizing) or in a gaseous medium (gas carburizing) at temperatures in the range 825-925°C. The hardening treatment may be carried out directly quenching from the carburizing media or by heating and quenching the carburized parts after first allowing them to cool to room temperature from the carburizing treatment. Regardless of what method is used, carburization always takes place via gaseous phases. Each of the above three methods has its own intrinsic characteristics which can produce differing case hardening results.

Pack Carburizing

The common commercial carburizing packs are reusable and contain charcoal or coke as the C source and 10-20 per cent alkali and alkaline earth metal carbonates

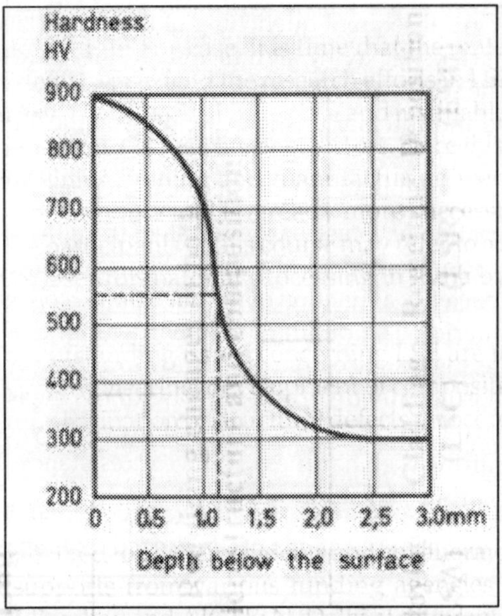

Figure 1.5: Schematic Diagram Showing Hardness Profile of Case Carburized Steel. Depth of case hardening, as shown in figure, is 1.15 mm (Thelning, 1984).

as the activator. The carbonates are usually bound to coke by oil tar or molasses. Barium carbonate (Ba_2CO_3) is the chief activator consisting of 50-70 per cent of the total carbonate content. The remainder of the activator is usually made up of calcium carbonate, although sodium carbonate may also be used. Hardwood charcoal is more reactive than coke as a source of C in pack carburizing. Nevertheless, coke offers certain advantages such as low shrinkage of the pack, good hot strength and good thermal conductivity. More active carburizing mixtures, therefore, contain both charcoal and coke as the C sources. Apart from the C source and the activator, certain non-burning compounds are also added to the pack to resist burning of parts when they come in contact with air at carburizing temperature.

Although the pack used for carburizing primarily consists of charcoal/coke as the C source and $BaCO_3$ as the activator, the carburization can take place with charcoal alone thanks to the presence of atmospheric oxygen enclosed in the carburizing box. Charcoal reacts with oxygen during heating-up period forming CO_2 which converts to CO on further reacting with charcoal as follows:

$$CO_2 + C_{charcoal} \rightarrow 2CO \tag{1}$$

As the temperature rises, the equilibrium is shifted to the right, *i.e.* more and more CO forms. At the steel surface, CO breaks down to CO_2 and C as follows:

$$2CO \rightarrow CO_2 + C_{nascent} \tag{2}$$

The atomic or nascent C thus released is readily dissolved by the austenite phase of the steel. The CO_2 produced in the above equation reacts again with charcoal according to eq. (1) and the cycle of reactions is repeated.

Since the amount of atmospheric oxygen can vary and may be insufficient to produce the carburizing gas, 10-20 per cent barium carbonate, is mixed with charcoal. The activator reacts during the heating-up period as follows:

$$BaCO_3 \rightarrow BaO + CO_2 \quad (3)$$

$$CO_2 + C_{charcoal} \rightarrow 2CO \quad (4)$$

The CO gas decomposes on steel surface in the same manner as described above (eq. 2) leading to absorption of C atoms by steel. All the reactions mentioned above are reversible in nature, *i.e.* they can proceed in both directions.

In pack carburizing, as in other carburization processes, the carbon concentration obtained in the case is a function of carbon potential of carburizing medium, carburizing temperature, time of carburizing and the chemical composition of the steel. The C potential of the atmosphere generated by carburizing medium, as well as the carbon content achieved on the surface of the work, increases directly with increase in the ratio of CO to CO_2. Thus, more C is made available by use of activators and carburizing materials that promote formation of carbon monoxide.

Carburization is effected at temperatures normally between 815 and 955°C during which the steel remains in austenitic phase. In recent years, the upper limits have been steadily raised, and carburizing at temperatures as high as 1095°C have been used. Steel making processes have improved to the extent that fine grain size is maintained at temperatures approaching or exceeding 1040°C. The rate at which the carburized case is formed increases rapidly with temperature. The exact carburization temperature and time are decided based on the austenitizing temperature of a steel as well as the case depth required.

Pack carburization of steel parts are usually carried out in boxes made of heat-resisting steel, type 25Cr-20Ni being suitable for this purpose. The parts are cleaned first and then kept completely immersed in the carburizing pack inside box as schematically shown in Figure 1.6(1). It is important to make lid of the box air-tight by sealing with clay or similar material in order to prevent ingress of oxygen which can affect the carburization process. The case depth is kept under control by inspecting test specimens or, if large parts are involved, by checking the actual parts.

Liquid Carburizing

Liquid or salt bath carburizing is also used for case hardening steel or iron parts. The parts are held at a temperature above Ac_1 in a molten salt that introduces carbon and nitrogen, or carbon alone, into the metal. Diffusion of the carbon from the surface toward the interior produces a case that can be hardened, usually by fast quenching from the bath. Carbon diffuses from the bath into the metal and produces a case comparable with one resulting from gas carburizing in an atmosphere containing some ammonia. However, because liquid carburizing involves faster heat-up (due to the superior heat-transfer characteristics of salt-bath solutions), cycle times for liquid carburizing are shorter than those for gas carburizing. Most liquid carburizing baths contain cyanide, which introduces both carbon and nitrogen into the case. The process in such cases is referred to as cyaniding. Certain types of liquid carburizing bath uses a special grade of carbon, rather than cyanide, as the source of carbon. These baths

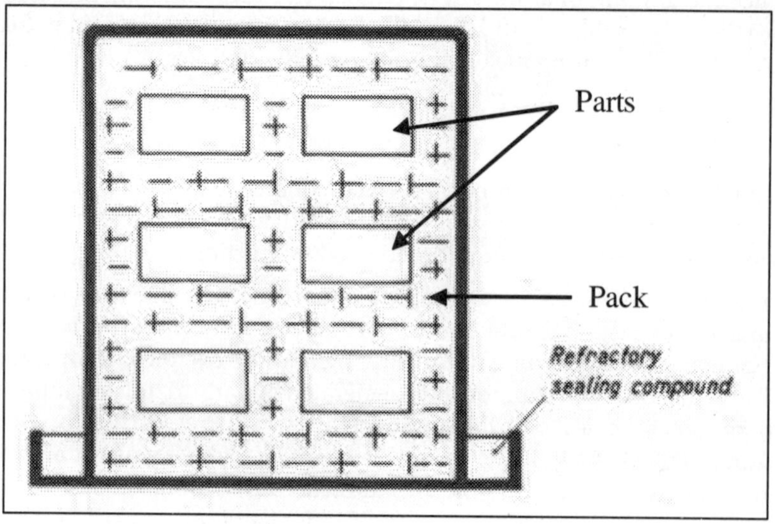

Figure 1.6: Section through Fully Packed Carburizing Box (1)

produce a case that contains only carbon as the hardening agent. The active carburizing agent in a common salt bath is sodium cyanide (NaCN) or potassium cyanide (KCN). It is believed that the carburization proceeds via a gaseous phase according to the following reactions:

$$2\,NaCN + O_2 \rightarrow 2\,NaCNO \tag{5}$$

$$4\,NaCNO \rightarrow 2\,NaCN + Na_2CO_3 + CO + 2\,N_{nascent} \tag{6}$$

$$2CO \rightarrow CO_2 + C_{nascent} \tag{7}$$

The first reaction takes place at the interface between the salt bath and the atmosphere; the other two reactions take place at the interface between the salt bath and the steel. Some of the nitrogen liberated by reaction (6) is also taken up by the steel. Salt-bath carburizing is usually carried out in the temperature range 850-950°C with the actual process temperature depending on the desired case depth. Liquid carburizing is advantageous in cases where very high case depths (up to 6 mm) are required.

Gas Carburizing

Gas carburizing is often referred to as "case carburizing" and, during last few decades, this method of carburization has become the most popular method of case hardening. Gas carburizing, in current commercial practice, uses carbon from hydrocarbon gases and easily vaporized hydrocarbon liquids. Gases most commonly used are natural gas, "manufactured" gas and certain propanes while butane is used infrequently. Natural gas and propane are the preferred sources when available in high purity forms. Liquids are also used as sources of carburizing gas. These liquids are usually proprietary compounds and range in composition from pure hydrocarbons such as terpenes, dipentene or benzene to oxygenated hydrocarbons such as alcohols, glycols or ketones. When a liquid is used, it normally is fed in droplet form to a target plate in the furnace, where it volatilizes almost instantaneously. The vapours dissociate thermally to provide a carburizing atmosphere containing carbon monoxide, carbon dioxide, methane and water vapour.

Most gas carburizing furnaces also employ one of several carrier gases to dilute and react with the hydrocarbon gas used as the principal source of carbon. Class 201, 202, 302 and 402 are the four common carrier gases used in gas carburizing industry. These gases are essentially mixtures of N_2, CO, CO_2, H_2 and methane (CH_4) in various proportions.

The mechanism by which C is transferred to steel surface during gas carburization is quite similar to that described in case of pack carburizing, *i.e.* decomposition of CO gas on steel surface as per the reaction in (Eq. 2) and subsequent absorption of active C atoms by the steel. However, in case of gas carburization, the CO gas is generated by reaction among the various gaseous species such as hydrocarbons and moisture present in the furnace atmosphere.

Vacuum Carburizing

Vacuum carburizing is a high-temperature gas carburizing process that is carried out at pressures below atmospheric pressure (below 100 kPa, or 760 torr). Vacuum carburizing temperatures typically range from 980 to 1050°C; in some cases, however, the range is extended from about 900 to 1095°C. The furnace atmosphere usually consists solely of an enriching gas such as natural gas, pure methane or propane. Nitrogen is sometimes used as a carrier gas. During the carburizing portion of the cycle, furnace pressures are maintained in the range of about 7 to 55 kPa (50 to 400 torr). Vacuum carburizing proceeds by the dissociation of hydrocarbon gas at the steel surfaces and by the direct absorption of carbon by the steel. The reaction with methane is

$$CH_4 + Fe \leftrightarrows Fe(C) + 2H_2 \tag{8}$$

Vacuum carburizing usually employs a two-step process. In the first step, carburizing is carried out. This step is followed by a diffusion step, in which the carbon diffuses into the steel at low pressure, causing the surface carbon to decrease and the case depth to increase.

In addition to reducing total processing time, which results primarily from operating at temperatures higher than those normally used in gas carburizing, vacuum carburizing offers several other benefits. Because the process can be carried out using only natural gas, the need for a carrier-gas is eliminated. Preheating, carburizing and post-carburizing may be done under vacuum, which results in highly clean parts and which may reduce or eliminate much of the usual pre-cleaning and post-cleaning. Alloying elements such as Cr, Mn and Si, are oxidized during long carburizing cycles in conventional carburizing atmospheres, resulting in a loss of surface hardening ability. This does not occur in vacuum carburizing, because of the absence of atmospheric constituents that causes oxidation. Exhausts from vacuum carburizing units contribute much less to environment pollution as compared to conventional gas carburizing. Process variables in vacuum carburizing are controlled with a precision that results in exceptional uniformity and repeatability of the hardened case.

Glow Discharge Carburizing

The glow-discharge or plasma carburizing process makes use of an ionized gas that serves as a medium for both heating and carburizing. This process takes place in a reduced pressure (0.1-2.7 kPa) of methane (CH_4). A potential of a few hundred volts is applied between the part to be carburized (cathode) and anode. This creates a thin envelope of plasma that closely follows the contours of even the most irregular part. This method of carburizing is said to have short process times and excellent process control.

Nitrided Coating

As discussed above, carburizing is an austenitic thermo-chemical treatment involving diffusional addition of interstitial alloying element C into the austenite phase, and relies on the subsequent transformation of austenite to martensite to produce high surface hardness. The process of nitriding, however, is a ferritic thermo-chemical treatment and takes place below A_1 temperature as shown in Figure 1.3. This process usually involves the introduction of atomic nitrogen into ferritic phase in temperature range 500-590°C and consequently no phase transformation occurs on cooling to room temperature. This process of hardening was first used at the end of 1920s and since then its application has continuously spread, due among other things to the fact that the process has been further developed and can now be applied to much larger varieties of steels than originally thought possible.

The beneficial properties imparted by nitriding to steels can be summarized as follows:

1. High surface hardness and wear resistance, together with enhanced anti-galling properties
2. Increased high temperature hardness
3. High fatigue strength and low fatigue notch sensitivity
4. Improved corrosion resistance for non-stainless steels
5. High dimensional stability

Basic Principles of Hardening

Figure 1.7 presents the equilibrium Fe-N phase diagram [1]. At the nitriding temperatures customarily used, nitrogen dissolves in Fe, but only up to a concentration of 0.1 per cent. As evident from the above figure, beyond the maximum solubility of nitrogen in Fe, γ'-nitride having a chemical formula of Fe_4N is formed. When nitrogen concentration exceeds about 6 per cent, the γ'-nitride begins to change to ε-nitride ($Fe_{2-3}N$). Below 500°C, ξ-nitride which has a chemical formula of Fe_2N forms. The nitrogen content of this phase is approximately 11 per cent.

When making observations in the metallurgical microscope, the γ' and ε-nitrides are seen as a white surface layer called 'white layer' or 'compound layer'. During

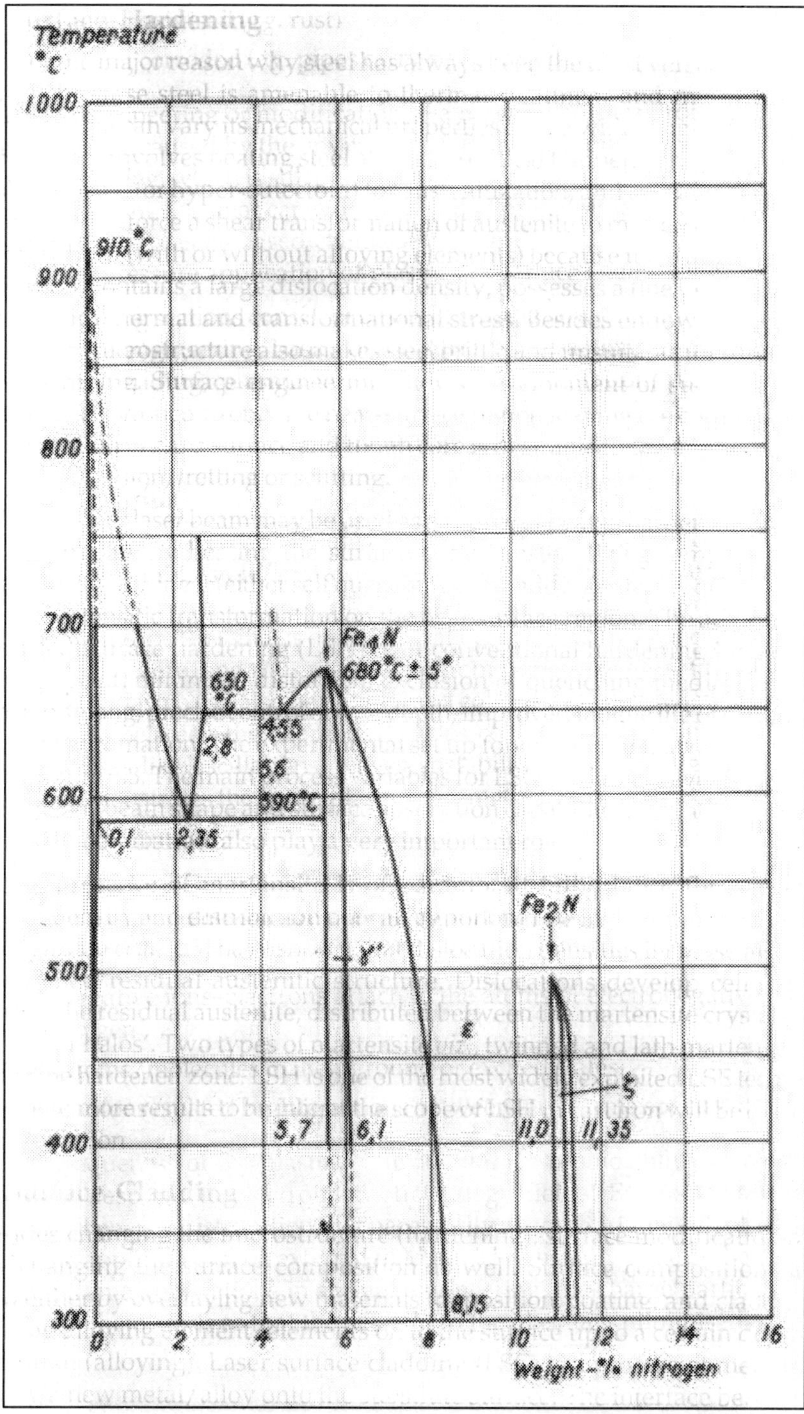

Figure 1.7: Iron-nitrogen Equilibrium Phase Diagram [1]

nitriding, simultaneous with the increase in thickness of white layer, the nitrogen diffuses further into the steel. When solubility limit is exceeded, nitrides are precipitated at grain boundaries and along certain crystallographic planes. Apart from nitrides, nitrogen together with C also forms carbonitrides. The high surface hardness which is obtained after nitriding is due to the formation of finely dispersed nitrides and carbonitrides which distort the ferrite lattice. After nitriding, the nitrided part sometimes is quenched in water (either cold or warm). This rapid cooling results in a supersaturated solid solution of nitrogen in α-Fe, which adds an extra contribution to the increase in fatigue strength which is a characteristic feature of nitriding. A typical nitrided coating microstructure is presented in Figure 1.8 [3]. All the above features are indicated in the figure.

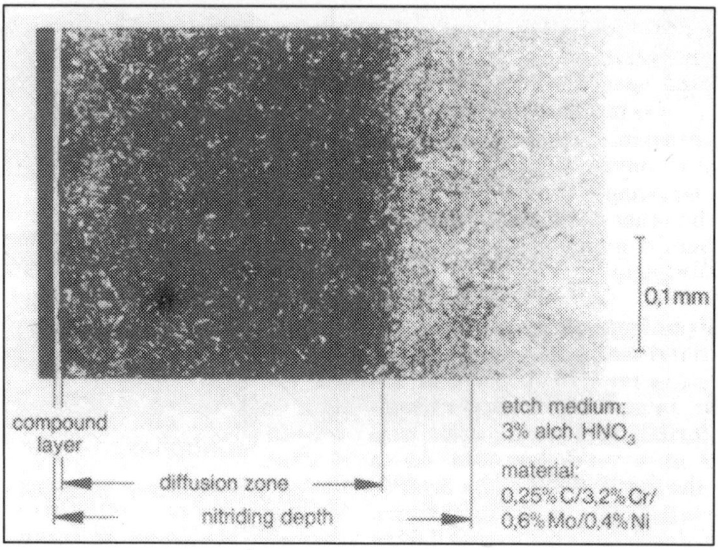

Figure 1.8: Microstructure of a Typical Nitrided Coating on Steels [3]

Although at suitable temperatures all steels are capable of forming iron nitrides in the presence of nascent nitrogen, the nitriding results are more favourable in those steels that contain one or more of the major nitride-forming alloying elements such as Al, Cr and Mo. Aluminium is the strongest nitride-former and steels containing about 0.85-1.50 per cent Al yield the best nitriding results [4]. Unalloyed carbon steels are not well suited to nitriding, because they form an extremely brittle case that spalls readily, and the hardness increase in the diffusion zone is small [4]. Nitrided steels are widely employed for industrial applications some of which are listed in Table 1.3.

Methods of Nitriding

The three main methods of nitriding are gas nitriding, liquid or salt-bath nitriding and plasma nitriding. There can be several variation within these methods. Nitriding by means of ammonia gas, which involves the absorption of nitrogen only, is usually referred to simply as nitriding or gas nitriding. When besides nitrogen, carbon is introduced the treatment is called nitrocarburizing, and this process can take place in a gaseous atmosphere or in a salt bath. It should be remembered that, since nitriding is a ferritic thermo-chemical treatment, all hardenable steels must be hardened and tempered before being nitrided. The tempering temperature must be high enough to guarantee structural stability at the nitriding temperature; the minimum tempering temperature is usually at least 30°C higher than the maximum temperature to be used in nitriding.

Table 1.3: Typical Applications of Nitrided Steels

Part	*Steel*
Hydraulic barrel	AMS 6470
Trigger for pneumatic hammer	AMS 6470
Tachometer shaft	AMS 6475
Helical timing gear	4140
Generator shaft	4140
Rotor and pinion for pneumatic drill	4140
Oil pump gear	4340
Loom shuttle	410 stainless
Aircraft cylinder barrel	AMS 6470
Clamp	4150
Spindle	4340

In the original method of gas nitriding, anhydrous ammonia is allowed to flow over the parts to be hardened, normally at about 500°C. The ammonia dissociates according to the equation given below:

$$2NH_3 \rightarrow 2N_{nascent} + 3H_2 \tag{9}$$

At the instant of dissociation, nitrogen remains in atomic form, and is absorbed by the steel as such. Gas nitriding is usually done for parts that require a case depth between 0.2 and 0.7 mm. The depth of hardening and extent of case hardness developed due to nitriding depends on several factors such as time-temperature schedule of the process, degree of dissociation of ammonia, concentration and nature of nitride forming elements in the steel etc. A typical hardness profile obtained across the nitrided layer on AMS 6475 steel is shown in Figure 1.9 [3]. The core hardness of this steel after nitriding is 41.5 HRC.

Liquid nitriding of steel, which is carried out in a molten salt bath, employs a similar temperature range as gas nitriding, *i.e.* 510-565°C. The case hardening medium in this method is a molten nitrogen-bearing fused salt bath containing both cyanides and cyanates. The typical commercial bath for liquid nitriding is composed of a mixture of sodium and potassium salts. The sodium salts, which comprise 60-70 wt. per cent of the total mixture, consist of 96 per cent NaCN, 2.5 per cent Na_2CO_3 and 0.5 per cent NaCNO. The potassium salts, 30-40 wt. per cent of the mixture, consist of 96 per cent KCN with the balance being small amounts of K_2CO_3, KCNO and KCl. The liquid bath is first aged at about 575°C for 12 h to increase the cyanate content to the

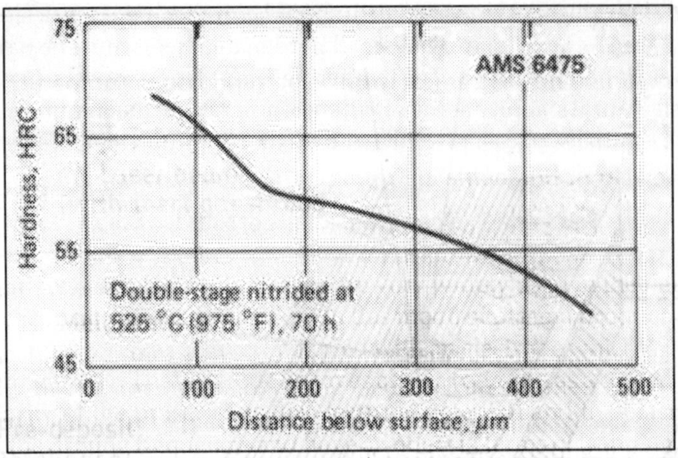

Figure 1.9: Harness Gradient Obtained for Nitrided AMS 6475 Steel [3]

desired level of ~45 per cent. Subsequently, steel parts are immersed in the liquid bath in the above mentioned temperature range (510-565°C). Nitriding time in the liquid process seldom exceeds 2 h. The salt bath gives off carbon and nitrogen in accordance with the equations described for liquid carburizing. The atomic nitrogen (and also carbon) is absorbed by the steel leading to surface hardening. During last decades, it has been found that by injecting air into the bath, better control of the cyanate content of the bath and consequent longer periods of nitriding can be achieved. In a modification to the conventional liquid nitriding bath, Na_2S is added as an additional component which leads to absorption of sulphur in addition of N and by the steel. This process is known as "Sulfinuz' treatment. Addition of S provides the steel with excellent anti-frictional properties.

Both gas nitriding and salt bath nitriding have certain inherent advantages and disadvantages. Salt bath nitriding has an obvious advantage over gas nitriding as treatment time is concerned. While nitriding time of less than 2 h is very common in case of salt bath process, it can be above 50 h in case of gas nitriding. Better cleanliness can be maintained with gas nitriding since there is no soiling of the work premises or of the nitrided components. However, with salt-bath nitriding, it is quite easy to spill salt on the top of the furnace and on the surrounding floor area. Formation of pores, which occurs during salt bath nitriding, increases with increasing treatment. This prevents carrying out nitriding for long durations in liquid bath, and nitriding period is usually restricted to 4 h. No such restrictions are imposed on gas nitriding although a nitriding time of 90 h is normally not exceeded for practical reasons. When a greater depth of nitriding is required than that obtainable with salt-bath nitriding, gas nitriding becomes extremely handy. For the same depth of hardening, gas nitriding causes less distortion of the of the hardened parts than liquid nitriding. In terms of wear resistance and toughness of the hardened steel, gas nitriding is said to have a slight edge over liquid nitriding.

The plasma or glow-discharge or ion nitriding process was patented at the beginning of the 1930s. For many years this was not used extensively on an industrial scale because of relatively expensive equipment and the diversified technology which is essential if the process is to be carried out correctly. Thanks to development work during recent years, the process has gained significant popularity. The glow-discharge or plasma nitriding process makes use of an ionized gas that serves as a medium for both heating and nitriding. In vacuum, high-voltage electrical energy is used to form a plasma, through which nitrogen ions are accelerated to impinge on the workpiece. This ion bombardment heats the workpiece, cleans the surface and provides active nitrogen for nitridation. The parts to be treated are charged into an airtight chamber, which constitutes the anode of the plasma nitriding unit. The parts making up the charge are so arranged that they are in electrical contact with the cathode, either by being placed on a bottom plate or suspended in a suitable manner. After being evacuated the chamber is filled with the process gas which, besides nitrogen, may contain hydrogen and methane. Thus, plasma nitriding may function both as conventional gas nitriding and as nitrocarburizing. The working pressure lies between 0-1 and 10 mbar. On applying a potential difference of 500-1000 V, the gas becomes ionized with the result that positive nitrogen ions having a high kinetic energy bombard the charge. Thereby the parts become heated to the required temperature and at the same time the nitrogen required for the nitriding process is introduced into the surface of the parts. All surfaces of the charge may be subjected to ion bombardment. This causes the formation of a blue-white glow discharge plasma around the components (Figure 1.10 [1]).

Metallurgically versatile, plasma nitriding provides excellent dimensional control and retention of surface finish. This process can be conducted at temperatures lower than those conventionally employed. Complete control of the nitrided layer leading to superior fatigue performance, wear resistance and hard layer ductility, can be achieved in glow-discharge nitriding process. Total absence of pollution,

Figure 1.10: During Plasma Nitriding a Blue-White Glow Discharge Plasma appears around the Components in the Charge [1]

efficient use of gas and electrical energy, total process automation, selective nitriding by simple masking techniques and reduced nitriding time are some of the other advantages of this process. The major limitations of plasma nitriding are high capital cost, need for precision fixturing with electrical connections, and lack of feasibility of liquid quenching for carbon steels.

Carbonitriding

Carbonitriding is a modified form of gas carburizing, rather than a form of nitriding. The modification consists of introducing ammonia into the gas carburizing atmosphere to add nitrogen to the carburized case as it is being produced. Nascent nitrogen forms at the work surface by the dissociation of ammonia in the furnace atmosphere; the nitrogen diffuses into the steel simultaneously with carbon. Typically, carbonitriding is carried out at a lower temperature and for a shorter time than gas carburizing, producing a shallower case than is usual in production carburizing. The treatment temperature is normally 800-900°C. In its effects on steel, carbonitriding is similar to liquid cyaniding. Because of problems in disposing of cyanide bearing wastes, carbonitriding is often preferred over liquid cyaniding. In terms of case characteristics, carbonitriding differs from carburizing and nitriding in that (a) carburized cases normally do not contain nitrogen, and (b) nitrided cases contain nitrogen primarily, whereas carbonitrided cases contain both.

Carbonitriding is used primarily to impart a hard, wear-resistant case, generally 0.075-0.75 mm deep. A carbonitrided case has better hardening ability than a carburized case (nitrogen increases hardening ability of steel; it also is an austenite stabilizer, and high nitrogen levels can result in retained austenite-particularly in alloy steels). Consequently, by carbonitriding and quenching, a hardened case can be produced at less expense within the case-depth range indicated, using either carbon or low-alloy steel. Full hardness with less distortion can be achieved with oil quenching or, in some instances, even gas quenching, employing a protective atmosphere as the quenching medium.

In spite of the increase in hardening ability that is achieved in case of carbonitriding, this process has certain inherent drawbacks. The applications of carbonitriding are more restricted than those of carburizing because this process is largely limited to case depths of about 0.75 mm or less, while no such limitation applies to carburizing. Lower processing temperatures and poorer control over nitrogen addition are the primary reasons for the above limitation of carbonitriding. The other major drawback of carbonitriding process is the potential of retaining larger quantities retained austenite in the hardened case. Nitrogen is known to lower the transformation temperature of austenite. Therefore, because of its nitrogen content, a carbonitrided case will, under identical post-treatment conditions, contain more retained austenite than a carburized case of the same carbon content. The low hardness resulting from the presence of retained austenite is undesirable in many applications. It can be extremely detrimental in components of closefitting assemblies. The delayed transformation of austenite to martensite also results in a volume increase that may cause moving parts to bind or "freeze" if it occurs in service.

Nitrocarburizing

Nitrocarburizing, or ferritic nitrocarburizing, as it is often called, is a thermo-chemical treatments that involves diffusional addition of both nitrogen and carbon to the surface of ferrous materials at temperatures completely within the ferrite-phase field. Thus, nitrocarburizing is a modified form of nitriding, rather than a form of carburizing. This process is usually carried out below 675°C. Carbonitriding, on the other hand, is performed with the steel in austenite phase field, *i.e.* above 760°C. Cycle times being usually less than 3 h; nitrocarburizing processes are termed "short-cycle" nitriding. The primary objective of such treatments is usually to improve anti-scuffing characteristics of ferrous engineering components by providing the surface with a compound layer exhibiting good wear/friction-resistant properties. In addition, fatigue characteristics can be considerably improved, particularly when nitrogen is retained in solid solution in the "diffusion zone" beneath the compound layer. Corrosion resistance provided by the compound zone is an important secondary benefit. This process produces a thin surface layer of iron carbonitrides and nitrides (the "white layer" or compound zone) with an underlying diffusion zone containing dissolved nitrogen and iron (or alloy) nitrides. The white layer enhances surface resistance to galling, corrosion and wear. The diffusion zone improves the fatigue properties.

Borizing or Boronizing

Borizing or boronizing is a thermo-chemical process in which boron is allowed to diffuse into steel leading to the formation of a surface layer containing metallic borides which possess extremely high hardness, *viz.* between 1600 HV and 2000 HV. This process may be carried out by means of gaseous, liquid or solid media. By employing commercially available compounds containing additions of activating agents, a surface layer of Fe_2B is formed. This layer has approximately the same coefficient of thermal expansion as steel.

In principle the method used for pack Borizing is the same as that employed for conventional pack carburizing. The process takes place at 800-1000°C. After cooling in the box the steel parts may be hardened as in case hardening. However, the surface hardness is the same whether the parts are subjected to a quenching operation or simply allowed to cool in the box. All steels can be borided. The thickness of the boride layer decreases as the content of alloying elements increases. Commercially available boriding packs such as EKABOR can provide a surface hardened layer of 0.01-0.06 mm when the treatment is carried out at 850°C for 1-8 h. The surface hardness achieved is about 1300 HV. The typical the microstructure of a boronized layer obtained on a steel is shown in Figure 1.11 [1].

Fundamentally, boride coating is different from its carburized and, to a certain extent, nitrided counterparts. It may be recalled that the coating forming element C remains in solid solution (in martensite phase) in the carburized coatings. In nitrided coatings also, N remains in solid solution in a-Fe although some of it forms nitride and carbonitride compounds. The hardening effect in carburized and nitrided coatings is achieved because of the distortion of the crystal lattice caused by the presence interstitial atoms such as C and N. In case of boride coatings, on the other

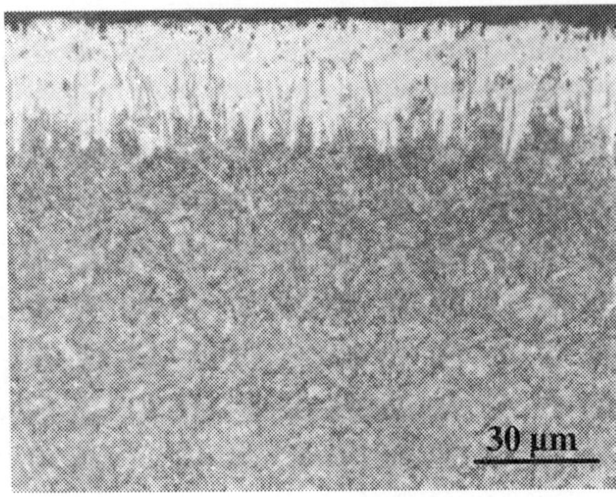

Figure 1.11: Microstructure of a Borided Coating on Steel SS 2225 (25Cr 4Mo) [1]

hand, B reacts with Fe to form a boride surface layer constituted of iron boride phases such as FeB and FeB_2 which are extremely hard. Thus, in boride coatings, the hard surface layer is the result of formation of compounds such as FeB and FeB_2 in the coating unlike in the case of carburized layer where the hardness is caused by distortion of crystal lattice.

Aluminized Coatings

Aluminized or diffusion aluminide coatings are formed by incorporating aluminium into the surface of a material by diffusion. Since these coatings provide protection against high temperature oxidation and corrosion, they are usually applied on Ni and Co-base superalloys which are used for elevated temperature applications. The components on which aluminide coatings are primarily applied are blades and vanes of gas turbine engines. Both aviation and land-based gas turbine components use these protective coatings. Aluminide coatings are also infrequently applied on steel components such as furnace parts which are used at moderately high temperatures. These coatings have also been used on high temperature Ti-base alloys for protecting them from oxidation damage.

Introduction of Al into the surface leads to the formation of aluminides of iron, nickel, cobalt and titanium depending on whether the substrate is a steel, Ni-base alloy, Co-base alloy or a Ti-base alloy. The aluminide phases that are normally formed in the coating on the above alloys are FeAl, NiAl, CoAl and $TiAl_3$, respectively. During service, Al present in the coating reacts with oxygen at high temperatures to form an alumina (Al_2O_3) layer on the surface of the coated alloy. The presence of the alumina layer greatly reduces further oxidation leading to increase in the life of the component. The aluminide phase of the coating acts as a reservoir of Al needed for continuous maintenance of the alumina layer on the surface. Once the coating becomes depleted of Al below a certain level, the desirable aluminide phases such as NiAl transform to lower aluminides like Ni_3Al which cannot sustain the surface alumina layer. In such a case, less protective oxide phases such as spinels form in the oxide layer which cause an increase in oxidation rate. Depletion of Al from the aluminide coating occurs due to its continuous use in maintaining the surface oxide layer and also because of its loss into the substrate under the Al concentration gradient between the coating and the substrate.

Mechanisms of Coating Formation

The mechanisms of formation of aluminide coatings have been widely reported on Ni-base alloys although the mechanisms in case of other alloys are expected to be similar. For the sake of convenience, these mechanisms are discussed here for Ni-base alloy substrates. Aluminide coatings are primarily formed using pack cementation methods. Depending on the Al potential or Al activity of the pack mixture, which depends on whether pure Al or an Al-alloy is being used as the Al source and its composition in the pack, aluminization is carried out either in two steps or in a single step. In the two-step process, aluminization is carried out at a relatively low temperature (700-850°C) in the first step. In the subsequent step, the aluminized specimen is removed from the pack and subjected to a post-aluminization diffusion treatment for a few hours at a temperature above 1000°C. The coating formation completes only after the second step. The two-step process is adopted when the activity of Al the pack is high, *i.e.* pure Al used as the Al source. This process is also referred to as low temperature high activity (LTHA) process.

When the activity of the pack is low, *i.e.* an alloy of Al such as Ni-Al, Cr-Al or Fe-Al instead of pure Al is used as the Al source, aluminization is carried out at a much higher temperature, usually above 1000°C. The coating formation in this case gets completed in the aluminization step and no post-aluminization treatment for the purpose of completing the coating development is usually provided. This aluminization process is often called as high temperature low activity (HTLA) process. The coating formation mechanisms in a two-step high activity process and a single-step low activity process are different and, therefore, discussed separately below.

Coating Formation in Two-Step Process

The coating formation mechanism depends primarily on the activity of Al in the pack at the aluminization temperature. In a two-step process, the coating formation during aluminization, which constitutes the first step of the coating process, occurs by the inward movement of the coating growth front (the interface at which Al reacts with Ni to form the coating). The nascent Al atoms that are released at the surface diffuse into the substrate and, in the process, react with the Ni atoms to form aluminide phases. This inwardly grown coating layer and its reaction front are denoted by INL and RFIN respectively in Figure 1.12(b). The reaction front for coating growth in this case lies below the initial sample surface ISS, as shown in Figures 1.12(a) and (b), and continuously moves inward into the substrate during aluminization. The presence of fine precipitates and substrate carbides lying fully or partially embedded in the coating layer is indicative of the inward coating growth during the aluminizing step [5-7]. The above mentioned precipitates form in the coating due to the lack of sufficient solubility of certain substrate elements like Cr, W, Mo, etc. in coating phases [7]. The aluminide phases that are usually formed in the coating after aluminization are Al-rich (hyperstoichiometric) β-NiAl and δ-Ni_2Al_3.

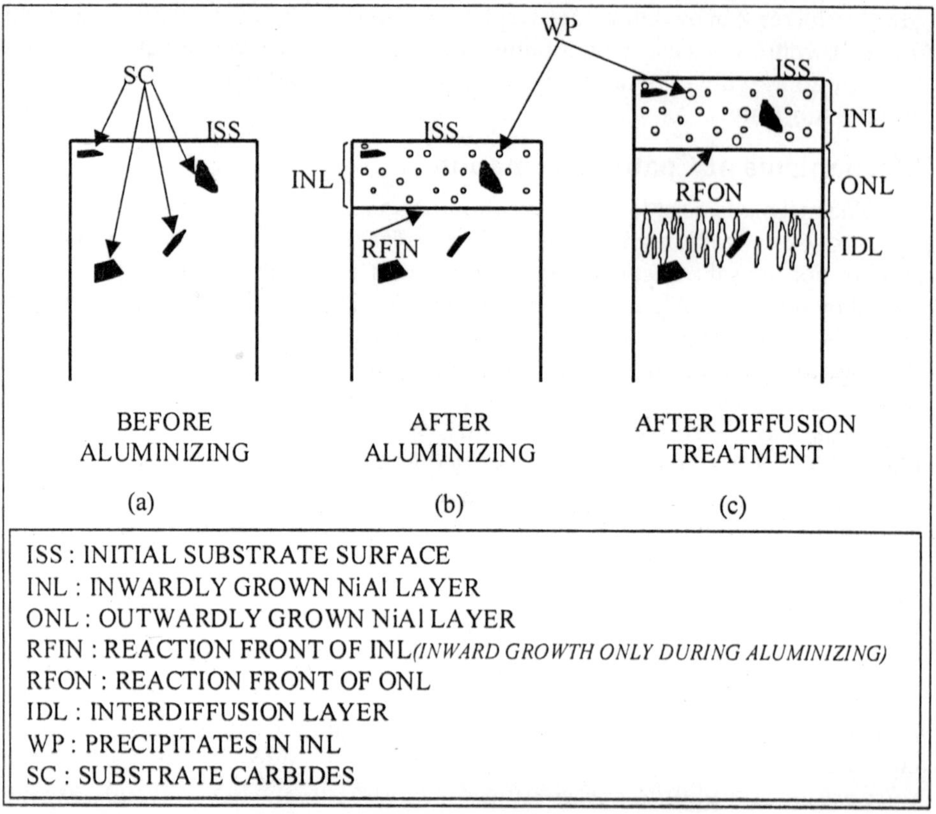

Figure 1.12: Schematic Representation of the Evolution of Coating Structure in a Two-Step High Activity Aluminizing Process [7]

The hyperstoichiometric β-NiAl and the δ-Ni_2Al_3 phases that constitute the as-aluminized coating are relatively unstable at high temperatures, typically above 1000°C, due to the high Al concentration gradient between the coating and the underlying unaffected substrate [6,8]. Therefore, prior to using the coating in actual operation, these phases are transformed to the more stable Ni-rich (hypostoichiometric) β-NiAl phase by subjecting the as-aluminized coating to a diffusion heat treatment. As mentioned earlier, this treatment is typically done at a temperature above 1000°C for a few hours. During the diffusion treatment, outward diffusion of nickel from the unaffected substrate takes place. which reacts with the Al provided by the hyperstoichiometric β-NiAl and δ-Ni_2Al_3 phases of the layer INL as shown in Figure 1.12(c). This results in the formation of additional β-NiAl by singular outward Ni diffusion leading to the development of an outwardly grown NiAl layer schematically shown as ONL in Figure 1.12(c). The reaction front RFON for the formation of the above outwardly grown NiAl layer coincides with RFIN at the start of the diffusion treatment and, subsequently, moves outward to form the layer ONL. Formation of the β-NiAl phase for the outwardly-growing NiAl layer continues until the decreasing flux of Al from the outer layer INL can no longer ensure adequate Al availability at the reaction front RFON. Subsequently, gradual conversion of INL to hypostoichiometric NIAl phase occurs. Thus, the layer INL shown in Figure 12(c), which consists of hyperstoichiometric NiAl and Ni_2Al_3 at the beginning of the diffusion treatment, eventually converts to be hypostoichiometric NiAl phase. However, this layer, unlike ONL, continues to contain the dispersed precipitates that existed after the aluminization step [Figure 1.12(b)]. In the final stages of the diffusion treatment, both INL and ONL layers are constituted of hypostoichiometric β-NiAl phase.

Simultaneous to the outward coating growth during diffusion treatment, an additional layer also develops below plane RFIN as shown in Figure 1.12(c). Removal of Ni from the substrate by outward diffusion causes an enrichment of Al in the substrate below RFIN. This enrichments of Al leads to the transformation of $\gamma+\gamma'$ phase structure of the super alloy substrate to β-NiAl [6]. Thus, a third coating layer constituted by β-NiAl phase forms which is referred to as the inter diffusion layer and schematically shown as IDL in Figure 1.12(c). This β-NiAl layer (IDL) is distinct from the two outer NiAl layers, *i.e.* INL and ONL, due to the presence of numerous complex precipitates in it. The formation of these precipitates in the inter diffusion layer is also ascribed to the loss of Ni from this layer. Most of the alloying elements, which usually remain in solid solution in the original $\gamma+\gamma'$ structure of the substrate, precipitate out as complex phases after the loss of Ni because of their inadequate solubility in the resultant β-NiAl matrix [6,7]. The structure and composition of the individual precipitates of the inter diffusion zone formed in the coatings on Ni-base super alloys vary from alloy to alloy although same basic structural features are always observed.

In summary, a typical two-step high activity coating is found to consist of three layers–an outer NiAl layer, an intermediate NiAl layer and an inner inter diffusion layer. These layers are schematically represented in Figure 1.12(c) as INL, ONL and IDL, respectively. The outer NiAl layer INL is clearly identified from the intermediate NiAl layer ONL because of the invariable presence of both substrate carbides as well as numerous other precipitates in it. The layer ONL, on the other hand, is usually devoid of such carbides and precipitates as shown in Figure 1.12(c). These layers can also be seen in the actual coating structure presented in Figure 1.13 [9].

Coating Formation in Single-Step Process

Unlike the high activity process which typically involves two steps, the entire aluminizing treatment in a low activity process is carried out in one single step at a temperature typically above 1000°C. The coating obtained on Ni base super alloys by adopting a low activity aluminizing process consists of two layers, namely an outer single-phase β-NiAl layer and the underlying inter diffusion layer, as schematically presented in Figure 1.14 [7]. An actual low activity aluminide coating, as reported by Goward and Boone [6], is reproduced in Figure 1.15 where the above mentioned coating layers can also be seen. The formation of the outer NiAl layer ONL involves its growth front RFON moving outward with respect to the initial substrate surface ISS, as indicated in Figures 1.14(a) and (b). The formation of NiAl phase in the outwardly growing layer ONL is caused by the reaction of outwardly diffusing nickel

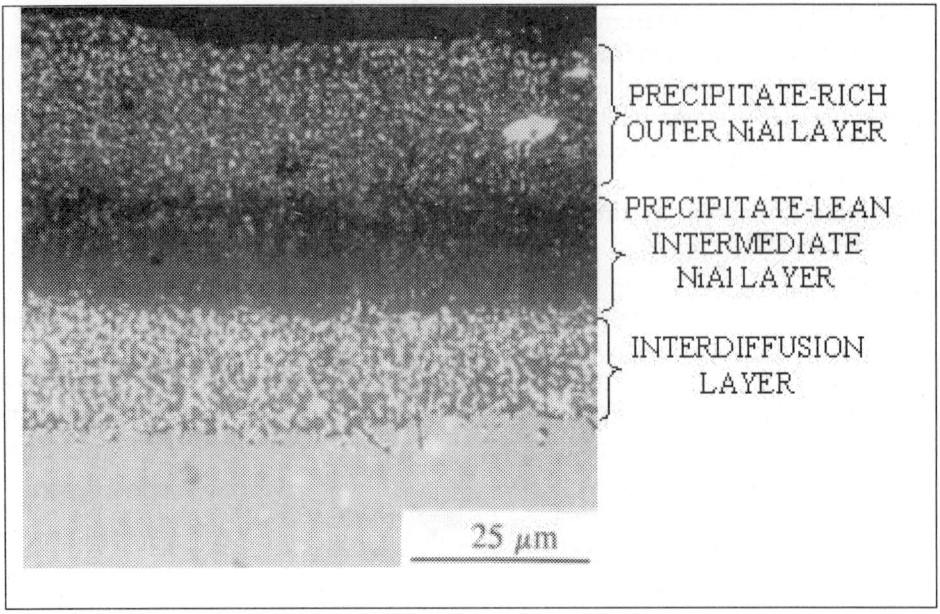

Figure 1.13: The Microstructure of a Typical Two-Step High Activity Aluminide Coating on a Ni-base Superalloy [9]

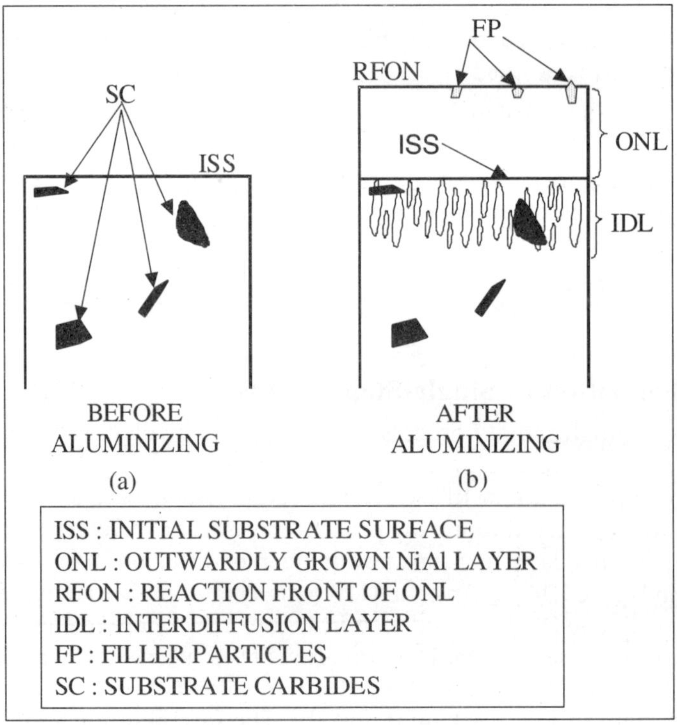

Figure 1.14: Schematic Representation of the Evolution of Coating Structure in a Low Activity Aluminizing Process [7]

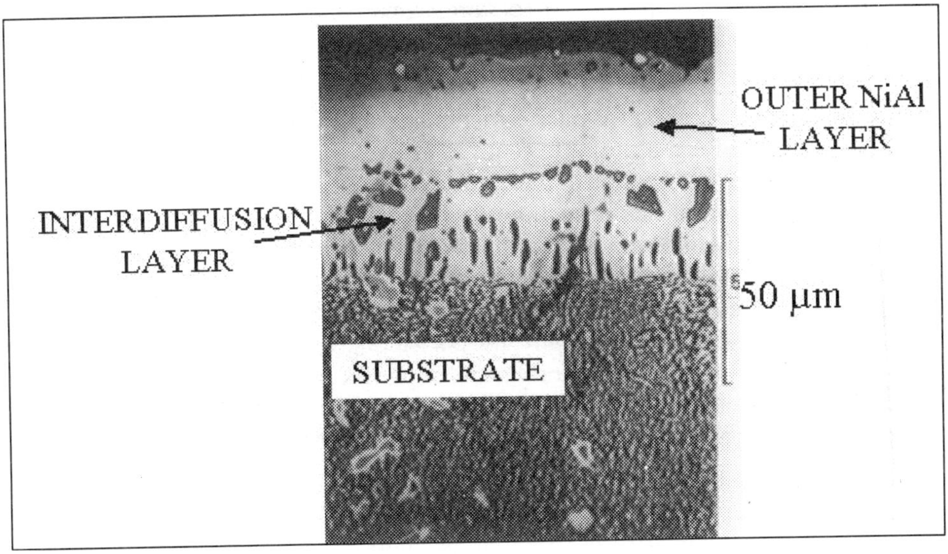

Figure 1.15: The Microstructure of a Low Activity Plain Aluminide Coating on IN-100 Superalloy [7]

from the substrate with Al from the pack at the reaction front RFON. The NiAl phase constituting layer ONL remains hypostoichiometric, *i.e.* Ni-rich, in composition throughout the aluminizing duration. Relatively less availability of Al in a low activity pack is responsible for outward movement of the coating growth front and formation of Ni-rich β-NiAl phase in the coating throughout the aluminizing process. Apart from the formation of outer layer ONL, an inter diffusion layer also develops in the coating because of the loss of Ni from the substrate in a manner as described earlier for a high activity coating. This layer is represented as IDL in Figure 1.14(b). A typical low activity coating on a Ni-base super alloy, thus, consists of two layers–an outer NiAl layer ONL which is devoid of any carbides and other precipitates, and an inner inter diffusion layer IDL, as schematically shown in Figure 1.14(b).

It is evident from Figure 1.14 that the nature of growth in a low activity process excludes the possibility of substrate carbides, the natural markers, getting embedded in the outer Ni-Al layer. Therefore, the substrate carbides in these coatings always lie at or below, and not above, the interface between the two layers of the coating which is, in fact, the initial substrate surface, *i.e.* ISS, as shown in Figure 1.14(b). Further, since substrate alloying elements do not diffuse outward in enough quantity along with Ni into the outer coating layer, this layer remains virtually precipitate-free. Since the outer surface of the coating, which is also the reaction front, moves towards the surrounding pack material, there is a likelihood of inert filler particles getting embedded in layer ONL of the coating, as demonstrated in Figure 1.14(b). This is one of the drawbacks of a low activity coating because such embedded particles in the coating can be potential weak sites during the use of the coating.

Methods of Aluminizing

Pack Aluminizing

Several techniques such as pack cementation, slurry coating and hot dip process have been used to form aluminide coatings [10-13]. However, among the various methods used for the formation of these coatings on steels and super alloys, the pack cementation technique has been the most commercially popular because of its simplicity and cost effectiveness. The pack cementation method of aluminizing, or simply "pack aluminizing" or "pack aluminization" requires the specimen/substrate to be kept immersed in a powder mixture or 'pack' which consists of an Al source, an activator and an inert filler. Normally, the pack along with the immersed specimens is kept in a heated retort for several hours, depending on the thickness and micro structural requirements of the coating, and subsequently cooled to room temperature. An inert/reducing atmosphere (Ar/H_2) is maintained throughout the coating process.

The various details pertaining to pack aluminizing process are well documented and available in form of several papers and reviews [14-17]. The steps involved in a typical pack aluminizing process and their sequence are shown in Figure 1.16. The Al source in an aluminizing pack can be pure Al powder or powder of an Al alloy such as Ni-Al, Cr-Al and Fe-Al having different concentrations of Al. The concentration of Al in the Al source determines the thermodynamic activity of Al in the Al-bearing vapour phases that form during the aluminizing process. Since the above Al activity directly decides the amount of Al that is transported from the pack to the substrate surface during coating formation at a given temperature, the composition of the Al source in terms of Al content is considered an important indicator of the nature of the pack.

The activator used in a pack is usually a halide salt such as NH_4Cl, NH_4F, NaCl, NaF etc. which is added in small quantities, typically 1-2 wt. per cent, to the pack mixture. This salt reacts with Al in the Al source during aluminizing to form Al-halides which remain in vapour form at aluminizing temperature. The Al halide vapour decomposes on the component surface to release nascent Al atoms which then react with the substrate to form the coating. The filler material that is most widely used in an aluminizing pack is alumina powder although mention of silica, kaolin and other such materials can also be found especially in the earlier literature [18].

After the coating and subsequent heat treatments (see Figure 1.16), the coated components are cleaned. Finally, they visually and/or otherwise inspected for coating quality, coating appearance etc. The extent of Al pick-up during the aluminizing may be determined from weight measurements before and after the treatment. Coating thickness is measured by a suitable destructive/non-destructive method, and, if needed, structural details of the coating are also evaluated by various micro structural characterization techniques such as metallography, X-ray diffraction etc.

Other Methods of Aluminizing

Although pack aluminization continues to remain the most favoured method industrially for aluminide coating formation, this method has all the drawbacks of a

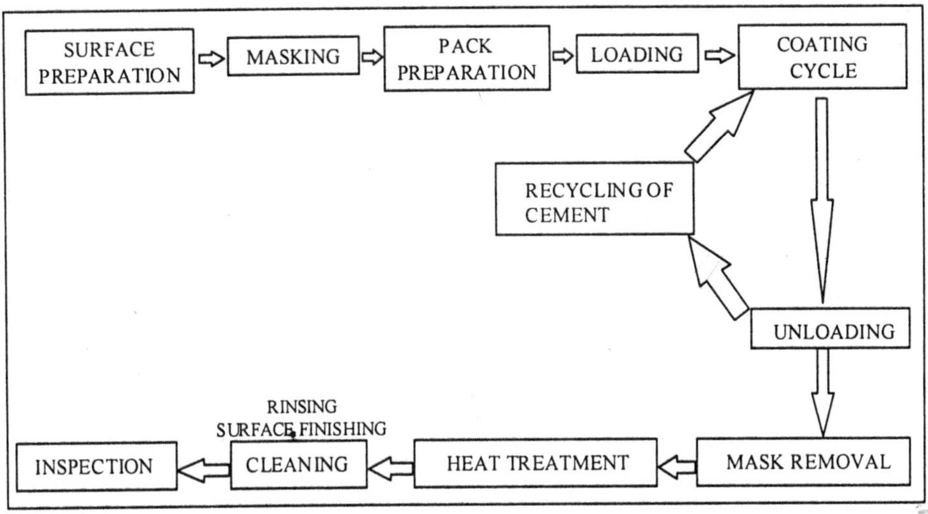

Figure 1.16: Various Steps Involved in a Typical Pack Aluminizing Process

typical pack cementation process. Additionally, since the pack material remains in physical contact with the substrate, pack aluminized coatings have the likelihood of having entrapped alumina filler particles which are highly undesirable for mechanically stressed components because they can act as weak spots during use of the coatings. Therefore, newer methods of aluminizing are increasingly being adopted in recent past for high-end aluminide coatings especially those for aircraft components. Among the advance processes that are being employed to achieved 'clean' and high performance aluminide coatings, the important ones are (a) vapour phase aluminizing and (b) aluminizing by conventional chemical vapour deposition (CVD) techniques [19-21]. Vapour phase aluminizing is essentially a pack aluminization process in which the substrate does not remain in contact with the pack material. In the CVD processes, Al is brought to the substrate surface in the form of gaseous species by a carrier gas where the nascent Al is released after decomposition of the Al bearing species. Subsequently, Al reacts with the substrate to form the aluminide coating. The above two methods are gaining popularity not only because of their propensity to produce clean coatings but also because of the excellent process control they offer.

Modified Aluminide Coatings

The coatings obtained by introduction of only Al onto the substrate surface by the aluminizing process are referred to as plain aluminide coatings. On superalloys, the β-NiAl/CoAl phase that constitutes these coatings acts as a reservoir of Al for the formation and sustenance of the protective alumina scale on the surface during high temperature exposure. The continued maintenance of a protective Al_2O_3 layer on the coating surface throughout the period of oxidation exposure is contingent on the availability of adequate quantity of Al from the coating. This is even more relevant under conditions where cyclic heating and cooling of coated components can cause severe spallation of the oxide layer and consequent accelerated consumption of Al from the coating. For making these coatings more effective against high temperature oxidation, it is desirable to enhance the adherence of the alumina scale to the coated substrate so that the scale spallation and consequent consumption of Al from the coating can be minimized. In order to achieve this, modification of aluminide coatings by introducing elements such as Ce, Y, Cr, Pd and Pt has been investigated [22-28], and it has been identified that Pt is most effective in this regard. Pt-modified aluminide coatings, or simply Pt-aluminide coatings, have been shown to possess much superior high temperature oxidation resistance compared to their plain aluminide counterparts [26,28]. Therefore, these coatings have been the subject of intense research in past few decades. The superior oxidation resistance of Pt-aluminide coatings as compared to their plain aluminide counterparts can be illustrated based on the data reported in several studies [26,28]. Figure 1.17, which shows the weight change plots during cyclic oxidation as reported by Farrell *et al.* [26], clearly demonstrates the above fact.

Enhancement in oxidation and hot corrosion resistance achieved by Pt modification [29-34] has led to widespread use of Pt-aluminide coatings on superalloy components of advanced gas turbine engines. While it is known that the presence of Pt in aluminide coatings enhances the adherence of the protective alumina scale on

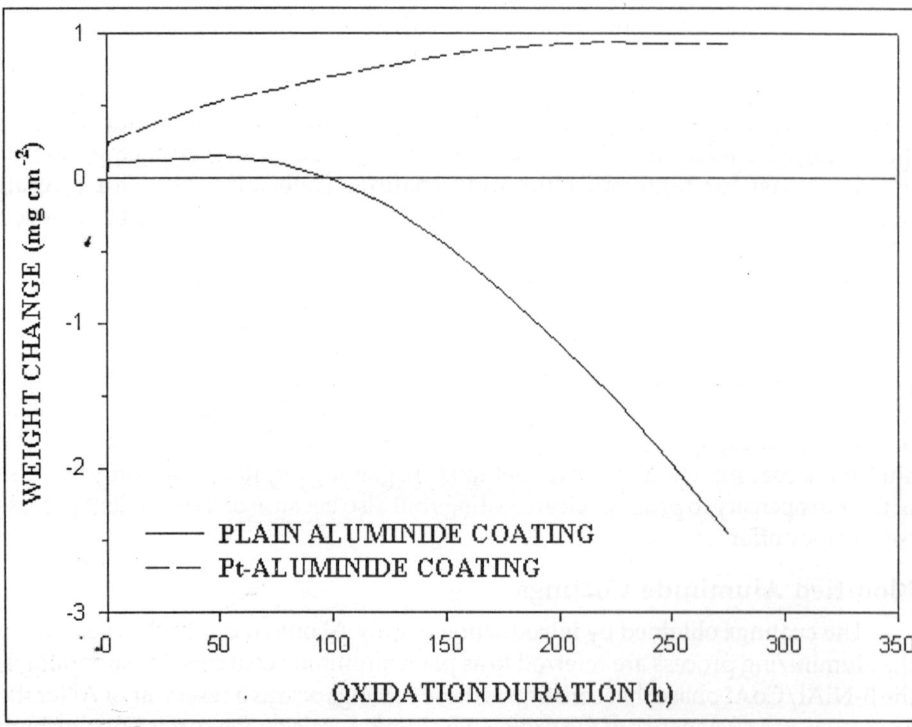

Figure 1.17: Weight Change vs. Time Plot for a Plain Aluminide and a Pt-aluminide Coating on IN-738 Superalloy during Cyclic Oxidation at 1100°C [26]. The superior oxidation resistance of the Pt-aluminide coating as compared to that of the plain aluminide coating is clearly evident from the figure.

the coating surface during high temperature exposure and, thereby, improves the oxidation resistance, the exact mechanism by which this improvement in adherence is achieved is not yet fully understood [29,34-40]. Several conjectures have been made in this regard which include the possibility of Pt reducing growth stresses in the alumina scale [34-37], and promoting the formation of oxide pegs for better anchoring of the scale on the coating surface [38,39].

The formation of Pt-modified aluminide coatings usually takes place in three distinct steps. Initially, a layer of pure Pt of thickness 7-10 μm is deposited on the superalloy substrate. Although the electroplating technique is most widely used for Pt deposition, techniques such as physical vapour deposition (PVD) [28,41] and fused salt electrolysis [42] have also been adopted for this purpose. In the second step, the plated sample is given a diffusion heat treatment, which is often called as "prior diffusion treatment", for improving the adhesion of the plated layer with the substrate. This treatment also causes certain amount of interdiffusion between the substrate alloy and the plated layer. The prior diffusion treatment is usually carried out in an inert/reducing atmosphere or in vacuum. The diffusion-treated samples are subsequently pack aluminized for developing the final coating. A typical Pt-aluminide coating microstructure reveals three distinct layers: (a) an outer Pt-rich two-phase layer having $PtAl_2$ phase distributed in a matrix of β-NiAl, (b) an intermediate β-NiAl layer having some amount of Pt in solid solution, and (c) the innermost interdiffusion layer which is typically found in any diffusion aluminide coating [43-46]. The above three layers are clearly evident from the microstructure shown in Figure 1.18. The outer Pt-rich layer in many currently available Pt-aluminide coatings is found to consist of single-phase β-NiAl which retains all the Pt in solid solution.

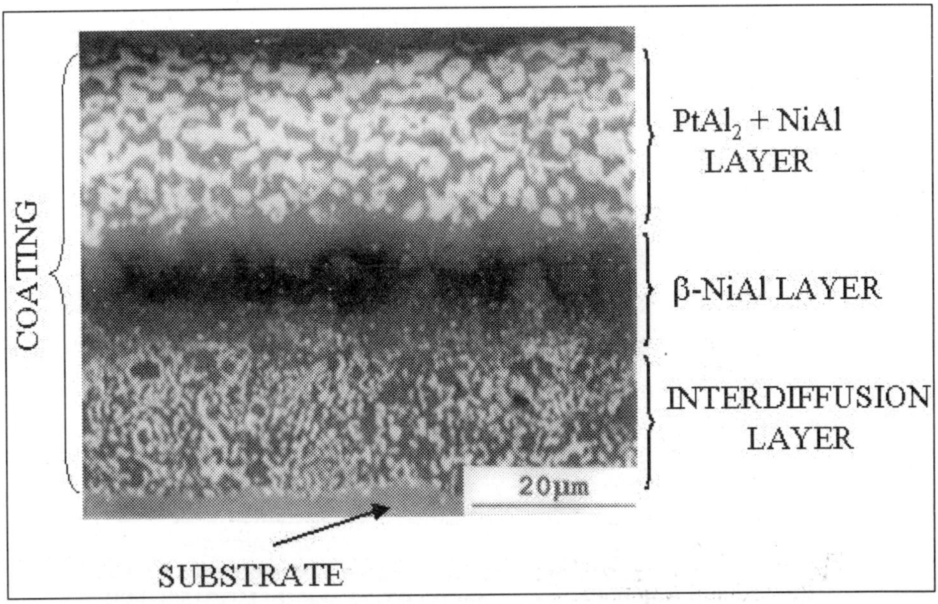

Figure 1.18: Typical Microstructure of a Pt-modified Aluminide Coating [43-46]

Currently, Pt-modified aluminide coatings are commercially available from several sources. One of the popular supplier is Chromalloy Ltd., U.K., which markets these coatings under brand name RT-22. Reports of other coatings such as JML-1, developed by Johnson Matthey Ltd., U.K., and LDC-2 can also be found in the literature [32,41,42]. More recently, Pt-aluminide coatings under the trade names SS82A and MDC150L are reportedly available on a commercial scale from Howmet Turbine Components Coating, U.S.A. and Thermatech, U.S.A., respectively [43]. Some of these coatings are presented in Figures 1.19(a)-(c) [43].

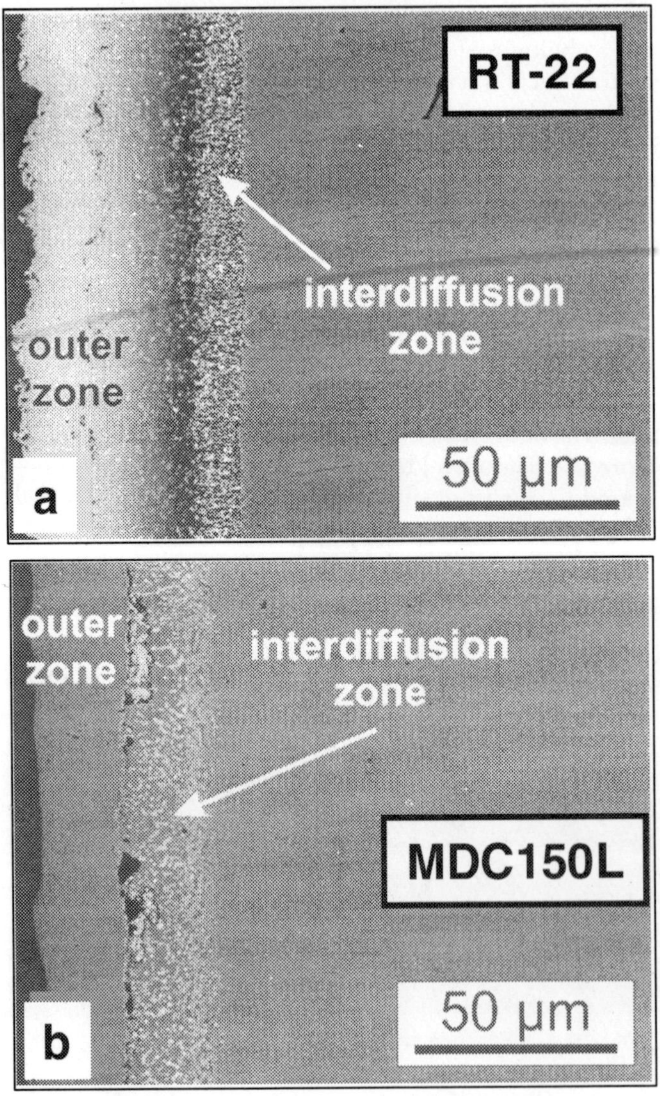

Figure 1.19: Typical Microstructures of Some Commercially Available Pt-modified Aluminide Coatings: (a) RT-22 and (b) MDC150L [43]

Chromized and Siliconized Coatings

Chromized coatings are produced by introducing Cr into the surface of super alloys and steels. The coating formation can be carried out by pack cementation method or by other conventional CVD methods. A typical chromizing pack consists of pure Cr powder as the Cr-source, a halide salt as the activator and alumina powder as the filler. The mechanism of chromized coating formation is very much similar to that of an aluminized coating. Siliconized coatings are also formed in a similar manner on super alloy substrates. The siliconizing pack contains pure silicon as the Si-source apart from containing a halide salt as the activator and alumina as the filler. While chromized coatings contains chromide phases, siliconized coatings are usually constituted of silicides (of Ni on Ni-base super alloys). Although both these two coatings are used for providing protection against high temperature oxidation and corrosion, the temperatures of their application are much lower than those for aluminide coatings. While the use of chromized and siliconized coatings are restricted below about 800°C, aluminide coatings, especially Pt-aluminide coatings, can be used up to 1200°C. During exposure to high temperature, chromized coatings form a chromia (Cr_2O_3) layer on the surface which reduces the further oxidation. Similar action is carried out by a silica (SiO_2) layer in case of siliconized coatings. Chromized coatings are said to possess superior hot corrosion resistance to aluminized and siliconized coatings.

References

1. Thelning, K.E., 1984. Steel and Its Heat Treatment, 2nd ed., Butterworths, London.
2. Leslie, W.C., 1981. The Physical Metallurgy of Steels. In. Materials Science and Engineering Series, McGraw-Hill, New York.
3. Metals Handbook. 1981. Vol. 4, 9th ed., ASM, Metals Park, Ohio.
4. Lakhtin, Y., 1983. Engineering Physical Metallurgy and Heat Treatment, Mir Publishers, Moscow, p. 263.
5. Goward, G.W., Boone, D.H., and Giggins, S., 1967. Trans. ASM, 60, p. 228.
6. Goward, G.W., and Boone, D.H., 1971. Oxid. Met., 3, p. 475.
7. Pichoir, R., 1978. Materials and Coatings to Resist High Temperature Corrosion. Applied Science Publishers, London, p. 271.
8. Duret, C., Mevrel R., and Pichoir R., 1984. Surface Engineering. NATO ASI Series, Martinus Nijhoff, Dordrecht, p. 452.
9. Das, D.K., Singh, V., and Joshi, S.V., 2000. Journal of Metals-e, 52, p. 1.
10. Puyear, R.B., 1962. Machine Design, July, p. 177.
11. Huminik, J., 1963. High Temperature Inorganic Coatings. Reinhold Publishing Corporation, New York.
12. Powell, C.F., Campbell, *I.E.*, and Gosner, B.W., 1955. Vapour Plating, John Wiley and Sons, New York.
13. Samuel, R.L., and Lockington, N.A., 1964. Chem. Proc. Eng., 45, p. 249.

14. Brill-Edwards, M., and Epner, M., 1968. Electrochem. Technol., 6, p. 299.
15. Goebel, J.A., 1979. Advanced Coating Development for Industrial Utility Gas Turbine Engines: Proc. 1st Conf. On Advanced Materials for Alternate Fuel Capable Directly Fired Heat Engines. U.S. Dept. of Energy and Electric Power Research Institute, p. 473.
16. Duret, C., and Pichoir, R., 1983. Coatings for High Temperature Applications. Applied Science, New York, p. 33.
17. Mevrel, R., and Pichoir, R., 1987. Mater. Sci. Eng., 88, p. 1.
18. Seigle, L.L., 1984. Surface Engineering. NATO ASI Series, Martinus Nijhoff, Dordrecht, p. 561.
19. Parzuchowski, R.S., 1977. Thin Solid Films., 45, p. 349.
20. Gauje', G. and Morbioli, R., 1982. High Temperature Protective Coatings. The Metallurgical Society of the American Institute of Mining, Metallurgical, and Petroleum Engineers, Warrendale, p. 13.
21. Sun, W., Lin, H.J., and Hon, M.H., 1987. Thin Solid Films, 146, p. 55.
22. Jedlinski, J., Godlewski, K., and Mrowec, S., 1989. Mater. Sci. Eng., A121, p. 539.
23. Tu, D.C., Lin, C.C., Liao, S.J., and Chou, J.C., 1986. J. Vac. Sci. Technol. A, 4, p. 2601.
24. Godlewski, K., and Godlewska, E., 1986. Oxid. Met., 26, p. 125.
25. Alperine, A., Steinmetz, P., Josso, P., and Costantini, A.F., 1989. Mater. Sci. Eng., A121, p. 367.
26. Farrell, M.S., Boone, D.H., and Streiff, R., 1987. Surf. Coat. Technol., 32, p. 69.
27. Twancy, H.M., Abbas, N.M., and Rhys-Jones, T.N., 1991. Surf. Coat. Technol., 49, p. 1.
28. Sun, J.H., Jang, H.C., and Chang, E., 1994. Surf. Coat. Technol., 64, p. 195.
29. Tatlock, G.J., and Hurd, T.J., 1984. Oxid. Met., 22, p. 201.
30. Gobel, M., Rahmel, A., and Schutze, M., 1993. Oxid. Met., 3/4, p. 231.
31. Patanaik, P.C., Thamburaj, R., and Sudarshsan, T.S., 1990. Surface Modification Technologies III. The Minerals, Metals and Materials Society, p. 759.
32. Bauer, R., Schneider, K., and Grunling, H.W., 1985. High Temperature Technol., 3, p. 59.
33. Wu, W.T., Rahmel, A., and Schorr, M., 1984. Oxid. Met., 22, p. 59.
34. Fountain, J.G., Golightly, F.A., Scott, F.H., and Wood, G.C., 1976. Oxid. Met., 10, p. 341.
35. Johnson, G.R., Cocking, J.L., and Johnson, W.C., 1985. Oxid. Met., 23, p. 237.
36. de Wit, J.H.W., van Manen, P.A., 1994. Mater. Sci. Forum, 154, p. 109.
37. Newcomb, S.B., and Stobbs, W.M., 1996. Elevated Temperature Coatings: Science and Technology II. Minerals, Metals and Materials Society, Warrendale, p. 265.

38. Allam, I.M., Akuezue, H.C., and Whittle, D.P., 1980. Oxid. Met., 14, p. 517.

39. Felten, E.J., and Pettit, F.S., 1976. Oxid. Met., 10, p. 189.

40. Felten, E.J., 1976. Oxid. Met., 10, p. 23.

41. Jackson, M.R., and Rairden, J.R., 1977. Metall. Trans. A, 8A, p. 1697.

42. Wing, R.G., and McGill, I.R., 1981. Plat. Met. Rev., 25, p. 94.

43. Angenete, J., 2002. Aluminide Diffusion Coatings for Ni Based Superalloys: Coating and oxide microstructure, Ph.D. Thesis, Department of Experimental Physics, Chalmers University of Technology, Göteborg University, Göteborg, Sweden.

44. Das, D.K., and Annapurna, J., 1996. Development of Pt-Aluminide Coatings on CM-247 Ni-Base superalloy : II. A Preliminary Study on Mechanism of Coating Formation, Defence Metallurgical Research Laboratory Technical Report, DMRL TR 96206, Hyderabad, India.

45. Krishna, G.R., Das, D. K., Singh, V., and Joshi, S.V., 1998. Mater. Sci. Eng. A, A251, p. 40.

46. Das, D.K., Singh, V., Joshi, S.V., 2000. Metall. Mater. Trans. A, 31A, p. 2037.

Chapter 2

Thermally Sprayed Coatings

Srikant V. Joshi
International Advanced Research Centre for Powder Metallurgy and New Materials
Opp. Balapur Village, R.R. District, Hyderabad – 500 005, India

Introduction

Most often, the degradation of a component during service, and its eventual failure, can be traced to material damage resulting from surface/environment interactions. Protective coatings are primarily sought to enhance the durability and/or performance of components employed in adverse operating conditions. Typically, these coatings are aimed at modifying the surface properties of critical components to provide enhanced resistance against deterioration due to mechanisms such as corrosion, oxidation, wear, or failure under an excessive heat load. In recent years, considerable advances in the field of coatings technology have accompanied the growing realization of the immense potential of surface engineering in the modern industrial world. Consequently, there are now available a number of methods for developing a wide variety of protective coatings.

Of all the coating methods, the thermal spray technique has gained by far the most widespread industrial acceptance to date. It has gradually emerged as the most industrially useful method of developing a variety of coatings, to enhance the quality of new components as well as to reclaim worn/wrongly machined parts. The availability of numerous variants of the thermal spray technique, the diversity of sprayable materials and continuing advances in spray-control systems have together created new opportunities for the thermal spray industry. Above all, the versatility of the thermal spray process is unmatched by any other surface modification technology.

What is Thermal Spraying?

The origin of the concept of thermal spraying actually dates back to the very early part of this century when a Swiss engineer, Max Schoop, and his associates developed a metal coating by blasting a surface with metal dust. Schoop and his team took another significant step forward in 1910, obtaining a metal coating by passing metal powder through an annular flame and propelling the near-molten particles onto a substrate by a blast of compressed air. Within a few years of the above work, the wire spraying process was developed which remains popular even today. To begin with, the above coating process became known as metallizing, since only metallic materials were deposited. The existing terms thermal spraying or flame spraying came later when the technological advances made it possible to deposit alloys, cermets and ceramics.

The present day status of modern thermal spraying as the major means of depositing protective coatings has been the result of the astounding progress made in several allied fields during the past three or four decades. The most significant development in the field was, for long, considered to be the introduction of the plasma spray torches which provided an impetus to accelerated growth of this technology– plasma arcs, and their variations, were the foundation of the thermal spray industry for more than two decades and widely used to obtain a range of protective coatings for numerous industrial applications. Among the relatively more recently developed techniques, high velocity oxy-fuel (HVOF) spraying first brought about a major revolutionary change in thermal spray technology. Coatings deposited by the HVOF methods were found to be comparable to those deposited by Detonation Gun (D-Gun), a proprietary process introduced in the late 1950s by Union Carbide for wear protection. Following the industrial acceptance of the first-generation gaseous fuel based HVOF systems, second generation HVOF systems capable of operating with liquid fuel (typically aviation turbine fuel) were soon introduced into the market in the late 1990s and found to be a resounding success. The proprietary D-Gun coatings, subsequently available from PRAXAIR Surface Technologies, had traditionally presented single sourcing concerns as these coatings were only available as service from the company with the associated equipment not available for sale. Consequently, the HVOF techniques offered a more convenient alternative and gained rapid popularity for depositing a wide range of wear resistant coatings. To further contribute to the growth of the thermal spray industry, the past few years have also seen some coating units based on the detonation spray process become available from sources within the erstwhile Soviet Union, Spain, USA and India. Although their cumulative current equipment sales are rather low pending appropriate benchmarking, the indigenous availability of detonation spray units in India is certain to provide a much needed impetus to the thermal spray market in the country.

More recently, the advent of the Cold Gas Dynamic Spray (CGDS) process has brought further excitement to the thermal spray industry. The CGDS technique eliminates most of the disadvantages associated with the high temperatures typical of all the above thermal spray variants *viz.* oxidation, thermal dissociation, phase change, elemental depletion etc. The advent of the CGDS method is also a logical progression in the direction that the thermal spray industry has sought to move in, namely reduction in thermal energy and increase in kinetic energy. Although the CGDS method has no widespread commercial applications established yet, it has opened up new vistas for overall thermal spray technology development.

Process Fundamentals

Thermal spraying, or flame spraying, are the generic terms used for processes wherein the deposited layer is developed by melting the material to be coated in a high temperature zone and propelling the molten droplets onto the substrate. In other words, all the variant processes of the thermal spray technique generally involve the following features:

- The material to be sprayed is heated in a flame or an arc so that it is substantially molten.
- The molten material is accelerated in a gas stream and propelled onto the surface to be coated.
- The partially/fully molten particles flatten upon impacting the substrate and subsequently cool and coalesce to yield the desired coating.

A graphical illustration depicting the above sequence of events leading to the formation of the coating by a thermal spray process is provided in Figure 2.1. The consumable coating material can be in the form of wire, rod or powder although use of feedstock in the form of spray-grade powders is most popular. Virtually any material that melts without decomposing can be sprayed. The sprayed coatings are built up layer by layer and, although the desired thickness of the deposit may vary depending upon the application, the protective coatings are typically 100-500 microns thick.

The mechanism of the bond between a sprayed coating and the substrate is mainly mechanical although there is some evidence that, in certain instances, a chemical interaction between the impacted particle and the substrate contributes to the adhesion. Apart from the need for adequate surface preparation prior to coating, it is known that the extent of particle melting and its velocity, at the moment of impact with the substrate to be coated, most crucially affect the coating characteristics. Complete particle melting is essential to minimize porosity in the coating while a high impact velocity leads to a well bonded coating. The gas temperature and gas

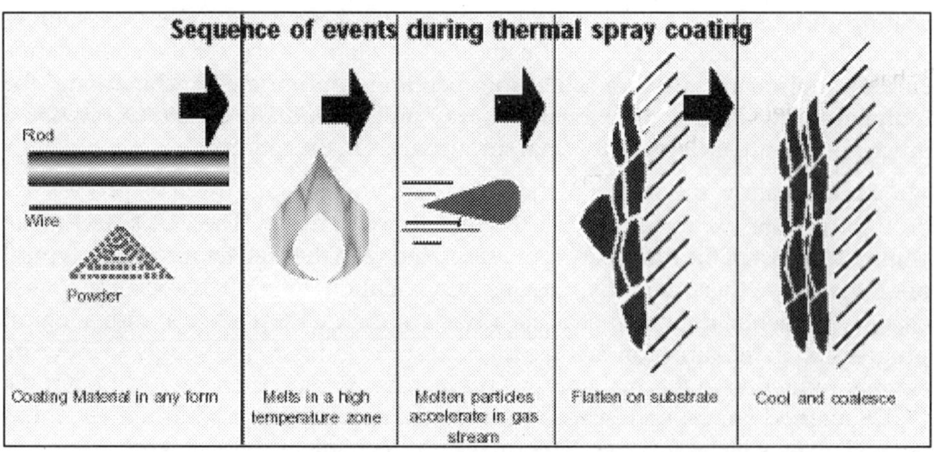

Figure 2.1: Sequence of Events during Thermal Spraying

velocity encountered by the powder particles during spraying are of considerable significance in determining the extent of particle heat-up and acceleration.

Of the three factors that primarily govern the coating-substrate bond strength *viz.* impact velocity of the particle, the physical state of the particle at impact and surface preparation, the first two are controlled by the coating process for any given powder or by the powder characteristics for any given coating technique. However, once the coating material and the method of coating have been chosen, substrate preparation is the single most important operation which determines the adhesion and warrants the same degree of control as the coating process itself. The primary requirement for achieving good adhesion is that the surface of the component to be coated must be clean, grease-free and rough. In this context, grit blasting of the surface provides certain unique benefits. Besides offering one of the best ways of cleaning the substrate and exposing fresh material, it also serves to roughen the surface, thereby increasing the surface area and ensuring satisfactory bonding of coatings.

Thermal Spray Variants

Table 2.1 presents the variety of thermal spray processes now available for coating purposes as a consequence of the above advancements in this technology along with their principal characteristics. While the process fundamentals and the surface preparation methods are similar for every variant, each of the processes included in Table 2.1 differs significantly in the manner in which the high temperature zone is generated and, consequently, in equipment design. These differences, along with the important characteristic features of all the variants are discussed below. The plasma spraying technique is dealt with in considerably more detail than any of the others because it is, by far, the most involved/complicated and, yet, the most widely used by virtue of its versatility.

Table 2.1: Principal Characteristics and Features of Important Thermal Spray Variants Employing Powders as Feedstock

	Flame	*Air Plasma*	*Detonation Spray*	*HVOF*	*Cold Spray*
Usual Coating Material Form	Powder/Wire	Powder	Powder	Powder	Powder
Heat Source	Oxy-fuel combustion	Plasma Flame	Controlled Detonation	Oxy-Fuel Combustion	Resistance heater to pre-heat gas
Flame Temp. (°C)	3,000-3,500	10,000-15,000	3,000-3,500	3,000-3,500	No flame; max. gas pre-heat 600 C
Gas Velocity (m/s)	< 300	400-500	3,000	1,500-2,000	2,500-3,000
Particle Temp. (°C)	1,500-2,200	2,700-3,500	1,500-2,000	1,500-2,000	400-600
Particle Vel. (m/s)	50-100	100-200	600-800	600-800	800-1000

Plasma Spraying

Most simply, a plasma may be defined as a partially ionized state of gas. It can be generated by imparting a sufficient amount of energy to strip off electrons from the gas molecules. Generally, a plasma is produced either by passing a plasma-generating gas through a high intensity arc struck between two electrodes (arc plasma) or by high radio frequency excitation of the plasma gas (RF plasma). A non-transferred DC arc plasma has been conventionally used for coating purposes and is most widely employed in the thermal spray industry. In this kind of torch, constriction of the arc and forcing it out through a nozzle enhances, respectively, both the arc temperature and the plasma gas velocity. The plasma flame basically serves as a heat source to melt the injected powder particles, which are then propelled onto the substrate where they deposit and form the coating. The quality of the developed coating depends, among other things, on the operation of the plasma gun, the gas-solid transport phenomena occurring as the particle traverses through the plasma flame and, finally, on the interactions between the impacting particles (which may be in a fully molten, partially molten or a solid state) and the substrate.

While the field of plasma coatings has significantly developed in several respects, particularly in the last twenty years or so, the basic design of the plasma gun has undergone little change during this period. A schematic diagram of a plasma gun is shown in Figure 2.2. Typically, the gun consists of a cylindrical anode (usually made of copper) and a cone-shaped cathode (usually made of thoriated tungsten). A high intensity arc struck between the two water-cooled electrodes ionizes the working gas to form a plasma. An increase in the operating power level increases the plasma volume but raises the arc temperature only slightly. An essential feature of a plasma torch, therefore, involves passing the arc through an anode nozzle of diameter smaller than that of the natural arc–this serves to increase the current density and, thereby, achieve significantly higher temperatures. Although some plasma torches have a power rating of up to 80 KW, a handling capacity of about 40 KW is usually adequate for most spraying applications.

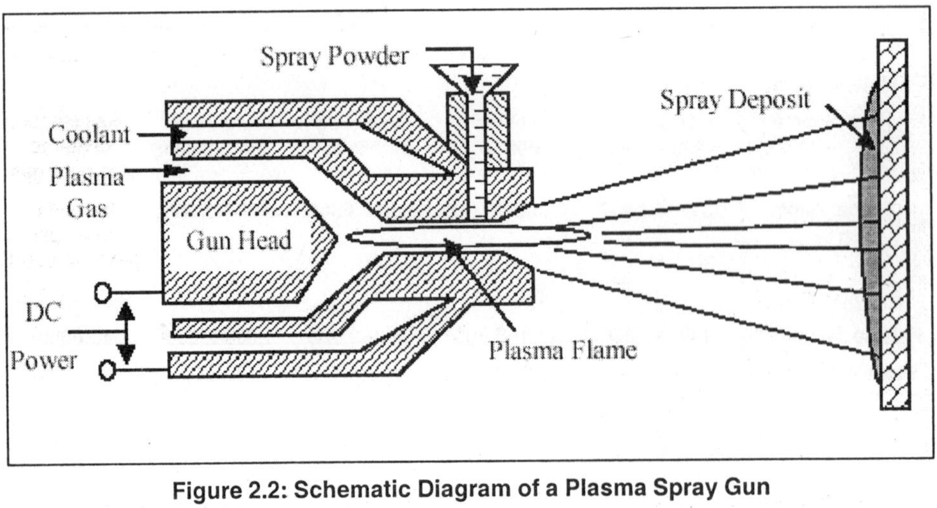

Figure 2.2: Schematic Diagram of a Plasma Spray Gun

The choice of the plasma arc gases, which is often dictated by their chemical reactivity, thermophysical properties and costs, affects electrode life, spray efficiency and deposit quality. Nitrogen, hydrogen, argon and helium are the commercially used plasma gases. In general, argon is the most suitable for routine applications with hydrogen being usually added to enhance the heat content of the plasma gas and thereby improve the gas-particle heat transfer to aid particle melting. Hydrogen also serves to provide a reducing atmosphere and somewhat suppress the oxygen contamination problem when the plasma coating is carried out under atmospheric conditions, as is often the case for normal industrial applications. However, only small amounts of hydrogen should be added, for it increases anode nozzle deterioration due to erosion.

A host of plasma variables influence coating quality and these need to be manipulated to maximize both the heat transfer to the particles as well as the particle velocity. In spite of the involved nature of the problem, the spray parameters should always be manipulated so as to achieve the highest possible particle velocity while assuring that the particle residence times are still sufficiently long to permit their complete melting. Usually, the most suitable spray parameters for a given powder size/composition are provided by the spray equipment supplier–in the absence of such information, the spray parameters have to be optimized in-house.

Equipment/Process Development in Plasma Spraying

In its most popular configuration, the plasma coating process is carried out under atmospheric conditions. However, as already illustrated in Figure 2.2, numerous variations of this process also exist with the Low Pressure Plasma Spraying (also known as Vacuum Plasma Spraying) being the most significant among them. In this technique, the plasma generator, the work piece and the powder port are all enclosed in a vacuum tank and the preheating, spraying as well as any post heat-treating operations are performed in a soft vacuum of about 50 torr. Although this technology requires a huge initial investment and is not yet indigenously available, plasma spraying in a low-pressure environment offers considerable advantages over the conventional air plasma spray process. It provides a much longer plasma jet, virtually eliminates the problem of possible oxidation due to incorporated air, improves the deposition efficiency of impacting particles and offers excellent dimensional thickness control besides reducing the porosity in coatings.

Another alternative to the existing high-energy coating systems involves operation of the plasma process under water. Materials with high melting points can be processed in the high-energy plasma jets operating under water. A reduction in the gas content in the coating compared to the atmospheric plasma spraying is one characteristic of this process. In addition, the substrate is constantly cooled by the surrounding water, allowing thermally sensitive materials such as plastics to be deposited. The water isolates the plasma under water and allows a controlled heat input into the base material. Dust or powder losses from the process are transferred to the water and a drastic reduction of radiation and noise emissions leads to a safer working environment. Applications of this variant of plasma spraying include wear and corrosion protection of surfaces located under water such as those in off-shore and marine installations.

Recently, radio-frequency (R.F.) induction plasma technology has gained wide acceptance for the preparation of plasma-sprayed coatings and structural free-standing parts of metals, ceramics and metal-matrix composites. Its principal advantages are the relatively large volume and low velocity of the discharge, which coupled with the ease of axial injection of the powder into the plasma, allows for the melting of relatively large particles at high throughputs. The absence of electrodes offers the added advantage of ease of operation under a wide range of conditions and also enables deposition of coatings, which are free of any metallic impurities from the electrodes. This is of particular significance when spraying coatings for medical applications or metal-sensitive, oxide-based high temperature superconductors.

Reactive plasma spray processes, where unique plasma chemistry can be used to form composites, ceramics and intermetallic materials, are also being developed. These processes introduce gases such as methane, propylene, nitrogen, oxygen, silane or borane into the plasma and, as the reacted compound particle film ruptures on impact, a dispersed composite material is created. This variant is acknowledged to have considerable potential but is yet to attract any significant commercial attention.

Combustion Flame Spraying

Prior to the acceptance of plasma spraying as a viable technique for developing protective coating on an industrial scale, the bulk of commercial coating production was based on combustion flame spraying and this deposition process is still widely used today. In this method, the material to be coated may be in the form of powder, wire or rod and the high temperature zone for melting the material is provided by the combustion of an oxygen-fuel gas mixture. The most common fuel gas is acetylene but propane or hydrogen can also be used.

In combustion flame powder spraying, the powder to be coated is normally introduced into the flame region by means of a carrier gas. However, torches employed for limited batch work have a hopper attached to the top of the torch for gravity feeding of the powder as shown in Figure 2.3. In case of the wire spray torch, the wire is gradually fed into the flame of a specially designed torch–the wire tip continuously melts and a supply of compressed air blasts the molten tip to produce a fine molten metal spray which is subsequently deposited on the substrate. In case of materials that cannot be drawn into a wire, fine powder contained in plastic tubes has also been used in conjunction with the wire spray torch.

As compared to a plasma spray system, a combustion flame spray unit is exceedingly simple and this translates into a considerably cheaper equipment and lower capital investment. The coating quality in combustion flame spraying depends mainly on the powder/powder injection variables and the gun-to-job distance. Furthermore, the velocity and temperature profiles in the combustion flame vary nowhere near as drastically with axial distance as they do in the case of a plasma flame. This makes the entire spraying process far easier to control by way of reproducibility of coating quality.

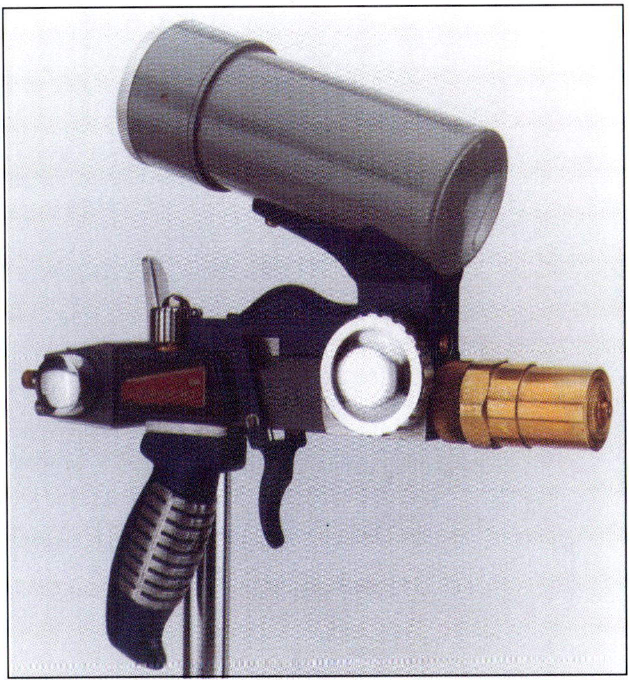

(a)

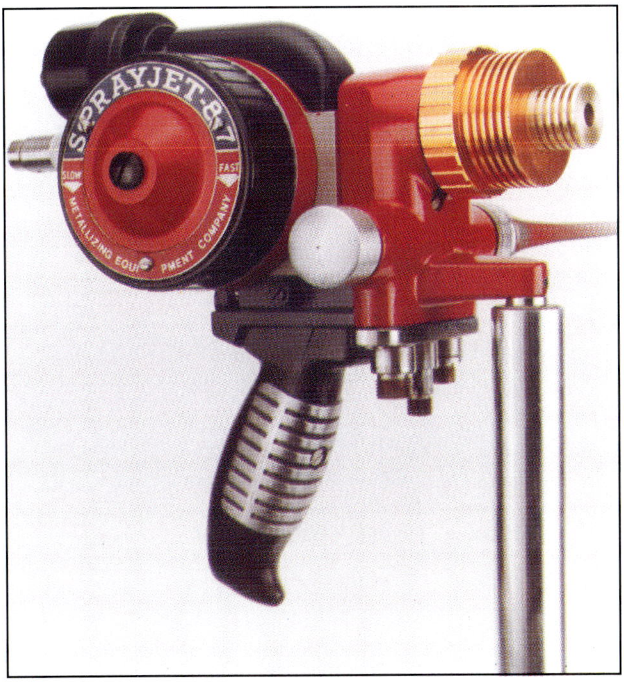

(b)

Figure 2.3: Combustion Flame Spray Guns
(a) With powder feeding and (b) With wire feeding

A major limitation of the combustion flame technique is that the maximum available flame temperature is only around 3000 C, which is insufficient to completely melt the powders of more refractory metals and many ceramics, unless the particle size is very fine. Also, the long flame tends to heat the substrate- besides causing possible workpiece distortion; this precludes the use of low melting point substrates such as plastics. As the sprayed particles are accelerated up to an impact velocity of only about 100 m/s and due to the oxidizing nature of the flame, this method can potentially yield coatings with a high oxide content and a high level of porosity, which may or may not be acceptable for the intended application of any given coating. However, use of compressed air as a propellant can substantially reduce the porosity. In some instances, using the torch to fuse the developed deposit can also increase the bond strength and coating density.

Arc Wire Spraying

Arc spraying differs from the combustion flame spraying process in that the heat source is a D.C. arc struck between two electrodes. The material to be sprayed, which must be available as rod or wire, forms the consumable electrodes of the gun. As in the case of combustion flame wire spraying, the molten trips of the electrodes are atomized and blown against the substrate by a blast of compressed air. Figure 2.4 shows an arc spray torch in operation. The temperature available in the arc is in excess of 4000 C and the heat transfer to the coating material is also extremely efficient in the case of an arc. This allows deposition of marginally denser coatings than the combustion flame method and, furthermore, enables use of high spray rates thus making the process well-suited for developing thick coatings on large components. An additional advantage is that arc spraying does not heat the substrate too much.

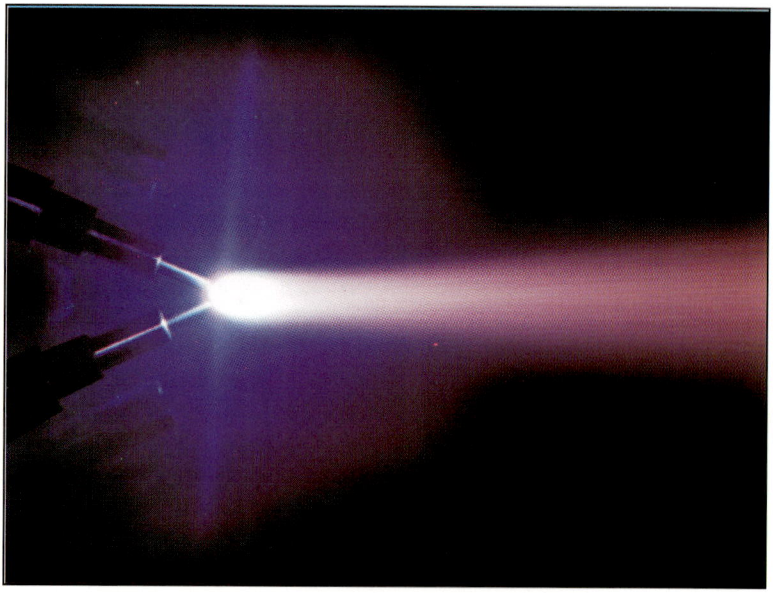

Figure 2.4: An Arc Spray Torch in Operation

However, a major shortcoming of this process is that only conductive metals can be applied by this method.

Detonation Spray Coating Technique

This was a coating technique developed almost simultaneously by the Linde Air Products Division of Union Carbide and by scientists in the erstwhile Soviet Union. Until recently, this process was strictly of a proprietary nature, widely known as the D-Gun technique as christened by Union Carbide, and available as a service only from the Coatings Service Division of Union Carbide U.K. Ltd. (now from Praxair Surface Technologies). However, detonation spray systems based on the original designs from the former Soviet Union have now become commercially available and shown to provide high-quality coatings. The detonation spray process basically depends on the controlled detonation of oxygen and acetylene in a specially designed combustion chamber to provide the high temperatures required for melting powders of the material to be coated. A schematic diagram of a detonation spray system is shown in Figure 2.5.

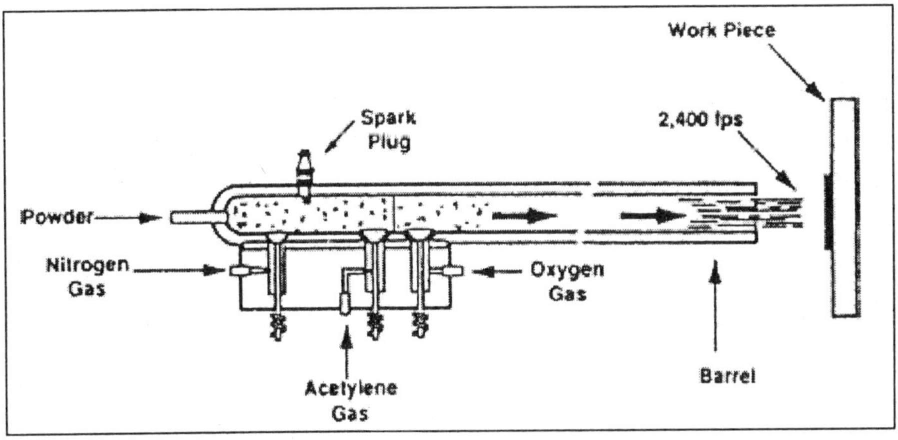

Figure 2.5: Schematic Diagram of a Detonation Spray System

As opposed to other flame deposition processes, which are continuous, the combustion process in the D-Gun technique is cyclic, with the oxy-acetylene mixture being detonated at precisely controlled intervals four or eight times a second by timed electrical pulses arcing across the gap of the spark plug. Following the detonation, the gas reaches a temperature of around 3000 C and a velocity as high as 3000 m/s. The powder particles are usually carried by nitrogen gas under pressure and suspended in the gas mixture prior to the detonation. The high temperature, high velocity gas products of the detonation melt the particles and accelerate them sufficiently so that they emerge from the gun at velocities as high as 1200 m/s depending upon the material being sprayed and its powder size characteristics. The resulting high velocity impingement of the particles on the substrate yields very dense coatings with excellent adhesion.

High Velocity Oxy-Fuel (HVOF) Spraying

User concerns about high coating costs and single sourcing of Union Carbide's D-Gun process provided a major impetus to evaluate alternate methods of generating similar high quality coatings and led to the development of the high-velocity flame spray (or HVOF) technique. HVOF thermal spraying was widely regarded as the most significant development in the thermal spray industry since the plasma gun was employed for coating applications. The growth of this technology, since the introduction of the first HVOF process called JET KOTE (a registered trademark of Stoody Deloro Stellite, Inc., U.S.A.) in 1982, has been phenomenal and numerous hypersonic combustion guns are now commercially available. The HVOF processes rely on the continuous internal combustion of a fuel gas with oxygen to produce the high temperature, high velocity exhaust gas stream essential for coating applications. Propylene, propane or hydrogen are the most commonly used fuel gases. While the flame temperature is close to 3000 C or only slightly above it, the flame velocity is hypersonic (in the range 1500-2000 m/s) and it is reported that uniform heat input and acceleration is available over almost a 12" distance. The powder material to be used for coating is injected in a carrier gas (nitrogen, argon or helium) and, owing to the high back-pressures created by the combustion process, a pressurized powder feeder is required. The powder is molten and propelled by the combustion gases to impact the substrate at very high speeds to provide well-adherent and dense coatings as in the case of detonation spraying. However, the relatively low gas temperatures available in the HVOF technique preclude the use of some refractory metals and many ceramics for spraying and represent a significant shortcoming compared to the detonation spray systems.

Many HVOF systems are commercially available today. Figure 2.6 shows the schematic diagrams of some of the gun chamber designs that have been adopted. Encouraged by the success of the 'first-generation' gaseous fuel based HVOF systems mentioned above, a second generation of liquid-fuel based HVOF units are also now in the market and being successfully employed in several industry segments.

Cold Gas Dynamic Spray Process

The development of the Cold Spray technique, also referred to as the Cold Gas Dynamic Spray (CGDS) process, represents the most significant advancement in the field of thermal spraying in recent years. As mentioned previously, the development of the CGDS technique is consistent with the desire of the thermal spray community to reduce the thermal energy content and increase the kinetic energy component during spraying. In the Cold Spray coating process, very fine powder particles of metals, alloys and composites (typically in the size range of 1-25 mm) are injected into a rapidly expanding supersonic gas jet at a temperature much lower than the melting point of the material. The simple process relies on a high-pressure compressed air/gas, which is accelerated by convergent-divergent de Laval type nozzle, to impart sufficient momentum to the injected powder particles. The powder particles are accelerated to supersonic velocities and propelled onto the part to be coated. Above a certain critical velocity that is characteristic of the coating material, particulate impact results in solid-state plastic deformation of the particles leading to formation of dense and highly adherent coatings. The processing gas can either be Compressed air,

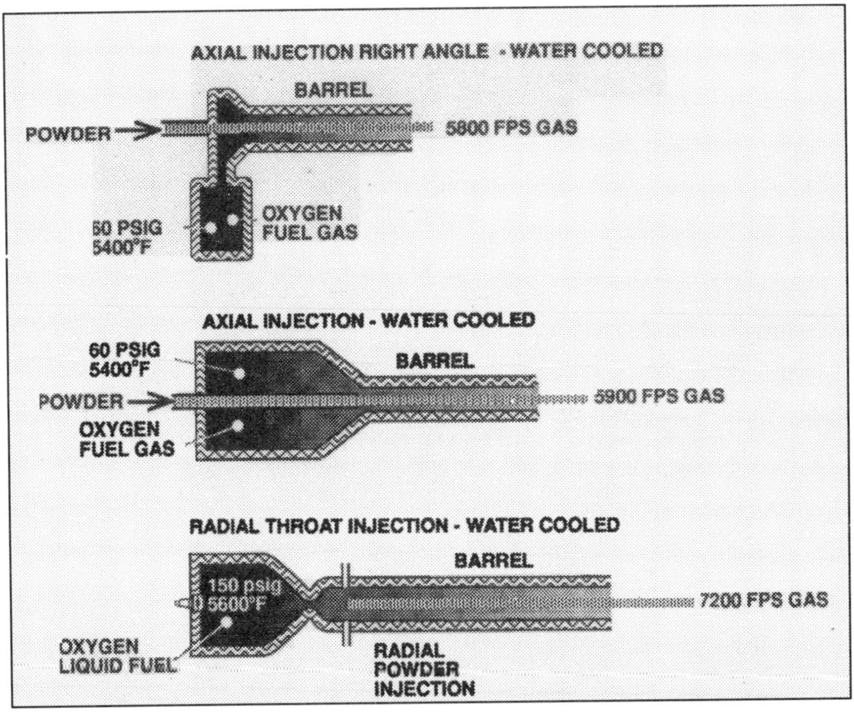

Figure 2.6: Schematic Diagrams of some Typical HVOF Systems

Nitrogen or Helium with the type of gas used also influencing the velocity of the gas jet. Occasionally, the gas is also preheated to temperatures not exceeding 500-600ºC to augment gas velocities as well as to aid plasticization of particles on impact. Figure 2.7 depicts a CGDS system in operation. The Cold Spray process has attracted widespread global interest since its emergence. This can be primarily attributed to the fact that it suppresses many of the detrimental aspects of traditional thermal spraying such as high temperature oxidation, thermal degradation, phase transformation, evaporation, melting, crystallization, residual stresses, gas entrapments etc.

As evident from the above discussion, each of the major thermal spray processes differ significantly in the manner in which the high temperature zone is generated and, consequently, in equipment design. Notwithstanding the fact that each of the above process variants has its advantages and limitations, all of them are now commercially well accepted and successfully employed for numerous industrial applications as detailed in a subsequent section.

Powders for Thermal Spraying

The versatility of the thermal spraying technique originates from the fact that any material that can be melted without decomposition can be thermally sprayed. Consequently, the range of materials addressed by thermal spraying includes metals, metal alloys, ceramics, cermets, composites, refractories and combinations of these. Some composite powders, which can yield deposits with unusual coating properties

Figure 2.7: Schematic Diagram of a CGDS System

by combining in an optimum manner the beneficial properties of their component materials, are also now available. As a result of the above diversity in spray-grade materials, coatings with wide-ranging functionalities can be developed enabling virtual tailoring of a surface to suit any prevailing operating environment. The various categories of coatings that can be developed by the thermal spray techniques include wear resistant, corrosion/oxidation resistant, thermal barrier, electrically conducting/ insulating, abradable and self-lubricating coatings. The powders normally used for each coating category and their typical areas of application are listed in Table 2.2.

The number of powders commercially produced for dedicated use in thermal spraying has also registered an impressive growth in recent times. The enhanced quality and the increased variety of spray grade powders available has made a significant contribution to the advancement of this technology and has been responsible, at least in part, for the widespread acceptance of the thermal spray technology on the manufacturing shop floor. As a result, the consumption of thermal spray powders has shown handsome growth in the last decade and this trend is likely to be maintained in the immediate future. For example, the estimated US market for spray-grade powders in 1990 was US$ 600-700 million and this had increased to over US$2500 million by the year 2002. Similarly, in Europe, the engineering (wear and corrosion) coating sector itself was valued at £4-5 billion in 1995, and this figure is predicted to increase to £8-9 billion by 2010.

Table 2.2: Popular Powders Used for Depositing Coatings to Achieve Different Functionalities

Property	*Powders Used*	*Applications*
Wear resistance	Al_2O_3, TiO_2, Cr_2O_3	Textile machinery parts, pumps components
	WC-Co, Cr_3C_2, TiC	Seals, jet-engine parts, paper industry
	Mo, Stellite, NiAl, NiCr	Bearings, automotive parts
Oxidation and corrosion resistance	NiCr, $MoSi_2$, Al, Al, AlZn, MCrAlY type (M=Fe, Ni or Co)	Electrodes, heating elements chemical industry, foundries, turbine blades, vanes
	S.S., Mo	Chemical and food industries
	Ta, Ti	Shipping industry
Thermal Barrier	ZrO_2 type, Al_2O_3, Mullite	Combustion systems, brazing and soldering equipment
Electrical insulation	Al_2O_3, Al_2O_3+TiO_2	High voltage components, computer and nuclear parts
Electrical conduction	Cu, Ag, Al	Electrical contacts, heating elements
Abradable and self-lubricating	Ni-graphite, NiCu	Turbine Seals for clearance control
	CaF_2-Ni	High temperature bearings

As a result of the availability of a wide array of powders with diverse chemistries for thermal spray applications, it is now possible to select very specific coating compositions that are best suited to combat any given surface degradation mechanism that is likely to be experienced by a part in service. Furthermore, numerous coating options is usually available to choose from based on the functionality required and the coating costs afforded by the intended application. Table 2.3 provides an example of the typical coating options available to combat different forms of wear.

Table 2.3: Coating Options to Resist Different Categories of Wear

Wear Process	*Typical Coating Options*
Abrasion	Al_2O_3, Cr_2O_3, NiCrBSiC, Wc-Co, TiC-Ni, Cr_3C_2–NiCr, Nano size Powders.
Cavitation erosion	NiTi, Cu-Ni, NiCrBSiCAlMo, 316S/S
Liquid impingement erosion	Stellite. Al_2O_3, WC- Ni
Fretting	Cr_3C_2–NiCr
Low amplitude	Cu-Ni, Cu-Ni-In
High amplitude	CoMoCrSi, Mo-NiCrBSi, T-800
Abradable	CoNiCrAlY, Ni-graphite, NiCrAl-Bentonite
Impact and sliding	WC-Co, T- 800, Cr_3C_2–NiCr, Nano size Powders.
Galling	Cr_3C_2–NiCr, T- 800
Solid particle erosion	WC- Co, Cr_3C_2–NiCr, WC- NiCrBSiC, Cr_3C_2- MCAlY, T-800
Adhesive	Al-bronze, Mo-NiCrBSi, Al_2O_3+ TiO_2, WC- Co
Scuffing	Mo, Mo-NiCrBSi,
Dry sliding	WC- Co, Cr_2O_3, Cr_3C_2–NiCr, NiCrBSi

Popular Methods of Manufacturing Spray-Grade Powders

There are numerous well-established methods of powder preparation like mechanical comminution, atomization, spray drying and sol-gel processing that have been routinely employed to manufacture spray-grade powders commercially. Each of these routes, lead to varied particle size morphology as well as particle size distribution as exemplified in Figure 2.8. In addition to the above methods, several unique methods of producing fine powders, including laser and plasma synthesis, have been recently developed. In addition to the vast number of metal, metal alloy and ceramic powders, there is now also emerging a growing demand for composite powders which can yield deposits with unusual coating properties by combining, in an optimum manner, the beneficial properties of their constituent materials. Compositing, which is the attaching of two or more dissimilar materials while permitting each to retain its own identity, is an effective way of forming a uniform multi-component coating. Ni-Al, Ni-Graphite, Ni-MoS_2, WC-Co, Ni-Cr-Cr_3C_2 and Ni-CaF_2 are some of the composite powders currently used for thermal spray applications. Compositing is normally achieved through chemical cladding, by one of the vapour/liquid/solid phase deposition techniques, followed by organic bonding and sintering.

(a) (b)

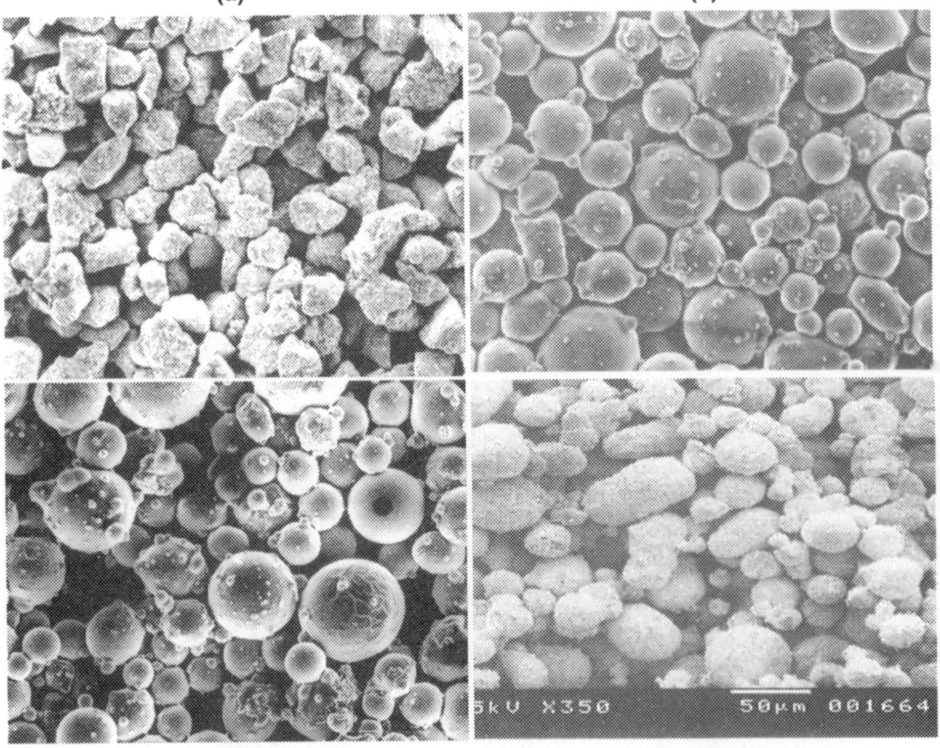

(c) (d)

Figure 2.8: Typical Morphologies of Spray-grade Powders Produced by Different Manufacturing Routes
(a) Sintering and crushing (b) Sol-gel derived (c) Atomized and (d) Agglomerated

Recently, a new route involving the micropelletization of the primary constituents by spray drying and their subsequent plasma densification has also been suggested.

While on the subject of manufacture of spray grade powders, it is pertinent to emphasize that similar chemical composition and particle size alone do not ensure identical powder behaviour in a plasma flame. There invariably results a wide variation in powder characteristics depending upon the method of manufacture employed, as illustrated in Figure 2.8. Such variations in powder morphology are known to influence the properties of developed coatings. From the standpoint of heat transfer alone, the grain structure (porous/dense) and grain shape (spherical/acicular) play important roles. Table 2.4, apart from providing another example of how the manufacturing technique influences the powder characteristics, also shows the variation in some mechanical properties of coatings developed using different powders.

Table 2.4: Dependence of Ni-5Al Powder Characteristics on the Manufacturing Techniques and their Influence on Plasma-Sprayed Coating Characteristics

Manufacturing Method	*Shape*	*Phase Present*	*Hall Flow (Sec)*	*Apparent Density (gm.cm^{-3})*	*Adhesion Strength (MPa)*	*Avg. M.H. (HV 300 gmf)*
Clad on Ni; Polyvinyl-phyrrolidon binder	Spherical	α Ni, Al			54.8	172
Al clad on Ni; Phenolic binder	Spherical	α-Ni, Al	16.0-17.9	3.7-4.0	55.3	176
Sintered	Spherical	Ni_3Al	17.5	3.9	46.8	195
Water atomized	Irregular	α-Ni	26.4-32.4	2.6-2.8	57.1	184
Gas atomized with argon	Spherical	α-Ni	13.8-15.0	43-45	65.3	203

It is evident from the above discussion that the quality of thermal sprayed coatings is governed significantly by the coating powder characteristics apart from other factors like spray process, operating variables and substrate surface preparation.

Factors Influencing Thermal Spray Coatings

As mentioned previously, the bond between a sprayed coating and the substrate is mainly mechanical. Thus, the extent of particle melting or heat-up and its velocity, at the moment of impact with the substrate to be coated, most crucially affect the coating characteristics. Complete particle melting is essential to minimize porosity in the coating while a high impact velocity also promotes plastic deformation of the impacting particles that is thought to be essential to obtain well-bonded coatings in case of the relatively low temperature thermal spray variants like Detonation Spray and Cold Spray.

The two factors that most critically influence the coating-substrate bond strength are the impact velocity of the particle and the physical state of the particle at impact. These are both controlled by the coating process for any given powder, or by the powder characteristics for any given coating technique. Thus, the powder characteristics play an important role in determining the quality of the developed thermal spray coating. While the thermal spray powders have to meet stringent particle size and particle size distribution requirements, their morphology, composition and chemical homogeneity also influence the resulting coating properties. Typically, the upper particle size limit in the case of plasma spraying is around 50 mm for ceramic powders and around 100 mm for metallic powders. For other thermal spray variants like Detonation Spray, HVOF and Cold Spray, the corresponding upper particle size limits are lower (approximately 25 μm and 50μm, respectively) because of the significantly lower hot zone temperatures available for melting or heating-up of particles.

The significant dependence of particle size on particle heat-up and acceleration during the spray process is exemplified in Figures 2.9(a) and 2.9(b), respectively,

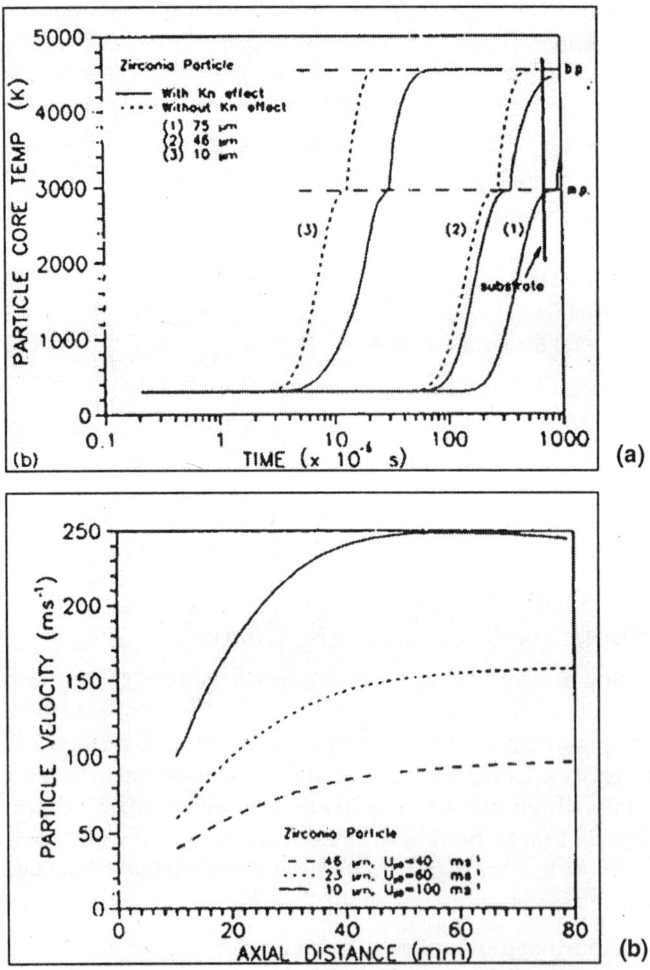

Figure 2.9: Estimates from a Theoretical Prediction Model Illustrating the Dependence of Size of Zirconia Particles on (a) Particle heat-up and (b) Acceleration during plasma spraying

which show theoretically predicted temperature and velocity profiles of zirconia particles in flight during plasma spraying. Due to the strong dependence of heat and momentum transfer rates on the particle size, it is often cost effective to size the powders to narrow limits since the expense incurred in this operation is more than compensated for, in the long run, by the enhanced quality of the protective coating developed.

Recent Developments in Thermal Spraying

Major thermal spray advances in the last 20 years represent over 80 per cent of the developments that have been made in the technology since Max Schoop introduced the wire spraying apparatus at the beginning of this century. Equipment and process advances have typically led in the technology with the advent of the cold spray technique, described earlier, representing the most significant development in recent times after the introduction of the HVOF systems specifically developed to match the coating quality provided by the detonation spray systems.

As previously discussed, numerous parameters influence the quality of thermal spray coatings. Improved control of these parameters has been the focus of many developments that have occurred during the past few years, and is also the subject of many ongoing efforts. These include incorporating empirical or real-time feedback looping, redesigning fundamental gun components, improving consistency of feedstock powders and rethinking of power supply designs. There also have been major changes in gas handling equipment. Mass flow control and metering has replaced traditional analog gages, enabling digital output with feedback potential. Data logging is gaining acceptance; flawed areas in a coating are now attributable to an excursion in gas flow. Similar control schemes have been adopted for the powder feed operation, including a variety of devices that display instantaneous powder feed rates. Powder feeders also have changed with fluidized bed feeders becoming common, permitting smooth flow for a variety of powders. The growing need for higher productivity and enhanced coating performance for critical components has lately been matched by progress in the area of equipment automation. A major step towards the end objective of obtaining reliable and reproducible high performance coatings has been the increasing use of computer control and robotics for the manipulation of plasma spray units.

For the past several years, research in thermal spraying in general and plasma spraying in particular has also focused on nanostructured materials. While the use of nanostructured feedstock has been widely explored, it has also been proposed that combining synthesis and consolidation processes into a single operation can be advantageous to realize nanostructured or at least finely structured materials in one operation. The oldest approach to achieve the above involved spraying conventional micron-sized spray grade powders comprising of agglomerated nanoparticles. The next advancement in this effort consisted of introducing the materials to be sprayed as a "liquid". The liquids used have been solutions (saturated or otherwise) as well as suspensions of micro- or nano-sized or sol-gel particles. Both processes have been shown to deliver fine splats and appear to hold promise to realize coatings with two or three different materials very deeply intricated (suspensions), or with new chemistry, because the molecular level mixing of constituents that is possible in solutions allows creation of new materials with metastable phases.

Applications of Thermal Spraying

Over the years, the areas of application of thermal spray coatings have grown in an impressive manner. Functionally, these coatings have been satisfactorily applied for

1. Wear protection
2. Thermal insulation/barriers
3. Oxidation protection
4. Corrosion resistance
5. Abradable seals
6. Abrasive surfaces/seals
7. Electrical conductance/insulation
8. Biomedical implants
9. Electronic and magnetic properties
10. Dimensional restoration/re-manufacture

In addition, thermal spraying is also being considered for emerging applications in metal, ceramic, intermetallic and composite forming. Given the diverse nature of surface characteristics that can be provided by the thermal spray techniques, the industrial utility of thermal spraying is wide ranging.

The portfolio of coating material-coating process options available within the thermal spray family to meet a specified industrial surface protection requirement is today impressive. As a result of the capability it offers to tailor the surface properties of components to provide greater resistance against degradation due to such diverse mechanisms such as corrosion, oxidation, wear or failure under an excessive heat load, the areas of application of the thermal spray coatings have grown in an impressive manner over the years. Some of the well-established areas of application of are listed in Table 2.5.

Historically, the gas turbine industry has been a major user of the thermal spray techniques. For example, the aero-engine alone requires thousands of parts to be coated by one or the other thermal spray variants. The aero-engine coatings can be classified into the following categories:

(*a*) Wear and erosion resistant coatings

(*b*) Abradable coatings for clearance control between rotating and stationary parts

(*c*) Oxidation and hot corrosion resistant diffusion and overlay coatings

(*d*) Thermal barrier coatings to insulate components from high temperatures

Table 2.5: Some Proven Thermal Spraying Applications

Area/Industry	*Partial List of Components that can Benefit from Surface Modification*
Aeronautics	Turbine blades, combustion chamber, nozzle guide vanes
Agricultural machines	Tractor parts, hydraulic rams, crusher rolls
Chemical/Petroleum	Reaction vessels, pump impellers, screw flights
Electronics	Computer and tape heads/guides
Food Processing	Shredder hammers, packaging knives, brewing containers
Internal combustion engines	Piston rings, combustion chamber, crankshafts
Pulp and paper	Pulper drive shafts, disintegrator hammers, chipper knives
Plastics and rubber	Pelletizer plates and cutters, mixer rotors
Power generation	Boiler tubes, induced draught fan blades, hydroelectric turbine blades
Steelmaking	Pickle line rolls, crane brake drums, ore crushers
Textile	Thread guides, winding spindles, flattening rolls
Transport	Brake drums, chassis, high speed train couplings

A majority of these coatings are applied on Titanium and Nickel alloys. Table 2.6 lists the different types of coatings and some of the actual aero-engine components on which these are applied.

Table 2.6: Some Prominent Thermally Sprayed Aero-engine Coatings and their Applications

Power to be Coated	*Function*	*Applicable Components*
IN-718	Dimensional Restoration	Compressor, low-pressure turbine parts
NiCrAlY	Oxidation Resistant Overlay/Bond Coat	Combustor, high-pressure turbine vanes, airfoils, casing
NiAl+Alumina	Labyrinth Knife Edge Abrasive Coating	All labyrinth seals–compressor to low-pressure turbine
CuNiln	Fretting Resistant Dovetail Coating	Fan dovetails, high-pressure compressor blade dovetails (front stages only)
Metcolloy 33	Fretting Resistant Soft Counterface	High-pressure compressor
NiAl+BN-Cermet	Abradable Coating	High-pressure compressor interstage seals
NiAl+Al-Si Polyester	Abradable Coating	Fan casing
WC-17% Co	Sliding/Impact Wear Resistance	Mid-span interlocks, shaft seal runners, support structures
NiCrAlY+YSZ	Thermal Barrier	High-pressure turbine airfoils, combustor, exhaust, shroud rings

The thermal spray industry has seen significant changes on several fronts during the past couple of decades. These have included a substantial growth in the range of industry sectors that it services. Largely restricted to gas-turbine industry initially, the thermal spray technique is rapidly gaining acceptance in the "front-end" design in other industry sectors as well. Several industrial applications are already well established and applications for thermal spray now range from bridge coatings to aircraft and aerospace engine components. It is expected that the aerospace sector will soon cease to be the dominant thermal spray market and that the automotive, industrial, chemical and infrastructure segments will grow.

An exhaustive discussion of the utility of thermally sprayed coatings on each of the components cited above is beyond the scope of this write-up. However, a few specific examples illustrating the benefit that can be derived by adopting this technology are listed below:

Example 1

Thermal barrier coatings offer a number of significant advantages to diesel operators. A 10-year study conducted in the U.S. under actual engine operating conditions has revealed, among others, the following benefits :

- Fuel savings, upto 11 per cent
- Longer engine life, upto 20 per cent
- Increased power, upto 10 per cent
- Reduced emissions, 20-50 per cent
- Reduced particulates, upto 52 per cent
- Reduced part temperatures, by about 100EC
- Lubricating oil savings, upto 15 per cent
- Reduced maintenance costs, upto 20 per cent

Example 2

In a compounding vessel used in the plastics industry, materials such as calcium carbonate and titanium dioxide are pre-blended before being final-formed. These abrasive materials lead to a high wear rate of expensive vessel liners. Plasma spraying of a layer of chromium oxide is found to double the life of the liners at about 25 per cent of the cost of a new vessel.

Example 3

Unprotected steel water-wall tubes fail rapidly under corrosive attack of sulfur-laden fuel in coal-fired boilers, especially when coupled with wear and erosion by fly ash. Coatings of NiCr, FeCrAlY or Al provide economical protection, eliminating costly replacement of tubes and unscheduled outages.

Example 4

The U.S. Navy combats corrosion aboard its ships with thermally sprayed aluminium. Off-shore oil structures, buoys, boat trailers and piers are also similarly protected. Al coatings provide effective protection in marine environments and can extend life five times longer than conventional paint systems.

National Status

Notwithstanding the fact that the thermal spray technologies have globally proliferated into virtually all major industry sectors, the industrial acceptance of these process in India has been sluggish. Thermal spray related R&D has also been on a very low key with only some research institutes devoting efforts to this subject. Most early research in the field was restricted to DMRL, Hyderabad, BHEL (R&D), Hyderabad and BARC, Mumbai apart from a few academic institutions. The establishment of the International Advanced Research Centre for Powder Metallurgy and New Materials (ARCI) in Hyderabad, which has Surface Engineering as one of its main areas of focus, has provided the much-needed momentum for R&D efforts in the field. The efforts at ARCI have also been devoted to indigenous equipment development and this has already benefited the Indian industry.

Table 2.7: Major Research Institutions Associated with Thermal Spraying

Institution and Spray Facilities	*Activity*
ARCI, Hyderabad Facilities: DSC, CGDS and Plasma Spray	DSC, CGDS and SPPS Coatings for All Applications: Basic studies, application development, industrial field testing of coatings, equipment development and building, transfer to industry
DMRL, Hyderabad Facilities: Plasma Spray	Thermal Barrier and Tribological Coatings: Basic studies and strategic applications
BHEL R&D, Hyderabad Facilities: Plasma Spray and HVOF	Plasma and HVOF Coatings for Land-Based Gas Turbines and Hydro-Power Generation: Application development, coating of turbine components
BARC, Mumbai Facilities: Plasma Spray	Plasma Spray: Basic studies, thermal plasma spherodization of powders
NAL, Bangalore Facilities: Plasma Spray	Thermal Barrier Coatings: Development of aerospace applications, formulation of speciality powders
CPRI, Bangalore	Thermal Barrier Coatings: Land-based gas turbine applications

The non-availability of indigenous sources for thermal spray equipment as well as feedstock materials has been one of the key factors that has hampered the growth of the thermal spray industry in the country. It would be appropriate to assume that there does not exist a single user industry sector in India where the full potential of the thermal spray technology has been realized. On the contrary, there are numerous sectors where the use of thermal spray technologies has barely been initiated in spite of the fact that proven case studies of similar applications are available from the developed countries. Prominent among such sectors are the automotive, petroleum, paper and metallurgical industries. So far, only the textile industry has seen commercialization of the thermal spray techniques to a significant extent.

However, the scenario in the country with respect to the thermal spray technologies is rapidly changing. Although the number of systems is still far lower than what a country of India's size and technological capability should realistically accommodate, it is a welcome change from the circumstances that prevailed not so long ago. A meaningful indicator of the promise of this technology is the spurt in the number of job shops that can now offer thermal spray coating service. There already exist nearly a dozen plasma spray job shops across the country, with a few of them also having HVOF coating units. The availability of indigenous detonation spray coating (DSC) systems from ARCI has also provided a major fillip to the thermal spray service capabilities in the country. In addition to these private jobbing facilities, the number of captive units meeting in-house coating needs has also increased in recent years. The emergent thermal spray variants such as cold gas dynamic spraying (CGDS) and solution precursor plasma spraying (SPPS) already form a part of the thermal spray canvas in the country and augment the now modest array of plasma, HVOF and DSC systems that are accessible to the thermal spray community. The above factors, along with the healthy growth trends observed recently, suggest that the future potential for the thermal spray technology in India is good.

Problem Areas in Thermal Spraying: Prospects for R&D Efforts

While the thermal spraying industry has seen significant growth in recent times, there is room yet to make substantial improvements in thermal spray technology. For example, in a plasma spray gun, only 3-5 per cent of the available thermal energy is used to melt the coating material; deposition efficiencies of some materials are 50 per cent or less; and the properties of the coatings are not yet equivalent to those of wrought materials. Also, many of the newer high-performance coating materials are relatively expensive. As a result, coating materials can account for 60-90 per cent of the cost of operating a thermal spray gun. Therefore, improving spray gun design for thermal efficiency, improving deposition efficiency and enhancing coating properties, as well as cost reduction, are major goals of ongoing studies by universities, research laboratories and supplier companies.

Thermal spray process development has, for long, sought a plasma spray gun, which allows axial, centre-line feeding of the feedstock powder into the core of the plasma jet, where the temperatures and velocities are highest. Various innovative approaches attempted for approximating central powder injection have shown promise and some have also been commercially adopted but these have not been very well received on account of concerns related to large power utilization and high costs. Consistency of powder feedstocks, control of powder feed rate and consistency of the heat region in plasma spraying are other areas demanding attention. As far as the relatively new Cold Spray process is concerned, the mechanisms associated with coating formation need to be better understood. Aspects related to nozzle design also need to be investigated to evolve an optimum configuration. Considerable efforts aimed at helium and powder feedstock recovery are also ongoing to make the CGDS process commercially more attractive.

Chapter 3

Detonation Spray Coating Technology: Capabilities and Applications

D. Srinivasa Rao
International Advanced Research Centre for Powder Metallurgy and New Materials (ARCI), Hyderabad – 500 005
E-mail: raods@arci.res.in, info@arci.res.in

ABSTRACT

Discusses in detail various aspects of Detonation Spray Coating (DSC) Technology. After a short historical introduction, examines its process fundamentals and equipment and apparatus used. Recounts the efforts done by ARCI in this direction. Further examines the working principles of detonation spraying and mechanism of coating formation. Describes DSC capabilities and coating characteristics. Looking into its applications, states that DSC technology is being widely used starting from aeronautical to agriculture industrial components. Of all the thermal spray variants, it is widely acknowledged to be the most superior by virtue of the very high particle velocities that it imparts, thereby leading to dense and well-bonded coatings which exhibit excellent tribological properties such as high abrasion and erosion resistance. Its use currently extends to all industry segments ranging from gas turbines to textile industries and from nuclear energy to data processing. Establishment of DSC technology at ARCI has recently culminated in successful development involving wholly indigenous fabrication of the coating system.

Keywords: *Detonation Spray Coating Technology, Corrosion resistance, Surface modification, Surface erosion, Alloys, Substrate materials, Coating materials.*

Introduction

In recent years, the surface modification approach has been attracting a great deal of attention as it presents a cost-effective way to combat degradation resulting from mechanisms such as wear, oxidation, corrosion, or failure under an excessive heat load without sacrificing the bulk properties of the component material. Various surfacing techniques are now available offering a wide range of quality and cost. Since a vast majority of industrial components deteriorate, and eventually fail, due to one of several wear phenomena that may be experienced during normal operation, considerable attention has been devoted to the development of coating materials and processes specifically to combat the routinely confronted wear modes, *viz.* erosion, abrasion and sliding. Among all the currently available coating processes, the thermal spray technique has gradually emerged as the most useful method of developing a wide variety of coatings to enhance the performance and durability of engineering components exposed to the above forms of wear.

Of all the thermal spray variants, the detonation spray coating (DSC) technology is widely acknowledged to be the most superior by virtue of the very high particle velocities that it imparts, thereby leading to dense and well-bonded coatings which exhibit excellent tribological properties such as high abrasion and erosion resistance. Initially developed specifically for depositing wear-resistant coatings of tungsten carbide, which undergoes substantial decarburization during plasma spraying, the use of the DSC technology currently extends to all industry segments ranging from gas turbines to textile industries and from nuclear energy to data processing.

The DSC process represents a frontier coating technology, hitherto unavailable in the country, and has wide-ranging applications in the engineering industry for enhancing the durability and performance of critical components. Establishment of DSC technology at ARCI has recently culminated in successful development involving wholly indigenous fabrication of the coating system. That encompasses substantial R&D and application development besides indigenous development of the technology that has already been transferred to three entrepreneurs, have ensured that the Indian industry reaps the benefits of this promising technology.

History and Success of Bringing DSC Technology into the Industrial Mainstream

The Detonation Spray Coating equipment was originally developed and patented in the United States by Union Carbide Corporation (*i.e.*, it has been renamed as Praxair Surface Coating Technologies) in 1955 and independently in 1969 at the Institute of Materials Science (Kiev, Ukraine). Since then, detonation coating of various materials has been applied in various industries in the United States, Japan, China, Korea, Germany and former Soviet Union. However, detonation-coating equipment generally has been unavailable for broad use in industry and research laboratories.

Widespread induction of any new technology into the regional industry invariably represents a significant challenge. Even if the technology in question is well accepted in the developed countries, the above task is particularly onerous if the technology is sourced from abroad, with the national institutes lacking 'readymade' relevant expertise, and make it further difficult if the industry has little prior experience or knowledge regarding the technology or any of its competing variants. In such a scenario, it is imperative that issues concerning process related R&D, technology demonstration, application development and market sensitization are all pursued simultaneously prior to technology commercialization. In this context, the successful incubation of the DSC technology by ARCI and its eventual transfer to the industry recently is a model case study and deserves to be recounted.

It was as early as 1990 when the DSC technology was first identified as being of significant promise for commercial exploitation in India in view of its ability to deposit high-quality coatings capable of enhancing the durability and performance of industrial components. Soon thereafter, a DSC system was acquired by ARCI from the Institute of Problems for Materials Sciences in Ukraine under the Integrated Long Term Program and the unit installed and commissioned in 1991. Although ARCI was then a fledging organization, R&D programs for optimization of coating quality were immediately initiated and, soon ARCI was ready to take up coating trails for the LPT Blades of the Jaguar aircraft's Adour Engine for HAL, Bangalore as well as on many other civilian components.

During the course of the job-work phase, it became abundantly clear that the DSC system was extremely versatile in its capability to deposit a large variety of coatings, virtually trouble-free during prolonged operation, could be operated with great ease by trained operators and commercially very attractive because of low operational costs. This only served to further reinforce the original conviction that the DSC technology was tailor-made for commercialization in India. As a consequence, ARCI decided to indigenize the DSC technology in collaboration with the Ukraine collaborating institute. Based on the long-term operating experience at ARCI, additional safety interlocks and some advanced operational features were incorporated in the refined version of the DSC system finalized for manufacture. Accordingly, fabrication of five DSC units was completed in August 1999 and, following extended operational tests, the units were made available in the year 2000 for acquisition by the private industries as part of a comprehensive technology transfer package. With prior efforts of ARCI having already enhanced national awareness regarding the DSC technology and created significant commercial interest among the industries, the technology has already been transferred to three private entrepreneurs.

With the considerable technical expertise developed by ARCI during the DSC technology's gestation period, ARCI has been able to provide wide-ranging services to the technology receivers. This has included extension of complete technical support for suitable coating material selection for any end application, process parameter development, coating characterization and certification of coating quality to potential customers, demonstration of coating on prototype components etc. In addition to routine assistance in the matter of installation and commissioning, rectification of operation related problems and quick supply of spare parts is also undertaken. As a result of the substantial created demand for the technology, ARCI has also had the luxury of ensuring that the DSC technology is transferred in a regionally exclusive manner so that all the technology receivers can flourish. This approach is being already seen to bear fruit as it has ensured significant growth of the DSC-based job shops in a very short time span.

Fundamentals of Detonation Spray Process

This DSC process is sometimes referred to as the D-gun process. It is different from the flame spray process. Instead of a continuous combustion process, it uses an intermittent series of explosions to melt and propel the particles onto the substrate. A barrel is filled with a small amount of powder and an explosive oxygen-acetylene mixture. With the use of a spark plug, the mixture is ignited. After ignition, a detonation wave accelerates and heats the entrained powder particles. The obtained particle velocities are high. Consequently, the coatings are dense and exhibit high bond strengths. After each detonation, the barrel is purged with nitrogen gas and the process repeats itself several times per second. This process is repeated over and over to achieve the desired coating thickness. The process produces noise levels that can exceed 140 decibels and requires special sound and explosion proof rooms. The coatings produced through this process are of excellent quality.

The most convenient form for gas mixture to explosion is a steel tube closed at one end, approximately 25mm internal diameter and 1350mm long (Figure 3.1). The tube is filled with an explosive mixture of gases such as methane, propane with oxygen or air and, in the case of the DSC, acetylene with oxygen. By means of an electric discharge or sparking plug as depicted in Figure 3.1, the mixture is ignited at the closed end. The gases bum for a short period of time, and the flame front accelerates, compressing and heating the un-reacted gas zone ahead of it. The velocity of this flame front is order of 10-15 m/s. Depending upon the gas mixture being compressed, a critical temperature is reached and self-ignition of the un-reacted gas mixture occurs, producing a strong light flash and a detonation or shock wave. The shock wave at the head of the detonation wave front compresses and raises the gas temperature, inducing

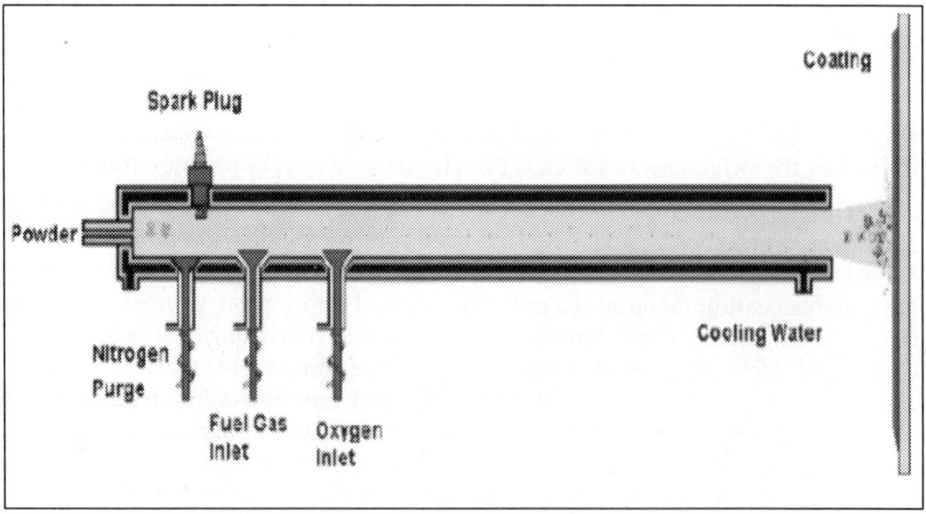

Figure 3.1: Schematic of Detonation Spray Barrel

violent chemical activity and reaction resulting in high gas pressures and temperatures of the order of 3890°C. The detonation wave front travels at a constant velocity, for a 1:1 ratio of oxygen/acetylene at 3500 m/s. The detonation velocity is the speed at which the phase boundary, between the completely reacted and un-reacted gas mixtures, moves within the tube.

Chemical reaction ceases when the detonation front passes through the flame front. The detonation front then continues to travel towards the closed end of the tube in the form of a shock wave only. The change from the combustion to detonation gas mixtures differs with respect to the composition of the gas mixtures, which is critical to upper and lower percentage limits of certain gases; outside these limits a detonation will not take place at all.

The detonation wave velocity varies with differing gas mixtures, and can be determined by the summation of the velocity of the gaseous combustion products in the detonation phase boundary plus the velocity of sound in the gaseous media.

For a gas mixture of equal volume of oxygen and acetylene in a tube of 1350 mm long, the detonation wave front travels through the entire tube length wherein the complete chemical reaction occurs within 0.5 milliseconds. In very simple terms the energy produced by the dynamic chemical reaction of the mixed gases maintains the shock wave until it passes out of the open end of the tube, leaving a tube full of reaction of the mixed gases at high pressure and temperature. They have only one way to go, quickly, of the open end expanding and accelerating but cooling down rapidly.

How to utilize this high temperature supersonic gas stream? By injecting and suspending the measured quantities of powder, the un-reacted gas mixture immediately prior to ignition provides an ideal medium for heating and hurling particles of a chosen material on to a prepared substrate which is placed in front of the open end of the gun barrel. As the detonation wave front passes through the suspended powder particles in the tube, they are heated, accelerated and decelerated. The calculated particle velocities induced by the chemical reaction zone are of the order of 40 m/s. However, the final high temperature high pressure system set up within the reacted gaseous products in the tube on completion of the detonation, picks up the molten particles and hurls them out of the open end of the tube at supersonic speeds.

The main differences between detonative conditions in the combustible mixture and thermal ones are known to be:

- The chemical reaction process caused by the contraction of the mixture due to shock wave when the energy released gives rise to the propagation of the shock wave itself;
- The energy, which is transmitted from the combustion products to a fresh mixture, is not associated with diffusion and heat transfer of work.
- Reactions in the detonation wave front proceed at temperatures, which are not attainable under normal conditions.

- The combustion products possess kinetic energy, their velocity and the detonation front propagation velocity having the same direction.

The integral characteristics of a stationary detonation wave in a gas (its velocity, the combustion products velocity, pressure, density, temperature) calculated from the one-dimensional model by Ya.B.Zetdovich are known to be in good agreement with experimental data. However, it is now established that the wave front of the self-sustaining detonation is three-dimensional and possesses various characteristics, which are not covered by the above theory. For example, laser technology used to investigate the structure of the detonation wave enabled a number of features to be determined, which are explained reasonably well by the kinetic theory of chemical induction. There is doubt that the concept of multi- wave structure of the detonation front will serve as the basis for developing more complete detonation theory in the future. Unfortunately, however, such theory is not yet available. Corrections for turbulence introduced in the equation for momentum and consideration for the effect of the boundary layer near the wall enables one only to refine the values of the detonation wave propagation velocity predicted by the one dimensional theory. As for the practical application of the detonative combustion process and the analytical tools, which are applicable for the above purpose, one must confess that of all the new theories, which have appeared, the one proposed by Ya.B.Zeldovich seems to be the most complete and to accord reasonably well with experiments.

The DSC Equipment

The DSC unit includes a barrel with a three-part combustion chamber and a gas distributor, which is connected to the first part of the combustion chamber. The gas distributor includes passages through which fuel gas and oxygen pass and are mixed. The first part of the combustion chamber is annular and includes a spark plug. The second part is cylindrical and has a plurality of peripheral channels, which is connected with the first part. The third part is cylindrical and is coupled to the second part, and to the proximal end of the main barrel. The combustion chamber and the main barrel are housed in a casing, which includes water-cooling passages as illustrated in Figure 3.2. A powder feeding tube passes axially through the first and second parts of the combustion chamber and terminates inside the third part. A first embodiment of the gas distributor includes two narrow annular passageways, which are fluidly coupled to each other by an axial opening and to an axial outlet. Fuel gas is mixed in one of the annular passageway and oxygen is injected into the other passageway. Mixed fuel gas and oxygen exits through the axial outlet to the first part of the combustion chamber. A second embodiment of the gas distributor has separate narrow diameter passages for fuel gas and oxygen. The dimensions of the passages are chosen according to a hydrodynamic Peclet criterion so that detonation wave damping is effected to prevent backfire.

The DSC equipment that is indigenously built at ARCI is shown in Figure 3.3. The key process parameters employed for spraying, like oxygen and fuel contents entering the combustion chamber in every cycle, no. of such cycles for second, the partial pressure of nitrogen etc., are controlled precisely through specially designed control console, which is illustrated in Figure 3.4. The control console has been

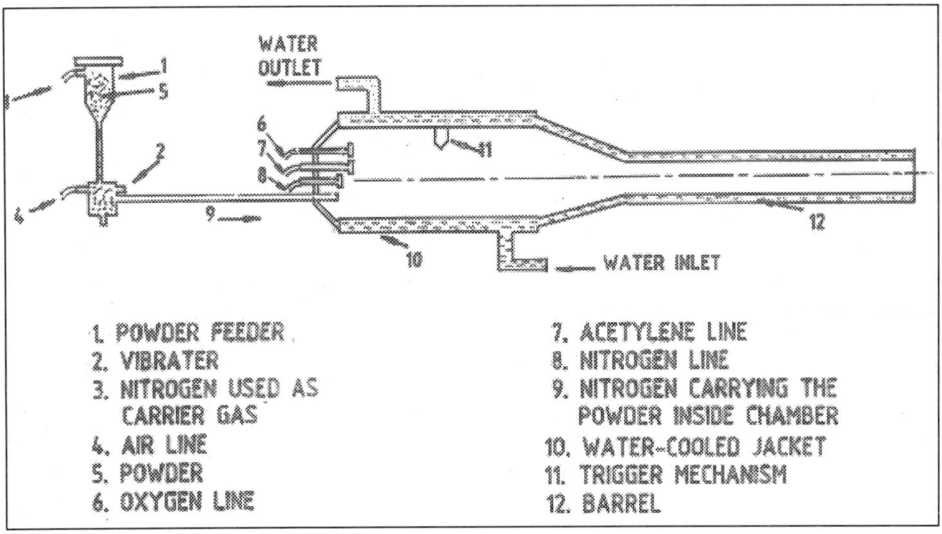

Figure 3.2: DSC Schematic Diagram

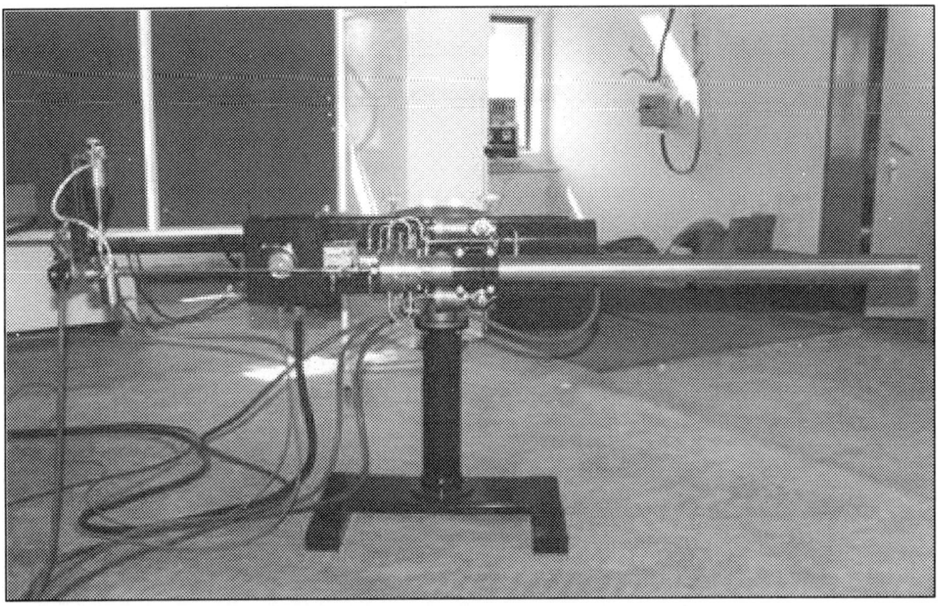

Figure 3.3: The DSC Apparatus

designed incorporating several operational safety features with cut off switches in such a way that in operation, maintenance and control is much easier for the operator.

In addition to the essential parts of DSC equipment as illustrated above, yet another important accessory of DSC unit, and of course, which is commonly required by the other thermal spray technologies, is the set of manipulators for controlling the

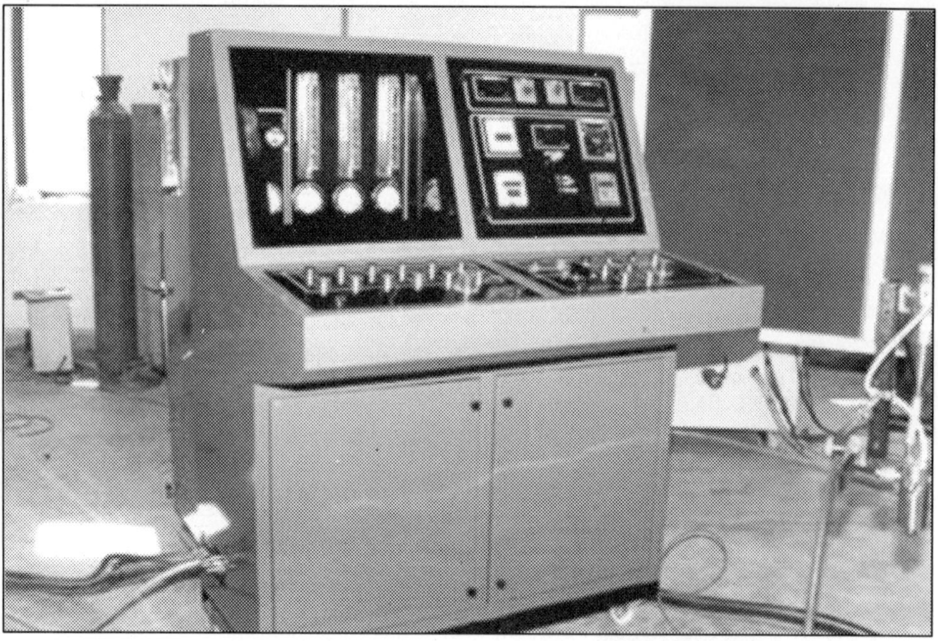

Figure 3.4: The Control Panel of DSC Unit

gun movements and movement of the parts to be coated so as to facilitate the uniformity in the coating obtained.

Working Principle of Detonation Spraying

The detonation spraying technique is cyclic. The structure of the working cycle and its individual step sequence is schematically shown in Figure 3.5. Although the Figure 3.5 depicts four different steps to facilitate the description and analysis of basic physicochemical phenomena involved in the working cycle, eight principal steps were illustrated.

1. Filling the detonation gun barrel with a fresh charge of combustible gas mixture and powder
2. Impulse bum-out of the fresh charge
3. Interaction of the powder particles with an impulse flow of combustion products
4. Change in the powder particles during their acceleration and movement in the flow of gas mixture combustion products
5. Interaction of the gas-powder stream with the substrate
6. Change in the powder particles during their interaction with the substrate
7. Change in the substrate surface during spraying of the coating and
8. Formation of the coating

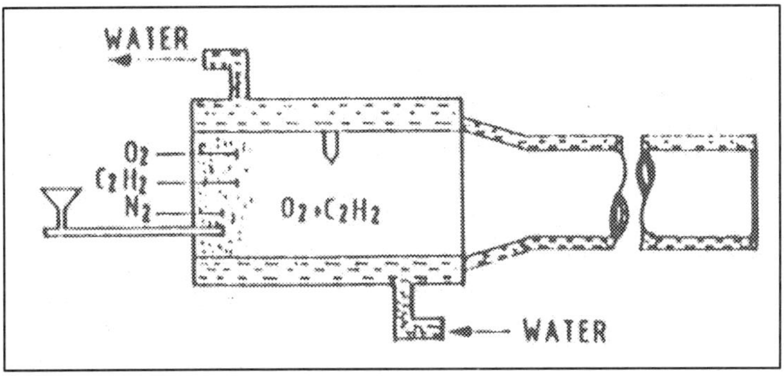

Filling chamber with O_2 and C_2H_2

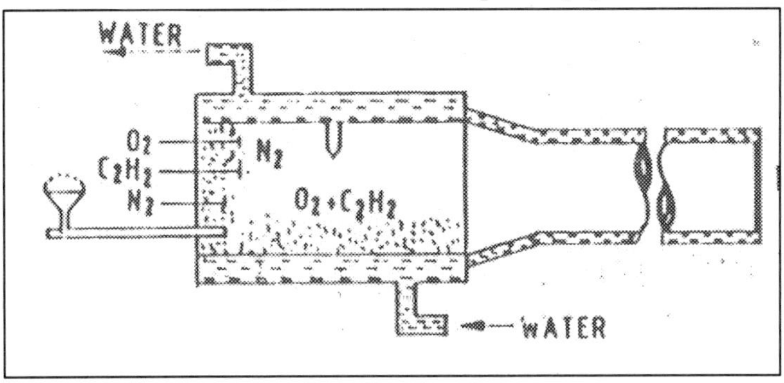

Powder charging

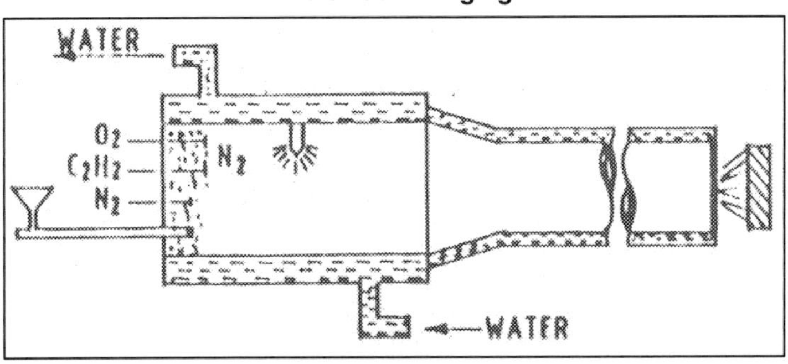

Sparking to initiate detonation

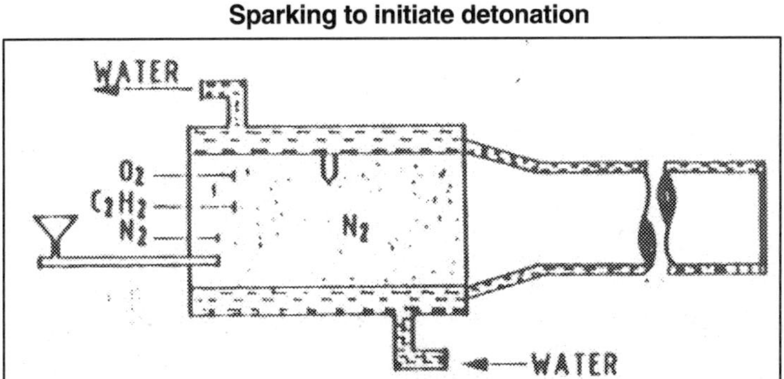

Purging with N_2 to clear gases and residual heat

Figure 3.5: Cycle of Operation in DSC Process

However, the last three steps should be considered jointly as directly interrelated and proceeds concurrently with the final step, the formation of a unit layer of coating. The steps of an impulse burn-out of the combustible mixture and formation of an impulse heterogeneous stream are governing steps and depend greatly on the quality of barrel filling.

Repetition of cycles causes the fresh charge to contain a certain amount of combustion products and powder from the preceding spray cycle. The gases and powder, on the one hand, reduce the mass of a fresh charge and, on the other hand, adversely affect the processes of burnout and formation of an impulse heterogeneous stream itself. It is therefore natural to attempt to reduce the fraction of residual gases and powder in the fresh charge. Also the flow of a working medium (purging gas, fresh combustible mixture or gas suspension of the initial powder) has a limited time to overcome the resistance of inlet paths, to contact heated surfaces etc. All this changes the density of the working medium and hence also the mass of charge capable of filling a barrel of predetermined dimensions, the degree of barrel filling, and the spatial distribution of fresh charge components in the barrel.

After the barrel has been filled, an ignition source operates, installed in the combustion chamber, or in an ignition chamber or in the combustible mixture feed piping. The mixture ignites, a flame front emerges, and burnout of the combustible mixture occurs accompanied by progressive acceleration of the flame until a stationary detonation mode arises. The pressure in the barrel during combustion rises; burnout is accompanied by ejection of incandescent combustion products jointly with suspended powder particles through the open end of the barrel. After the discharge, the pressure in the barrel first drops below atmospheric pressure (because of the inertia of the flow) and then equalizes. The combustion chamber is again filled with a fresh charge, and the process is repeated. Thus, vibrational burning with discontinuous oscillations (relaxation or non-resonance burning) occurs in the barrel. The exhaust of gas from the barrel is caused by a detonation wave moving inside the barrel, in which the gas temperature and pressure decrease and the velocity increases.

Pre- and Post-Coating Operations

Preparation of the parts for the application of thermal spray coatings generally falls into four categories, not including dimensional measurements etc. These categories are surface cleaning, grit blasting, masking and fixturing, the brief explanation pertaining to each category is as follows:

Cleaning

The cleanliness of the surface to be coated is essential. The solid material like oxide scales and other foreign bodies but also oils, machining fluids, lubricants, rust inhibitors etc must be cleaned. De-scaling is usually done by machining but not by heavy grit blasting which otherwise leads to the transfer of foreign material. The part should be degreased and then not handled by bare hands before being grit blasted. Grit blasting should immediately follow the last degreasing step.

Grit Blasting

Though grit blasting is mandatory for majority of the coating techniques, the DSC technique does not require this step prior to coating deposition if the substrate is not very hard. This is because, in DSC, the very high kinetic energy of the powder particles either of carbide or oxide particle, or perhaps, extremely cleans the surfaces due to the impact of the powder particles. Normal surface roughness required is about 4 microns (Ra). Excessive grit blasting also may create a topography that is not amenable to bonding as compared to that offered by light grit blasted surface. Choice of the grit depends on the substrate to be coated and in some cases on the final application of the part. Alumina, one of the expensive and most inert grits available is most commonly used. For some applications, hardened steel grit can be used, but care must be taken not to induce galvanic or other corrosion situation. For harder surfaces silicon carbide is preferred. Grit should not be used continuously without resizing and cleaning. Same grit blasting cabinet should not be used for both thermal spaying surface preparation and general shop cleaning.

Masking

This step is very commonly used to limit the area being coated. It is usually far less expensive to use masking than to remove excess coating over spray by machining or grinding. There are many types of tapes and oxide loaded paints or lacquers that are in usage. The later can only be used for relatively low velocity plasma deposited coatings, because they can not withstand the impact energy experienced in many of HVOF or detonation spray coating techniques. For the higher velocity deposition processes, masking materials such as fiberglass reinforced high temperature tape, adhesive backed steel or aluminum foil, or sheet metal masking may be required. More substantial masking is required for detonation spray coatings.

Masking can be of significant cost in any coating process. Efficiently designed and fabricated masking is usually well worth the time and effort involved in designing and applying it. Masking should be designed such that no bridging adjacent to the coated areas occur that would subsequently cause chipping when the masking agent is removed. Expensive masking can also be produced using materials from which the coating can be stripped so that the mask can be reused several times. In other cases, relatively inexpensive, but precisely cut and formed masking can be made that can be disposed of after each or repeated use.

Fixturing and Tooling

Like masking, efficient fixturing and tooling can have a significant impact on the total cost of thermal spray coating. Obviously, safety is the prime consideration particularly when the part must be rotated or moved at high velocities and/or the part is of heavy weight. Otherwise, higher the volume of parts to be coated of the same type, more expensive would be the fixturing and tooling to improve the efficiency of the coating process. Frequently, fairly elaborate fixturing and tooling is used to coat a large number of parts mounted on an annulus plate, a cylinder, or on a traverse and index fixture. The extensive use of robotics to control the movements by the part and the torch or gun, can reduce the need for elaborate fixturing and tooling and still maintain a highly efficient and very flexible control on the coating process.

Finishing

This step is essentially a post-coating operation. Although the detonation spray coatings are used in the as-coated condition for many applications, most applications probably require a finishing operation of some kind. Finishing techniques vary from non-dimensional brush finishing to simply remove the highest peaks thereby create a semi-nodular surface to the diamond grinding and lapping to produce a surface that has mirror-like qualities with a roughness of less than 0.05 micrometers *Ra*. Although machining can be used on a few metallic coatings, hard coatings usually require grinding with silicon carbide or diamond. Diamond is preferred by many, in particular for cermet and oxide coatings of the highest density. Specific finishing parameters depend on the coating composition, its mechanical properties, and the surface finish required. Occasionally, surface topography parameters in addition to *Ra* (average peak height) are specified to ensure performances for a given application.

Mechanisms of coating formation

The basic features of formation of detonation coatings are generally attributed to a high velocity particle collision against the substrate. A relation for an approximate estimation of the velocity Vp (m s^{-1}) of particles with a condition when the thermal energy and the kinetic energy of the particles are equal is given as:

$$Vp = 100\,(Cp.Tp + L/1{,}195)^{0.5}$$

where,

Cp (cal/g/°C) is the specific heat capacity of the particle material, Tp (°C) is the temperature of the particles and L (cal/g) is the latent heat of melting of the particle material. The higher the specific heat capacity of the particle material, the larger are the velocities of particle collision with the substrate needed for equal contributions of the thermal and kinetic energies to the coating formation.

Figure 3.6 schematically illustrates the coating formation mechanism in DSC process. Substrates surface, commonly in the grit blasted condition usually offers a rough surface, allows the molten/semi-molten or plasticised particles to get distributed owing to the high kinetic and thermal energy of the sprayed particles. High impulse pressure encourages mechanical activation of the contact surfaces concurrently with thermal activation of the contact area of the substrate; the interaction of the particle and substrate materials in general will be considered as a mechano-chemical reaction. As a result of contact friction, particles spreading over the substrate under the action of a high impulse pressure clean the contact surface by destroying surface films and forcing them to the periphery of the contact zone. At the same time, a particle plastically deforms the metallic substrate, become active centres, within whose limits a chemical interaction between the materials being bonded proceeds. The formation of defects in the contact zone is also promoted by contact friction. The rate of formation of active centres in the contact zone and the time of action of the high impulse pressure govern the strength of the bond being formed between particles and the substrate.

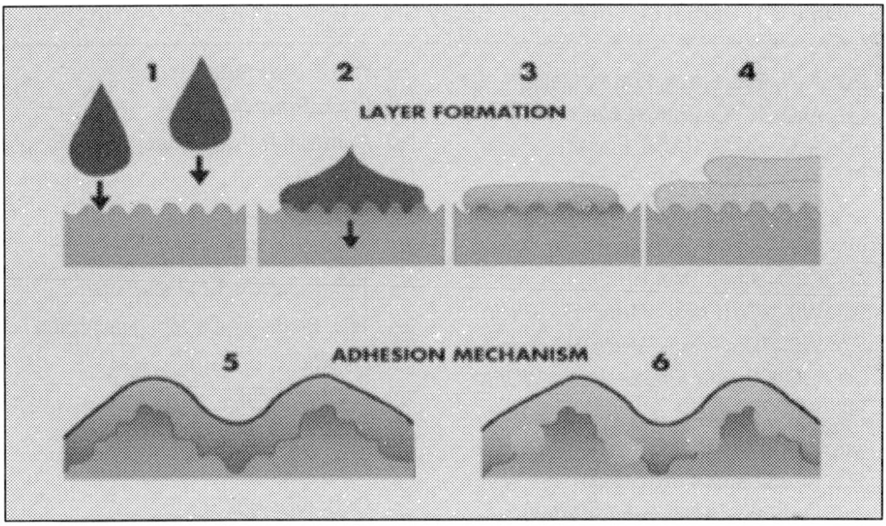

Figure 3.6: Schematic of Coating Formation in DSC Process

Changing the detonation-spraying process variables affects the degree of heating and the velocity of the particles. With insufficient acceleration of the powder while the other conditions unchanged, the strength of the bond of coatings to substrates sharply decreases, while their porosity increases to 5-8 per cent. The porosity of coatings sprayed under the optimum conditions is usually not more than 1.0 per cent. However, the powder temperature also substantially affects the coating porosity. The presence of over 10 per cent liquid phase (melt) in the material being sprayed at the moment of coating formation as well as the heating of powder below the melting point causes relatively large individual pores in the coating. However, the bond strength, micro-hardness, microstructure and phase composition of coatings depends on the geometric and production process variables like coating thickness, explosive mixture composition, spraying distance, powder feed method etc.

DSC Capabilities and Coating Characteristics

Feedstock

The DSC is a versatile process and can be used for almost all types of thermal spray grade powders except high temperature materials such as Zirconium Oxide with partially stabilized Yttrium or Magnesia. The list of powders that are being used in ARCI developed DSC is given in Table 3.1.

Surface Roughness

The DSC technique produces smoother surface finish than the other thermal spray variants such as Plasma, Flame, and Wire Arc spray coating technique. However, HVOF sprayed coatings will exhibit near to the DSC coatings smoothness. Generally, the surface roughness range of the detonation spray coatings is 3-5 mm Ra. Many of the DSC deposited components are being used in as coated conditions.

Table 3.1: Partial List of Popular Spray Grade Powders Being Used in DSC

Al_2O_3 (HC. Starck, Praxair, Metco)	WC-10.5Ni(Praxair)
Al_2O_3-40TiO_2 (Praxair, Metco, HC.Starck)	Cr_2O_3 (Praxair)
Al_2O_3-13TiO_2 (CUMI, Praxair)	Aluminum bronze(Praxair)
Al_2O_3-3TiO_2 (CUMI, Praxair)	NiCr-5Al (HCST)
Cr_3C_2-25NiCr (Praxair, Metco, HC.Starck)	Stellite-6(Delero)
Cr_3C_2-10NiCr (HC,Starck)	Cr_2O_3-20 Al_2O_3(Praxair)
Cr_3C_2-20NiCr (Metco)	Ti(C,N)-38%(Ni, Co)(Fraunhofer)
WC-12Co (Praxair, Metco, HC.Starck)	Ti(C,N)-38%(Ni, Co, Mo) (Fraunhofer)
WC-17Co (Praxair, Metco, HC.Starck)	Fly ash
WC-10Co-4Cr (Praxair, Metco, WOKA)	AlN (from SHS)
WC-17Co-8FEP	Fe-Cr
WC-10.5Co (Praxair)	Cr_3C_2-NiCr-TiB2
Ni-20Cr(Metco, HC.Starck)	Cr_3C_2-NiCr-TiB2
Co-Cr-Al-Y & Ni-Cr-Al-Y (Praxair)	Fe-SiC
SS316-Martensitic, Austenitic (Metco)	$Al_{65}Fe_{20}Cu_{15}$ (Quasi-Crystalline coatings)
Fe, Ni and other pure metals	

Deposition Rate

The deposition rated DSC is usually referred to as coating thickness per shot. The variation in density of the coating material does affect deposition rate and is closely related to the powder feed rate for each particular coating. However, in general terms a single detonation produces a 15–22 mm diameter coated area with 5-10 mm thickness buildup. The deposition rate can be improved with appropriate selection of process parameters such as oxygen to acetylene ratio, gas volume, spray distance and etc. As exemplified in Figure 3.7, changing the oxygen-fuel ratio and volume in case of Cr_3C_2-25NiCr coatings has considerably changed the deposition rate per shot.

Microhardness

The hardness of DSC coatings is generally higher than that of equivalent coatings produced by the other flame spraying techniques. Higher micro-hardness is due to the dense structure achieved by the high impact velocity of the particles. The typical hardness for tungsten carbide cobalt is 1250HV. The micro hardness of various coatings obtained by detonation spray is given in Table 3.2. The hardness of similar coating produced by plasma spraying would be order of 800-1000HV. Those of coatings produced by combustion spraying would be still lower.

Microstructure

The microstructure of a DSC is illustrated in Figure 3.8. The figure shows a typical cross-section of a WC-12Co coating. At the substrate/coating interface, there is no separation or porosity. The laminar structure is well illustrated, showing the cross-section of the platelets formed by the high velocity impact of particles against substrate and particles. The structure of DSC is very uniform as shown in Figure 3.8.

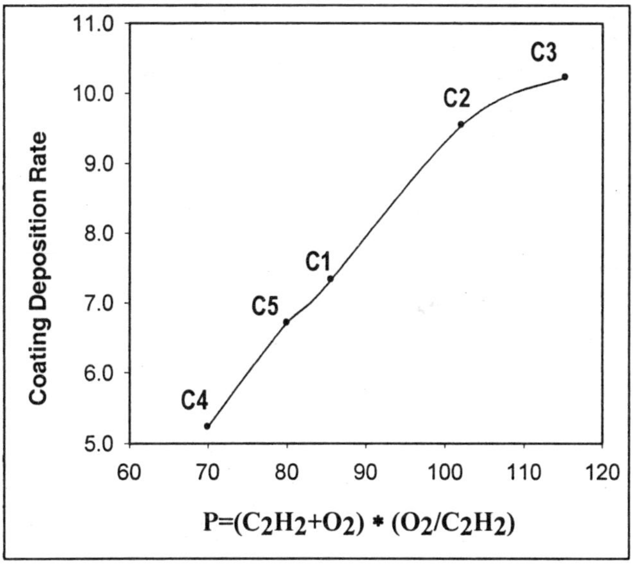

Figure 3.7: The Variation of Coating Deposition Rate of Cr_3C_2-NiCr Coating

Table 3.2: Detonation Spray Coating Characteristics

Coating Composition	*Hardness $HV_{0.2}$*	*Porosity %*	*Bond Strength Psi*	*Max. Operating Temp., °C*
WC-10.5Co	1374	0.8	>10,000*	480
WC-12Co	1160	0.6	>10,000*	480
WC-17Co	1095	0.7	>10,000*	480
WC-10.5Ni	1040	0.6	>10,000*	480
WC-19.5Cr_3C_2-6.5Ni	1060	0.2	>10,000*	760
WC-10Co-4Cr	1086	1.0	>10,000*	480
Cr_3C_2-20NiCr	1042	0.6	>10,000*	820
Cr_3C_2-25NiCr	940	0.4	>10,000*	820
Al_2O_3	1160	1.1	6000	1000
Al_2O_3-13TiO_2	1076	0.7	7000	700
Al_2O_3-40TiO_2	1040	0.9	7000	700
Ni20Cr	379	0.4	>10000*	900
NiCrAlY	523	0.6	>10000*	900

Porosity

The porosity of detonation gun coating is very fine (Figure 3.8). The overall mount is between 0.25-1 per cent, evenly distributed low level of porosity can be compared values of several per cent for most plasma sprayed coatings and even higher level porosity for combustion sprayed coatings.

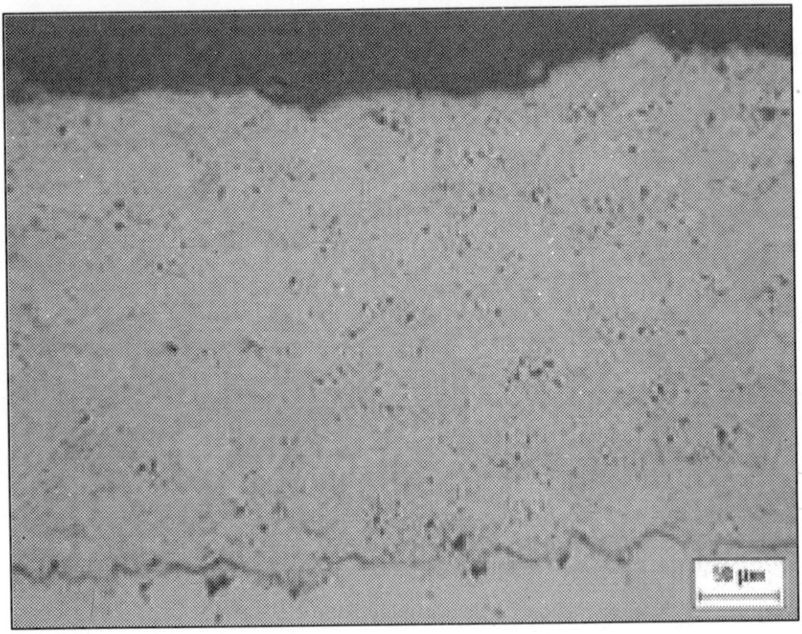

Figure 3.8: Dense Structure of DSC Deposited WC-12Co Coating

Decarburization

Appropriate selection of oxygen and acetylene gas ratio and its volume plays a vital role in controlling the decarburization rate and it is well known that the decarburization or free oxide content in detonation sprayed coatings is negligible and particularly for carbide based wear resistant coatings. Limited studies for investigating the above have been conducted by various institutes and recommended that the O_2 to C_2H_2 gas ratio should be less than the 1: 1.17 for WC based powders. In case, O_2 to C_2H_2 gas ratio more than 1: 1.17, WC phase will transform to W_2C or other complex compounds with the presence of Co or Ni such as $Co_3W_3C_4$/NiWC. Apart from the above gas parameters influence on decarburization of coatings other parameters like spray distance and powder charging depth are also important. For example, experiments were conducted for evaluating the four grade of WC-12Co powders decarburization rate by varying the spray distance from 130 to 200 mm and shown in Figure 3.9.

Adhesion Strength

The bond strength of detonation is a mechanical interlocking between the coating and substrate. The typical method of measuring the bond strength of DSC coatings is by the ASTM recommended practice C633. In this technique, the end of a 25,4 mm diameter bar is coated and then a mating bar bonded to it, usually with an epoxy or double side adhesive tape. The test is limited by the strength of the epoxy or tape. For many of DSC coatings, the bond between the coating and the substrate exceeds the strength of the epoxy. Generally, the DSC deposited wear resistant carbide coatings

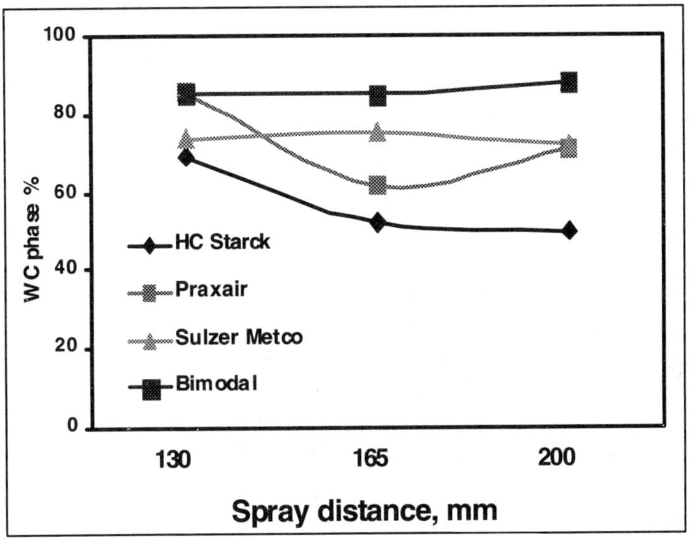

Figure 3.9: Influence of Spray Distance on the WC-Phase per cent in the Coatings

always exceeds 10000 psi and very rarely the plasma spray coatings exhibits this strength.

Residual Stress

Depending upon the type of thermal spray technique, either tensile or compressive residual stresses are formed in coating. Generally, the low particle velocity thermal spray techniques will result in formation tensile residual stress like in the case of APS, Wire Arc, flame spray techniques that is a most prevalent state for this kind of processes. In case of high particle velocity techniques such as DSC, HVOF and Cold spray will produce desired compressive residual stresses. The residual compressive stress is certainly beneficial to the substrate if the fatigue is of concern. The tensile residual stress exceeds the strength of the coating, the coating tends to crack and may spall from the substrate or it may bend the substrate in case of stress developed is sufficient. In some cases, the coating may also detract from the bond strength and other mechanical properties and lead to chipping of coatings at sharp edges.

For example, several experiments conducted to determine the residual stresses in coatings and the results were illustrated in Figure 3.10 which represents two types of residual stresses obtained from DSC and APS deposited coatings. The DSC deposited coating exhibit completely compressive residual stress and APS coating resulted in tensile residual stress.

Tribological Performance

This study presents a comparison of the tribological behaviour of some popular wear resistant coatings, for example WC-12Co, Al_2O_3 and Cr_3C_2- NiCr deposited by

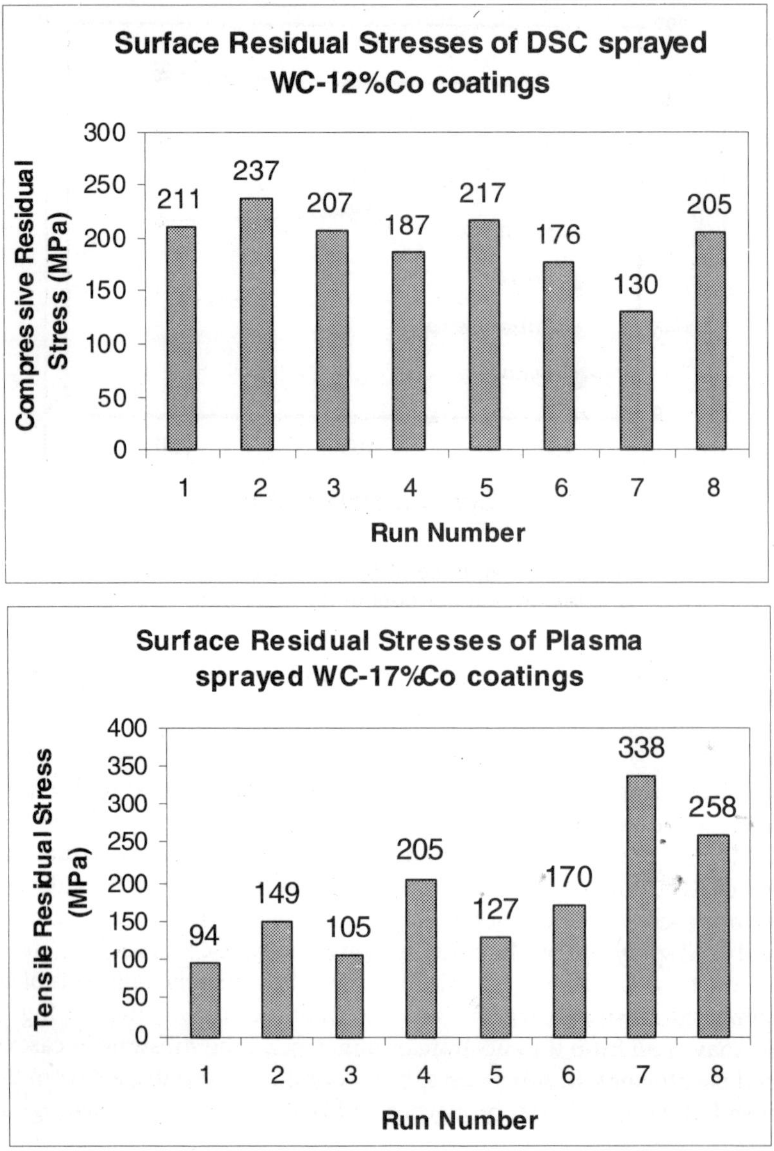

Figure 3.10: Residual Stress Comparison of DSC and APS Deposited WC-Co Coatings

the DSC and APS techniques. To enable a comprehensive comparison of the above indicated thermal spray techniques as well as coating materials, the deposited coatings have been characterized for tribological performance like solid particle erosion tests, rubber wheel sand abrasion tests.

The sliding and abrasion wear rate of DSC and APS deposited WC-12Co, Al_2O_3 and Cr_3C_2-NiCr was illustrated in Figure 3.11. It is very clear that the DSC coating properties are several folds better than the APS deposited coatings.

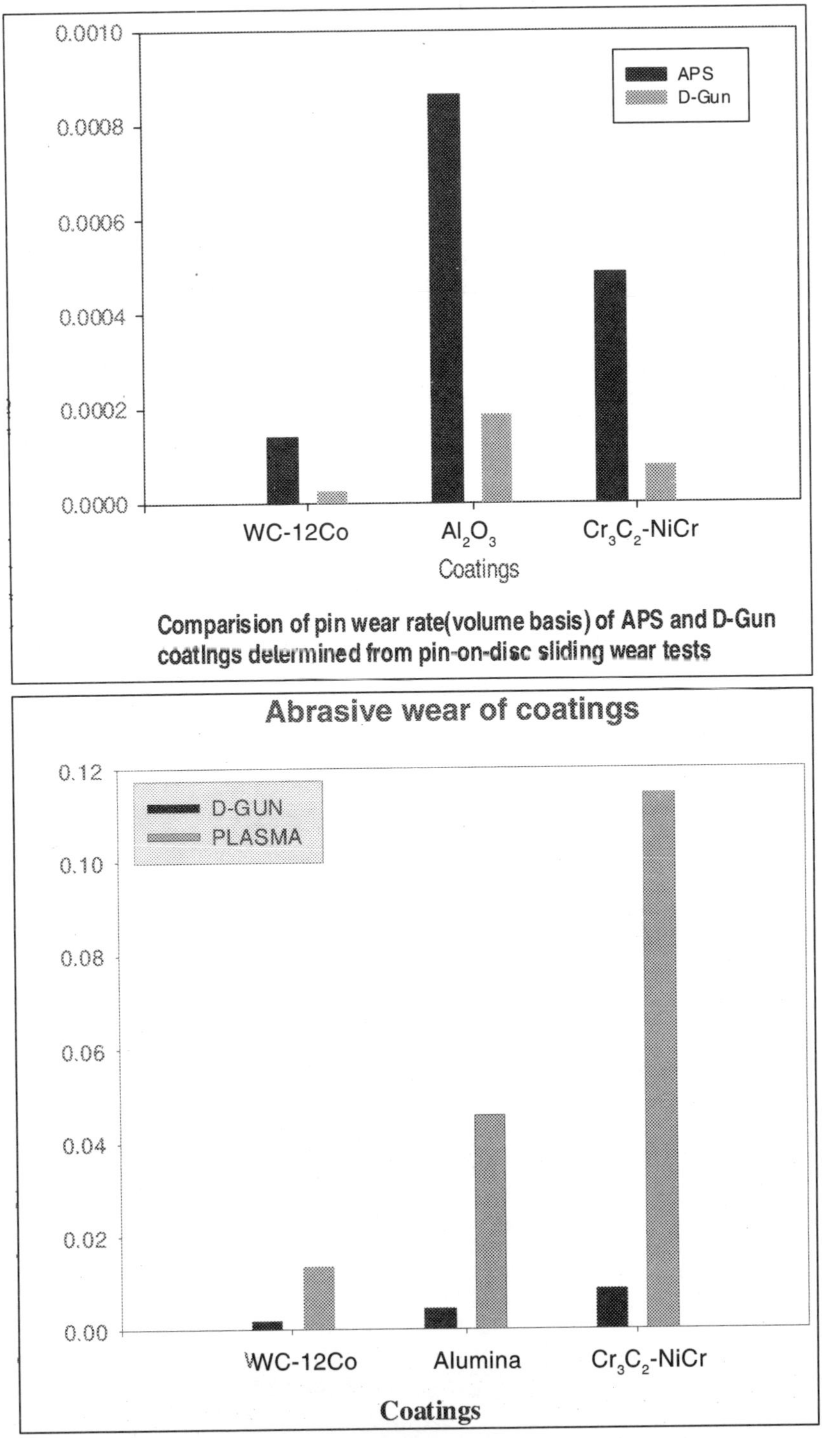

Figure 3.11: Sliding and Abrasion Wear Rates of DSC and APS Deposited Coatings and Pin on Disk Sliding Wear

Influence of Binder on Properties of WC-Based Coatings: A Case Study

Recognizing the considerable potential of detonation sprayed Tungsten Carbide (WC) coatings in improving the tribological behaviour of parts exposed to aggressive wear prone environments, one of the recent developments in the field have focused on formulating new grades of spray-grade WC powders with different types of binders with varying content Co, Ni, Cr and their combinations are the more popular binders employed among the commercially available powder grades, the choice of a specific binder being largely dependent on the operating conditions to which the coated component is eventually exposed. However, the effect of binder on the performance of the WC-based coatings has not yet been fully investigated.

In the present study, WC-11Co, WC-10.5 Ni and WC-10Co-4Cr powders were selected to investigate the influence of binder type on the properties of detonation sprayed WC-based coatings obtained on mild steel substrates. Prior to deposition, these powders were characterized in terms of their particle size distribution, morphology (Figure 3.12) and phase content. While depositing each powder, key process variables like oxy-fuel (O_2/C_2H_2) gas ratio, volume of the O_2+C_2H_2 gas mix

WC-10.5Ni WC-10Co-4Cr

WC-10Co

Figure 3.12: Morphology of Carbide Powders

and spray distance were varied over a wide range with a two-fold objective *viz.*, to investigate the influence of the above process variables on coating properties and to identify the optimum process parameters for coating each powder.

Coatings obtained were comprehensively characterized for surface roughness, microstructure, micro-hardness and phase composition. In addition, tribological performance of these coatings was also evaluated using dry sand abrasion test, pin-on-disc sliding wear test and solid particle erosion test.

A three phase experimental test matrix was designed to optimize the three key variables, namely spray distance, oxy-fuel ratio and the gas volume, for detonation spraying of each of the three different WC grades of powder. All the powders were thoroughly investigated prior to taking up the coating exercise. Results of the particle size analysis confirmed that the particle size distribution exhibited by each powder was suitable for use in detonation spraying. SEM examination of these powders revealed the WC-11Co and WC-10.5Ni powder particles to be angular and irregular while the WC-10Co-4Cr powder particles were seen to be spherical in shape. The quantitative analysis of XRD patterns obtained for the three powder grades showed that, in all cases, WC was the predominant phase present as expected.

The coated samples were sectioned, metallographically polished and observed under SEM to assess the micro-structural characteristics. Figure 3.13 SEM images of WC-11Co, WC- 10.5Ni and WC-10Co-4Cr coatings deposited with the optimized parameters clearly revealed that the substrate-coating interface was clean and the coating well bonded with the substrate. Image analysis of these microstructures also confirmed that the porosity in these coatings was less than 0.3 per cent. The influence of spray distance, oxy-fuel ratio and total volume of the gas mix on the micro-hardness, porosity and surface roughness of coatings was carefully investigated. For example, the results clearly revealed that a spray distance of 130 mm, an oxy-fuel gas ratio of 1:1.21 and a volume of oxy-fuel gas mix equal to 5440 slph yielded highest hardness and lowest porosity in detonation sprayed WC-11Co coatings. Studies on abrasion, erosion and sliding wear performance of WC-11Co coatings also indicated that the above values of spray distance, oxy-fuel ratio and its volume led to best results from the standpoint of tribological behavior. Thus, the lower level settings of spray distance, oxy-fuel ratio and gas volume were considered optimum for detonation spraying of the WC-11Co powder. A similar procedure was also adapted to optimize parameters for depositing the WC-10.5Ni and WC-10Co-4Cr powders by detonation spraying.

In order to correlate the tribological behaviour observed with the phase constitution of the coatings, a detailed XRD analysis of all the coating specimens generated was also carried out. Results revealed that a higher WC phase content consistently led to superior wear performance of the coatings under all the wear modes investigated. At higher oxy-fuel ratio and spray distance, the extent of decarburization of the WC-phase was found to increase, thus leading to the formation of brittle phases like W_2C and other complex phases. The influence of binder type on coating properties could be clearly assessed following optimization of the detonation spray variables for depositing WC-11Co, WC-10.5 Ni and WC-10Co-4Cr coatings. Results as depicted in Figure 3.14 illustrate that the micro-hardness of WC-based

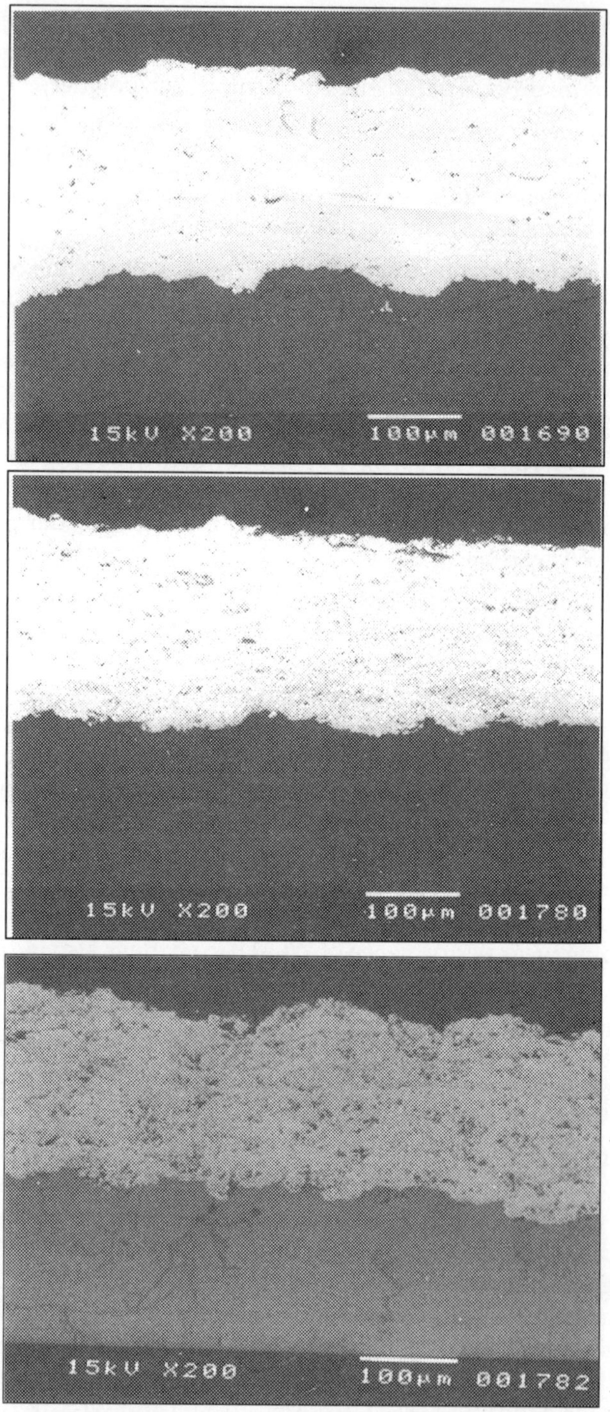

Figure 3.13: The Cross Sections of the WC Coating Microstructures (I.e WC-11Co at 130 mm, WC-10.5Ni at 165 mm and WC-10Co-4Cr 165 mm spray distance)

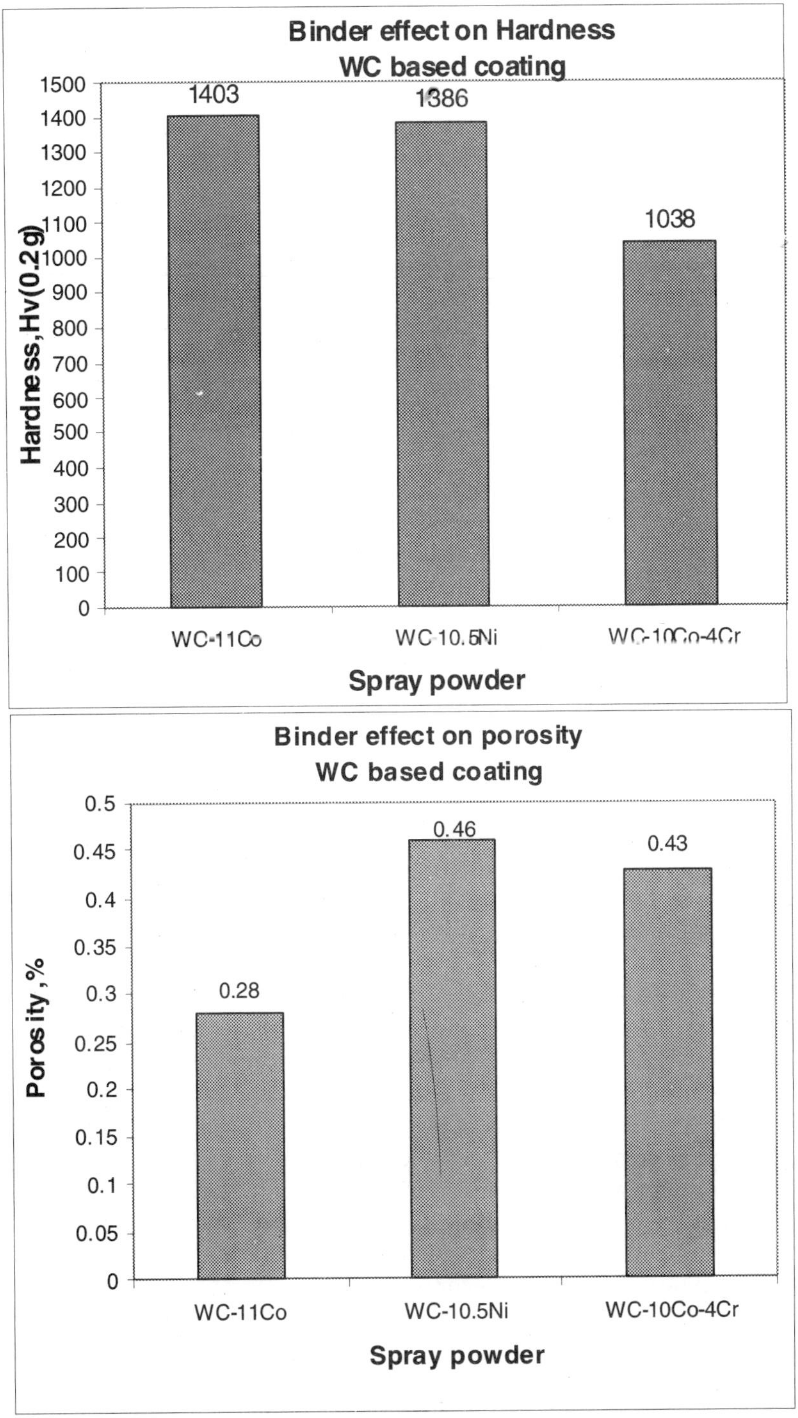

Figure 3.14: Influence of Binder on Micro-hardness and Porosity of WC Coatings

coatings depends on the type and extent of binder phase used. While the optimized WC-11Co and WC-10.5Ni coatings yielded micro-hardness values of about 1400HV, the hardness of the WC-10Co-4Cr coating was under 1050HV. However, the porosity levels in all the three coatings were consistently low and varied in the narrow range of 0.25-0.50. Interestingly, the WC coatings with cobalt and chromium as binders exhibited superior abrasion and erosion wear resistance compared to coatings with Ni-binder, while the sliding wear performance of coatings with Co binder was significantly superior to that of coatings with Cr and Ni binders.

Investigations are currently ongoing to explain the above observations. The three coatings deposited with the optimized detonation spray parameters were subjected to XRD analysis to determine their phase constitution. Results have revealed that the WC-10Co-4Cr powder yielded by far the highest WC phase content and this presumably accounts for the superior performance of this coating under abrasive and erosive wear conditions as illustrated in Figure 3.15. However, its relatively poor performance under sliding wear conditions remains yet to be explained. The WC-11Co coating yields good performance under all wear modes while its erosive and abrasive wear behavior is only marginally inferior to that exhibited by WC-10C0-4Cr, its sliding wear behavior is exceptionally good. This may be attributed to the high WC to W_2C ratio combined with the presence of ductile Co-containing phases in the coating.

Applications

The detonation spray coatings are being used starting from aeronautical to agriculture industrial components. The partial list of popularly used powders and their applications are given in Table 3.3. For example, the DSC can produce coatings of various carbide based coatings like WC-10Co on the high strength alloy gears of turbines, for helicopter engines, with 0.3 per cent porosity, >70 MPa adhesion and 1392 HV average micro-hardness. This enhances the surface characteristics thereby increasing service life of the components to several folds. Some of the proven areas of DSC applications, which are being used on regular basis, are provided in Table 3.4.

Advantages and Limitations

Gas detonation is capable of producing the highest pressure, velocity, and density in the gas flow, which is not achievable by all other spraying techniques. The typical values of gas temperature, substrate temperature, particle velocity and the porosity, bond strength and the oxide content in the resultant coatings deposited by different thermal spray process has been presented in Table 3.5. As illustrated in Table 3.5, the detonation coatings are characterized by extremely high density, micro-hardness, and low porosity, and are suitable for applications requiring extremely good mechanical and tribological characteristics. In addition, a great variety of materials can be sprayed by using the DSC technology, and a broad spectrum of substrates can be coated. In addition to premium coating quality, the DSC technique offers the following benefits:

Figure 3.15: Influence of Binder Content on Tribological Properties of WC Based Coating Properties

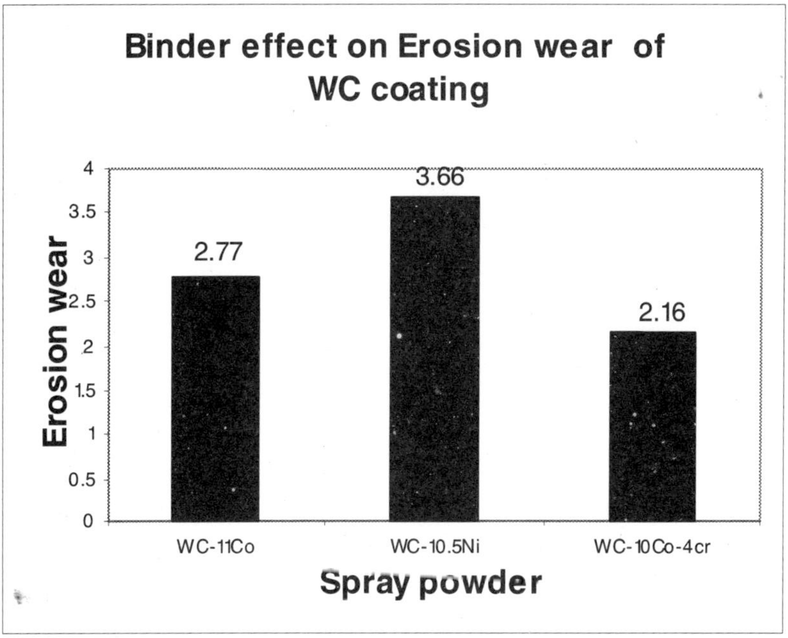

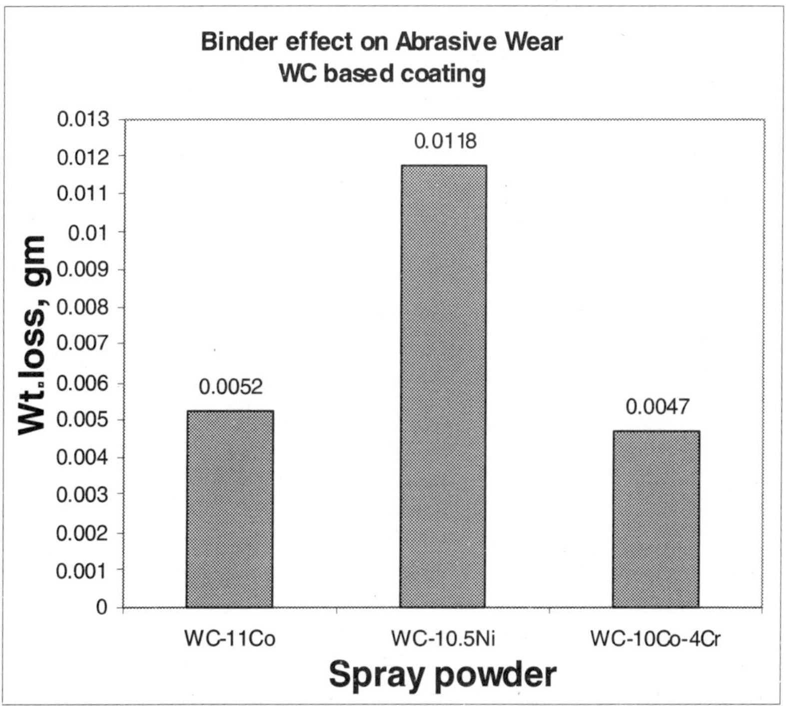

Contd...

Figure 3.15–Contd...

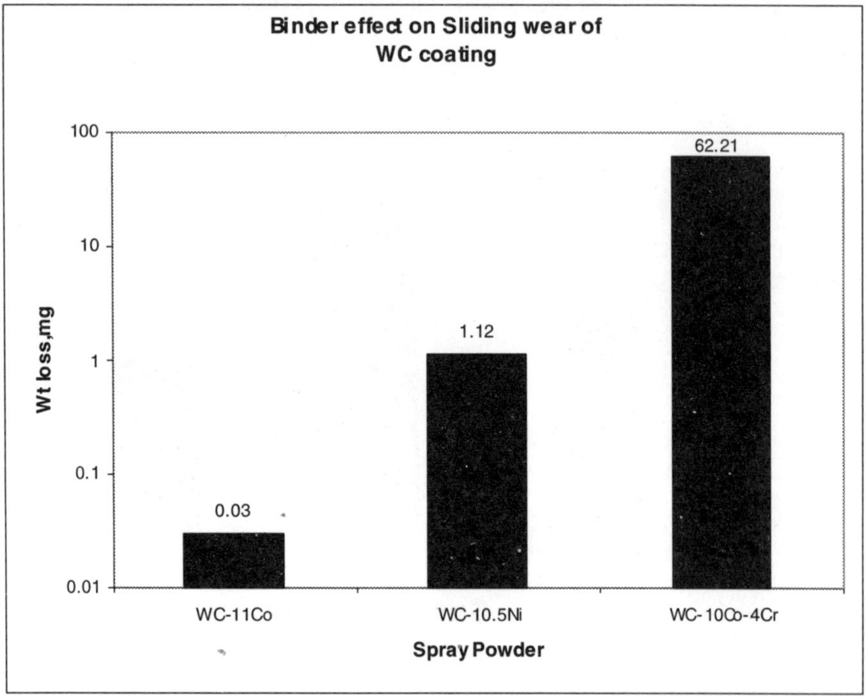

Table 3.3: Detonation Spray Coatings and their Applications

Material	*Typical Coating Hardness (Hv)*	*Max. Operating Temperature (°C)*	*Field of Application*
100Al		400	Corrosion resistant Electroconductive coatings
100Cr	500	1000	Refractory, corrosion resistant coatings
100Fe	400	500	Reconditioning
100Ni	400	800	Corrosion and wear resistant coatings, pre-coatings
100Cu		450	Electroconductive coatings
100Mo	169	400	Antifriction coatings, intermediate non-machined layers of fit seats
Alumina	1100	980	Heat and wear resistant high temperature protective coatings injuring high chemical stability and oxidation resistance
Cr_2O_3	1350	510	High refractoriness, erosion resistant at high temperature, protection of surface against oxidation
$60Al_2O_3$– 40 TiO_2	950	700	Heat and wear resistant coatings
WC-Co(88-12)	1200	560	High wear resistance, improved mechanical, impact and heat shock-resistance

Contd...

Table 3.3–Contd...

Material	*Typical Coating Hardness (Hv)*	*Max. Operating Temperature (°C)*	*Field of Application*
WC-Ni	1200	700	High wear and erosion-resistance, impact strength
$75Cr_3C_2$-15NiCr	800	980-1000	High wear resistance at high temperature, corrosion and erosion resistance
NiAl (95-5), (80-20)	200	1000	Corrosion-resistant bond coat
NiCrAlY (Ni-22Cr-10Al-1Y)	400	1300	Corrosion resistant bond coat for high temperature application
CoCrAlY	400	1200	Corrosion resistant bond coat for high temperature applications
NiCr	200	1350	Corrosion and oxidation–resistant bond coat

Table 3.4: Proven Applications of DSC Coatings

Nomenclature	*User Industry*	*Coating Material*	*Purpose*
Ink Metering Roll	Printing	Al_2O_3-TiO_2/Cr_2O_3	Woar and Corrosion Resistance
Tension Roll	Steel	WC-12Co	Smoother hard surface and Improved frictional Coefficient
Wire Drawing Blocks and Pulleys	Wire	Al_2O_3, Al_2O_3-TiO_2, WC-12Co	Abrasion and Sliding Resistance
Fibre Contact Surface	Textile	Al_2O_3, Al_2O_3-TiO_2	Abrasion and Corrosion Resistance
Extruder Screws (Flight Tips and Flaws)	Plastic	WC-Co	Abrasion Resistance
Pump Impeller Neck Rings, Rotor Shafts, Pump Sleeves and Mechanical Seal Faces, Pressure Control Valves	Chemical	WC-Co-Cr, Al_2O_3-TiO_2, Cr_3C_2-NiCr WC-Ni	Corrosion Resistance
Valve Stems, Seats and Gates, Rods and Plunger	Oil	WC-Co	Sliding Resistance and Friction Coefficient Abrasion Resistance
Brake Drums	Crane	WC-Co	Wear Resistance

Low Substrate Temperature

Due to the intermittent nature of the process, the amount of heat transferred to the substrate remains low. The substrate temperature, during spraying, does not exceed 150°C and can also be maintained close room temperature, if necessary. This makes possible the spraying of critical components such as crankshafts, or turbine blades, without the danger of causing thermal deformations, or chemical changes. It is also possible to spray low melting point materials, such as plastics.

Table 3.5: Typical Characteristic Features of Different Thermal Spray Processes

Process Name	*Gas Temperature °C*	*Substrate Temperature °C*	*Particle Velocity m/sec*	*Porosity %*	*Bond Strength psi*	*Oxide Contents %*
DSC	2500-3500	20-140	Up to 1000	0.1-1.0	Extremely high	0.1
HVOF	2500-3500	500-700	500-1000	0.1-2.0	Extremely high	0.2
Plasma Spray	5500-8300	700-1000	100-300	1.0-10.0	Very high	0.1-1.0
Cold Spray	30-700	30-70	600-1200	<0.5	Extremely high	0.0
Wire Arc	4000-6000	500-800	240	10-20.0	High	0.5-3.0
Flame Spray	2500	500-700	30-180	10-30.0	High	4.0-6.0

Low Cost of Operation

The practical working experience of the DSC customers, demonstrates that the use of DSC process in production, is very cost-effective; thus results in substantial operational and maintenance cost reduction as compared to the typical HVOF and plasma spray systems. This is due to the following five principle factors:

Less Gas

A high-pressure fuel and oxygen supply is not needed. The gases are fed into the combustion chamber at low pressure, slightly exceeding atmospheric pressure. The total amount of gases consumed per unit mass of powder sprayed, is about 10 times less than the typical HVOF system.

Less Water

Due to the minimal amount of heat transferred to the gun barrel, a special cooling system is not needed. Water consumption is very low.

Less Electrical Energy

Typically, the DSC unit consumes a total of 500W of electrical power, which is comparable to a regular electric bulb. A water chiller is not needed, which requires a lot of electricity.

Less Downtime

The process is simple to operate and very reliable. There are literally no components that can wear out likewise to De Laval nozzle used in the HVOF gun.

Less Scrap

There are no nozzles that can melt, or disintegrate, and spit copper on the expensive part being sprayed.

No Expensive Proprietary Powders

An additional bonus of DSC technology is the reduced cost for powder, as superfine powders, or powders of special composition are not needed to produce a superior coating.

In addition to the advantages listed above, there exists few more. The gun can be mounted at any angle, or can be fixed in the arm of a robot-manipulator. Multiple gun barrels can be installed to satisfy process requirements for high volume rate of deposition in a continuous production line and barrels can be manufactured to the exact length, and bore, to optimize a specific coating. The gun design allows the use of same unit as a grit blaster, without modifications of the barrel, or the gas supply system. This eliminates the need for a separate grit-blasting booth, and it allows for a combination of surface grit blasting and further spraying, in one continuous operation. These are generic examples–your exact results will depend on spray duration, spray distance, process settings, other parameters, and the substrate material

However, the following substantial deficiencies are inherent in this method.

1. The cyclic nature of the spray coating process impedes stabilizing and monitoring the production parameters of the process.
2. The high level of noise and other harmful effects necessitate isolation of the processing zone from the environment.
3. The use of a high temperature stream with detonation products of a complex composition, containing CO_2, CO, H_2O. H_2, O_2, N_2, OH, H, O, NO, N and other gases, as the working medium imposes certain production restrictions on producing coating from the materials that contain elements which actively interact with the gases.
4. The light level of impulse pressure, under interaction of the two-phase stream with the substrate, imposes certain production restrictions on the spraying of non-rigid surfaces.

The above listed advantages and deficiencies of the method show up differently in solving specific engineering problems of spraying coatings. A correct selection of the field of application of the method allows its advantages to be utilized most efficiently and the manifestation of its shortcomings to be minimized.

Current Status of DSC Technology

Currently, the Praxair Surface Coating Technology Inc (PST), USA and few institutes from CIS countries are actively pursuing the DSC related development works to enhance the performance of the system. The PST have developed an advanced unit named as Super D-Gun wherein the particle velocity and temperature were further enhanced by introducing the fuel gas as powder carrier instead of Nitrogen. M/s. Aerostar, Spain which has been recently taken over by PST had developed a semi automatic high frequency pulsed DSC system named as HFPD wherein oxygen-propane gas mix is used for combustion.

Wherein in Russia, an automated DSC unit that has multiple revolving gun barrels was developed for enhanced coating productivity. Apart from India, some of the institutes form China, Korea, Germany, France and Finland also have obtained DSC units from CIS countries and they have been using for R&D studies as well as for in-house limited applications. However, the DSC systems may be obtained from CIS institutes or from A Flame Corporation and Praxair Surface Coating Technologies

USA. Recently, ARCI has developed indigenous detonation spray coating unit in collaboration with IPMS, Ukraine and three units have been transferred to private entrepreneurs.

Future R&D Prospects

The R&D works related to development of DSC technology or understanding of the process thoroughly have been carried out by the patent holders from USA and CIS countries, and most of the important data was not revealed. The available literature on DSC technology is limited to only on coating properties of various powders and its applications. The R&D studies accomplished at ARCI were meant for optimizing the coatings process parameters of various types of powders by varying only the process variables and wherein the variables of basic gun design were kept constant. Based on results obtained at ARCI and also information available from literature, it was found that the maximum particle temperature and velocity, by changing the process variables, yielded the best quality of the coating. It is expected that the gun/ equipment variables will also affect the particle temperature and velocity and hence coating properties. In order to get maximum temperature and velocity, it is necessary to vary the equipment variables at different levels. Due to the lack of information related to the gun variables in English journals and proprietary of the process, the correlation between the gun variables and the coating properties were not well understood.

The effect of equipment variables like powder charging depth, internal diameter of the powder nozzle pipe, the size of the combustion chamber, the length of the barrel and diameter, etc plays a vital role on velocity and temperature of powder particle which in turn effect the tribological properties of the coating apart from coating adhesion strength, thickness build-up and deposition efficiency rate. Depending upon the density of the powder, these variables are fixed at different levels so that the maximum particle temperature and velocity can be obtained for achieving the best quality of the coating. Another area of study for understanding the process is validation of the above properties with the theoretical modeling.

For example, two types of the density of the powders such as low density Aluminium Oxide and high-density Tungsten Carbide Cobalt powders, which are being used widely industry will be of the correct choice to understand the intricacies of the design and for enhancement of the design. The studies may be conducted by varying the gun design variable like powder charging depth, internal diameter of the powder nozzle pipe, barrel diameter and length, the size of the combustion chamber and etc, and on the other hand coating process variables like Acetylene gas flow rate, Oxygen gas flow rate, Acetylene/Oxygen gases ratio and its volume, Powder feed rate, Spray distance, Detonation frequency, Barrel temperature/water temperature cut-off, Speed of job motion, Extent of overlap and etc.

The above studies may result in clear understanding of the particle behaviour in side the gun barrel like melting, motion, acceleration, temperature and coating properties for each set of the variables. In turn it would help in improving the gun design for enhanced performance of the Detonation Spray Coating Technology. Another area of the study may also be conducted to generate wide spectrum of database that is useful for development of expert system to "choose" an optimum parameter set for depositing any given coating material. In order to determine the optimum parameter set, appropriate consideration of the following major steps of the detonation spray process is essential.

- Detonation of oxy-fuel mixture (Detonation wave parameters, chemical reactions during combustion, constitution of combustion products, velocity and temperature profiles)
- Gas-particle transport phenomena (heat transfer, momentum transfer, particle velocity and like intra-particle heat transfer, non-continuum effects etc.).
- Interaction between the impacting powder particles/droplets and the surface being coated (Wetting at substrate-coating interface, heat transfer to the substrate, hydrodynamic behaviour etc,).
- Coating structure (dependence on splat formation)
- Desired coating properties (structure property relationships)

References

1. V.Kadyrov, Margarita, Y., Sen, D.Srinivasa Rao, K.P.Rao and A.V.Saibaba, *Detonation Coating Process*, Transactions of the Powder Metallurgy Association of India, P. Ramakrishnan (ed.), 1993, Vol.20, p. 1-5
2. G. Sundararajan, D. Srinivasa Rao, D. Sen, K.R.C. Somaraju, *Tribological Behaviour of Thermal Sprayed Coatings*, Proceedings of the XIth Surface Modification Technologies, Eds. T.S. Sudarshan, K.A. Khor and M. Jeandin, Pub. The Institute of Materials, London, 1998,p.872-886
3. D. Srinivasa Rao, D.Sen, K.R.C.Somaraju, S.Ravi Kumar, N.Ravi and G.Sundararajan, *The influence of powder particle velocity and temperature on the property of Cr_3C_2-25NiCr coating obtained by detonation-gun*, Proceedings of the 15th International Thermal Spraying Conference, Nice, France, 1998, p. 385-393
4. Sundararajan, G., K.U.M.Prasad, D. Srinivasa Rao, and S.V.Joshi, *A Comparative Study of Tribological Behaviour of Plasma and D.Gun Sprayed Coatings under Different Wear Modes*, Journal of Materials Engineering and Performance, volume 7(3), June 1998, p. 343- 351.

Chapter 4

Cold Gas Dynamic Spraying

G. Sivakumar and S.V. Joshi

International Advanced Research Centre for Powder Metallurgy and New Materials, Hyderabad – 500 005

ABSTRACT

This paper deals with the exciting technology of Cold Gas Dynamic Spraying (CGDS) or Cold Spray Coating Technology (CSCT), which was originally conceived in mid 1980s at Institute of Theoretical and Applied Mechanics, Russia by Alkhimov and colleagues and later patented in United States in 1994 and in Europe in 1995. It is considered to be a logical step in the effort to reduce the thermal energy content and increase the kinetic energy content in thermal spray techniques. Points out that ARCI has now joined select band of institutions worldwide that are equipped to pursue research pertaining to this new and exciting field that has numerous promising applications. As in the case of DSC technology already commercialized by ARCI, the Centre proposes to eventually build completely indigenous Cold Spray systems for technology transfer to Indian industries so that a frontier coating technology becomes available to Indian entrepreneurs at financially attractive terms.

Keywords: Protective coatings, Surface engineering, Wear and tear, Environmental corrosion protection, Cold gas dynamic spraying.

Introduction

The quest for development of protective coatings to combat material degradation under aggressive environments has led to the advent of advanced coating technologies, which provide value-addition through tailored surface properties. Over the years, the surface degradation problems have been addressed by different coating techniques. Among the various surface modification methods, techniques such as flame spray, arc spray, detonation, plasma, HVOF spray which are categorized as 'thermal spraying', have been well-established for forming coatings using wide range of materials for a variety of applications, ranging from aviation to electronics industry. Applications include protection from different modes of wear, high temperatures, chemical attack, and the more mundane uses of environmental corrosion protection in infrastructure maintenance engineering.

Although the art of thermal spraying started way back in early 1900s', the last decade has seen a virtual revolution in the capability of the technology to produce truly high performance coatings of a great range of materials on many different substrates. This enhancement of the thermal spray domain has been achieved largely through introduction of new spray variants, the enhancement of spray process controls, employment of state-of-the-art methods for feedstock materials production, and use of modern techniques of quality assurance. The latest technological advance in the field pertains to the development of the Cold Gas Dynamic Spraying (CGDS) or Cold Spray Coating Technology (CSCT). The method was originally conceived in mid 1980s at Institute of Theoretical and Applied Mechanics, Russia by Alkhimov and colleagues and later patented in United States in 1994 and in Europe in 1995. Although relatively new, the CGDS technique is acknowledged to have significant promise and is attracting increasing attention both in the research community as well as in potential user-industry segments. This paper is devoted to this exciting technology which is considered to be a logical step in the effort to reduce the thermal energy content and increase the kinetic energy content in thermal spray techniques.

CGDS Process Description

CGDS is a process of applying coatings by exposing the metallic or non-metallic substrate to a high velocity (300 to 1200 m/sec) jet of small particles (1 to 50 μm) accelerated by a supersonic jet of compressed gas (Air/Nitrogen/Helium). While the specifics of the bonding of the metal particles to the substrate material are the subject of ongoing studies, it is thought that the high velocity impact of the particles disrupts the oxide films on the metal particle and provides intimate conformal contact under high local pressure, thus permitting bonding to occur by allowing a kind of "explosive welding" of the materials. This hypothesis is consistent with the fact that wide ranges of ductile materials, such as metals and polymers, have been cold spray deposited. On the other hand, experiments with non-ductile materials, such as ceramics, have not been successful unless they are co-deposited along with a ductile matrix material. A suitable combination of particle velocity, temperature and size leads to the deposition of reliable coatings. As discussed in an earlier chapter, the principle behind the formation of thermally sprayed coatings is to melt material feedstock (wire or powder) with a heat source (electric/gas) and then accelerate the molten droplets to impact onto the substrate, where rapid solidification at the rate of 10^4 to 10^8 K/sec occurs. In contrast to the conventional thermal spraying processes, the particles are accelerated by a supersonic gas jet at a temperature much lower than the melting point of the material in the CGDS technique, enabling only solid-state deposition of coating materials. The relative comparison of particle velocity and gas temperatures typical of the CGDS process with that of other different thermal spray processes is shown in Figure 4.1.

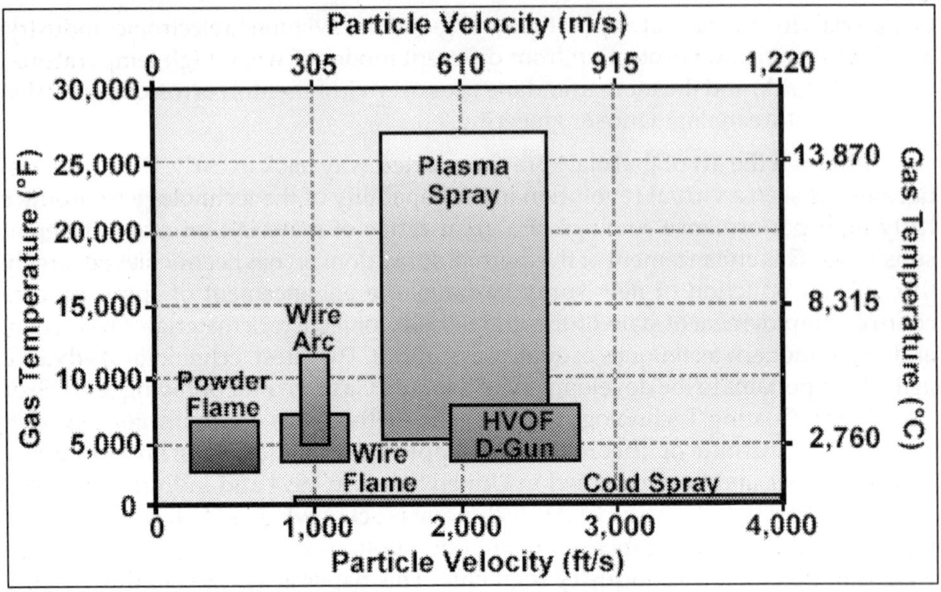

Figure 4.1: Comparison of Different Spray Process Based on Particle Velocity and Gas Temperature

Many of the problems associated with conventional thermal spray coating methods arise because of the high temperatures required to heat or partially melt the coating material. The high deposition temperatures often preclude the spraying of temperature-sensitive materials like Titanium onto low melting point materials such as plastics. Moreover materials prone to phase transformations, excessive oxidation, evaporation, and/or crystallization also cannot be satisfactorily coated. Additional problems arise due to possible residual stress effects and deformations induced by the thermal expansion mismatch between the coating and substrate. Even if the coating remains bonded to the substrate, the inherent residual stresses may cause unacceptable distortions to weaken the bond strength and accelerate coating failure. Many of the above mentioned problems can be suppressed by using the low temperature CGDS process.

Figure 4.2 shows a schematic of the CGDS system. The operating gas (Compressed Air/Nitrogen/Helium) at an elevated pressure is introduced to a manifold system containing typically a ceramic catridge gas heater, where the heating of the gas is accomplished electrically. A metered quantity of powder is fed by a high-pressure powder feeder to an intermediate mixing chamber, from where the heated operating gas carries the powder particles through a de Laval type converging-diverging supersonic nozzle. The compression of the heated gas and the corresponding powder particles happens at the throat region. It may be pointed out that the particles are significantly accelerated as they pass through the nozzle. Figure 4.3 illustrates the typical variation in particle velocity with axial position in the nozzle. As may be expected, the extent to which any particle is accelerated is governed by the material

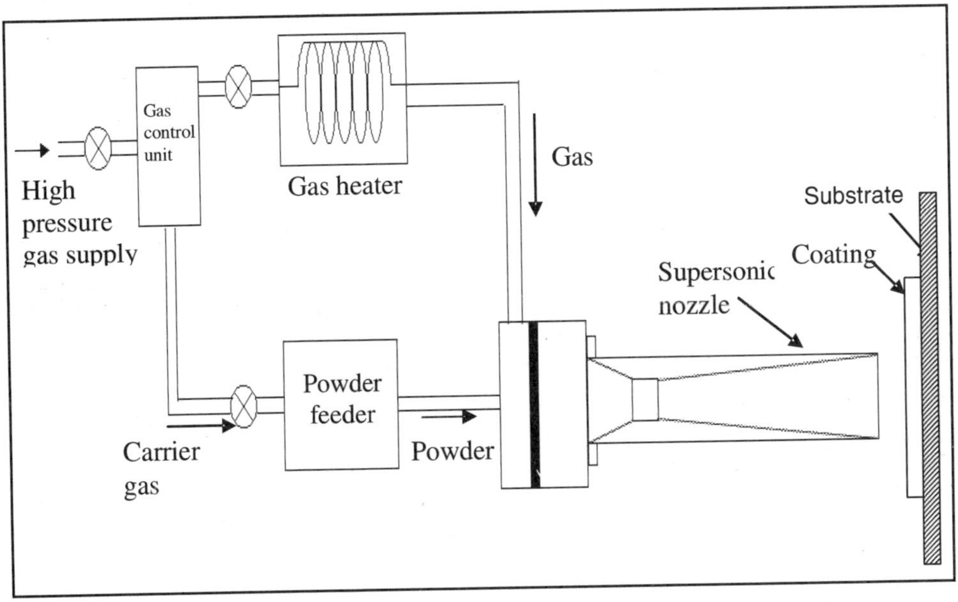

Figure 4.2: Schematic of the Cold Spray Process

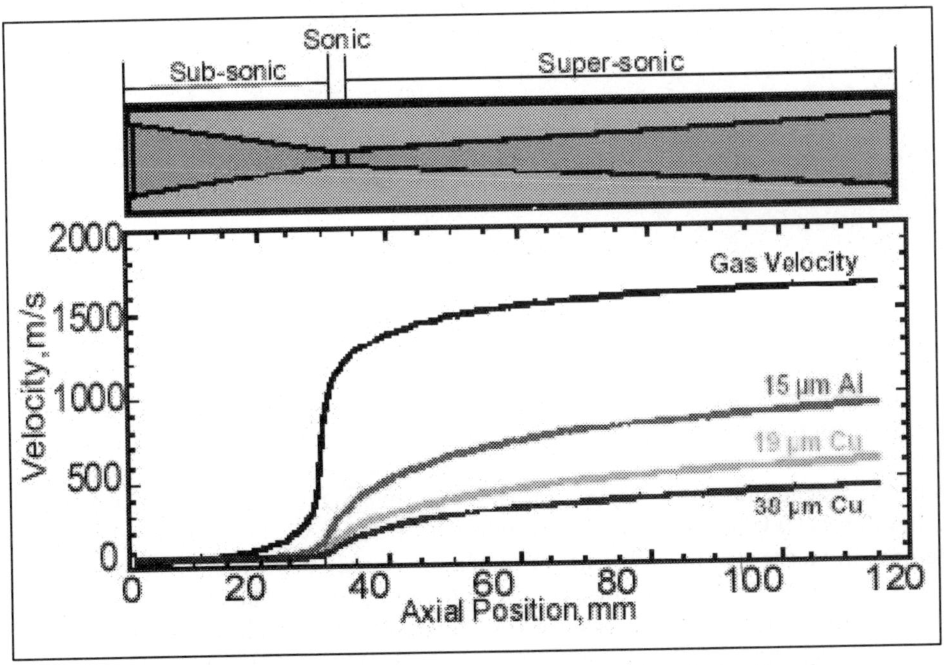

Figure 4.3: Schematic of Particle Acceleration Inside the Nozzle

density as well as the size of the particle as evident from the results shown in the above figure.

The warm/hot particles that are accelerated to supersonic velocities by the time they exit the nozzle proceed to impact the part to be coated, where they flatten. Manipulation of the gun and the job and the deposition of successive layers are eventually responsible for the formation of a coating of desired thickness on large areas and on complex geometrical shapes. Figure 4.4 shows a typical cold spray gun in operation.

Figure 4.4: Cold Spraying Process in Operation

Process Fundamentals

Generally, when a particle-laden gas jet impacts the surface of the substrate, one of three different phenomenon can occur depending on the particle velocity, V_p. (a) When V_p is low, the particle simply ricochets from the surface leaving craters (b) When V_p reaches moderate values, solid particle erosion occurs resulting in substrate material loss (c) When V_p exceeds a certain critical value V_c, the particles plastically deform and adhere to the substrate or onto a previously deposited layer to form an overlay deposit, analogous to thermally sprayed coatings. The critical velocity to allow coating formation varies with powder material as well as its particle size characteristics and typically falls in the range of 500-700 m/s. During CGDS spraying, the particle velocity at impact depends on the operating gas pressure, pre-heat temperature and marginally on standoff distance apart from the particle size and material density as mentioned earlier. Since particles with velocity higher than the critical velocity form the deposit, the deposition efficiency increases with particle velocity. This is clearly evident from the results obtained during spraying of copper on an aluminium substrate with different particle velocities as shown in Figure 4.5.

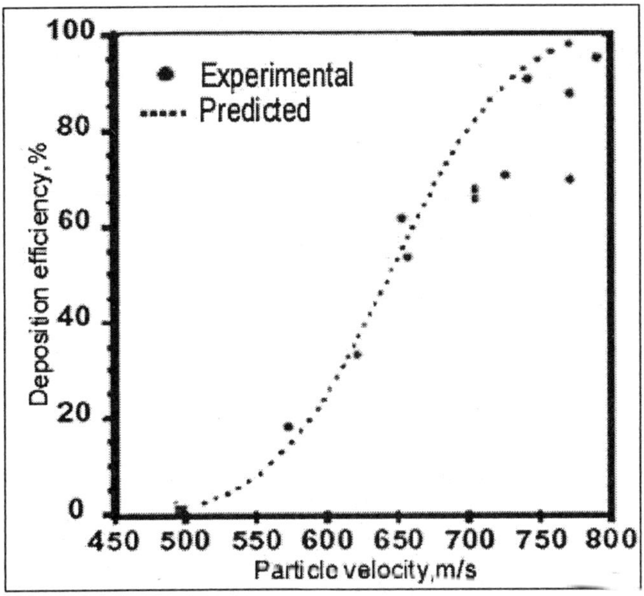

Figure 4.5: Relation between Deposition Efficiency and Particle Velocity [R.C. Dykhuizen *et al.*, 1998]

The velocity of the gas at the throat, V_t, is given by,

$$V_t = (\gamma R T_t)^{0.5} \quad (1)$$

In equation 1, γ is the ratio of specific heats (approximately 1.7 for monatomic gases and 1.4 for diatomic gas), R is the specific gas constant (universal gas constant divided by gas molecular weight), and T_t is the gas temperature at throat. The gas velocity at nozzle exit is given by,

$$A/A_t = [2M^{-1}/(\gamma+1)\,(1+(\gamma-1)\,0.5\,M^2)]\,\hat{}\,(0.5\,(\gamma+1)/(\gamma-1)) \quad (2)$$

where, A_t is the area at throat and M is the mach number. The powder particles are accelerated and reach a velocity V_p given by,

$$V_p = V\,(C_D A_p \rho x / m)^{0.5} \quad (3)$$

where,

V is the gas velocity, C_D is the drag coefficient, A_p is the area of the particle, ρ is the gas density, x is the distance traveled by the particle and m is the particle mass.

In the cold spray process, deposition of powder particles is largely governed by their kinetic energy, which is dependent on the gas velocities for a given powder feedstock. The gas velocities, in turn, strongly depend on the nozzle geometry, gas temperature, and type of gas. The above equations also explain the dependence of particle velocity on particle size and material density previously illustrated in Figure 4.3. Reported results also clearly reveal that the deposition efficiency increases with particle velocity and gas temperature as illustrated in Tables 4.1 and 4.2 respectively. It is pertinent to mention here that the velocity of particles is usually determined by laser doppler anemometry and deposition efficiency by weighing methods.

Table 4.1: Typical Dependence of Deposition Efficiency on Particle Velocities during Cold Spraying of Different Materials [A.P. Alkhimov *et al.*, 1994]

Powder Material	*Particle Velocity, m/s*	*Deposition Efficiency, %*
Copper	650 ± 10	10
	800 ± 10	30
	900 ± 10	40
	1000 ± 10	80
Aluminium	650 ± 10	40
	1000 ± 10	60-70
	1200 ± 10	80-90
Nickel	800 ± 10	10
	900 ± 10	40
	1000 ± 10	80
Vanadium	800 ± 10	10
	900 ± 10	30
	1000 ± 10	80
Alloys	700 ± 10	10
	800 ± 10	20
	900 ± 10	50

Table 4.2: Deposition Efficiency as a Function of Inlet Air Temperature during Cold Spraying of Different Materials [A.P. Alkhimov *et al.*, 1994]

Powder Material	*Air Temperature, °C*					
	10	*20*	*100*	*200*	*350*	*400*
Aluminium	0.1–1 per cent	1–1.5	10	30–60	90–95	–
Zinc	1–2	2–4	10	50–80	–	–
Tin	1–30	30–40	40–60	–	–	–
Copper	–	–	10–20	50	80–90	90
Nickel	–	–	–	20	50–80	80–90
Titanium	–	–	–	50–80	–	–
Iron	–	–	–	20–40	60–70	80–90
Vanadium	–	–	–	20	40–50	60–70
Cobalt	–	–	–	20	40–50	50–60

In practice, stagnation jet pressures of 1-3 MPa and stagnation jet temperatures in the range of 100-600°C are employed during cold spraying. Various gases such as air, nitrogen, helium etc can be used to provide the necessary 'critical' particle velocity for coating formation. The sonic velocity increases with increase in ratio of specific heats and also by reducing the molecular weight of the gas. Thus, the use of helium which has a higher γ value (1.66) compared to nitrogen (1.4) and a substantially lower molecular weight (4:28) results in much higher jet velocity as illustrated in Figure 4.6 and, in turn, leads to a substantially higher deposition efficiency. It is pertinent to point out that various metals and alloys can be sprayed even at room jet stagnation temperature (without any heating whatsoever), if the necessary particle velocity is achieved.

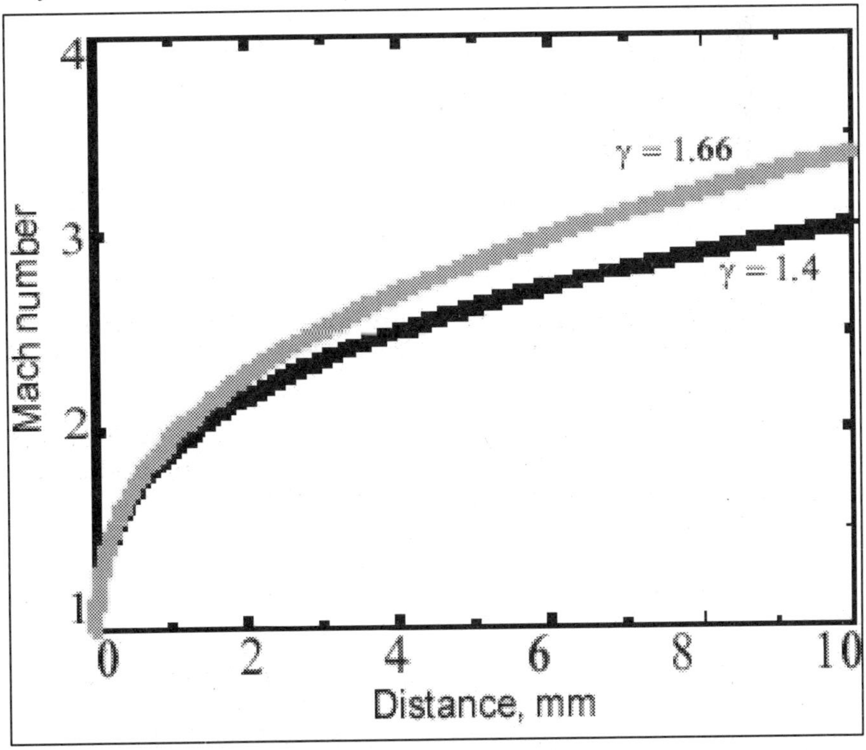

Figure 4.6: Influence of Specific Heat Ratio of Process Gas on Exit Gas Velocity [R.C. Dykhuizen *et al.*, 1998]

Mechanism of Coating Formation

Before discussing the coating characteristics, it is meaningful to understand how the coating builds up during cold spraying. The overall coating development process can be divided into four distinct stages as per a recently proposed mechanism. These four stages are depicted in Figure 4.7 with the transition between individual stages being governed by the incident particle velocity. The four stages can be briefly listed as follows:

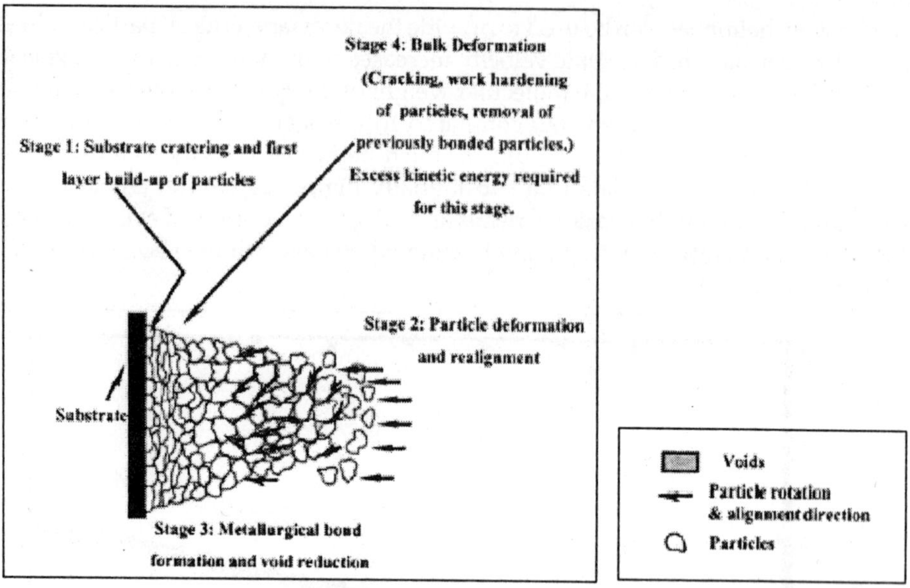

Figure 4.7: Coating Formation Stages in the Cold Spray Process [T.H. Van Steenkiste, 2002]

Stage 1

Substrate cratering and first layer build up of particles (Initial cratering, deformation of substrate, splat formation, fracturing of surface oxides on metal particles and substrates and first particle layer formation).

Stage 2

Particle deformation and realignment (multilayer coating build-up) as described in Figure 4.8.

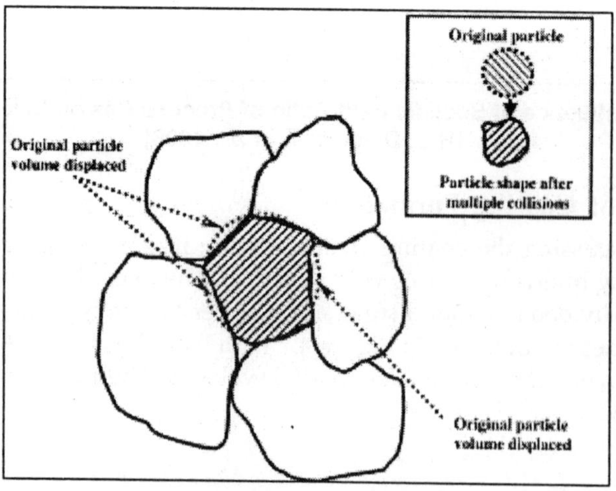

Figure 4.8: Schematic of Particle Deformation Under Incident Particle Collision [T.H. Van Steenkiste, 2002]

Stage 3

Metallic bond formation between particles (particle–particle bonding) and void reduction.

Stage 4

Further densification and work hardening of the coating.

The basic coating formation arises from a transformation of the kinetic energy of the particles into mechanical deformation and thermal energy. This transformation occurs rapidly, on the order of 10^{-7} s. The overall process involves a succession of changes involving compaction, realignment and localized plastic flow of material under high contact pressures to remove surface oxide layers and allow metallic bonds to form. The micrographs of cold sprayed aluminium and copper coatings shown in Figure 4.9 depict some of the above features of cold spraying. As compared to the

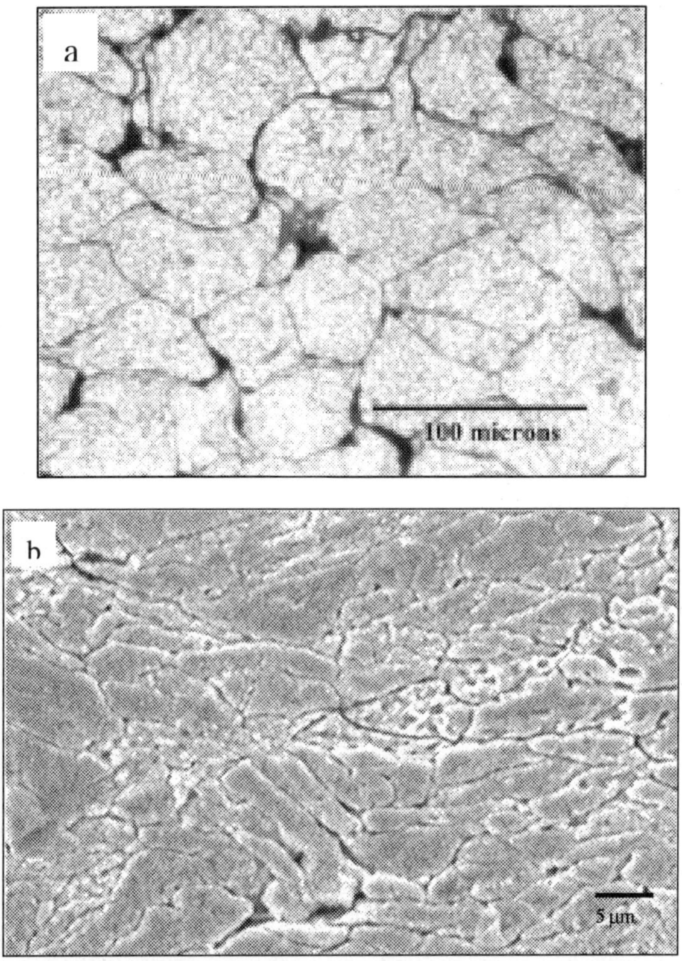

Figure 4.9: Optical Micrograph of Etched (a) Aluminium [T.H. Van Steenkiste, 2002] and (b) Copper Coating Deposited by CGDS Technique

original powder grain structures, these micrographs are indicative of substantial grain deformation consistent with plastic deformation of the particles on impact with the substrate and a peening effect due to impact of subsequent particles.

The bonding process during cold spraying is a complex particle–particle and particle–substrate interaction. The coating formation process is generally found to be anisotropic, with metallic bonds forming primarily along the direction of the incident particle velocity and splats flattening parallel to the coating surface. For achieving thicker coatings, a higher kinetic energy is desirable so that the coatings transform from a non-bonding compacted state to an increasingly metallurgically bonded coating.

Influence of Process Parameters

The typical operating regimes for obtaining cold-sprayed coatings are:

- Gas Pressure: 14-20 kg/cm^2
- Gas Temperature: 100- 600°C
- Gas Flow Rate: 10-50 cfm
- Powder Feed Rate: 1-15 kg/hr
- Spray Distance: 8-30 mm
- Powder size: 1-50 μm

The influence of each individual process variable on the quality of cold sprayed coatings has been the subject of considerable research interest. Nozzle design, too, has attracted significant attention. Nozzles made of high strength die steels are commonly used. While different nozzle shapes have been studied, few results reported in the literature suggest that the rectangular nozzles produce higher deposition efficiency than other shaped nozzles. The influence of nozzle size on coating formation has also been a subject of investigation.

Effect of Gun Variables

Different studies have been made to assess the influence of process variables on coating characteristics. On the basis of these studies, many supporting results have been published on the subject, providing useful insights into the mechanisms associated with coating formation and factors critical to the selection of suitable parametric regimes. Based on the various research investigations carried out till date, the following process variables are deemed to most significantly influence coating quality: (*a*) Gas Pressure, (*b*) Gas Temperature, (*c*) Gas Flow Rate, (*d*) Powder Feed Rate, (*e*) Spray Distance. Efforts in the area of mathematical modelling have also allowed estimating the threshold values of key parameters. Some of the predicted results are shown in Figure 4.10 indicating the effects of gas temperature and the standoff distance on particle velocity under different conditions.

The effect of varying key process parameters on coating quality can be most instructively visualized by examining the coating microstructures. This is illustrated in Figure 4.11 which depicts the coating microstructures obtained during cold spraying of NiCr-Cr_3C_2 coatings by employing different gas temperatures of 400 and

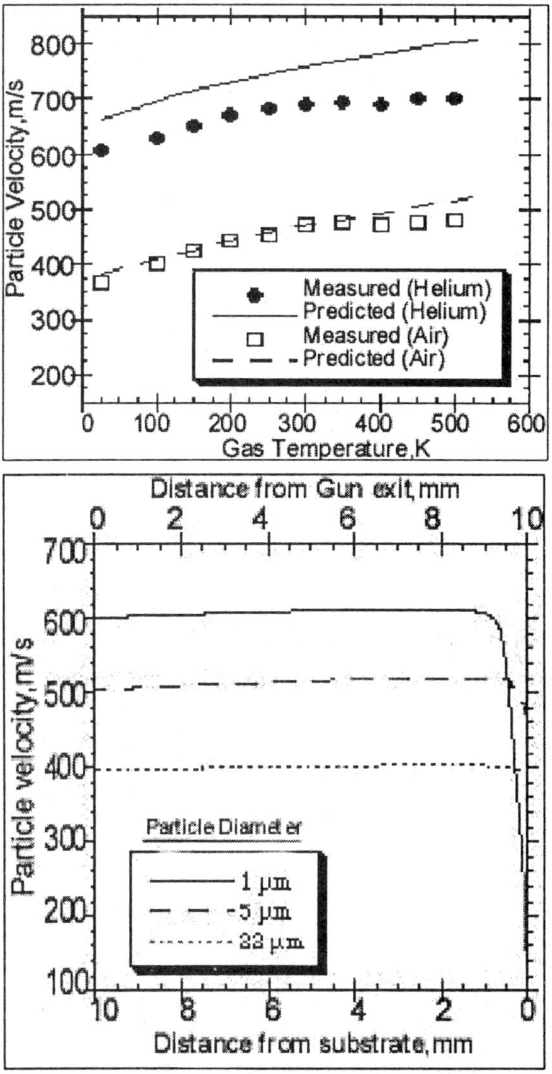

Figure 4.10: Effect of Gas Temperature and Standoff Distance on Particle Velocity [D.L. Gilmore *et al.*, 1999]

600°C. As shown in the figure, the coatings deposited at a lower pre-heat temperature of 400°C exhibited a poor microstructure with relatively high porosity levels in the coating, which is an indicator of lower incident particle velocity. The inter-particle bonding was also seen to be very poor at low pre-heat temperature. In contrast, pre-heating the gas to a temperature of 600°C is seen to significantly improve the coating microstructure.

The coating thickness built up per pass and deposition efficiency is also significantly influenced by varying the process parameters. For example, Figure 4.12

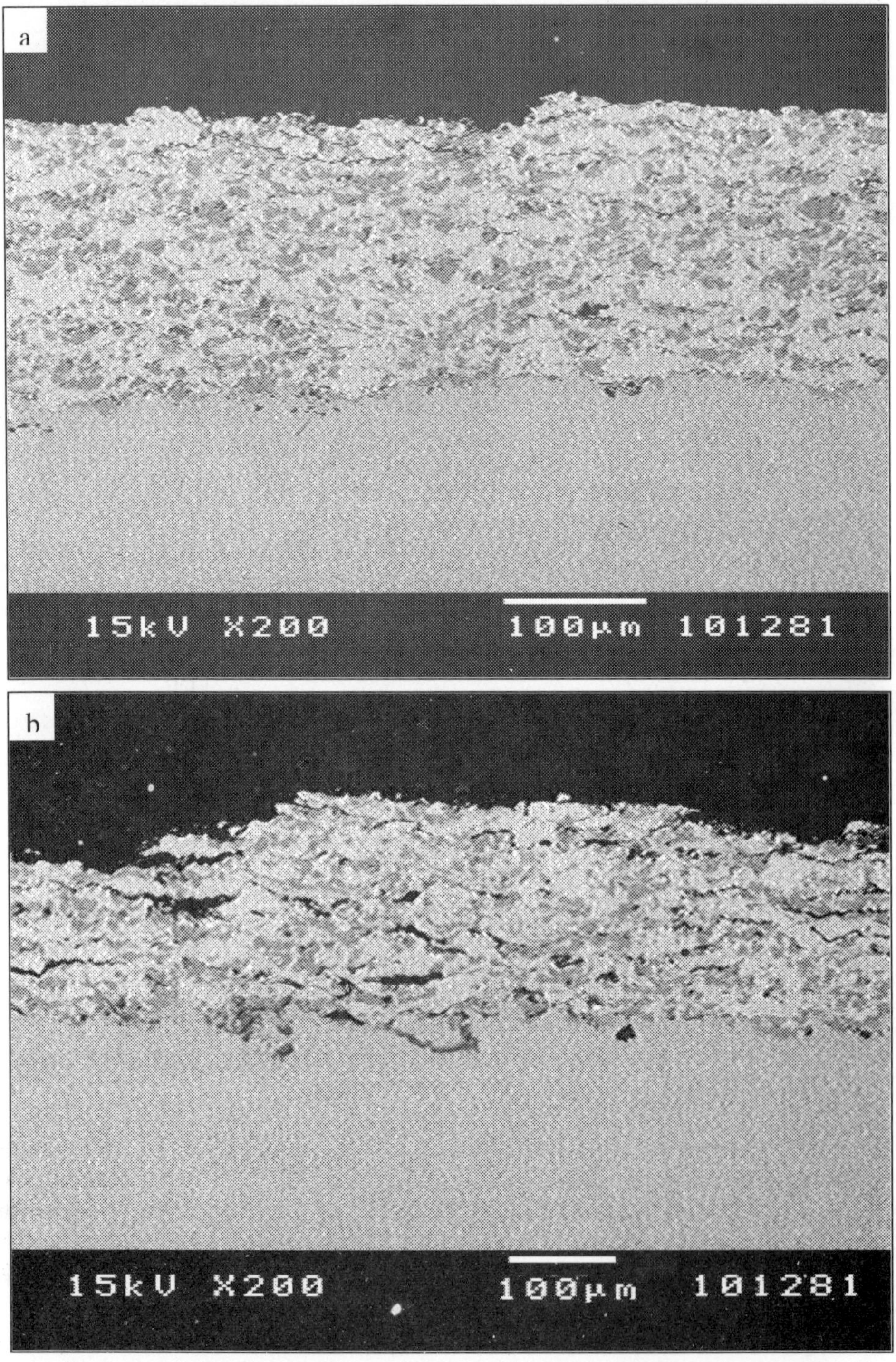

Figure 4.11: Cross-sectional Microstructure of NiCr-Cr_3C_2 Coating with Air Pre-heat Temperature of (a) 600ºC, (b) 400ºC

depicts the variation in deposition efficiency with gas temperature and powder feed rate during spraying of Titanium. Incidentally, because of high temperatures involved and the rapid oxidation that ensues with conventional thermal spray variants, the industrial potential of CGDS for spraying reactive and oxidation prone materials such as titanium has been attracting increased attention. Special expensive accessories such as inert shrouds or high vacuum can be avoided with cold spraying. As evident from Figure 4.12, typical deposition efficiencies in the range of 50-70 per cent can be achieved by employing gas temperatures higher than 250ºC using nitrogen as the process gas.

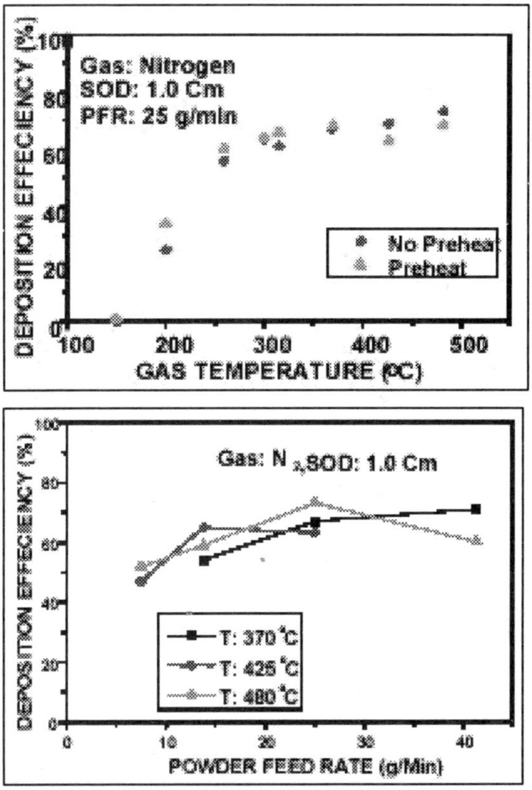

Figure 4.12: Variation in Deposition Efficiency of CGDS Titanium Coatings with (a) Inlet gas temperature and (b) Powder feed rate [J. Karthikeyan *et al.*, 2000]

Similarly, Figure 4.13 shows the effect of various operating parameters on the relative coating thickness per pass during CGDS spraying of aluminium, copper and iron coatings on two different substrates, namely Al and brass.

Influence of Type of Processing Gas

As discussed earlier, the type of gas employed in the CGDS technique plays a very prominent role in determining the quality of the deposited coating. Air is most preferred because of economic considerations while results till date have amply demonstrated that Helium produces superior coatings. Hence, a proper selection of the operating gas needs to be done substantially based on desired coating properties, functional requirements and cost considerations.

The spraying of NiCr-Cr_3C_2 coatings was also tried with both air and helium and coatings obtained from both compared for properties. The results shown in

Figure 4.13: Influence of Various Process Parameters on Cold Spraying of Cu, Al, and Fe Coatings on Al and Brass Substrates: (a) Nozzle air inlet pressure (b) Nozzle air inlet temperature (c) Stand-off distance and (d) Powder feed rate [T.H. Van Steenkiste, 1999]

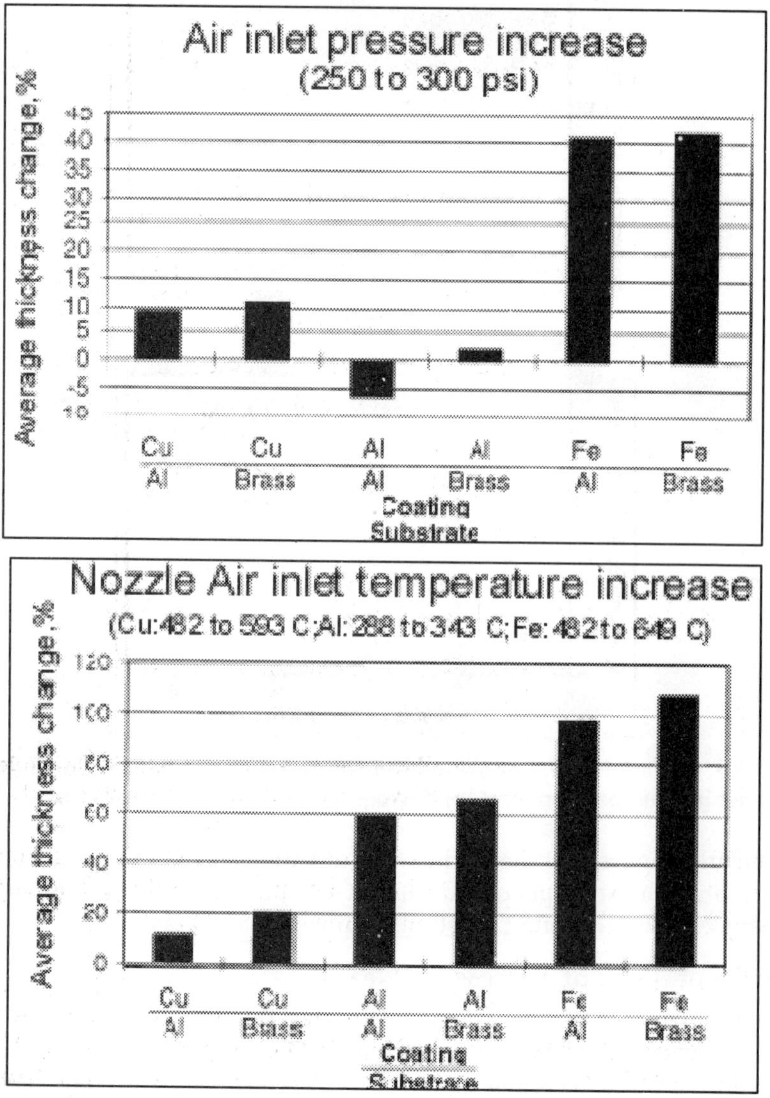

Contd...

Figure 4.13–Contd...

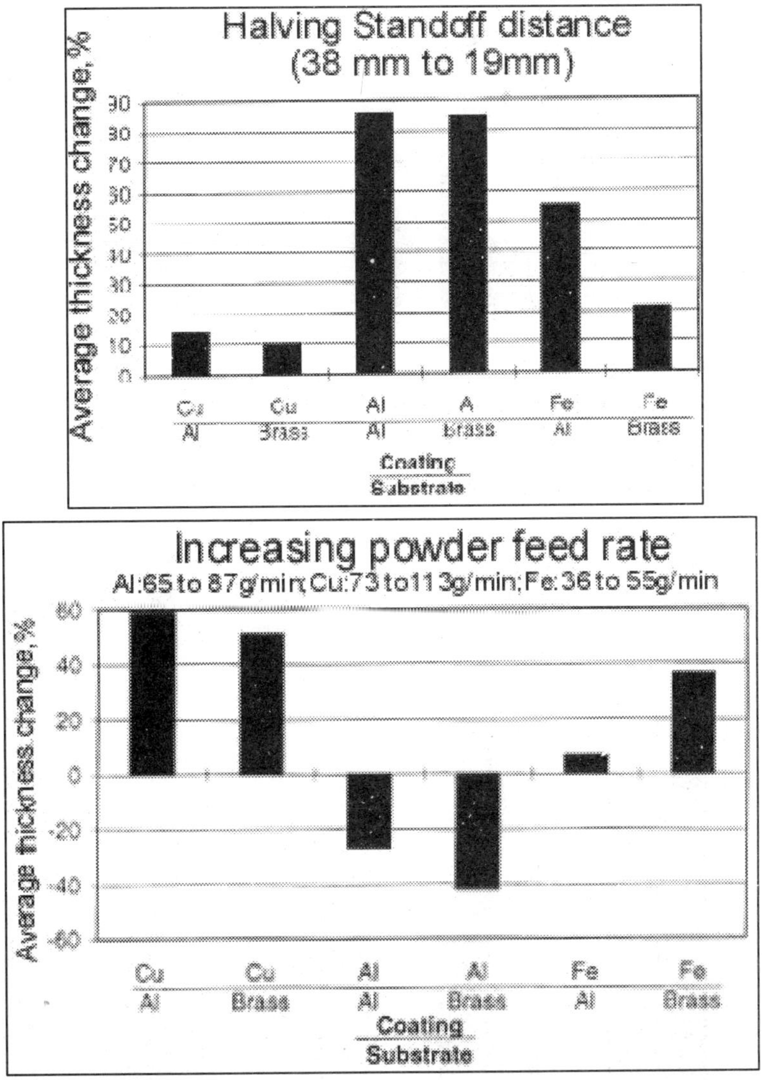

Figure 4.15 clearly indicate wide ranging improvements in all coating properties, when Helium is used as the operating gas.

Influence of Nozzle Geometry

The most widely studied nozzle geometries have included circular, rectangular and elliptical exit types. Among these nozzles, the rectangular shaped nozzles have been reported to produce substantially higher particle velocities compared to the circular nozzle as shown in Figure 4.16. Hence the rectangular nozzles are considered to favor higher deposition efficiencies.

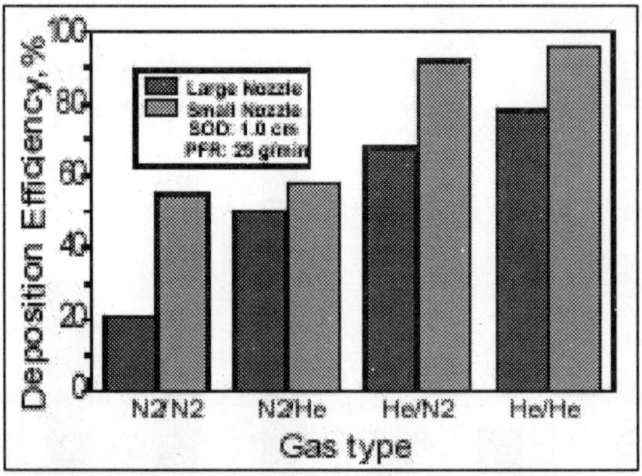

Figure 4.14: Illustrates the Influence of Gas Type Employed as Process as well as Carrier Gas on Deposition Efficiency, Using Cold Spraying of Titanium as a Case Study. The use of helium for both operation as well as powder feeding is seen to yield best results.

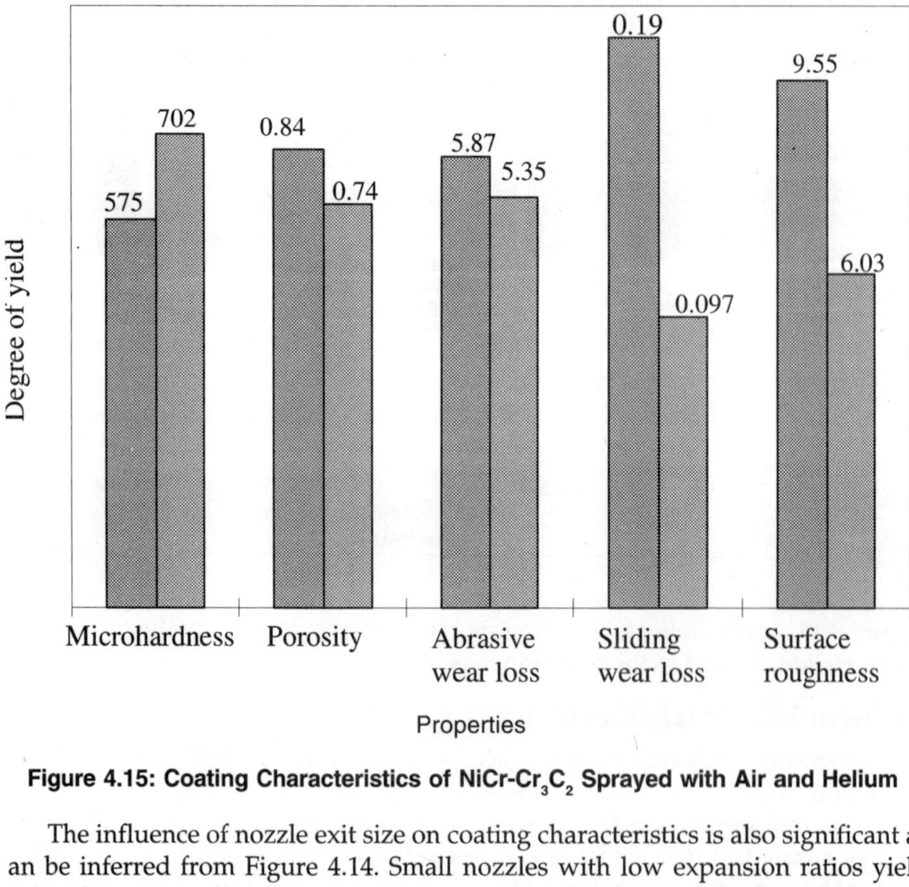

Figure 4.15: Coating Characteristics of NiCr-Cr_3C_2 Sprayed with Air and Helium

The influence of nozzle exit size on coating characteristics is also significant as an be inferred from Figure 4.14. Small nozzles with low expansion ratios yield ıigher deposition efficiencies than large nozzles with higher expansion ratio for any ;iven processing gases.

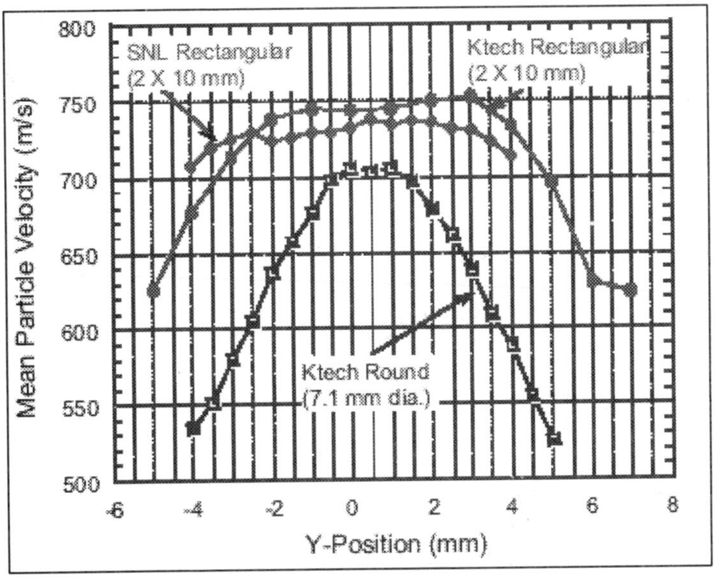

Figure 4.16: Influence of Nozzle Geometry on Particle Velocity [W.M. Robson, 2002]

Versatility of CGDS Technique

Powders of varying chemistry have been successfully deposited by the CGDS process till date. A partial list of coatings deposited by the CGDS route is provided below and provides ample evidence of the versatility of the technique. The starting morphologies of some of the different powders widely used with cold spraying are also shown in Figure 4.17 and this suggests that powders with varying morphology (and manufactured by different routes such as water/gas atomized, blend, fused) can be accepted as feedstock. Different feedstock studied include,

Aluminium	Ni-5Al	NiCr-Cr_3C_2
Copper	SS304	WC-Co
Iron	SS316	StelCar
Zinc	SS420	
Titanium	IN718	
Molybdenum	Al-12Si	
Chromium	Cu-50Ni	
Cobalt	Ni-50Co	
Tantalum	Fe_3Pt	
Tin	Ni-20Cr	
Polymers	NiCrAlY	
	Active Braze alloy	
	Aluminium Bronze	
	Cu-50Zn	

Characteristics of CGDS Coatings

Microstructural Aspects

The microstructural features of CGDS deposited coatings can be studied similar to the traditional thermal sprayed samples. Samples are cut by a low speed cutter and hot mounted and polished by subjecting to standard metallographic preparation. The cold sprayed coatings are generally observed to be dense and well-adherent to the substrate as seen in Figure 4.18. It is especially interesting to point out that efforts to deposit carbide coatings have also yielded exciting results as exemplified by the

Figure 4.17: Powder Morphologies of (a) Aluminium Bronze, (b) Nickel, (c) NiCrAlY, (d) NiCr-Cr_3C_2

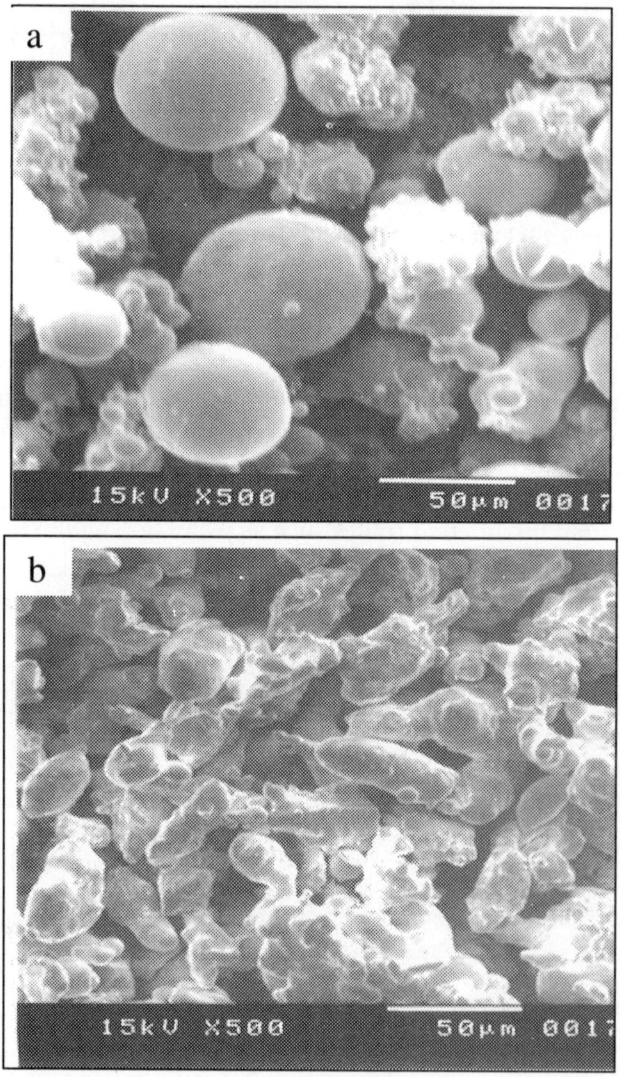

Contd...

Figure 4.17–Contd...

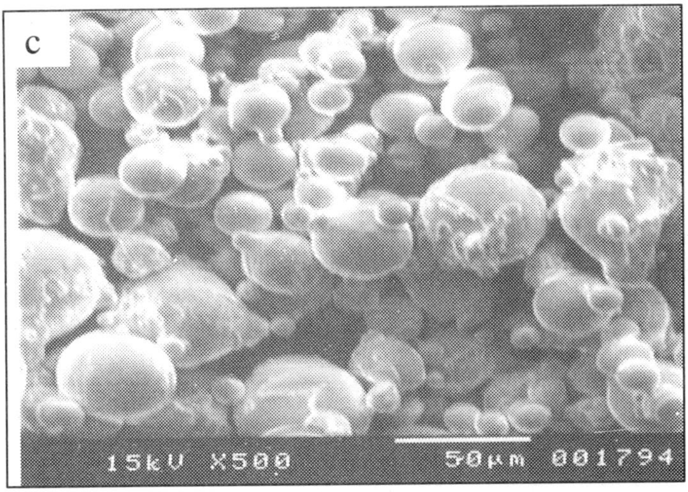

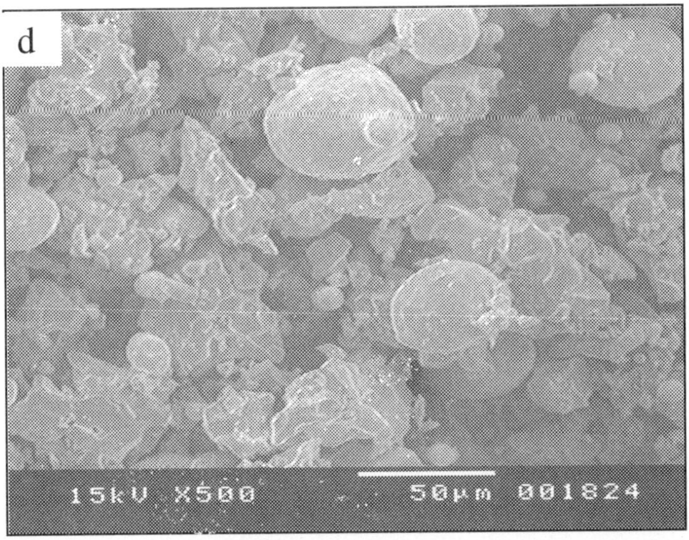

secondary image of cold sprayed NiCr-Cr_3C_2 coating of 700 µm as illustrated in Figure 4.18 (d) which reveals the formation of an extremely dense layer. Various inherent advantages of the CGDS process such as suppression of *in situ* oxidation, retention of phases etc, are also manifested in the microstructures noted in Figure 4.18. These important aspects are, however, also separately discussed below.

Suppression of Oxidation

The possibility of *in situ* oxidation during spraying metals and alloys by the traditional thermal spray techniques represents a significant problem. The low temperatures to which the feedstock powders are exposed during CGDS process successfully suppresses this problem and potentially opens up avenues for special applications. For example, use of cold sprayed copper for electrical conductivity

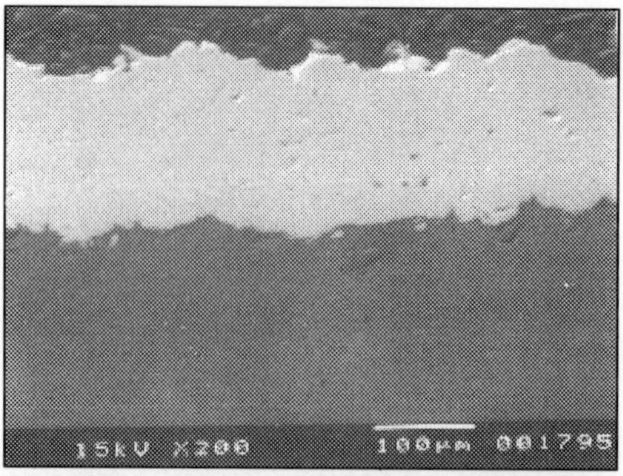

Figure 4.18(a): Aluminium Bronze Coating

Figure 4.18(b): Nickel Coating

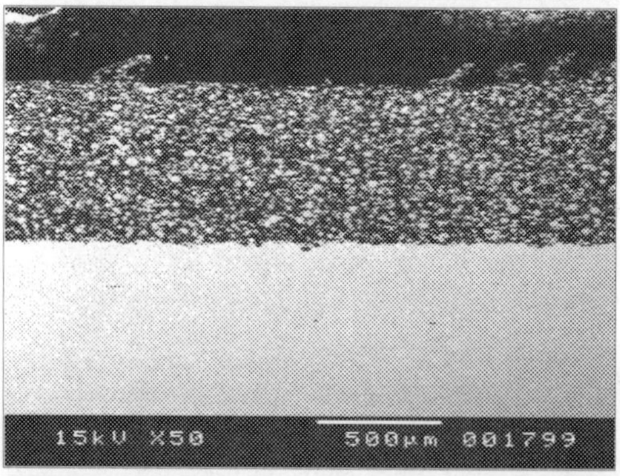

Figure 4.18(c): NiCrAlY Coating

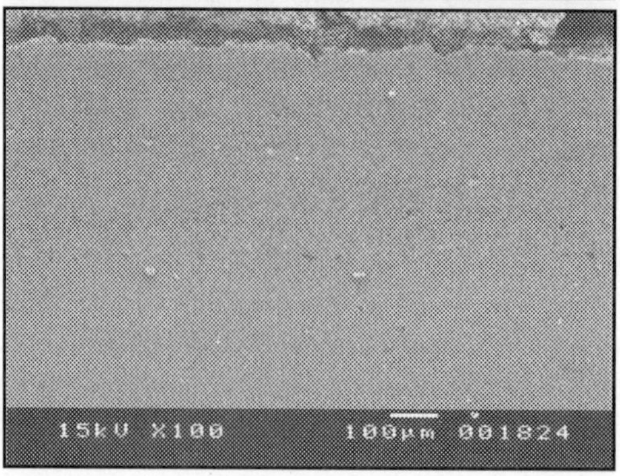

Figure 4.18(d): NiCr-Cr_3C_2 Coating

applications is considered most promising. The typical cold sprayed coating cross-section is shown in Figure 4.19 exhibits a dense and homogenous microstructure. The oxygen content in the starting powder and coatings were determined to be 0.3 per cent and 1 per cent respectively. The EDAX spectrum of the CGDS copper coating is shown in Figure 4.20 and reveals negligible oxygen presence. The XRD pattern confirms the purity of cold sprayed copper coating as shown in Figure 4.21.

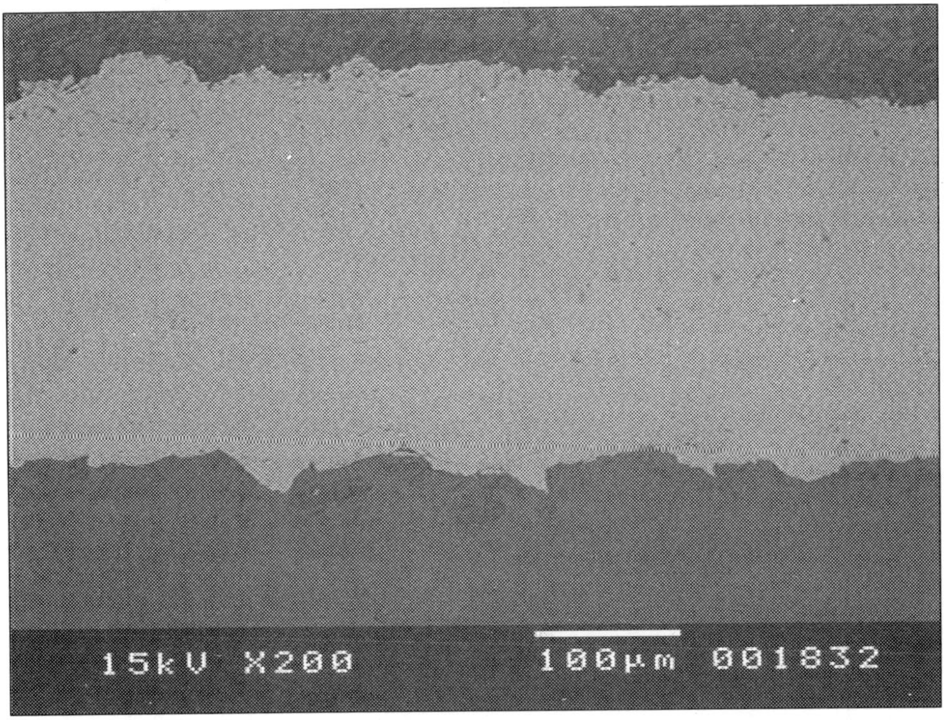

Figure 4.19: Dense Microstructure of Cold Sprayed Copper Coating

The above results clearly suggest that the currently employed thermal spray methods of applying copper such as wire-arc spray, plasma spray etc can be successfully replaced by cold spraying due to relatively higher conductivity that can potentially be offered by cold sprayed copper coatings. A comparative study of electrical conductivity achieved in copper coatings deposited by different techniques is shown in Figure 4.22. This clearly establishes the advantages of the CGDS method for electrical conductivity applications, since this route yields coatings with properties closest to those of wrought copper. Consequently, there is wide scope for application of above coatings on aluminium lugs for increased electrical conductivity and this is already being adapted commercially in Russia and in some western countries.

Retention of Phases

By virtue of the inherently low operating temperatures, the cold spray technique also provides the ability to retain the starting phases, any oxidation or decarburization has prompted cold spraying of carbide coatings. While spraying of NiCr-Cr_3C_2 has

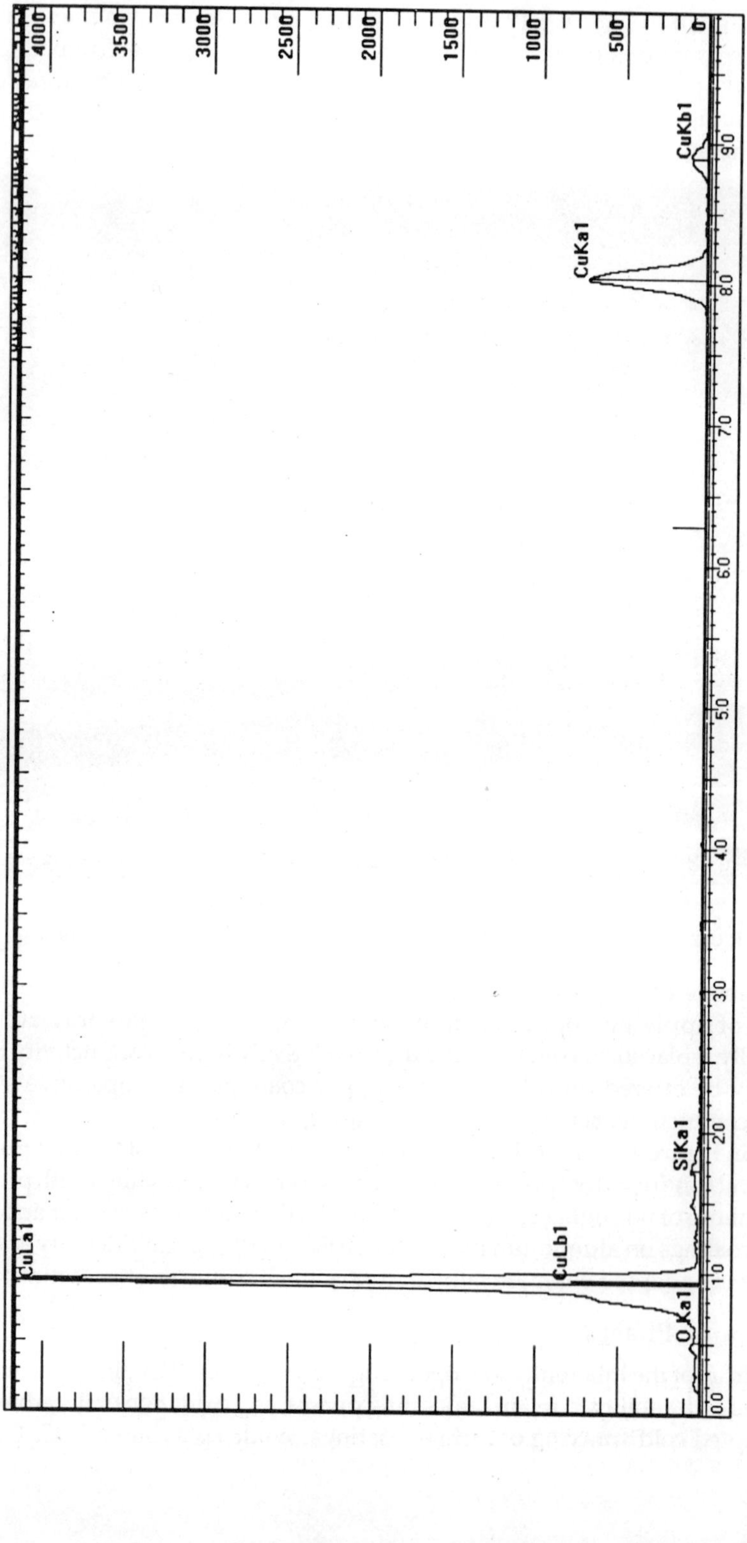

Figure 4.20: EDAX Spectrum of a CGDS Copper Coatings

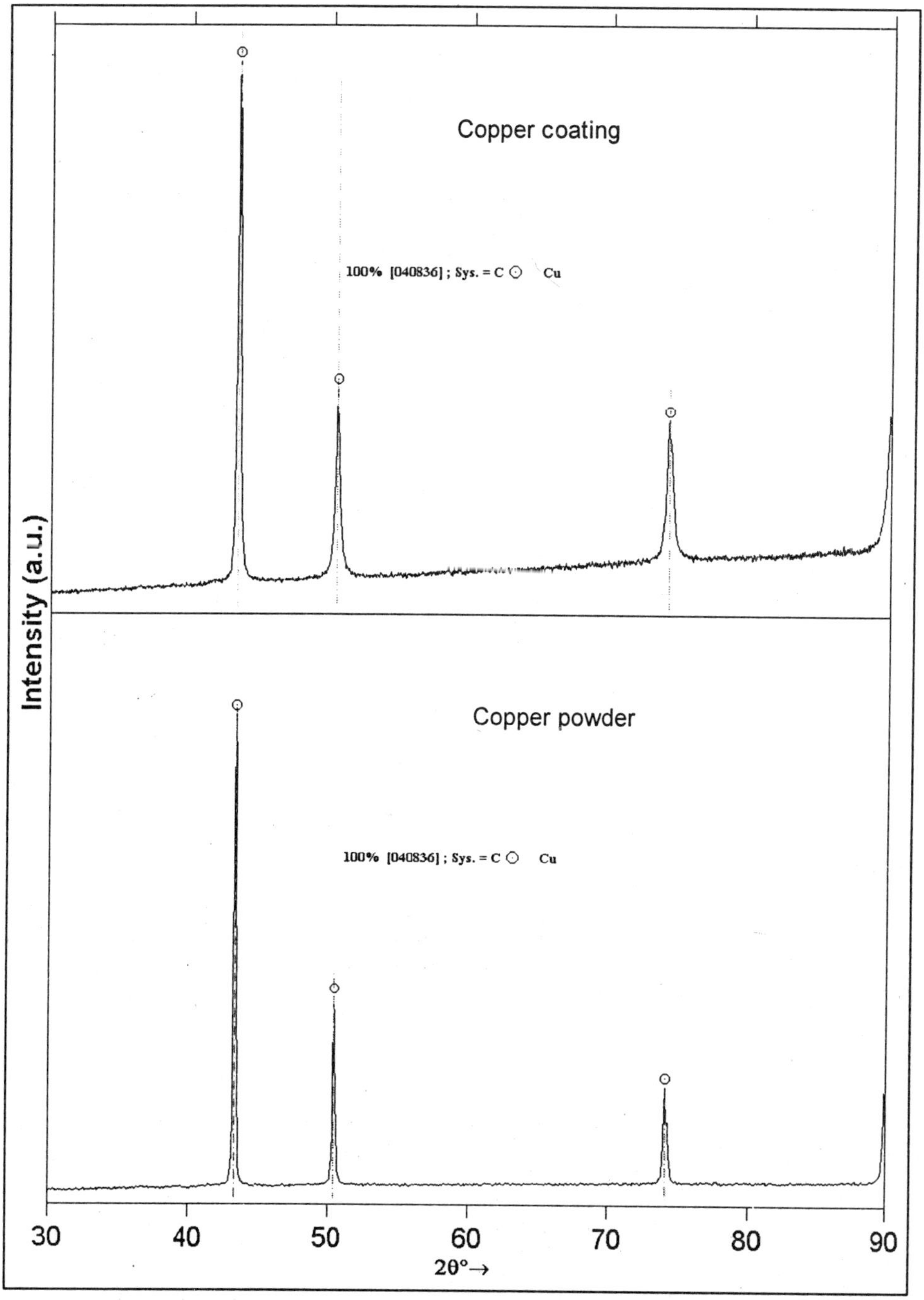

Figure 4.21: Comparison of XRD Patterns of a CGDS Copper Coating and the Starting Powder

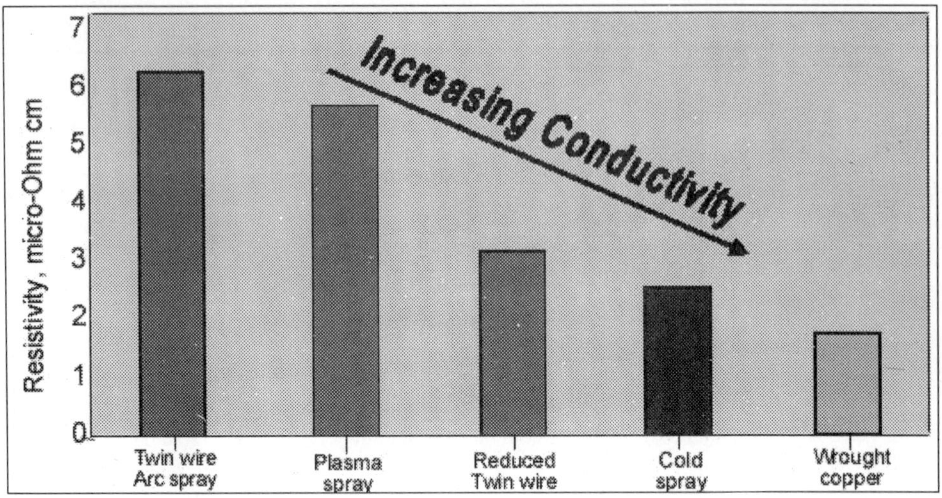

Figure 4.22: Electrical Conductivity of Copper Coatings Deposited by Different Thermal Spray Process [Mark F. Smith, 2002]

been accomplished successfully as discussed earlier, spraying of WC-Co was also attempted in view of its well-proven wear resistant applications. The most striking feature of cold sprayed WC-Co compared to conventional thermal spray variants is the complete absence of any decarburization. The XRD patterns shown in Figure 4.23 exemplify the success of the CGDS in retaining the phases originally present in the powder. However, most attempts reported on cold spraying of WC-Co till date have indicated that the thickness that can be developed is limited to 25 mm because of the inherent particle hardness. This limitation may have to be overcome before cold spraying of WC-Co coatings becomes practically attractive.

Splat Formation

Investigation of splat morphologies resulting from particle impact on the substrate is most informative, since it can provide insight into the coating formation mechanism as well as particle behavior under different processing conditions. Figure 4.24 shows the typical splat patterns of cold sprayed particles under 2000X magnification. Unlike the characteristic splash type splats with the high temperature thermal spray variants like plasma spray, which suggest a completely molten state, the cold sprayed splats are typically non-splashy type. The figure also reveals that the splat patterns vary with the material being sprayed. It may also be pointed out that splats obtained with NiCr-Cr_3C_2 reveal that the hard carbide particles penetrate the substrate, while the soft binder phase stays at the periphery of the carbide particle.

The extent of particle flattening (ratio of particle diameter before and after impact) is also a meaningful aspect to investigate since it can be correlated to the deposition rate and efficiency. Hence, during development of new coatings by cold spraying, it is always advisable to study the splat morphology by varying the process parameters and determine the flattening ratio. The variation in the splat patterns with pre-heat temperature during CGDS coating of SS316 is shown in Figure 4.25 as an example.

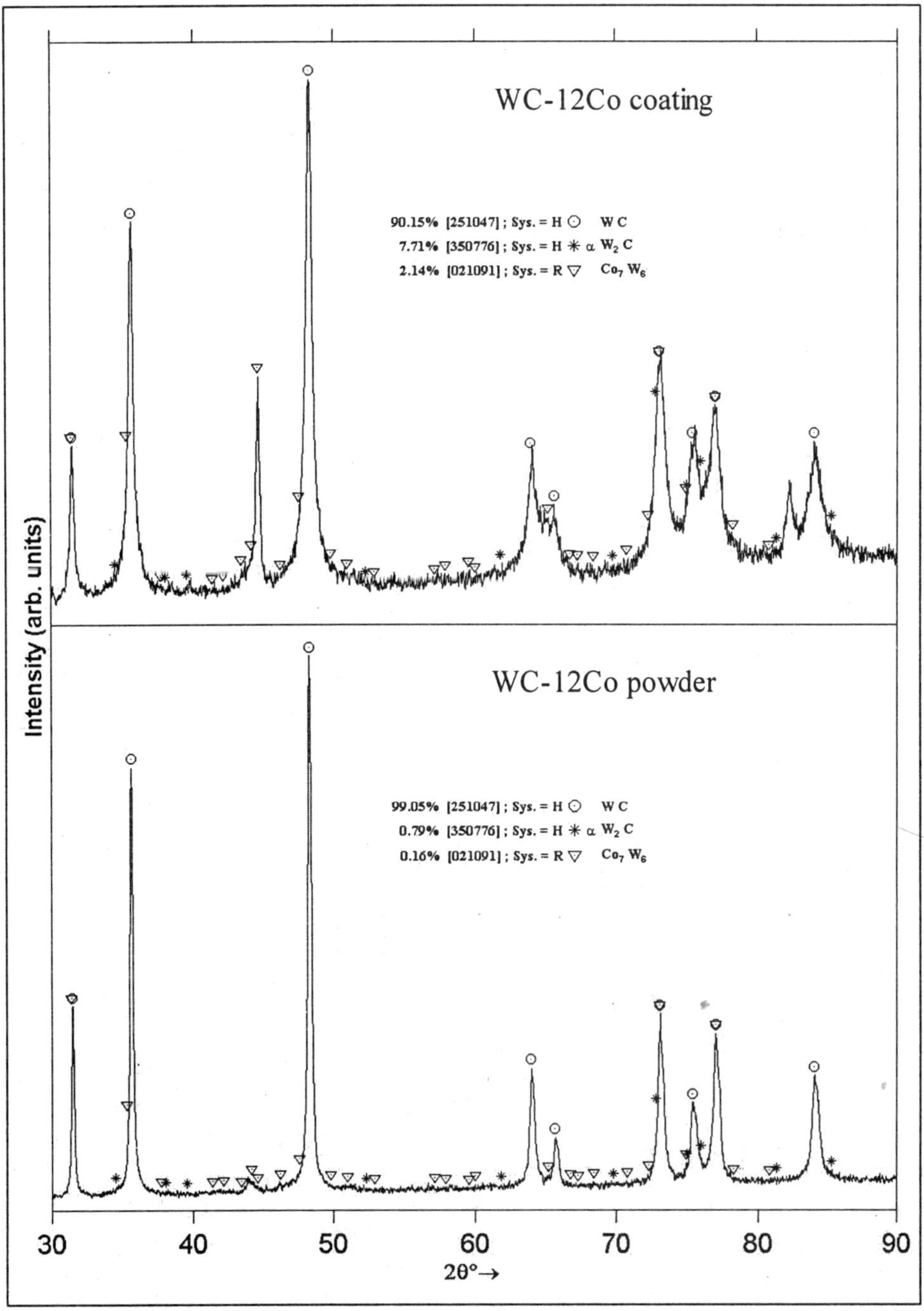

Figure 4.23: XRD Pattern of WC-Co Powder and its Cold Sprayed Coating

Figure 4.24: Splat Morphology of Cold Sprayed (a) NiCr-Cr_3C_2 (b) SS316, (c) Al-12Si, (d) Ni-5Al powders

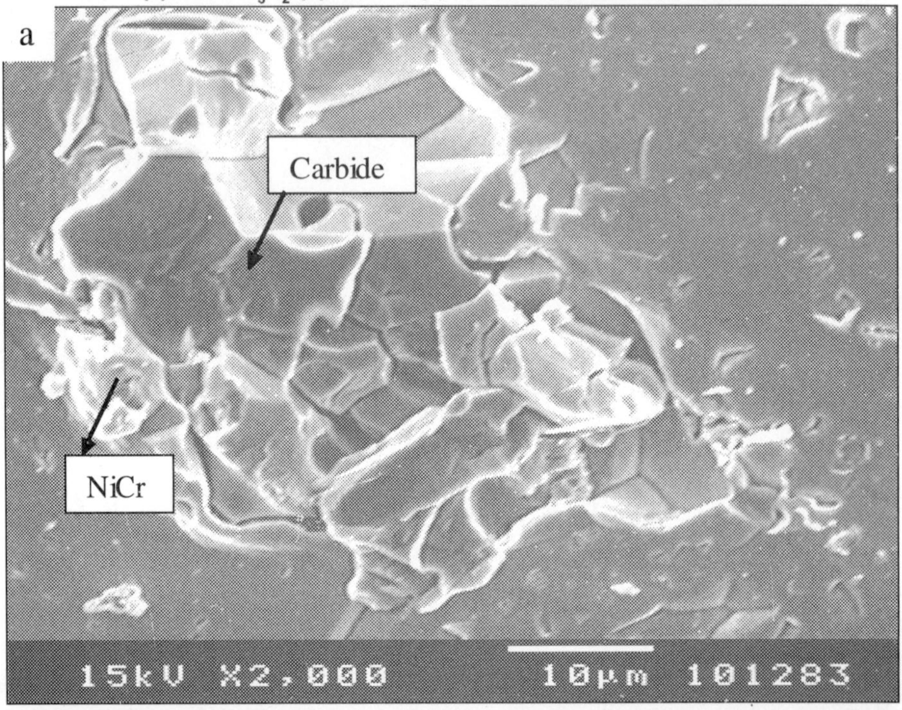

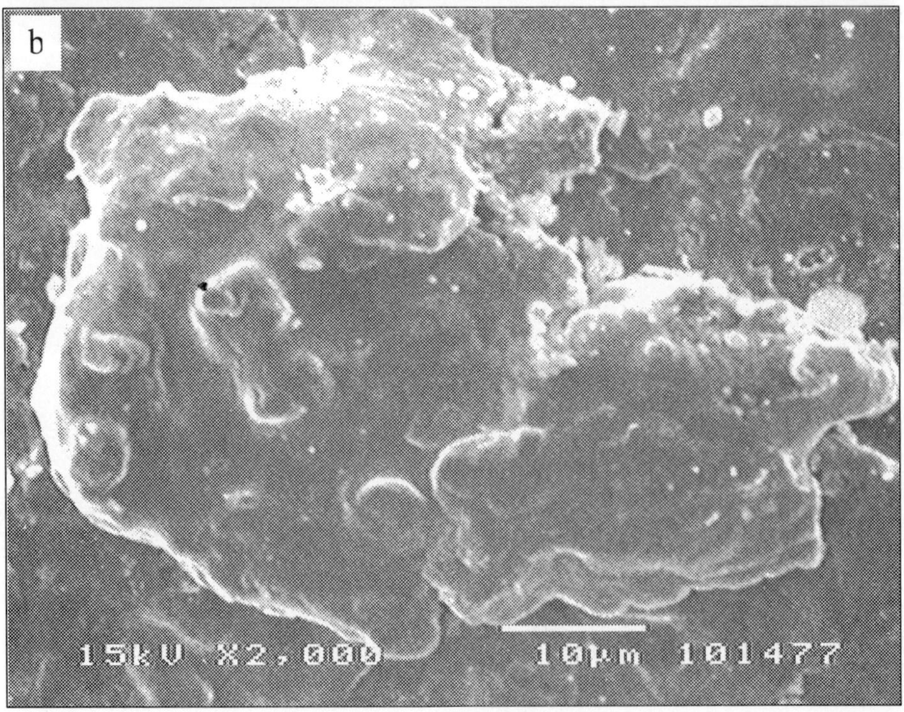

Contd...

Figure 4.24–Contd...

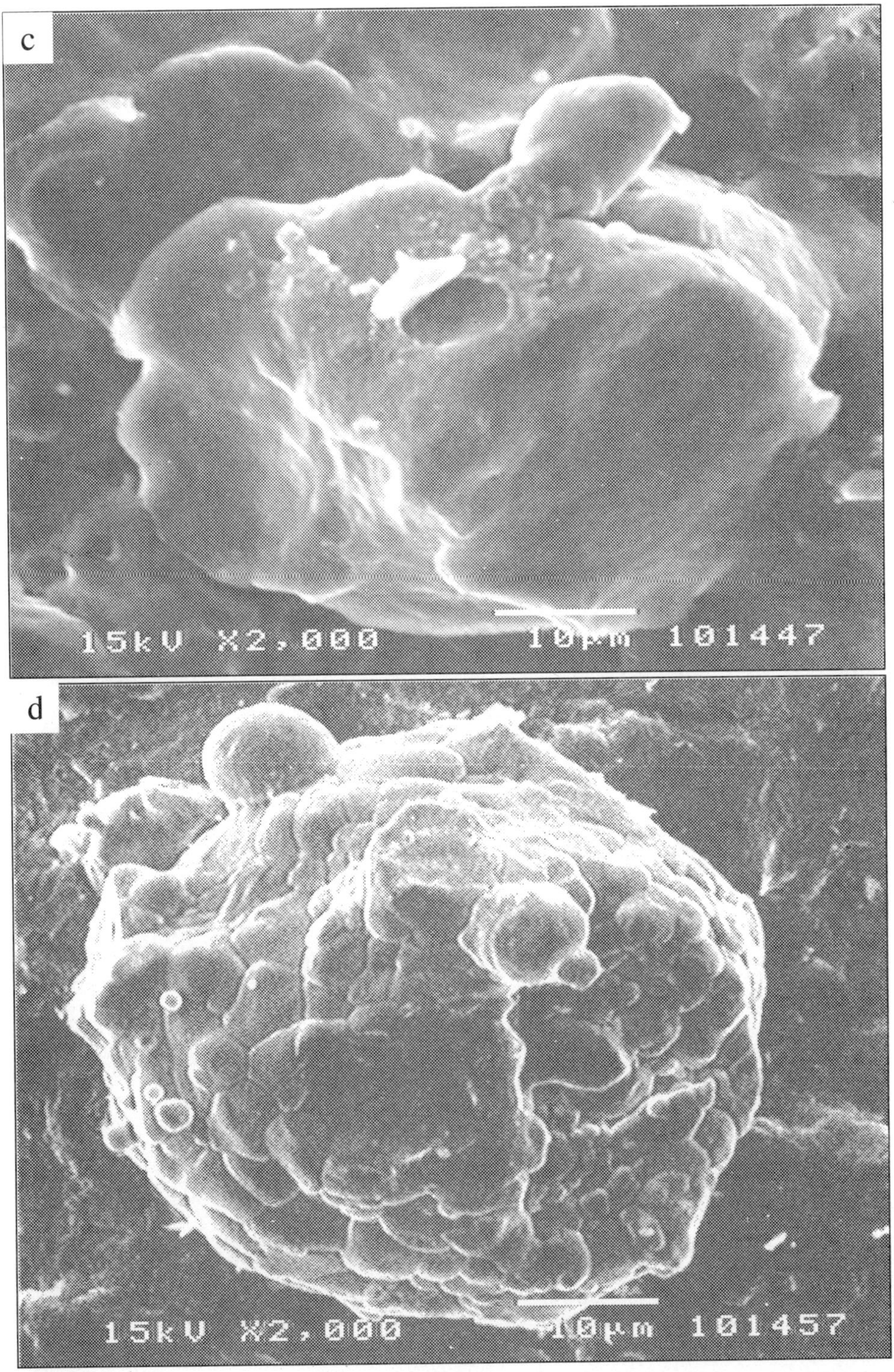

Figure 4.25: Variation of Splat Patterns with Gas Pre-heat Temperature During Cold Spraying of SS316 (a) 500ºC, (b) 550ºC, c) 600ºC

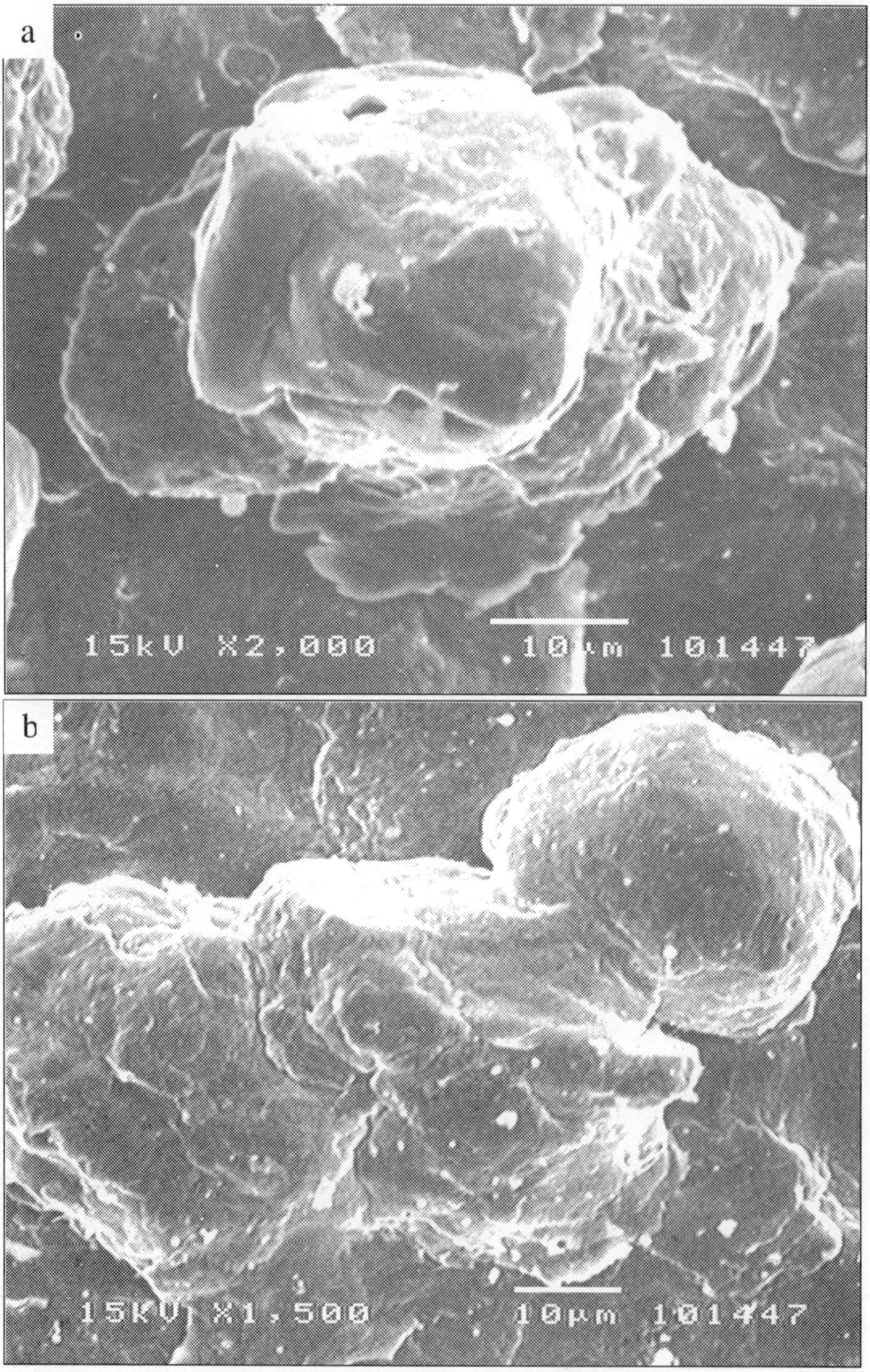

Contd...

Figure 4.25–Contd...

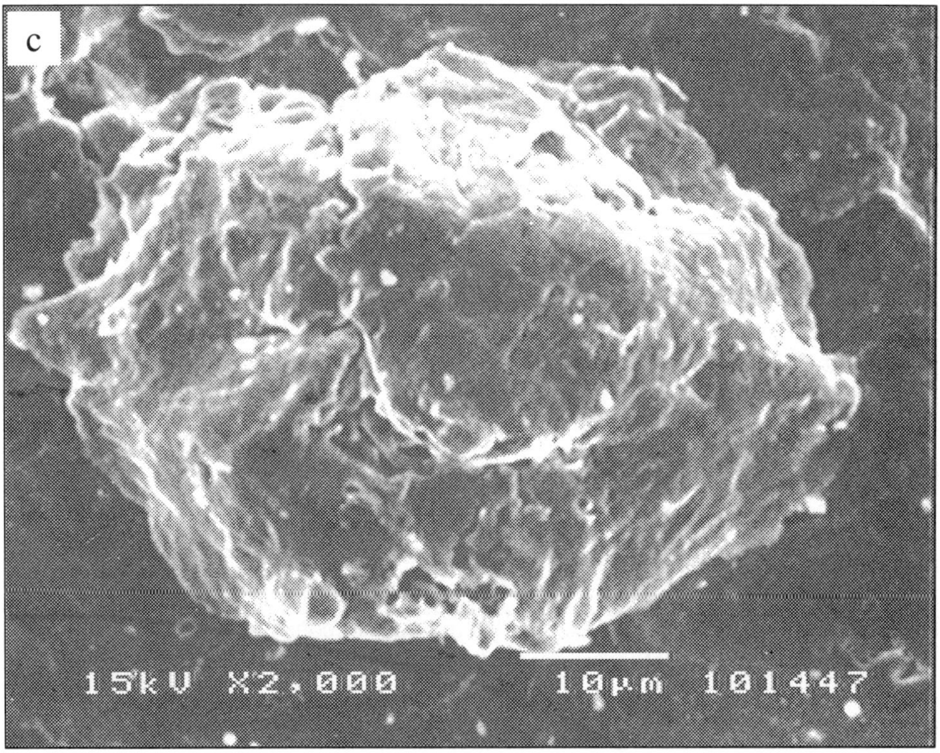

Advantages and Limitations

By eliminating the deleterious effects typically associated with the high temperatures that are characteristic of most thermal spray variants, the CGDS technique offers significant technical superiorities and new possibilities. The prominent advantages are listed below:

Coating Characteristics

- Retains properties and chemistry of initial particles
- Low oxide with no oxidation
- Low residual stress level and mostly compressive
- High thermal/electrical conductivity
- High density deposits
- High hardness, cold-worked microstructure
- Avoids any phase transformation

Deposition Capabilities

- Thermally sensitive materials like Titanium, etc
- Powders with a particle size < 5-10 µm (such as nanocrystalline)

- Highly dissimilar materials without undesirable interactions
- Deposits some cermets in solid state
- Spraying of deposits on glass/ceramics as well

Operational Efficiency

- Minimum surface preparation/masking, short standoff distance
- High productivity due to high powder feed rate (up to 15 kg/hr)
- High deposition rates and efficiencies for most materials
- Potential to Recycle Process Gas and Feedstock
- No significant heating of the substrate
- Increased operational safety because of the absence of high- temperature gas jets, radiation, and explosive gases

The fact that CGDS is a "cold" process and the coating is formed by a high velocity jet of solid phase particles also lends certain limitations to the technique which are listed below:

1. The process deposits only ductile materials; otherwise, the hard phase needs to be mixed with an additional softer ductile phase to achieve coatings
2. The hardness effect of the substrate should be considered as it also involves in the particle deformation. Hence the substrate should be hard enough to deform particles.
3. The critical velocity requirements of certain feedstock cannot be achieved with air or nitrogen, forcing the use of expensive helium.
4. The characteristic cold working of particles does not allow the much-needed ductility to the coatings. Consequently, post-treatment of coatings is required in many cases.
5. The use of inert atmospheres for highly reactive materials like titanium adds marginally to the operational cost.
6. The current status of the technology is still immature with limited coating property data and field experience available.

Applications

The versatility of the CGDS makes it promising for a very broad range of potential applications in the energy, automobile, aerospace, shipbuilding, farming machinery, electronic and instrument industries. CGDS offers attractive economic and energy related advantages over competing thermal spray methods. For instance, the typical deposition efficiency achievable with CGDS systems far exceeds that obtained by conventional thermal spray systems, especially while coating metals and alloys. Moreover, this improved deposition efficiency can be attained at much lower temperature resulting in lower energy consumption. As the powder is relatively unscathed during the process, great savings can be potentially achieved by recycling the unused powder in subsequent coating cycles. It is pertinent to mention that most

CGDS coatings are generated with typical gas pre-heat temperature range of 150-500ºC. Consequently, the power requirement in CGDS is about 5-8 kW for a powder feeding rate of 3-5 kg/hr whereas, for same spray rate, the typical power requirements in a plasma spray system are about 30 kW.

Apart from the economics, the major attraction of the CGDS route is its ability to eliminate or suppress "defects" attributable to high temperature in conventional thermal spray processes. For example, Figure 4.26 shows the typical microstructure

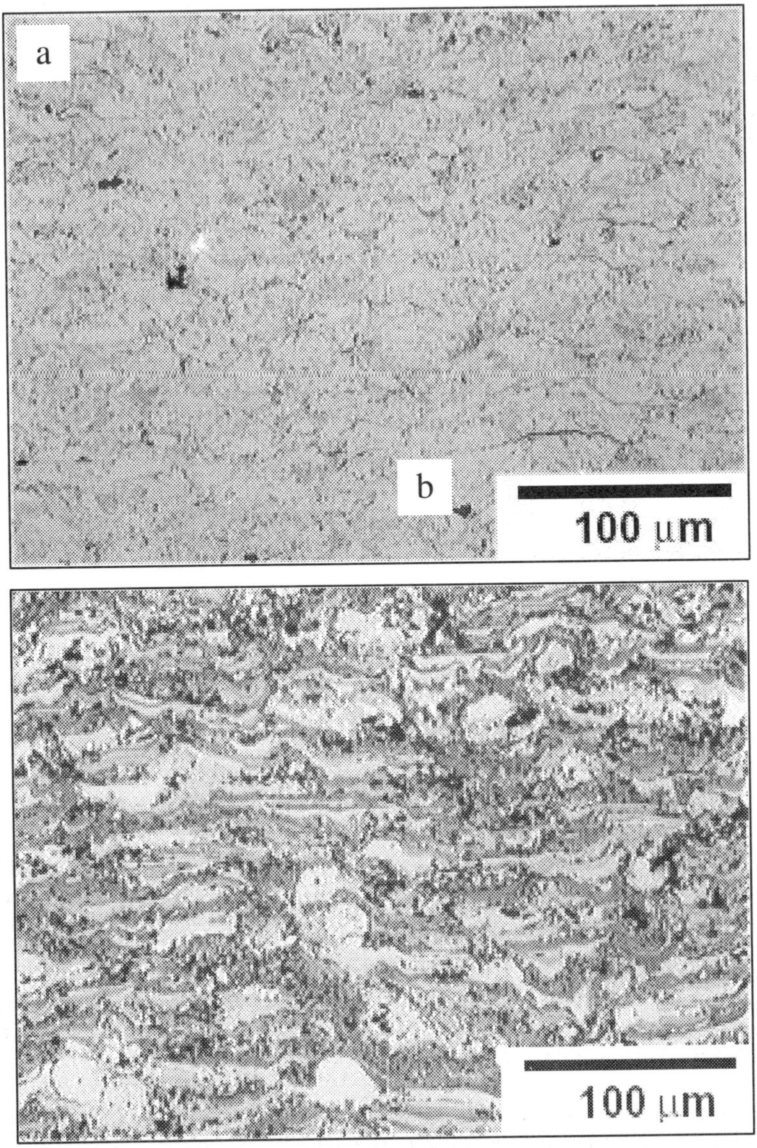

Figure 4.26: Copper Coating Produced by (a) Cold spray and (b) Plasma spray

of copper coating sprayed by CGDS and plasma spraying. The oxygen content in the CGDS coatings is 0.4 per cent with porosity levels less than 0.5 per cent, while the corresponding values in case of plasma spraying are 1.3 per cent and 5 per cent respectively. Accordingly, the electrical conductivity is as high as 85 per cent of conductivity of OFHC in CGDS copper coating but only 15 per cent of conductivity of OFHC for plasma sprayed copper coating.

Owing to the above benefits that the CGDS method offers over other thermal spray variants, it is attracting increased industrial attention. Cold Spray technology can be used to produce and repair a wide range of industrial parts–for example, engine parts, turbine blades, rocket nozzles, pump parts, and others. The unique qualities of cold spray technology can add immediate value to existing manufacturing processes. Some of the promising areas in which cold spray application are being explored are listed below:

- *Wear and Corrosion resistance*: Owing to the dense microstructures that can be produced and high phase retention ability of the process, wear resistance applications of cold sprayed coatings for many engineering components are conceivable. The non-porous and pure coatings that can be deposited can also provide good corrosion resistance.
- *Biomedical applications*: The porosity levels can be easily controlled with cold spraying and by imparting high porosity levels in implant biomaterials like titanium coatings, the potential of CGDS can be explored in biomedical applications.
- *Electrical coatings*: Pure coatings with reduced oxide content can be effectively used for conductive applications in electrical conductors (printed circuit boards), electrical contacts, etc. The application regime can also be extended for providing shielding for electromagnetic radiation and metallization of plastics (protection from static electricity, electrostatic painting).
- *Dielectric coatings*: Ability to spray polymer coatings without any volatile organics emissions can enable cold spraying of non-conductive dielectric coatings on industrial components.
- *Fabrication of prototype components*: The thickness that can be built up by cold spraying extends upto 10mm and such capability can be utilized for direct fabrication of some components and dimensional restoration of worn and damaged metallic, alloy, cermet parts. Additionally, spraying of narrow areas without masking is possible, which lowers the cost of masking.
- *Strategic applications*: Spraying of specialty alloys or metals that are used in strategic components without decomposition or phase change is an attractive application. Since the gun is relatively small in dimension, the ability to spray internal structures can be utilized for cylinder bore coatings in internal combustion engines and internal pipes for corrosion protection.

Although, there are no currently established applications of commercial interest, the technology is considered to have substantial promise for its adoption in a wide spectrum of industry segments. Some of the numerous envisaged applications include the following:

- Copper on Aluminum lugs for enhanced electrical conductivity,
- Zinc coatings for electrical and corrosion protection,
- NiCr coatings for corrosion protection,
- MCrAlY coatings for hot-gas corrosion protection, especially in TBC's,
- Aluminum Bronze for bearings,
- Nickel coatings for refurbishment applications,
- Stainless steel on mild steel for corrosion resistance.

Thus it is clear that the technique can be utilized for meeting a wide range of functionalities. The process is also considered to offer a good alternative to technologies such as electroplating, soldering and painting.

The fabrication of near net shapes has been an area of worldwide research and being pursued by various thermal spray methods. The fabricated shapes are increasingly finding applications in strategic and novel applications like MEMS. The ability to produce thick coatings in the order of 10 mm makes cold spraying ideal for near net shape fabrication as shown in Figure 4.27.

Helium and Unused Feedstock Recovery

The important parameters governing the economics of the CGDS process are the powder cost, cost of operating gas and deposition efficiency. The cost model recently proposed for the CGDS technology is shown in Figure 4.28 and reveals that the gas cost is a dominating factor especially in case of use of compressed nitrogen and

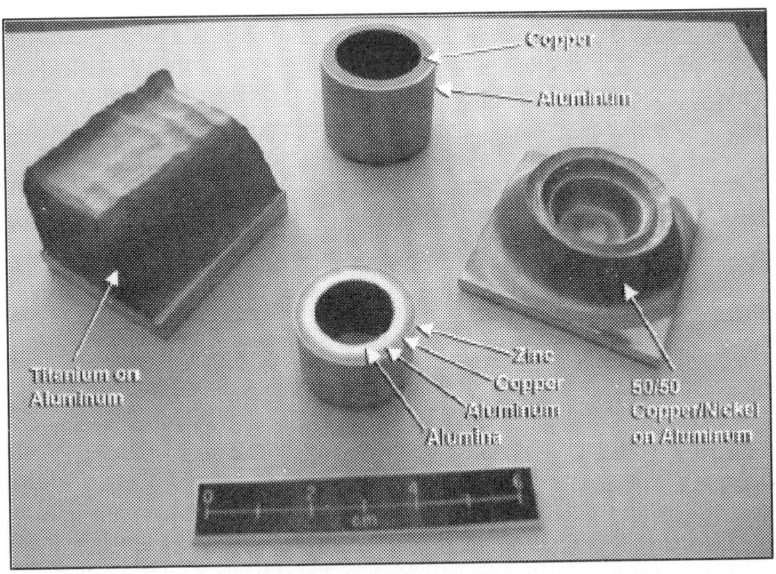

Figure 4.27: Ability of Cold Spray to Generate Near Net Shapes

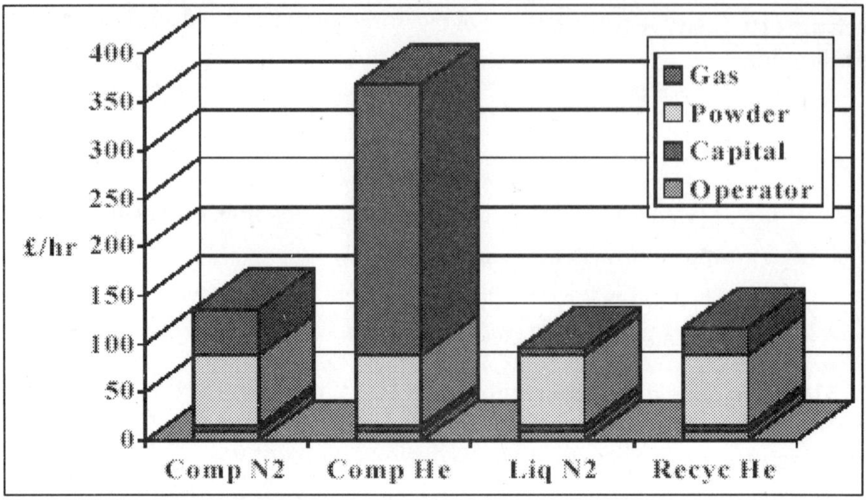

Figure 4.28: Cost Model for Operating Process Variables with CGDS

helium. When quality considerations demand the use of Helium, the component of gas cost can be nearly 6 times that of the powder cost. This provides ample motivation for assessing the practicality of recovering the Helium gas as well as the unused feedstock.

The cold spray system uses large quantity of helium gas to produce high quality coatings. The major factors favoring helium are its inertness and high thermal conductivity (He: 0.086, N_2: 0.0149, Ar: 0.01 BTU/hr Ft ºF). To lower operating costs and conserve this limited resource, adoption of helium recycle technology being actively considered to make the process more prudent. A schematic of such a recovery system conceived by Praxair is shown in Figure 4.29.

Apart from helium, the spray grade fine powders generally used for cold spraying are also expensive. Since these are almost unscathed during the low temperature CGDS process, they can be reused after proper collection. Efforts are already underway to engineer schemes that will enable effective feedstock recovery.

National Status of the Technology

Since the emergence of the CGDS technology in the world scene, many research laboratories and manufacturing firms have been investing efforts to commercialize CGDS system. To enable faster transition of the technology from the laboratory to the industry, a consortium of leading thermal spray research and manufacturing companies comprising of Sandia National Laboratories, Daimler Chrysler, Ford Motor Company, Ktech Corporation, Pratt and Whitney, Praxair and Siemens Westinghouse has been formed and taken lead in addressing several immediate research and manufacturing requirements. However, in the South Asian region, the International Advanced Research Centre for Powder Metallurgy and New Materials (ARCI) has been the first institute to set up a cold spray facility. Consistent with ARCI's continuing efforts to bring emergent surface engineering solutions to the door step of Indian

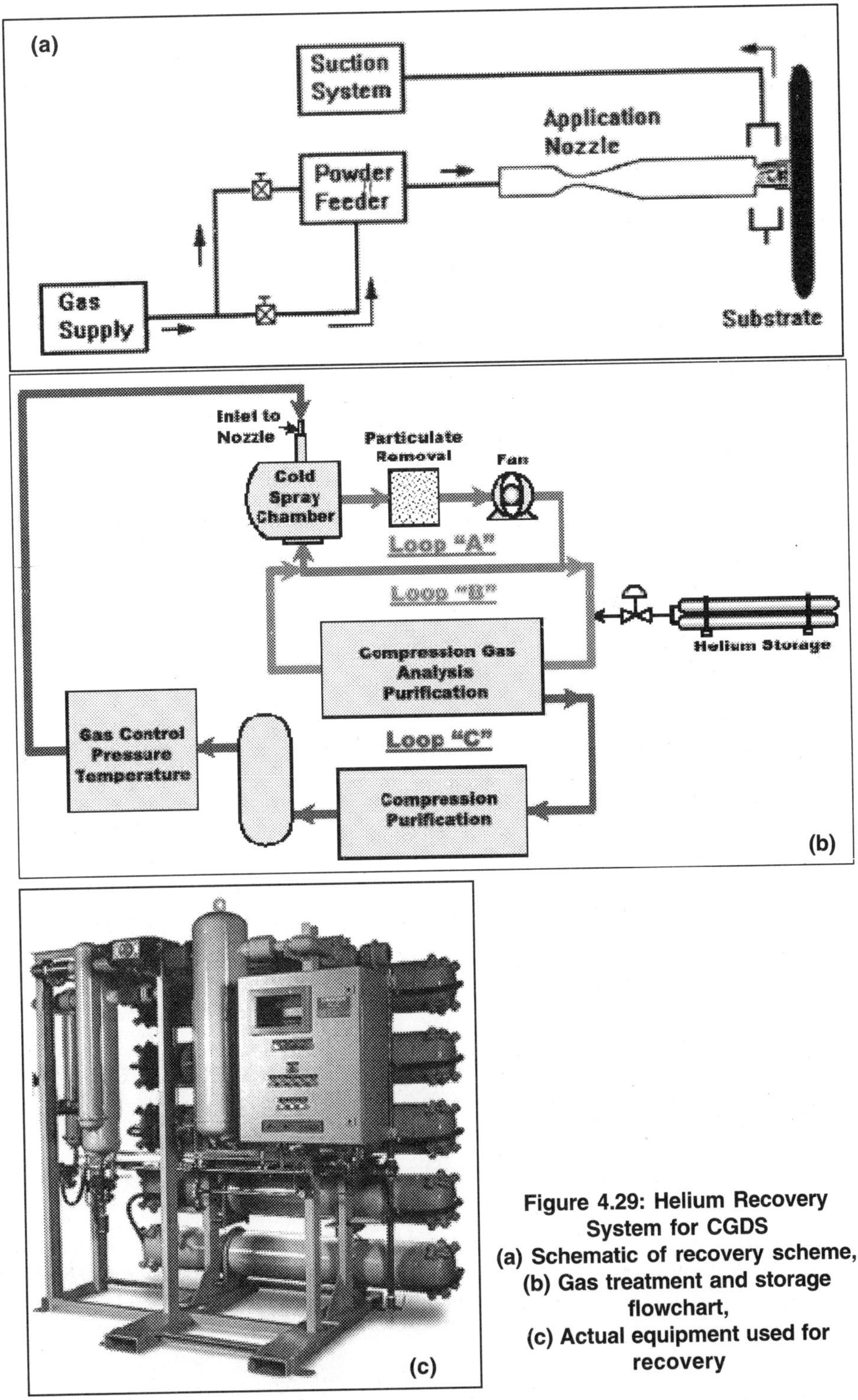

Figure 4.29: Helium Recovery System for CGDS (a) Schematic of recovery scheme, (b) Gas treatment and storage flowchart, (c) Actual equipment used for recovery

industry, a CGDS system has been recently established at the centre in collaboration with the Institute of Theoretical and Applied Mechanics (ITAM) of the Russian Academy of Sciences. As a result, ARCI has now joined select band of institutions worldwide that are equipped to pursue research pertaining to this new and exciting field that has numerous promising applications. As in the case of DSC technology already commercialized by ARCI, the Centre proposes to eventually build completely indigenous Cold Spray systems for technology transfer to Indian industries so that a frontier coating technology becomes available to Indian entrepreneurs at financially attractive terms.

Scope for R&D Efforts

Notwithstanding its immense application potential, the commercialization of the CGDS technique is hampered by various manufacturing and technical barriers. Some of these can constitute ideal subjects for inter-disciplinary research in this exciting field. The prominent technical issues are related to the following:

1. Process fundamentals are poorly understood
2. Capability to satisfactorily predict particle behavior leading to coating deposition is still lacking
3. Reduction/elimination of the use of expensive helium is essential
4. *In-situ* or post-deposition heat treatments to enhance performance must be assessed
5. Use of novel methods like liquid feedstock can be evaluated to extend the applicability of the CGDS technique and
6. Nozzle design optimization needs further research attention

Besides the above, problematic manufacturing issues pertain to the limited sources of commercial coating equipment and coating suppliers. Since the CGDS technique typically demands relatively fine powders, improved feed technology needs to be developed. There is also a need for development of an articulated robot-compatible spray gun. The absence of coating property data/field experience is also presently a problem but this is certain to be eliminated as these coatings are increasingly utilized for industrial applications.

References

1. A.P.Alkhimov, A.N.Papyrin, V.F.Kosarev, N.I. Nesterovich, and M.M.Sushpanov, "Gas Dynamic Spraying Method for Applying a Coating", U.S.Patent 5,302,414, 12 April 1994.
2. R.C.Dykhuizen and M.F.Smith, Gas Dynamic Principles of Cold Spray, *J. Therm. Spray Technol.*, Vol 7(2) (1998), p 205-212.
3. Mark F. Smith, Richard A. Neiser, and Ronald C. Dykhuizen, Introduction to Cold Spray and New Opportunities It Offers, in *Cold Spray: New Horizons in Surfacing Technology*, 9-10 September 2002, Albuquerque, New Mexico, USA.

4. T.H. Van Steenkiste, J.R. Smith, R.E. Teets, J.J. Moleski *et al.*, Kinetic spray coatings, *Surf. Coat. Technol.*, Vol. 111 (1999), p. 62-71.

5. Technical presentations in *Cold Spray Workshop*, 14-15 July 1999, Albuquerque, New Mexico, USA.

6. J.Karthikeyan, C.M.Kay, J.Lindeman, R.S.Lima, and C.C.Berndt, in Thermal Spray 2000, (Ed.) C.C.Berndt, ASM International, Ohio, USA, 2000.

7. T.H. Van Steenkiste, J.R. Smith, R.E. Teets, Aluminum coatings via kinetic spray with relatively large powder particles, *Surf. Coat. Technol*, Vol 154 (2002) p.237–252

8. D.L.Gilmore, R.C.Dykhuizen, R.A.Nieser, T.J.Roemer, and M.F.Smith, Particle Velcoity and Deposition Efficiency in the Cold Spray Process, *J. Therm. Spray Technol.*, Vol 8(4) (1999), p. 576-582.

9. M.K.Dekker, R.A.Nieser, D.Gilmore and H.D.Tran, in Thermal Spray 2001:New Surfaces for a New Millennium, (Ed.) C.C.Berndt, K.A.Khor and E.F.Lugscheider, ASM International, Ohio, USA, 2001.

10. A.Papyrin, Cold Spray Technology, *Adv. Mater. Process.* September (2001) p. 49-51.

11. R.C.McCune, W.T.Donlon, O.O.Popoola and E.L.Cartwright, Characterization of Copper Layers Produced by Cold Gas-Dynamic Spraying, *J. Therm. Spray Technol.*, Vol 9 (1) (2000), p. 73-82.

12. W.M. Robson, R.E. Blose, T.J. Roemer, R.T. Nichols, A.J. Mayer, D.A. Beatty, and A.N. Papyrin, Cold Spray Systems for Industrial Surface Engineering Applications, in *Cold Spray: New Horizons in Surfacing Technology*, 9-10 September 2002, Albuquerque, New Mexico, USA.

Chapter 5

Sol Gel Coating: Basic Science, Processing and Applications

U.S. Hareesh
Centre for Ceramic Processing,
International Advanced Research Centre for Powder Metallurgy and New Materials
Hyderabad – 500 005, India

ABSTRACT

Sol gel coatings, the wet chemical approach to surface engineering, have been gathering momentum over the past few years through a variety of applications in fields ranging from medical science to textile chemistry. Coatings with diverse functionalities imparted through incorporating nano powders are already been brought to market. The major advantage that enabled this is the ease with which the coatings are applied as compared to traditional coating techniques. A close control of chemistry of precursors and tailoring of rheological properties and shelf life of sol are the key parameters that promote the use of sol gel coatings for industrial applications. A review of the process has been attempted here with an introduction to basic science of process, the techniques with which coatings are applied and the various applications in which the coatings are currently being proved useful.

Introduction

Sol Gel Process

The Fundamental Concepts

SOLS

A sol is a stable dispersion of colloidal solid particles in a liquid phase where the solid particles denser than the surrounding liquid are small enough for the forces of dispersion to be greater than that of gravity. Colloidal particles are in a finely divided state and hence have a high surface s to volume v ratio according to the equation

$$\frac{s}{v} = \frac{3}{r}$$

where,

r is the radius of the particle. As r tends to zero the ratio goes to infinity and hence colloidal particles have high total surface energy. Therefore sols are thermodynamically unstable. Hence, with no constraints applied, the system can spontaneously tends towards a state with lower surface to volume ratio with particles bonded together. This process is called flocculation if it is reversible and coagulation when irreversible.

Sols can be made kinetically stable by preventing aggregation of the colloidal particles by three mechanisms;

1. Imposition of a surface charge by either the preferential dissociation of one of the lattice ions of the sol particles or adsorption of a charged species on to sol particles.

   Peptisation of boehmite sol by dilute nitric acid at pH 2.5 to 3.5 when H+ ions are adsorbed, is an example
2. Steric interaction between organic macromolecules adsorbed on the surface of sol particles and solvent molecules.

   Use of Darvan C (Ammonium polyacrylate) for stabilisation of metal oxide sols is an example
3. Solvation stability effective in aqueous based systems where the energy of hydration required to dehydrate the layers of water molecule surrounding the sol particles contribute to the imposition of an energy barrier.

Gels

A gel can be defined as a three-dimensional solid network structure completely impregnated with a liquid. There is high degree of homogeneity in the dispersion of liquid within the solid. The important criteria for sol to gel transformation is the attainment of a thermodynamic equilibrium between the solid and liquid phases arising from strong particle–solvent interaction

Gels are designated basically on the type of dispersed liquid phase. The gel is called hydrogel (or aquagel) when it is mostly water. When the liquid phase is mainly alcohol, the gel is alcogel. The inorganic sol or gel obtained from a chemical reactant, containing the cation of the ceramic is called the precursor. Depending on the precursor type two types of sol gel processes can be distinguished; an aqueous based process that start from the solution of the metal salt and an alcohol based process that start from a metal alkoxide.

Aqueous Based Process

The general formula of a metallic salt is M_mX_n where M is the metal, X an anionic group and m and n are stoichiometric coefficients. In the aqueous based process the metal salt dissociate into ions in solution resulting in the formation of a solvated cation represented as $\{M(H_2O)_N\}^{z+}$ where z^+ is the charge of the cation and N is the number of water molecules in which the cation is entrapped. This is then followed by a hydrolysis reaction in which deprotonation of the solvated cation takes place transforming the aquo ligand molecule H_2O into an hydroxo ligand OH^- or into an oxo ligand O^{2-}depending on the number of detached protons. The gel formation is then accomplished by condensation reactions in which complexes of the metal atom M react with one another in a polymerisation reaction. This can proceed either by the construction of an 'ol' bridge where one or more hydroxo ligands are caught between two metal atoms as in 'Olation' or an 'oxo' bridge where an oxo ligand is caught between two metal atoms as in oxolation[6].

When the sol preparation is done by condensation, the particles are prepared by slow, controlled nucleation and growth of crystals at an elevated temperature. It can also be prepared by a dispersion method in which the metal salt is hydrolysed rapidly to a gelatinous precipitate by the addition of a base followed by washing to remove excess electrolytes and peptisation by an acid around pH 3 [20]

$$MX \longrightarrow M^{z+} + X^{2-} \text{ (in solution)}$$

$$M^{z+} + N\,(H_2O) \longrightarrow [M\,(OH_2)_N]^{Z+}$$

Solvated Cation

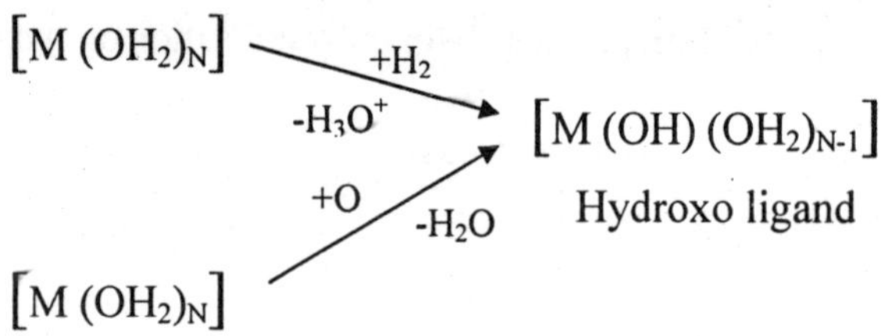

$$[M\,(OH)\,(OH_2)_{N-1}]^{(Z-1)^+} \longrightarrow [M\,(OH_2)_{N-1}]^{(Z-2)^+} + H_3O$$

Oxo ligand

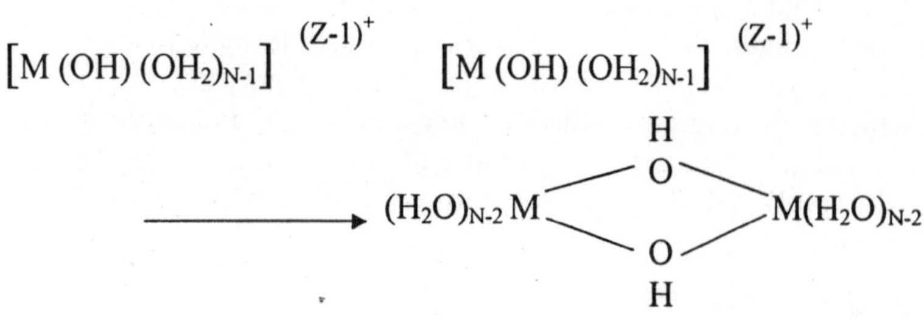

Olation

Alcohol Based Process

The precursors in this process are alkoxides which are compounds with a chemical formula $(MOR)_z$ formed as the result of the direct/indirect reactions between a metal M and alcohol R-OH. In this process there is no distinct sol formation step. A simultaneous hydrolysis and condensation reaction proceeds ultimately leading to the formation of a gel. The reaction can be represented as:

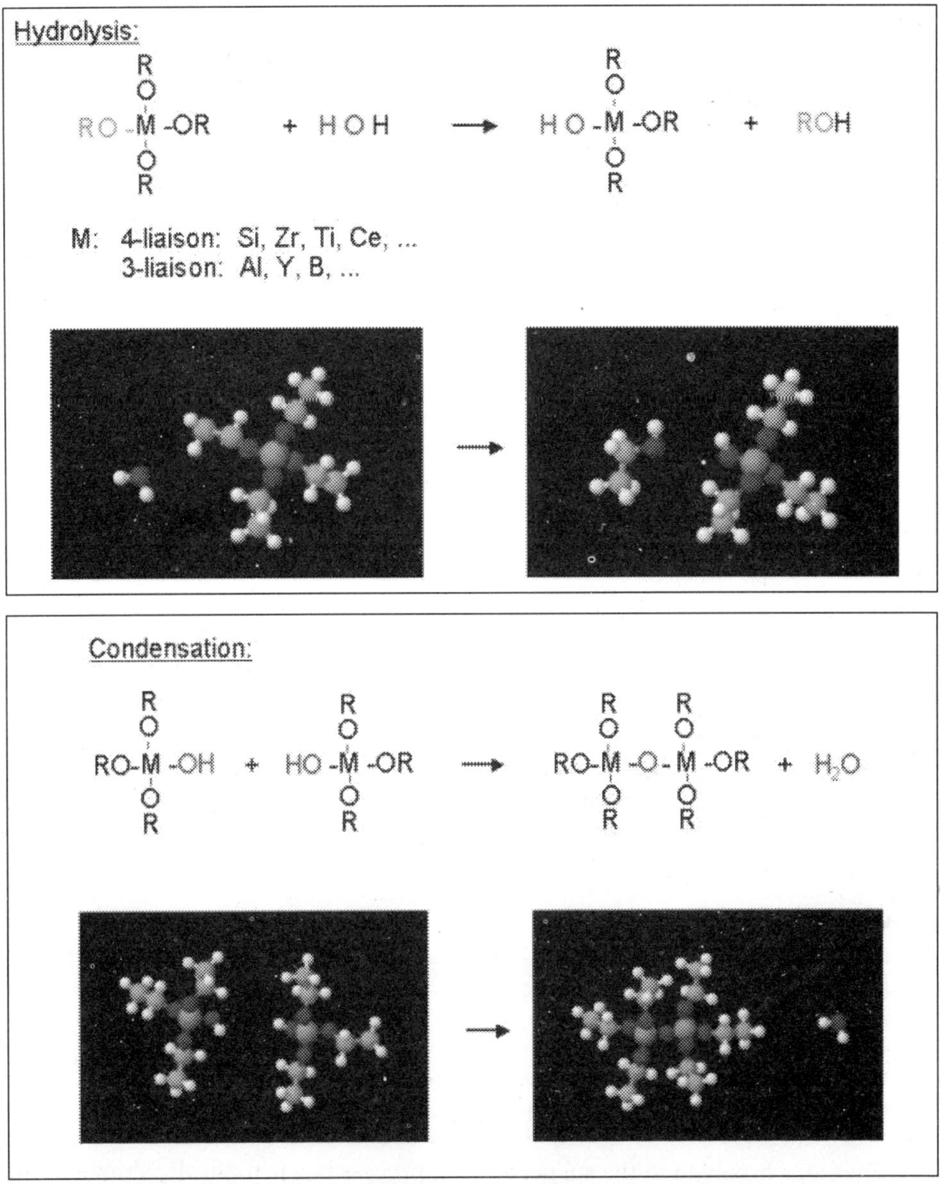

Figure 5.1: Model Showing Alkoxide Based Sol Gel Process

Outline of the Process

Figure 5.2 represents the general outline of the sol gel process and compare the gel formation from solution versus sols.

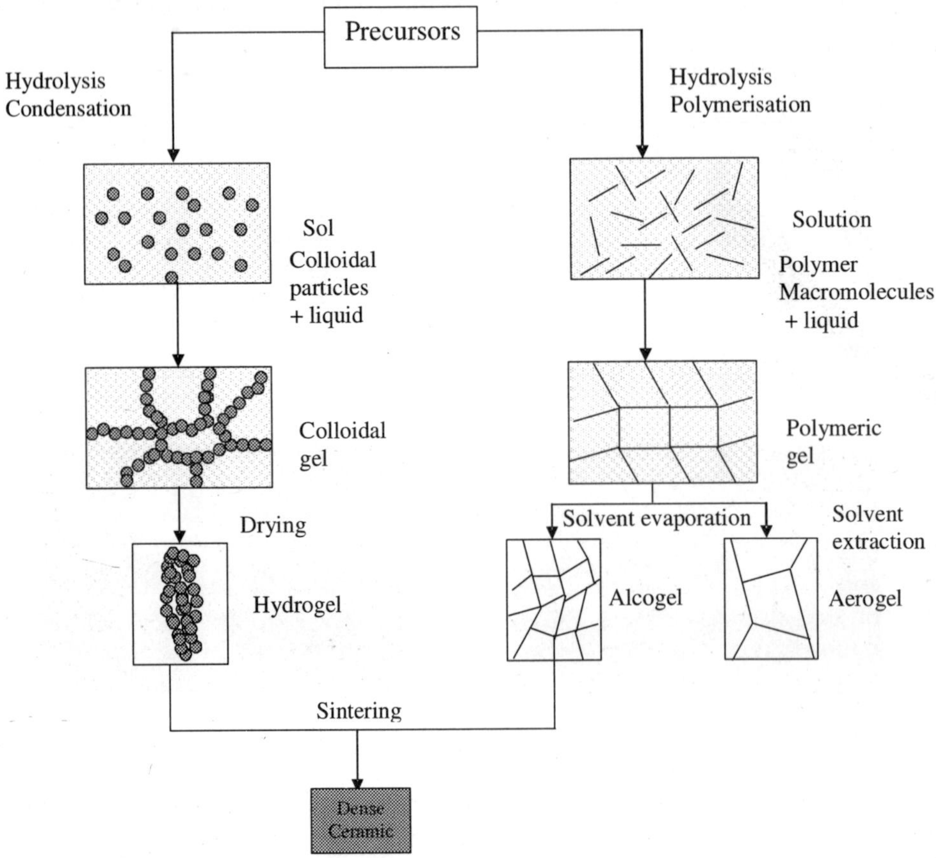

Figure 5.2: Schematic Outline of Sol-gel Process and Comparison of Colloidal and Polymeric Routes

Sol Gel Coatings

Coating Techniques

The selection of coating technique in the case of sol gel derived thin and thick films depends on the following factors:

- Thickness of coating
- Coating area
- Complexity and type of substrate

The steps involved in the sol gel coating process is schematically shown in the flow chart below:

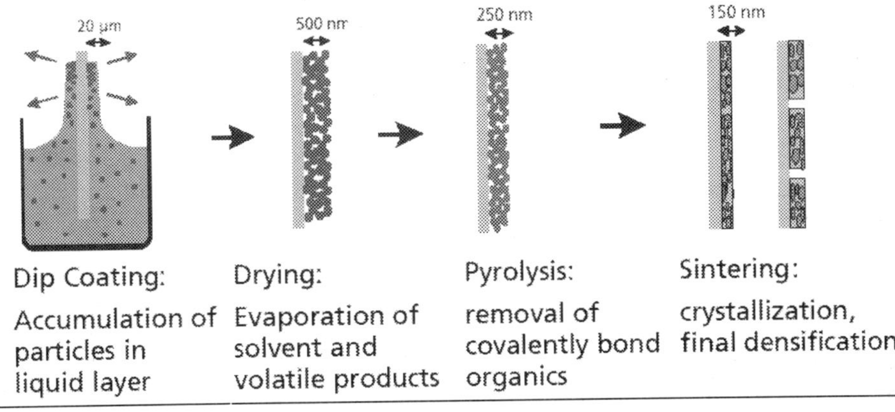

Figure 5.3: Process Flowchart for Deposition of Sol Gel Films

The most commonly used techniques are discussed as follows:

Dip Coating

Substrates with curved or flat surfaces without many complexities can be coated using a dip coating process where in the substrate to be coated is immersed in a sol and withdrawn with a well-defined withdrawal speed under controlled conditions of temperature and atmosphere. The thickness of the coating is defined by the parameters like withdrawal speed of substrate, solid content and viscosity of sol. If the sheer rate generated during withdrawal can be kept in the Newtonian regime, the coating thickness is given by the Landau-Levich equation

$$h = \frac{0.94.(\eta.v)^{2/3}}{\gamma_{LV}^{\ 1/6}(\rho.g)^{1/2}}$$

where,

h is the coating thickness, η is the viscosity; γ_{LV}–surface tension (liquid-vapour), ρ is the density and g-acceleration due to gravity.

Hence, it can be seen that by choosing an appropriate viscosity, which in turn depends on the solid content, the coating thickness can be precisely varied by a control over the withdrawal speed. Thickness in the range 20 nm to 50 µm can thus be achieved.

As shown in Figure 5.5 the dip coating process involves the dipping of substrate in to the sol. As the substrate is withdrawn (speed of which is determined by the desired thickness) a wet layer of sol particles with solvent is formed on the surface. This is followed by gelation of the layer through solvent evaporation. Sol particles, stabilized by surface charges, approach as a result of evaporation of solvent to distances less than the repulsion potential leading to gelation. The resulting gel layer requires

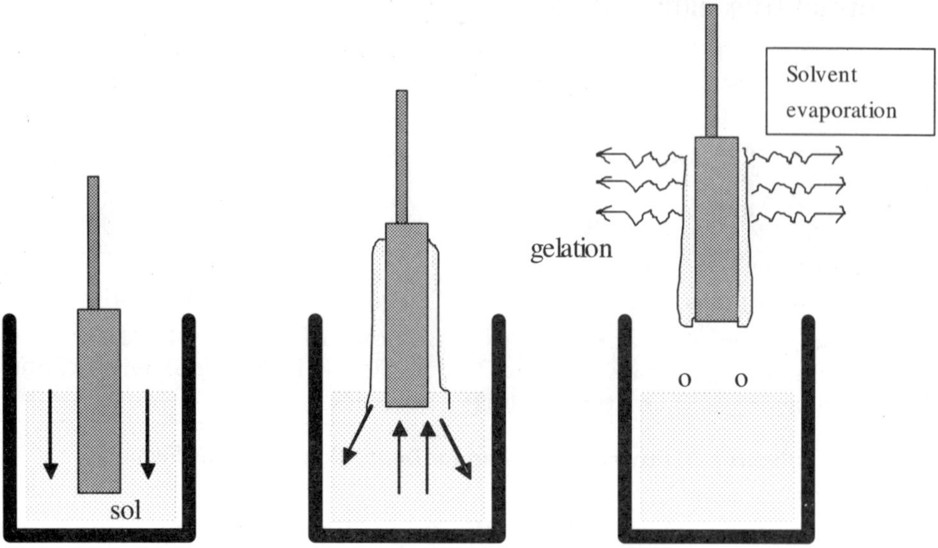

Figure 5.4: Schematic Showing the Dip Coating Process

Figure 5.5: Photograph of a Dip Coating Set Up

thermal treatment, temperature of which is determined by the composition of layer. However, since the gel particles are extremely fine (in the order of a few nanometers) possessing high surface energy the films can be densified at remarkably reduced temperatures compared to bulk structures.

Dip coating process is extensively used for wet coatings on glass for a variety of applications like antireflection coatings, solar energy control systems a few of which will be dealt later in applications.

Angle Dependent Dip Coating

A variation of the dip coating, in which the angle dependency is also taken into account, to have a better control on the thickness, has been developed as an alternative for dip coating. The coating thickness is dependent on the angle at which the substrate is dipped in to the sol surface.

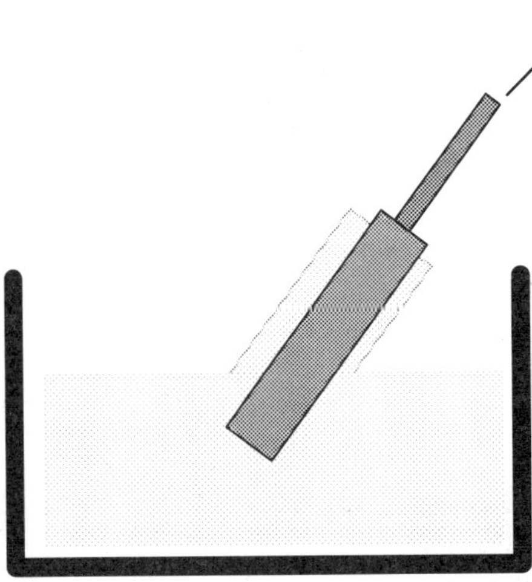

Figure 5.6: Schematic of Angle Dependent Dip Coating

It has been shown that angle dependent coating can be used to coat layers with varying thickness between the top and bottom of the substrate and therefore has been used extensively in optical coatings where the number of layers and thickness are to be precisely controlled to obtain the desired properties. In a slight modification of the procedure, bottles can be coated by a simultaneous revolution of the bottle during the withdrawal process.

Spin Coating Process

In the spin coating process, the substrate is made to spin around an axis perpendicular to the coating area.

A sol is deposited on the substrate through a feeder. The substrate is then allowed to spin at a fixed rpm, which helps in the formation of a layer on the substrate. Subsequent drying and gelation will provide the coating on the substrate.

The main parameter that determine the quality of coatings is the rheological parameters of the coating sol and should necessarily be in the Newtonian regime. Reynolds number of the surrounding atmosphere also plays a role in determining

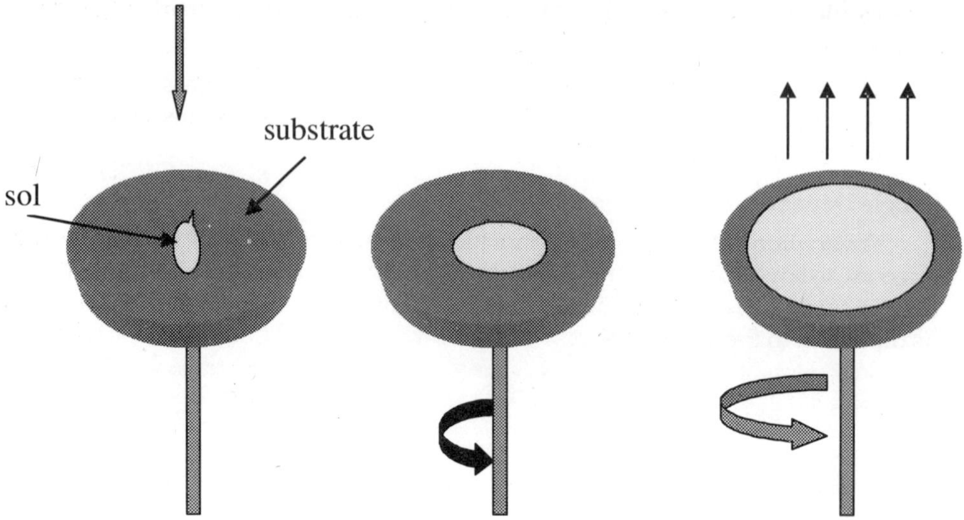

Figure 5.7: Schematic Showing the Stages of the Spin Coating Process

Figure 5.8: Photograph of Spin Coating Set Up

the quality of coatings, particularly for optical components. Atmospheric friction influences the rotation velocity and high turbulences can lead to defects in optical coatings.

The dependence of the final thickness of spin coated layer on the processing and materials parameters can be semi empirically quantified as

$$h = (1 - \rho_A / \rho_{A0}).\left(\frac{3\eta.m}{2\rho_{Ao}.\omega^2}\right)^{1/3}$$

where,

ρ_A is mass of volatile solvent/unit volume

ρ_{A0} is initial value of ρ_A

h–final thickness

η is viscosity and ω=angular speed

m-evaporation rate of the solvent

Spray Coating

This is the simplest and most practical method for depositing coatings on to large area substrates and substrates with complex shapes. Use of an automatic flat spraying equipment in combination with high, volume, low pressure nozzles has been shown to be very effective for the deposition of thin coatings (100-200 nm) with a thickness accuracy of 5-10 per cent. Compared to dip coating procedures spray coating is much faster and can be fully automated to be an inline process. Sols with shorter pot life can be used and the wastage is minimum.

Wire Coating

The wire unrolled from a bobbin is going through a vertical furnace, which after cooling down is submerged in the sol bath. The coated wire, after drying is heat-treated for phase formation and densification of the coatings and is further enrolled on a second bobbin. This apparatus allows the deposition of several layers because the bobbin's motors can be automatically permutated.

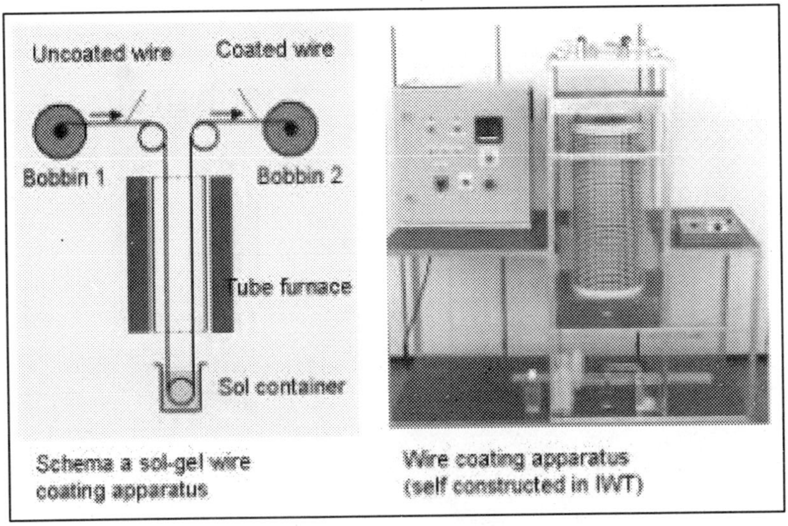

Figure 5.9: Sol Gel Wire Coating Techniques

Casting

This coating technique is widely used for thicker coatings, often in the range 5-50 microns, and can be simply applied by means of an application bar with precise thickness control. Most commonly organic–inorganic hybrids with higher viscosities are applied by this technique. Few drops of the coating slurry/sol is placed on the pretreated substrate and the application bar is used to spread and wipe down the slurry in the form a layer over the substrate.

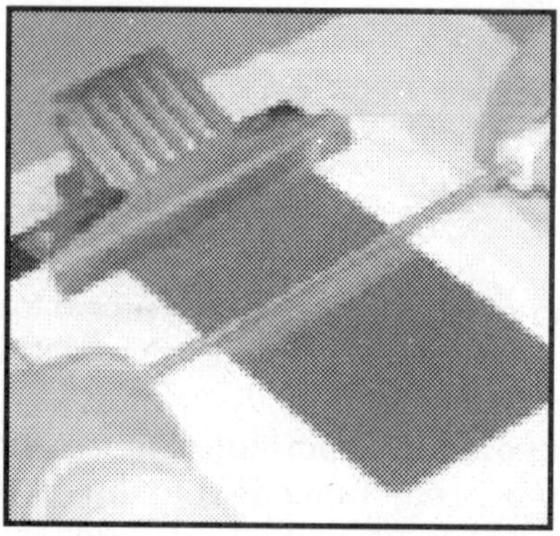

Figure 5.10: Showing Application of Sol Gel Coatings by Casting

Flow Coating Process

In the flow coating process, the sol to be deposited is poured over the substrate and the thickness of the coating is dependent on the angle of inclination of the substrate, the coating liquid viscosity and the rate of evaporation of solvent. The advantage of the coating process is that large substrates can be coated very easily. Additionally, the spinning of the substrate (rather for small substrates) helps to attain homogeneity. The process is shown schematically as below:

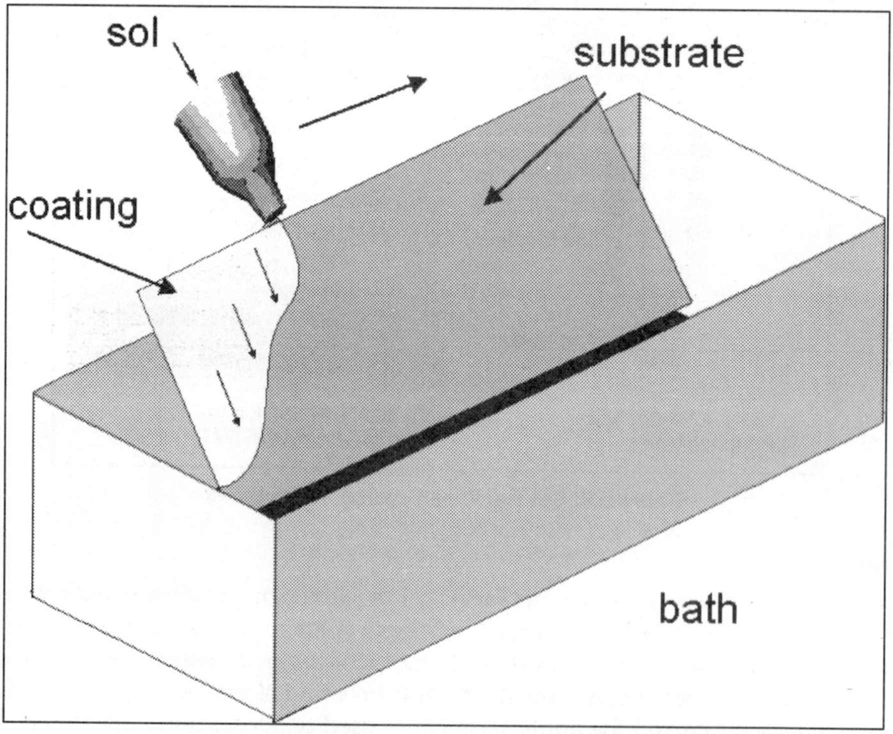

Figure 5.11: Set Up of a Sol Gel Flow Coating Process

Applications of Sol Gel Coatings

Table 5.1 summarises the spectrum of applications where sol gel coatings are currently being employed. A wide range of functionalities can be obtained by a right selection of the composition. A few of the applications are dealt in detail as following.

Table 5.1

Application	*Material*
Protective coatings	
Corrosion resistance	Hybrid, oxides
Diffusion barrier	Silica, hybrid
Thermal barrier	β-Al_2TiO_5, Porous Silica
Hard coat and abrasion resistance	Hybrid and Oxides
Ion bombardment protection	Silica-alumina
Adhesion and Bonding	Silica, titania, hybrids
Optical coatings	
AR coating	Oxide multilayer
Waveguide and optical device	Doped WO_3 and other oxides
Transperent conductive	ITO, Tin oxide, vanadium based
White coat	Alumina based
Surface modification	
Hydrophobic, water-repellent	Hybrid
Anti-fog	Hybrid
Ferroelectric devices	
Ferroelectric capacitor, Non-volatile memory, sensors and actuators, MEM's, ultrasound imaging, piezoelectric motor, pyrodetector	PZT, BT, BST, etc.
Low dielectric materials	Porous materials (pure silica or hybrid)
Sensors based on encapsulation of sensing element	Flouresence chemistry, enzymes, protein, functional reagents
Others	
Nanomembranes	Alumina. Titania, zirconia
Biomaterial	Calcium phosphate

Organic–Inorganic hybrids

Unique Combination of the interesting physical properties of an inorganic material like strength, hardness with that of an organic material possessing properties of flexibility, polymerization etc has lead to the generation of an interesting class of materials called organic–inorganic hybrids. The following schematic shows the method of preparation of the same.

A suitable selection of surfactant (mostly with multiple functional groups) to make it compatible with the inorganic as well as organic materials is a prerequisite

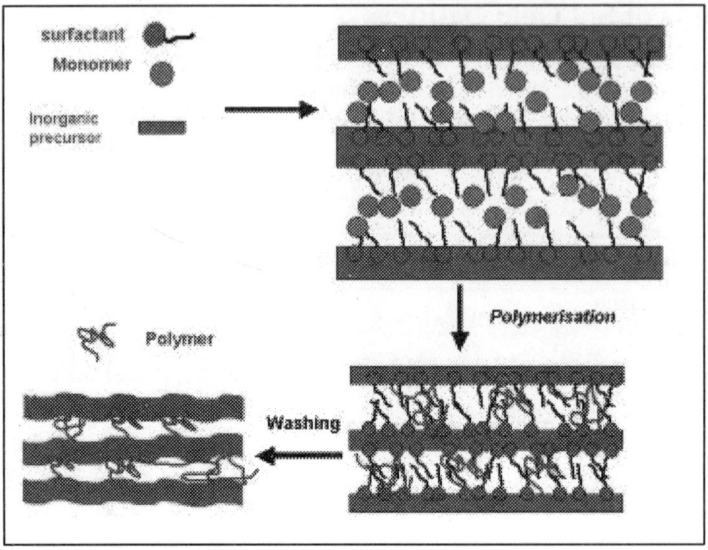

Figure 5.12: Synthesis of an Organic-Inorganic Hybrid Material

for the preparation of such materials. Molecular level mixing of the two components followed by polymerization will lead to the hybrid material possessing properties of both. The mixed solution leading to the formation of a hybrid is a very useful material for coating applications. Depending on the application suitable compositions can be formulated by incorporating powders of the required functionality.

A specific example for the preparation of easy to clean coatings is mentioned below. This is based on the introduction of hydrophobic components with low free surface energy (perfluorinated organic compounds) into the basic binder composition to result in an organic–inorganic nanocomposite which can then be applied by any of the normal wet chemical deposition techniques like spraying, dipping, casting etc. The coatings are subjected to polymerization by UV irradiation or by thermal treatment

A friction reducing coating on an elastomeric support is provided by the application of a sol having a polymerisable silane containg a glycidyloxy functional residue or a mercapto functional residue. An epoxy resin hardener is used to improve further the cross limking of the siloxane skeleton. The mechanical stability of the coatings are improved further by incorporating nanoparticulates suitably surface modified to make it system compatible.

Hydrophilic Coatings

Coatings with good wetting behaviour to water are useful for outside surfaces as well as for inside surfaces (antifogging properties). In addition to this specific functionality the coatings would also require sufficient mechanical properties like scratch and abrasion resistance, stability against wet climate conditions and UV radiation. Organic–Inorganic hybrid coatings functionalised for the said purpose

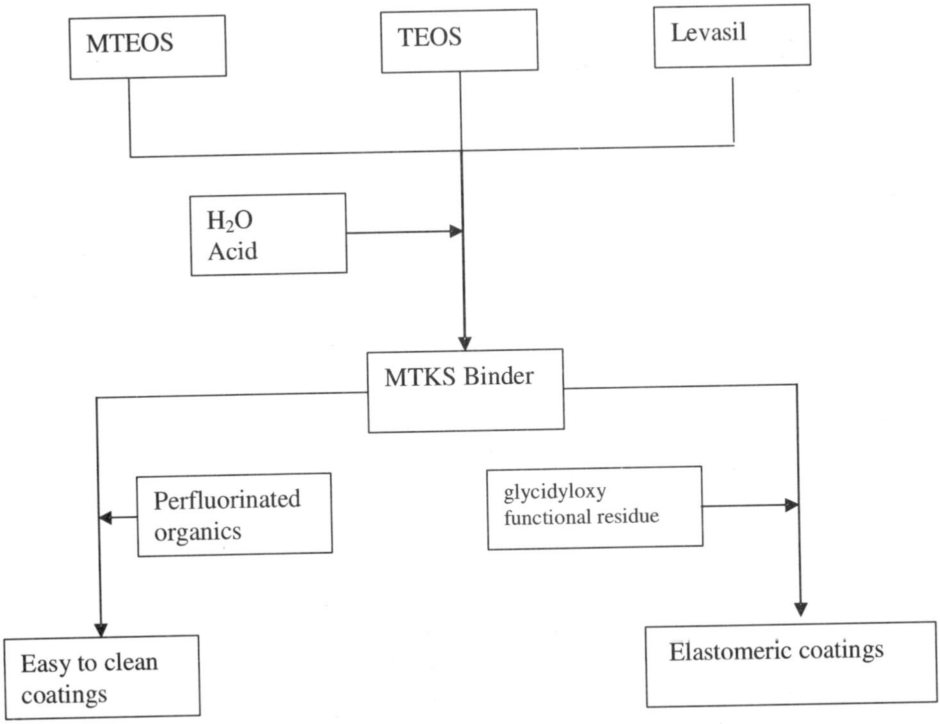

Figure 5.13: Flowchart Showing the Preparation of Organic-Inorganic Hybrid Coatings

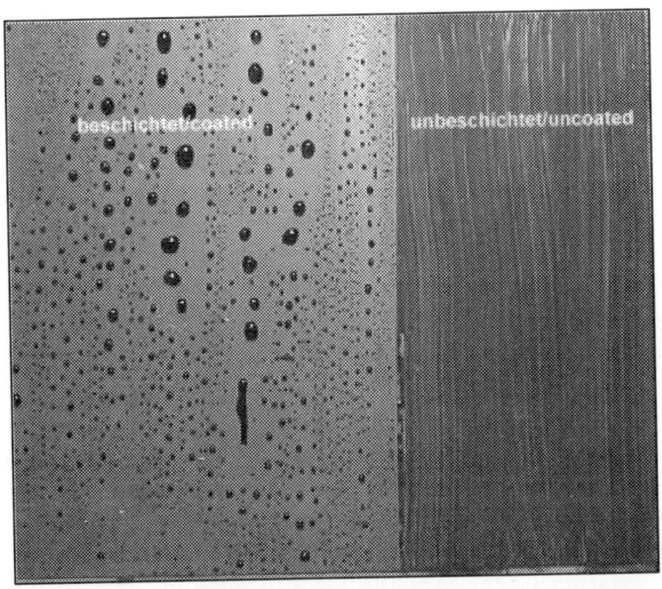

Figure 5.14: Photograph Showing the Efficiency of Easy to Clean Coatings Against Paint on Metallic Substrates

has been developed. (Nanomers) developed at Institute for New Materials, Germany is a typical example.

AR Coatings

Inorganic multilayered optical coatings with varying refractive indices are deposited on to glass substrates. Nanosized particles of SiO_2, ZrO_2 and TiO_2, surface modified with organic silanes are applied as coating to substrates of glass, sapphire etc. Polymerisation/Poly condensation lead to organically cross linked network which on thermal treatment gives densified coatings. Stacks of optical layers thus prepared with refractive indices varying between 1.46 and 2.2 on glassby dip coating and subsequent UV curing or heat treatment at 450°C gives rise to anti reflection coatings.

Micropatterning with "moth eye" structures on plate glass can be used to develop angle-dependent anti refective systems. Such systems are currently being developed for solar collector systems.

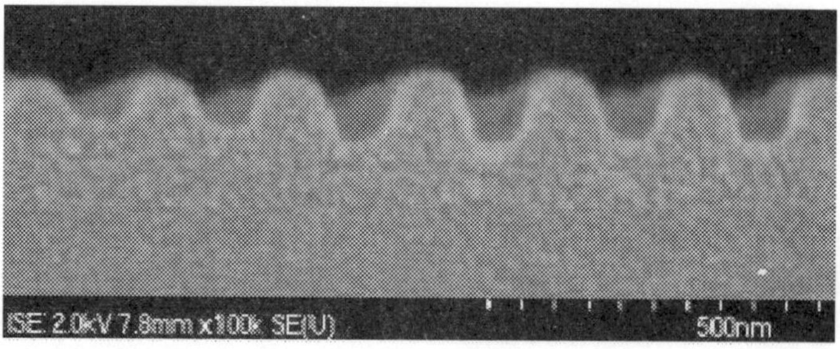

Figure 5.15: Micropatterning with Moth Eye Structures Embossed on Sol Gel Films by Polymeric Stampers

Photocatalytic Coatings

The mechanism of photocatalysis is schematically shown in Figure 5.16. Photons of light are absorbed by the semiconductor (anatase titania) which promote electrons from valence band to conduction band creating an electron-hole pair. The necessary condition for a material to be a photocatalyst is that the energy gap between valence and conduction bands be narrow and is capable of absorbing a photon having energy more than or equal to its band gap. Titania with a band gap of 3.2 ev combined with the high oxidation potential of the valence bond, of approx. VVB=3.1eV meets this prerequisite.

The resulting free-radicals are very efficient oxidizers of organic matter. Photocatalytic activity in TiO_2 has been extensively studied because of its potential use in sterilization, sanitation, and remediation applications. The ability to control PCA is important in many other applications utilizing TiO_2 including paint pigments and cosmetics that require low PCA.

Sol Gel method has been extensively used in the development of titania coatings. Alkoxide precursors like isopropoxide or inorganic salts like titanyl sulphate can be

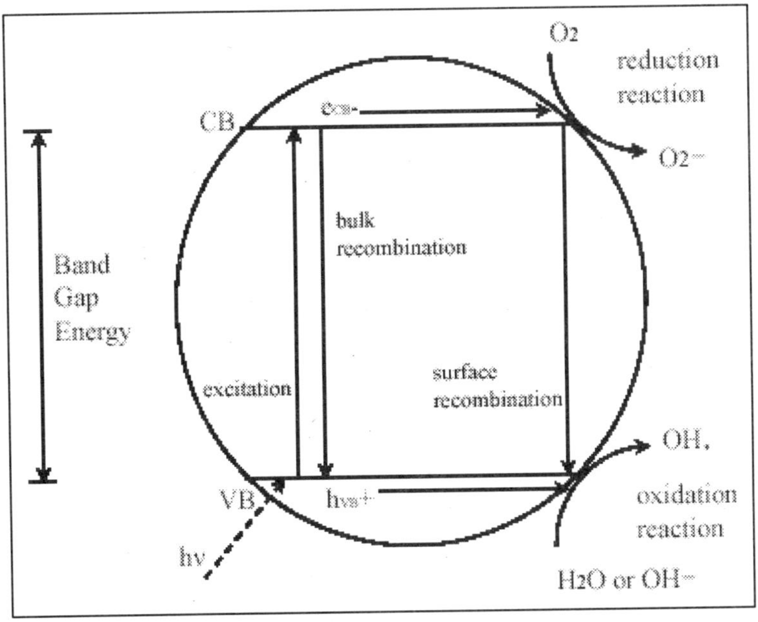

Figure 5.16: Principle of Photocatalysis

used to prepare a titania sol which forms the photocatalytic form of anantase. The sol can be applied by any of the techniques mentioned above depending on the application and substrate.

Antimicrobial Coatings

The photocatalytic activity of titania results in thin coatings of the material exhibiting self cleaning and disinfecting properties under exposure to UV radiation. These properties make the material a candidate for applications such as medical devices, food preparation surfaces, air conditioning filters, and sanitary ware surfaces.

The use of titania coatings for various applications are summarized in the Table 5.2 below.

Table 5.2

Property	*Application*
Photocatalytic Super hydrophylic	Self cleaning tiles 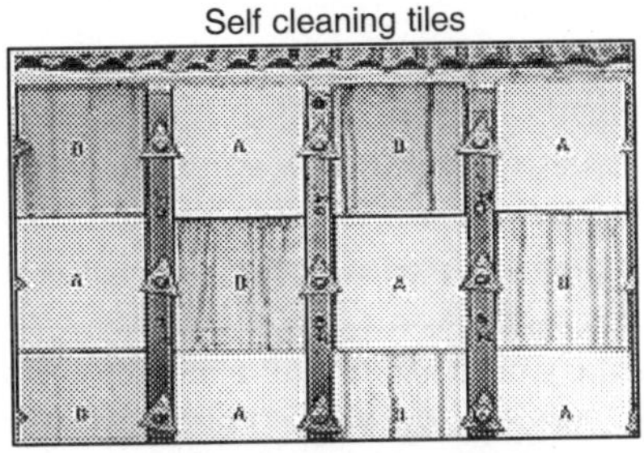 A-Titania coated B Uncoated

Contd...

Table 5.2–Contd...

Property	*Application*
Antimicrobial TiO_2 coated tiles in an hospital environment showed the surface bacteria on the wall surfaces were reduced to zero, plus airborne bacteria counts were reduced.	Coating of tiles in hospital rooms 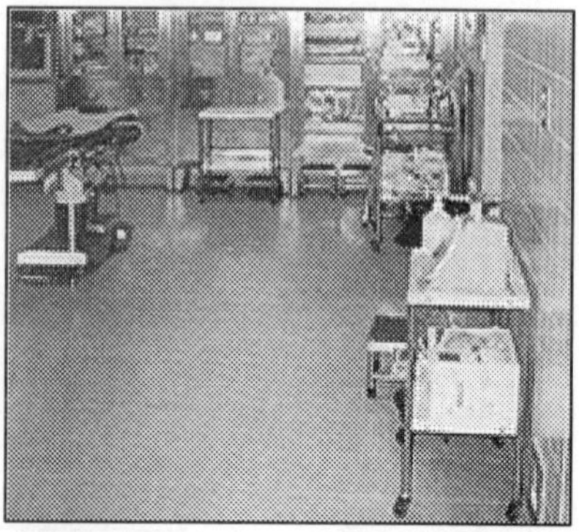
Antifogging Superhydrophylic property of titania coatings prevents the formation of water droplets by spreading it as a film on glass	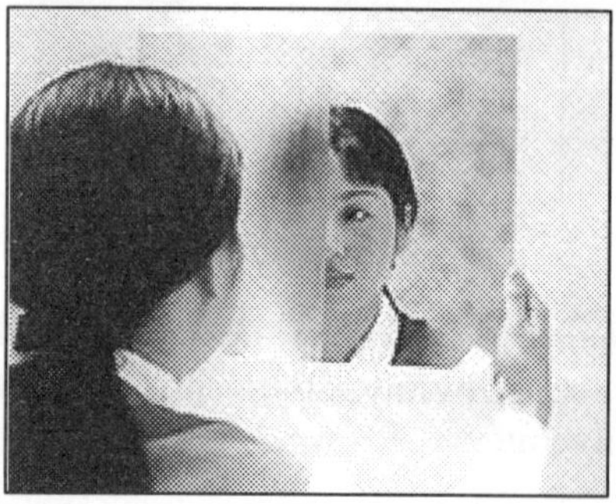

Ceramic Membranes

By virtue of their thermal and chemical stability ceramic membranes are being preferred to polymeric counterparts. In order to deposit membrane layers over porous inorganic supports, sol gel methods are extensively used. One such activity is currently being pursued at ARCI where alumina membrane layers are deposited on honeycomb supports to obtain an asymmetric membrane having final pore sizes of the order of 200 nm.

Conclusions

Sol Gel process is very conducive to develop coatings of various functionalities and applications. Application techniques are simple and cost effective. Large area substrates and substrates of complex shapes can be suitably coated by a right choice of coating technique. Availability of nanopowders will aid in the development of coatings with improved quality by aiding low temperature densification and thereby improving adhesion characteristics. Surface modification of nanopowders will ensure maximum dispersion as well as compatibility to the matrix. Coatings of Inorganic-organic hybrids combine the beneficial effects to result in nanocomposites with improved properties of strength, hardness and flexibility.

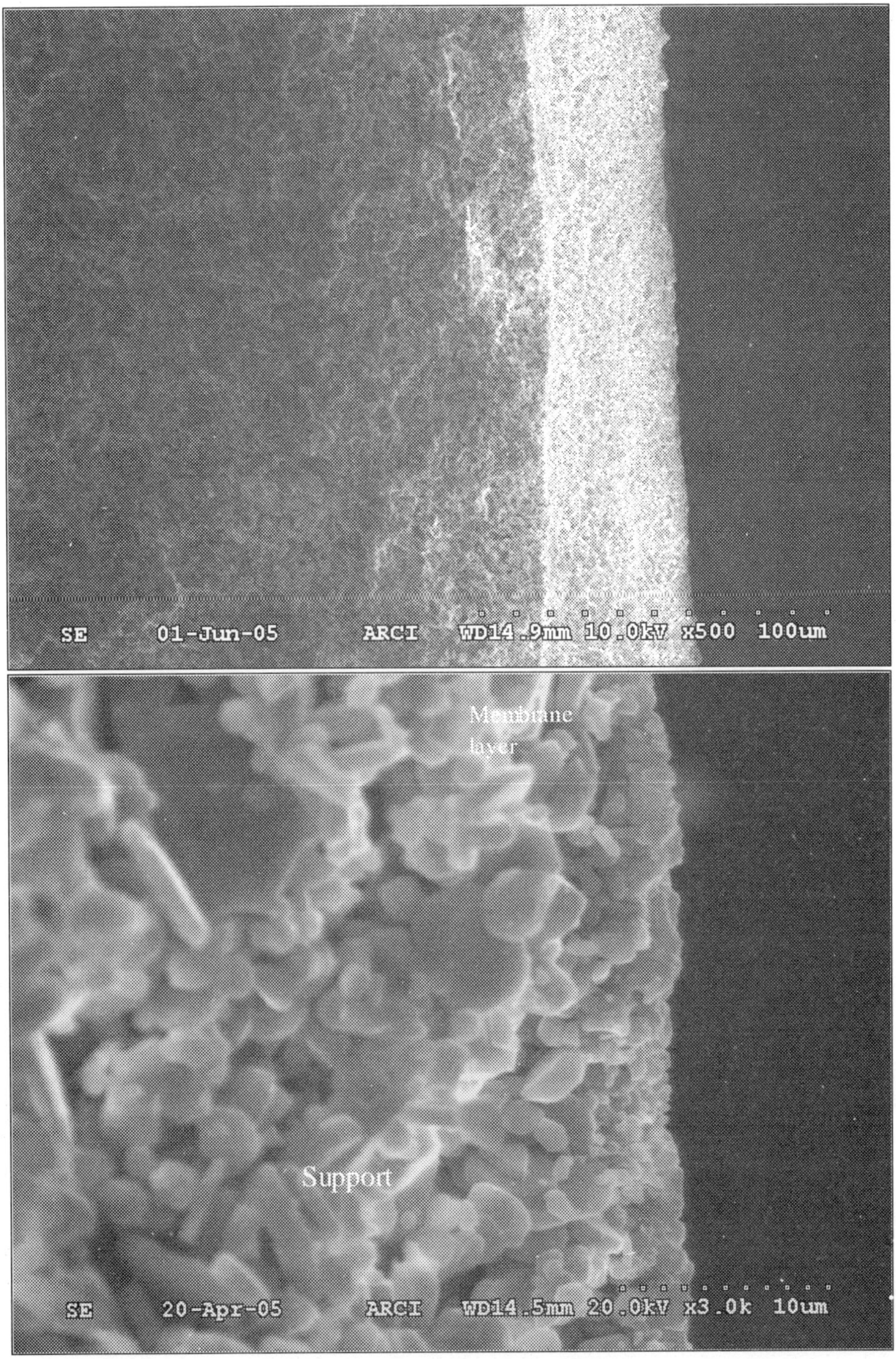

Figure 5.17: SEM Fractograph of Alumina Based Membrane Layers Adhering onto a Porous Support

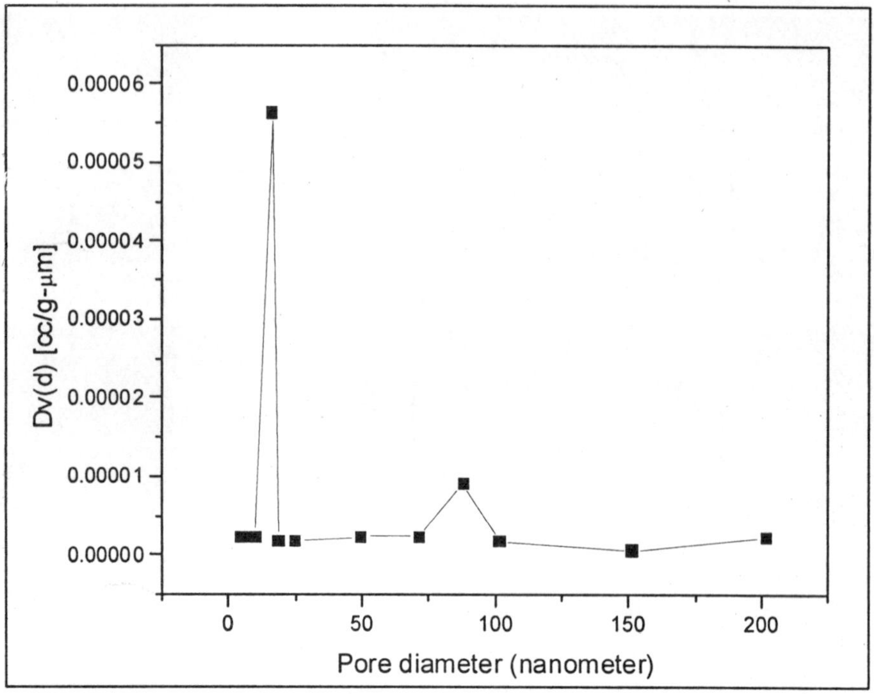

Figure 5.18: Pore Size Distribution of Nanoporous Membrane Layer

References

1. Introduction to Sol Gel processing, A. C. Pierre, Kluwer Academic Publishers, Massachussets, USA, 1998
2. Sol Gel Science, The Physics and Chemistry of Sol gel Process, C. J. Brinker and G. W. Scherer, Academic Press, NewYork, 1990
3. Wet coating technologies on glass, M.Mennig and H.Schmidt at www.solgel.com
4. Tutorials available at www.solgel.com
5. www.IWTBremen.sol-gel-verfahren.
6. www.titaniumart.com
7. www.nano-x.de

Chapter 6

Laser Assisted Surface Modification of Engineering Metals and Alloys

Jyotsna Dutta Majumdar and Indranil Manna*

Department of Metallurgical and Materials Engineering,
Indian Institute of Technology, Kharagpur – 721 302, India
**E-mail: imanna@metal.iitkgp.ernet.in*

ABSTRACT

Light Amplification by Stimulated Emission of Radiation (laser) is a coherent and monochromatic beam of electromagnetic radiation that can propagate in a straight line with negligible divergence and deliver controlled quantum of energy/ power in a given beam-mode/configuration and wavelength. As a result, laser finds a wide-ranging applications in different material processing like machining, joining, manufacturing and surface modification. In the present lecture, fundamentals and a brief overview of the current status of laser assisted surface modification of engineering materials will be discussed with specific examples from the work by the author and his collaborators. The processes covered are broadly divided into the following major categories: laser surface cleaning, laser assisted coloring, laser surface texturing, laser surface hardening, laser surface cladding and laser surface alloying. The materials considered are mostly metallic, but the principles of treating non-metallic components are also briefly touched upon. With a brief introduction on lasers, laser-matter interaction and classification of laser surface engineering, a major part of the discussion will focus on laser assisted surface modification of metallic materials to enhance resistance to wear, corrosion and oxidation. A list of recent studies of importance in these areas is included for the benefit of the participants of this course. Finally, a brief comment is made on the scope of laser surface engineering in future.

Keywords: *Laser, Wear, Corrosion, Oxidation, Microstructure, Surface engineering.*

Introduction

Surface composition and microstructure play a crucial role in determining important surface dependent properties like color, appearance, wear, corrosion and oxidation. The engineering approach to tailor these properties lies in modifying the surface composition and/or microstructure of the near surface region of a component without affecting the bulk [1]. In this regard, the more commonly practiced conventional surface engineering techniques like galvanizing, diffusion coating, carburizing, nitriding and flame/induction hardening possess several limitations like high time/energy/material consumption, poor precision and flexibility, lack in scope of automation/improvisation and requirement of complex heat treatment schedule. Furthermore, the thermodynamic and kinetic constraint of restricted solid solubility limit and slow solid state diffusivity imposes additional limitations of these conventional and near-equilibrium processes.

In contrast, surface engineering based on application of electron, ion and laser beams are free from many of the limitations of equilibrium surface engineering methods [1-6]. A directed energy electron beam is capable of intense heating and melting the surface of even the most refractory metals and ceramics [7]. However, an electron beam produces a gaussian energy deposition profile in the irradiated zone, and hence, is more suitable for deep penetration welding or cladding of similar or dissimilar solids. Moreover, the scope of generation of x-ray by rapid deceleration of high-energy electrons impinging on a solid substrate poses additional disadvantage of a possible health hazard. As an alternative, ion beam processing offers practically an unlimited choice and flexibility of tailoring the surface microstructure and composition with an implant which has otherwise no or very restricted solid solubility in a given substrate [8,9]. However, the peak concentration of implanted species, like the energy deposition peak in electron beam irradiation, lies underneath and does not coincide with the surface. Furthermore, the requirement of an expensive ionization chamber, beam delivery system and ultra-high vacuum level are serious deterrents against large-scale commercial exploitation of ion beam assisted surface engineering. In comparison, laser circumvents majority of the limitations cited above with regards to both conventional and electron/ion beam assisted surface engineering methods and offers a unique set of advantages in terms of economy, precision, flexibility, novelty of processing and improvement in surface dependent properties concerned [6,7,10].

Among the notable advantages, laser surface engineering (LSE) enables delivery of a controlled quantum of energy (1-30 J/cm^2) or power density (10^4-10^7 W/cm^2) with precise temporal and spatial distribution either in short pulses (10^{-3} to 10^{-12} s) or as a continuous wave (CW). The process is characterized by an extremely fast heating/cooling rate (10^4-10^{11} K/s), very high thermal gradient (10^6-10^8 K/m) and ultra-rapid re-solidification velocity (1-30 m/s) [1,4,5]. These extreme processing conditions often develop a meta stable microstructure and composition in the near surface region with a large extension of solid solubility and formation of amorphous phases in a crystalline matrix.

History of Laser and its Application

LASER stands for light amplification by stimulated emission of radiation. Following the initial foundation laid by Einstein, Maiman [11] developed a ruby laser for the first time in 1960. Almost all important lasers including semiconductor lasers, Nd:YAG lasers, CO_2 gas lasers, dye lasers and other gas lasers were invented during 1962-68. Incidentally, CO_2 laser was invented by C. K. N. Patel, an Indian working in the Bell Laboratory, USA [12]. By mid 1970s, more reliable lasers were made available for industrial applications such as cutting, welding, drilling and marking. Surface engineering activities (cladding, alloying, glazing and deposition) picked up from 1980 onwards.

Table 6.1 summarizes commercially available lasers and their main areas of application. The laser medium could be solid, liquid or gaseous offering a wide range of characteristic wavelength and power/energy output. Lasers are generally named after the physical state and chemical identity of the active medium. Consequently, we have crystal, glass or semiconductor, solid state, liquid medium, and gas lasers. The latter (gas lasers) can be further subdivided into neutral atom lasers, ion lasers, molecular lasers and excimer lasers. The typical commercially available lasers for material processing are (a) solid state crystal or glass laser–Nd:YAG, Ruby, (b) semiconductor laser–AlGaAs, GaAsSb and GaAlSb lasers, (c) dye or liquid lasers–solutions of dyes in water/alcohol and other solvents, (d) neutral or atomic gas lasers–He-Ne laser, Cu or Au vapor laser, (e) ionized gas lasers or ion lasers–argon (Ar^+) and krypton (Kr^+) ion lasers, (f) molecular gas lasers–CO_2 or CO laser, and (g) excimer laser–XeCl, KrF, etc. The wavelengths of the presently available lasers cover the huge spectral range from the far infrared to the soft x-ray. For details about the history of laser and its application, one of the earliest textbooks on laser material processing by Steen [6] is a useful guide.

Table 6.1: Commercially Available Lasers and their Industrial Applications

Laser	*Discovery*	*Commercialization*	*Application*
Ruby	1960	1963	Metrology, medical applications, inorganic material processing.
Nd-Glass	1961	1968	Length and velocity measurement.
Diode	1962	1965	Semiconductor processing, bio-medical applications, welding.
He-Ne	1962		Light-pointers, length/velocity measurement, alignment devices.
Carbon Dioxide	1964	1966	Material processing–cutting/joining, atomic fusion.
Nd-YAG	1964	1966	Material processing, joining, analytical technique.
Argon ion	1964	1966	Powerful light, medical applications.
Dye	1966	1969	Pollution detection, isotope separation.
Copper	1966	1989	Isotope separation.
Excimer	1975	1976	Medical application, material processing, coloring.

Generation of Laser

Laser is a coherent and amplified beam of electromagnetic radiation or light. The key element in making a practical laser is the light amplification achieved by stimulated emission due to the incident photons of high energy. A laser comprises three principal components, namely, the gain medium (or resonator), means of exciting the gain medium into its amplifying state and optical delivery/feed back system. Additional provisions of cooling the mirrors, guiding the beam and manipulating the target are also important. The laser medium may be a solid (*e.g.* Nd:YAG or neodymium doped yttrium-aluminum-garnet), liquid (dye) or gas (*e.g.* CO_2, He, Ne, *etc.*). For gas and diode lasers, the energy is usually introduced directly by electric-current flow, whereas, an intense flash of white light produced by incandescent lamps introduces the excitation energy in solid state crystal lasers. The sudden pumping of energy causes the laser medium to fluoresce and produce intense monochromatic, unidirectional (parallel/convergent) and coherent rays [1,2]. Among the commercially available lasers, CO_2-laser seems one of the earliest developed and most popular lasers for material processing because they are electrically more efficient (15-20 per cent) and produce higher powers (0.1–50 kW) than other lasers in the continuous mode. Despite being less efficient in energy coupling with metals due to longer wavelength (10.6 $_m$m), the higher wall plug (~ 12 per cent) and quantum (~ 45 per cent) efficiency and output power level of CO_2 lasers more than compensate for the poor laser-matter energy coupling capability. On the other hand, Nd:YAG and Ruby lasers possess shorter wavelength and are more suited to pulsed mode of applications requiring deeper penetration, smaller area coverage and precision treatment for a specific purpose.

CO_2-laser device consists of three main parts - a gain or laser medium, an optical resonator or cavity with two mirrors, and an energizing or pumping source that supplies energy to the gain medium. The chemical species in the gain medium determines the wavelength of the optical output. Between the two mirrors, one is a fully reflecting and the other a partially reflecting one. From the quantum mechanical principle, when an external energy is supplied to an atom, the irradiated atom attains an excited state. The excited atom spontaneously returns to the ground state (E_1) from the higher energy state (E_2) by emitting the energy difference as a photon of frequency (ν):

$$\nu = (E_2 - E_1)/h \tag{1}$$

where,

h is the Planck's constant. This phenomenon is known as spontaneous emission. A spontaneously emitted photon may in turn excite another atom and stimulate it to emit a photon by de-exciting it to a lower energy level. This process is called stimulated emission of radiation. The latter is coherent with the stimulating radiation so that the wavelength, phase and polarization between the two are identical. A photon interacting with an unexcited atom may get absorbed by it and excite it to higher energy state. This situation, called 'population inversion' is created by the pumping source. The photons moving along the optic axis interact with a large number of excited atoms, stimulate them and by this process get amplified. They are reflected back and forth by the resonator mirrors and pass through the excited medium creating more photons. In each round trip, a percentage of these photons exit through the partially transmitting mirror as intense laser beam. Finally, the laser beam is either guided on to the work-piece by using reflecting mirrors or delivered at the desired site through optical fibers. For details on laser production and manipulation, the reader may refer to the textbooks by Steen [6] and Draper [7].

Laser Material Processing

The domain for different laser material processing techniques as a function of laser power and interaction time is illustrated in Figure 6.1 [1,6]. The processes are divided into three major classes, namely involving only heating (without melting/ vaporizing), melting (no vaporizing) and vaporizing. Obviously, the laser power density and interaction/pulse time are so selected in each process that the material concerned undergoes the desired degree of heating and phase transition. It is evident that transformation hardening, bending and magnetic domain control which rely on surface heating without surface melting require low power density. On the other hand, surface melting, glazing, cladding, welding and cutting that involve melting require high power density. Similarly, cutting, drilling and similar machining operations remove material as vapor, hence need delivery of a substantially high power density within a very short interaction/pulse time. For convenience, a single scalar parameter like energy density (power density multiplied by time, J/mm^2) is more useful for quantifying different laser assisted processes. However, the practice is not advisable as the specific combination of power and time (rather than their product) can only achieve the desired thermal and material effect. In the present lecture, we will focus our attention only to laser assisted surface engineering, which normally needs low power and higher interaction time (Figure 6.1).

Laser-Matter Interaction

The energy-deposition process from a pulsed/continuous wave laser beam into the near-surface region of a solid involves electronic excitation and de-excitation within an extremely short period of time [8,9]. In other words, the laser-matter interaction within the near-surface region achieves extreme heating and cooling rates (10^3–10^{10} K/s), while the total deposited energy (typically, 0.1-10 J/cm^2) is insufficient to affect, in a significant way, the temperature of the bulk material. This allows the near-surface region to be processed under extreme conditions with little effect on the bulk properties. In general, laser matter interaction mainly depends on the optical properties of the laser and bonding/electronic structure of the substrate. More precisely, the energy gap between conduction and valence bands dictates the ease of incident energy coupling to the irradiated surface. Thus, a metal having overlapping conduction and valence bands reflects majority of the incident laser, while a polymer or semiconductor shows variable energy coupling efficiency depending on the bonding and physical properties of the material. In this lecture, however, we will mostly consider interaction of laser with metallic substrates, as that is most common in laser surface engineering (LSE).

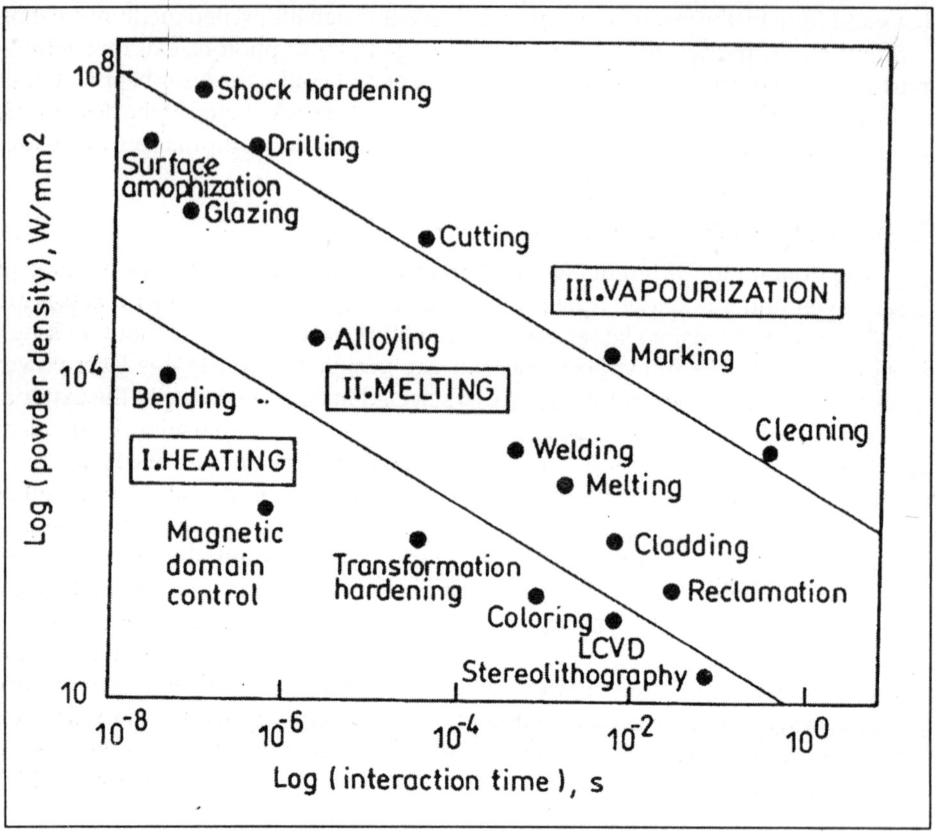

Figure 6.1: Scope of Laser Assisted Material Processing as Determined by the Power Density and Interaction Time Combination

Lattice Heating

The initial stage in all laser-metal interactions involves coupling of laser radiation to electrons on the substrate surface. This interaction first occurs by the absorption of photons from the incident laser beam promoting electrons from lower to higher states of energy. Electrons that have been excited in this manner can divest themselves of their excess energy in a variety of ways. For example, if the photon energy is large enough, the excited electrons can be removed entirely from the metal. This is photoelectric effect and usually requires photon energies greater than several electron volts. Most laser processing applications, however, utilize lasers emitting photons with relatively low energy. Electrons excited by absorption of laser radiation do not, therefore, have enough energy to be ejected from the metal surface. Such electrons must, nevertheless, lose energy to return to an equilibrium state after photon excitation. This occurs when excited electrons are scattered by lattice defects like usual non-crystalline regions in a crystal such as dislocations and grain boundaries. The overall effect is to convert electronic energy derived from the beam of incident photons into heat. It is this heat that is useful (indeed necessary) in all surface treatment applications.

Energy Absorption

Figure 6.2 describes the process that is important in electron excitation and excited carrier relaxation process involved during laser-matter interaction [8]. Photon interaction with matter occurs usually through the excitation of valence and conduction band electrons throughout the wavelength band from infrared (10 μm) to ultraviolet (0.2 μm) region. Absorption of wavelength between 0.2-10 μm leads to intra-band transition (free electrons only) in metals and inter-band transition (valence to conduction) in semiconductors. Conversion of the absorbed energy to heat involves (a) excitation of valence and/or conduction band electrons, (b) excited electron-phonon interaction within a span of 10^{-11}-10^{-12} s, (c) electron-electron or electron-plasma interaction, and (d) electron-hole recombination within 10^{-9}-10^{-10} s (Auger process). Since free carrier absorption (by conduction band electrons) is the primary route of energy absorption in metals, beam energy is almost instantaneously transferred to the lattice by electron-phonon interaction.

Spatial Distribution of Deposited Energy

The spatial profile of deposited energy from laser beam is illustrated in Figure 6.3. For laser irradiation, the beam intensity I at a depth z for the normally incident beam of initial intensity I_o (in W/m^2) is given by [8]:

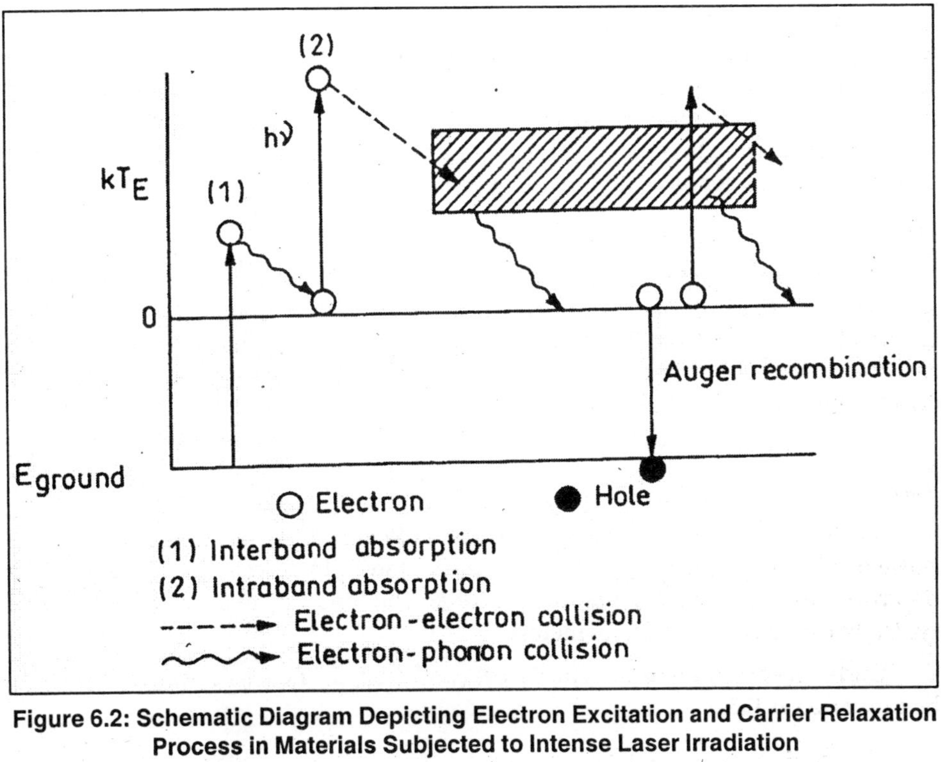

Figure 6.2: Schematic Diagram Depicting Electron Excitation and Carrier Relaxation Process in Materials Subjected to Intense Laser Irradiation

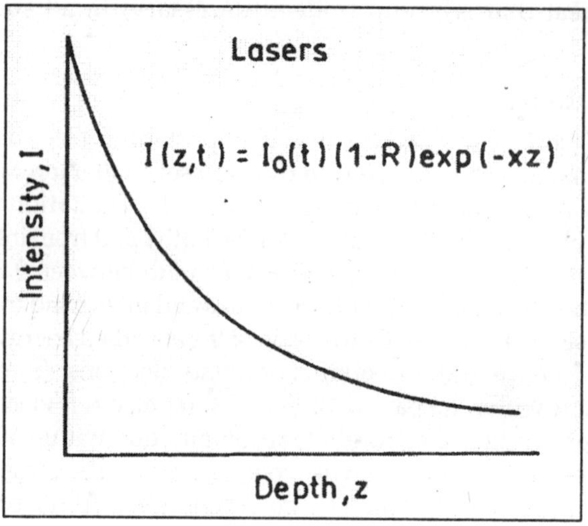

Figure 6.3: Spatial Profile of Deposited Energy Following Laser Irradiation of Solid Matter

$$I(z,t) = I_0(t)\,(1-R)\exp(-\alpha z) \tag{2}$$

where,

I_0 is the incident intensity, t is time, R and α are the reflectivity and absorption coefficients, respectively. Since α is very high ($\sim 10^6$ cm^{-1}) for metals, light is totally absorbed within a depth of 100-200 Å. The efficiency of optical coupling is determined by the reflectivity (R). R for metals is relatively low at short wavelengths, rises abruptly at a critical wavelength (related to the plasma frequency of the free electron plasma), and then remains very high at long wavelength [8].

Heating due to Laser Irradiation

Usually, the deposited energy of laser irradiation is converted into heat on a time scale shorter than the pulse duration or laser interaction time [8]. The resulting temperature profile depends on the deposited energy profile and thermal diffusion rate during laser irradiation. Thermal diffusivity (D) is related to thermal conductivity (k) and specific heat (CP) as follows:

$$D = k/(\rho\, C_p) \tag{3}$$

where,

ρ is the density. The vertical distance (z) over which heat diffuses during the pulse duration (t_p) is given by, $z = (2Dt_p)1/2$. Here, z in comparison to absorption depth (α^{-1}) determines the temperature profile. The condition of $\alpha^{-1} << z$ is applicable typically for laser irradiation of metals.

Under the one dimensional heat flow condition, the heat balance equation may be expressed as [13]:

$$\rho c_p \frac{\partial T(z,t)}{\partial T} = Q(z,t) + \frac{\partial}{\partial z} k \frac{\partial T(z,t)}{\partial z} \tag{4}$$

where,

T and Q are the temperature and power density at a given vertical distance of depth (z) and time (t), respectively. Q follows a functional relation with z same as equation (2). The heat balance equation (4) may be solved analytically if the coupling parameters (α and R) and materials parameters (ρ, k and c_p) are not temperature and phase dependent. However, phase changes are unavoidable except in solid state processing. Thus, the heat balance equation is solved by numerical techniques like finite difference/element methods. Details on mathematical modelling of heat transfer in laser material processing may be obtained in several textbooks [6,13].

Laser Surface Engineering

Figure 6.4 presents a brief classification of different laser surface engineering (LSE) methods that involve mainly two types of processes. The first type is meant for only micro-structural modification of the surface without any change in composition (hardening, melting, shocking and texturing), while the other involves both micro-structural as well as compositional modification of the near-surface region (alloying, cladding, *etc.*). In this section, we will review the basics of the LSE processes, discuss methodology involved and highlight the scope of applications. However, we will begin with a process that essentially involves controlled material removal, but is included in this section because it concerns only the surface and not the bulk.

Laser Surface Cleaning

Cleaning of metal surface can be carried out by selected area laser irradiation at an optimum combination of incident power, interaction/pulse time and gas flow rate (that sweeps the dislodged atoms from the surface) leading to surface evaporation and removal of small particles or continuous layer from the metal surface. Figure 6.5 shows the schematic mechanism of laser cleaning. At the initial stage, a plasma plume is formed due to ionization of the atoms vaporized from the surface and blocks the beam-surface contact (Figure 6.5a). As the irradiation stops, the temporary compression on the surface changes into tension and causes spallation of the oxidized layer (Figure 6.5b). A dramatic improvement of cleaning efficiency in terms of area and energy is possible when laser irradiates the work-piece at an oblique or glancing-angle rather than normal incidence. Furthermore, substrate damage is greatly reduced and probably eliminated at glancing angles.

This process is a contact less, fast and flexible method that can remove any material (oil, paint, polymer, oxide, metal, ceramic) with a precision of less than a micrometer. Hence, laser cleaning is of immense importance to clean priceless paintings, archaeological exhibits, semiconductor devices and even dead or alive animal/human tissues. Let us now examine a few practical examples.

Psyllaki and Oltra [14] have investigated the influence of pulsed laser irradiation on removal of oxide layer from the surface of stainless steels developed by high

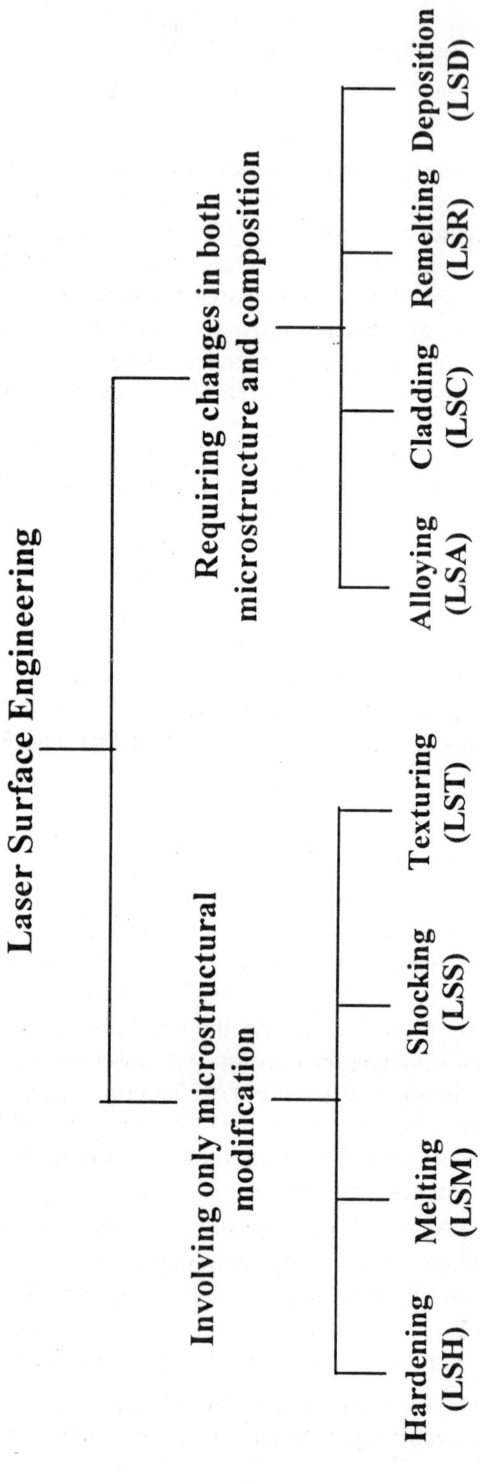

Figure 6.4: General Classification of Laser Surface Engineering

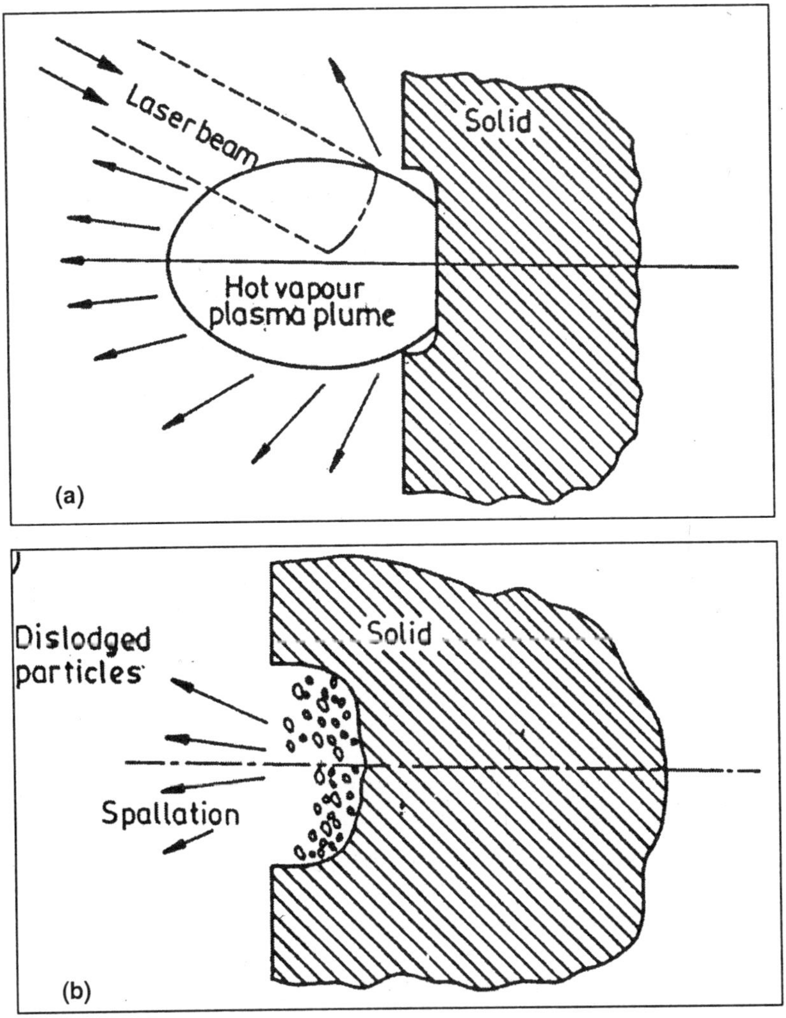

Figure 6.5: Schematic Diagram Showing the (a) Principle and (b) Methodology of Laser Surface Cleaning

temperature oxidation using a pulsed Nd:YAG laser. At 1.0-2.0 J/cm^2 energy density, irrespective of the composition and thickness of the surface layer, YAG laser irradiation resulted in expulsion of the oxide layer without any material removal from the underlying metal.

The demand for new wafer cleaning technology after plasma etching increases as the industry enters into submicron or nanotechnology processes. The success of low-resistance interconnecting high-density ultralarge-scale integrated devices depends on the cleanliness of via holes. The side wall and bottom polymers resulting from reactive ion etching of via holes can be removed by a non-contact dry laser-cleaning technique using pulsed excimer laser irradiation [15]. Similarly, laser cleaning is capable of removing the polymers by sub-threshold ablation, even at fluences limited by the damage threshold (= 250-280 mJ/cm^2) of the underlying Al-Cu metal film with titanium nitride (TiN) antireflective coating. Between laser ablation by Nd-YAG and excimer laser, it appears that although the shorter 7 ns Nd-YAG laser pulse gives a greater etch thickness than the 23 ns excimer laser pulse, the former also tends to damage the metal films and the silicon substrates of the via wafers more easily. Thus, excimer lasers are more suited to clean silicon and other semiconductor devices.

Tsunemi *et al.* [16] have demonstrated that pulsed laser irradiation of oxidized metallic surfaces in an electrolytic cell under proper voltage conditions could be a promising new approach for effective removal of oxide films. Systematic measurements on simulated corrosion-product films by optical reflectance profile and energy dispersive X-ray spectroscopy showed that the utilization of a basic electrolyte solution and imposition of a certain cathodic potential prior to laser irradiation were essential for high removal efficiency.

In general, laser cleaning should find potential applications in paint removal metallurgy, semiconductor fabrication technology, de-contamination of nuclear power plants and mask-less patterning of oxidized surfaces.

Laser Colouring

The colour we see on a metallic material is mostly due to the composition, thickness and optical properties of the thin oxide layer present on its surface. Using this principle, a particular colour or lustre on the surface of a metal can be developed by a controlled laser irradiation and formation of a thin and desired oxide layer. Lu and Qiu [17] have investigated laser-assisted colouring/discolouring and bleaching of amorphous WO_3 thin film during pulsed laser deposition. The original films could be coloured from light brown to purple by a single pulse of KrF excimer laser irradiation at 248 nm and subsequently bleached to brown by a single pulse of Nd-yttrium–aluminium-garnet laser at 1.06 μm in air. It is suggested that colouring is due to polaron transition or photochemical activation, while photo-thermal oxidation is responsible for the bleaching process. In stainless steel, a thermo-chemical reaction between oxygen and stainless steel is believed responsible for colouring during excimer laser irradiation in air [18]. With increasing laser fluences, the temperature rise in the irradiated area of stainless steel surface increases, which enhances oxygen diffusion into the surface and oxidation reaction within the irradiated area. Thus, laser irradiation in vacuum is an easy way to avoid discoloration of stainless steel.

In order to identify the electronic process involved in colouring, Seifert *et al.* [19] have carried out single-colour femtosecond pump-probe experiments to investigate the time dependence of laser-induced ultra fast desorption and deformation processes of silver nano-particles in glass. Muggli *et al.* [20] have reported that a single-color illumination of a copper surface by a red or an ultraviolet femtosecond laser pulse yields a three-photon (red) or a two-photon (UV) photoemission process. On the other hand, a multicolour and multi-photon process ensues when the red and the UV pulses overlap both in space and in time on the photocathode. It is shown that this emission process results from the absorption of one red and one UV photon by an electron.

Besides aesthetics, laser assisted colouring has serious applications in photo-chemical reactions, photo-electric devices, and several other optical devices like reflectors and absorbents.

Laser Surface Texturing

Texture or preferred orientation plays a very important role in determining the mechanical, electrical, magnetic and many other bulk material properties of crystalline solids. Surface texture is also an important parameter for many manufacturing technologies and in many cases, modification of surface topography is essential. Since laser irradiation can supply sufficient thermal activation to the surface atoms to align along certain crystallographic planes and directions as per the anisotropy constant, laser can be a useful tool in developing the desired texture on the surface of bulk solids or thin films. Existing methods of surface texturing include roll surface texturing, shot blasting, electric discharge texturing, etc. On the other hand, surface texturing using concentrated energy beams (laser, electron, etc.) has significant potential to replace the existing technique of surface texturing. Focused laser or electron beams produce surface depression surrounded by a smooth ring or solidified metal without any physical contact. The topography of the surface texture is reproducible and can be controlled by varying the process parameters (laser power, scan speed, beam diameter, etc.). Specific advantages of laser assisted surface texturing over electron beam texturing include the amenability to process in air (no vacuum), low cost, flexibility and better control over the process.

Figure 6.6 shows the schematic of roll surface laser texturing unit [21]. The beam of a continuous wave CO_2 laser is modulated by a chopper and then focused onto a surface of the rotating roll. The focusing/modulating unit is translated along the roll axis to provide the texturing of roll surface. The interaction of the focused laser beam

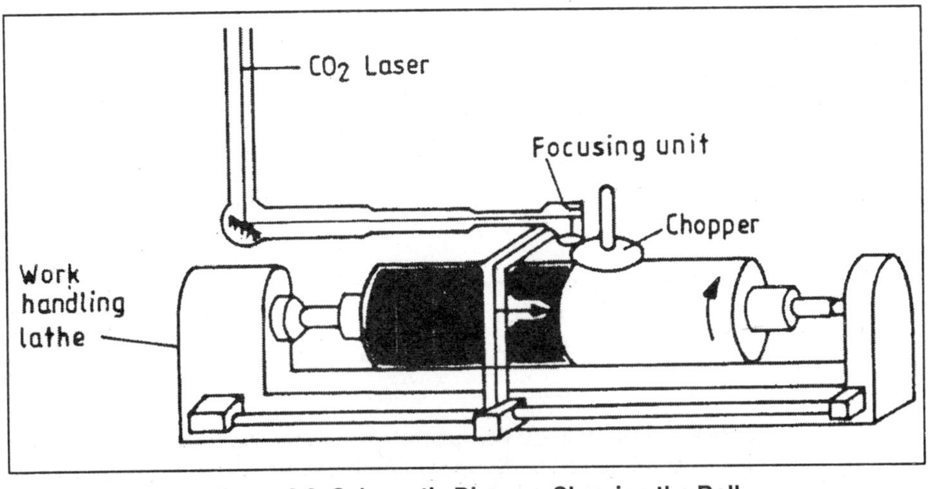

Figure 6.6: Schematic Diagram Showing the Roll Surface-Texture Development Unit

with a material surface is accompanied by melting, melt motion, evaporation, sublimation, and solidification, resulting in change of the surface topography. Different regimes of interaction zone may be produced by variation of the interaction parameters, which determine the physical nature of the process.

Figure 6.7 shows the parameter region corresponding to the different texture of the surface of electrolytic copper as a function of laser beam intensity and number of pulses applied [22]. In region I, unorganized structure of the surface with 2-5 μm droplets and thermal cracking was found. Both unorganized and mono dimensional structures following surface grain boundaries were present in region II. In region III a bi-dimensional surface structure, represented by regularly distributed conical formations, was observed. The bi-dimensional structure was observed above the threshold of plasma formation and is explained by evaporation front instability. It may be noted that, the regime varied with materials, its surface conditions and with the types of laser used. Under certain conditions of lasing smoothening of metal surface may be achieved. Smoothening requires low beam intensities just capable of melting surface. A typical application of laser assisted texturing currently in practice is texturing of the head landing zone on computer hard disk surfaces.

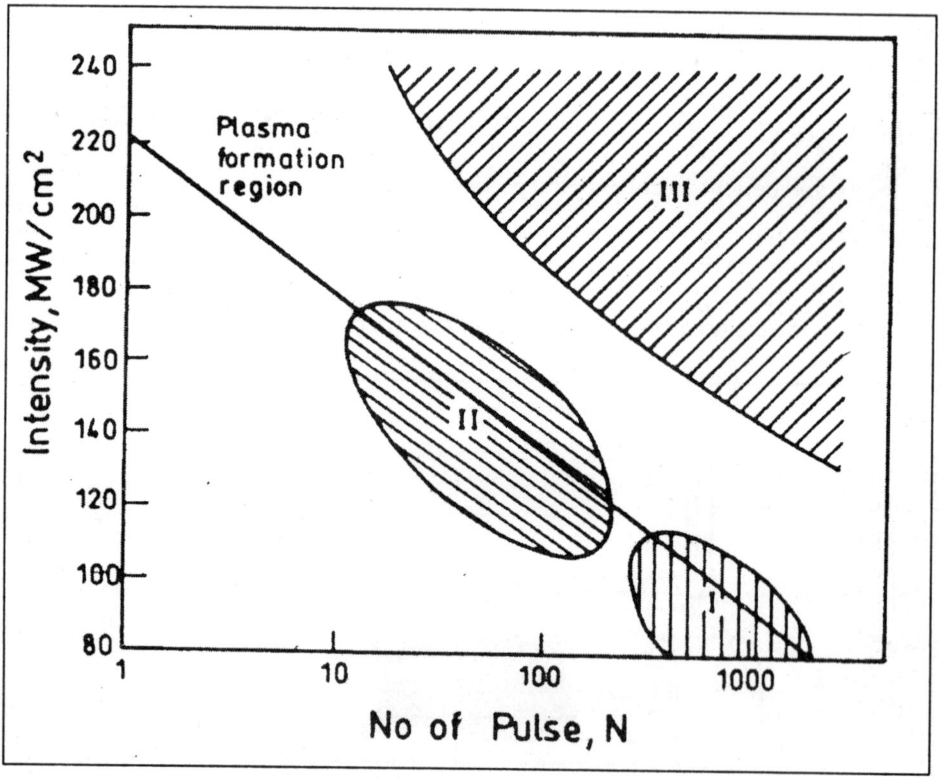

Figure 6.7: An Intensity Versus Pulse-Number Diagram for Electrolytic Copper in Ambient Air

Laser Surface Hardening

One of the major reason why steel has always been the most versatile structural material is because steel is amenable to thermal treatment and micro-structural modification that can vary its mechanical properties over a wide range. Hardening is a treatment that involves heating steel above the critical temperature (AC_3 for hypo-eutectoid and AC_1 for hyper-eutectoid) for austenitization, and cooling at faster than a critical rate to enforce a shear transformation of austenite to martensite. Martensite is extremely hard (with or without alloying elements) because it is a supersaturated solid solution, contains a large dislocation density, possesses a fine grain size and is associated with thermal and transformational stress. Besides endowing significant strength, such microstructure also makes steel brittle and unsuitable for direct use or further machining. Surface engineering allows development of such martensitic microstructure confined to only the near-surface region leaving the core unaffected. This combination of hard surface and tough core is desirable for a good resistance to wear, fatigue, erosion, fretting or scuffing.

A high power laser beam may be used as a source of heat to harden surface of the Fe-based substrate by heating the surface above austenetizing temperature and subsequently quenching it (either self quenching or by adding external quench media) to induce martensitic transformation on the near surface region. The advantages of laser assisted surface hardening (LSH) over conventional hardening include high processing speed, minimum distortion, exclusion of quenching media, possibility selective hardening, precise control of case depth, improved fatigue life of components and scope of automation. The experimental set up for temperature controlled LSH is shown in Figure 6.8. The main process variables for LSH are laser power, scan speed, beam diameter, beam shape and surface absorption capability. The thermo physical properties of the substrate also play a very important role.

While presence of martensite is of paramount importance for hardening, dislocation density and distribution play an important role in determining the strain and strength. Lu *et al.* [23] have shown that dislocation densities increase also in the laser hardened residual austenitic structure. Dislocations develop cellular sub-structure in the residual austenite, distributed between the martensite crystals in the form of 'lunar halos'. Two types of martensite *viz.*, twinned and lath martensite may co-exist in the hardened zone. LSH is one of the most widely exploited LSE techniques to date. Some more results to highlight the scope of LSH in cast iron will be discussed in the next section.

Laser Surface Cladding

Besides changing the microstructure (hardening), surface modification might involve changing the surface composition as well. Surface composition can be changed either by overlaying new materials (deposition, coating, and cladding) or introducing alloying element/elements on to the surface up to a certain depth and concentration (alloying). Laser surface cladding (LSC) is a process of melting and depositing a new metal/alloy onto the substrate surface. The interface between the deposited layer and substrate is usually sharp with minimum penetration or mass transfer at the interface. Minimum porosity, good interfacial bonding and better control

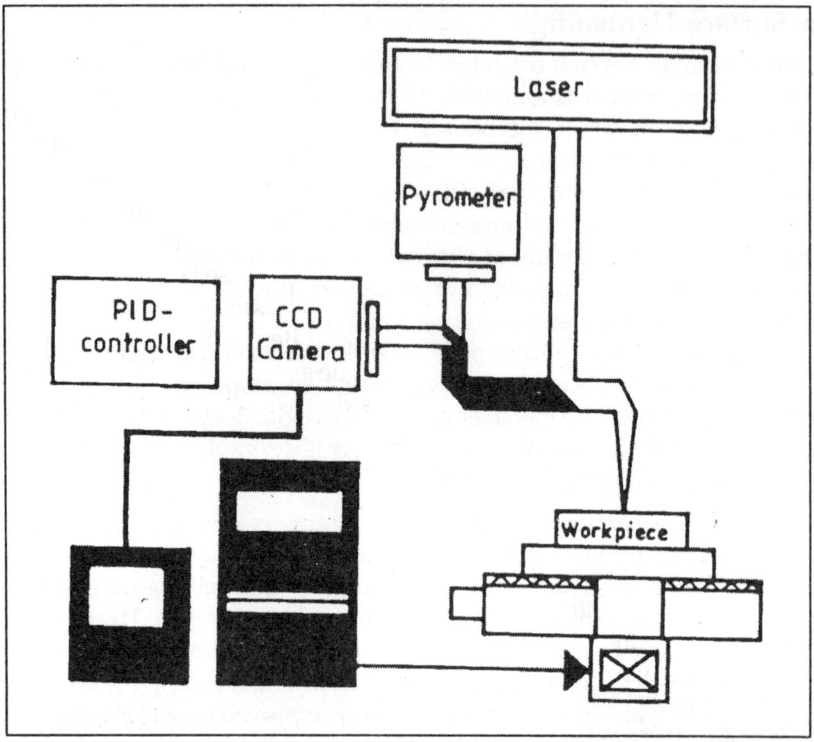

Figure 6.8: Schematic of Laser Surface Hardening Set Up

over the process, development of metastable microstructures are the major advantages of LSC over competitive coating methods like plasma spraying, arc welding, *etc.* At times, LSC is adopted as a secondary process of improving the coating quality (reduce porosity, roughness, heterogeneity) of a plasma sprayed surface. The properties of the clad layer are critically dependent on its microstructure and chemical composition. The microstructure of the clad layer, in turn, is closely linked to the rate of cooling associated with the process and hence, depends on laser power, scan speed and thermo-physical properties of the clad material. The bonding and properties at the interface on the other hand, is dependent on the degree of intermixing at the interface and difference in coefficient of thermal expansion.

In the recent times, LSC of ceramic-metal composite coatings on hard metals and high temperature components (turbines, tubes, heat exchangers) has made significant progress in view of the vast potential of its application in tribology. Different ceramic phase particles such as WC, TiC, SiC and $Cr_{23}C_6$ and diverse matrix alloys have been used to generate such composite clad layers for studies on wear resistance [24]. NiCrMo layer with varying ZrO_2 contents has been successfully deposited by this method [25]. LSC with pneumatic powder delivery has been claimed to provide a very effective technique for binding near net shape metallic deposits to the substrate [26]. The major obstacle in popularizing LSC over the conventional gas and plasma based coating methods lies in cost factors (installation), inertia to change the existing technology and requirement of skilled manpower.

Laser Surface Alloying

Among the various LSE methods, Laser Surface Alloying (LSA) involves melting of a deposited layer along with a part of the underlying substrate to form an alloyed zone for improvement of wear, corrosion and oxidation resistance. Figure 6.9a illustrates the scheme of LSA with a continuous wave laser. It includes three major parts: a laser source with a beam focusing and delivery system, a lasing chamber with controlled atmosphere and a microprocessor controlled sweeping stage where the specimen is mounted for lasing. The process includes melting, intermixing and rapid solidification of a thin surface layer with pre/co- deposited alloying elements (Figure 6.9b). The coating material may be pre-deposited by any of the conventional means like electro-deposition, plasma spray and physical/chemical vapour deposition or may be injected in the form of powder or powder mixture into the melt at the time of laser treatment and is termed as co-deposition. In LSA with a pre- or co-deposition, a 20-30 per cent overlap of the successive melt tracks is intended to ensure micro-structural/compositional homogeneity of the laser treated surface. The sweeping stage (x-y or x-y-z-θ) allows laser irradiation of the intended area of the sample-surface at an appropriate rate and interaction time/frequency. The irradiation results into transient melting of the deposit with a part of the underlying substrate, rapid mass transfer by diffusion/convection in the melt pool, and ultra-fast solidification to form an alloyed zone. The depth, chemistry, microstructure and associated properties of the alloyed zone depend on the suitable choice of laser/process parameters *i.e.* incident power/energy, beam diameter/profile, interaction time/pulse width, pre or co-deposition thickness/composition and concerned physical properties like reflectivity, absorption coefficient, thermal conductivity, melting point and density. In the next section, we will discuss a number of examples to explain the scope of LSA to enhance resistance to wear, corrosion, oxidation, etc.

Specific Studies on LSE of Ferrous Alloys

Having understood the basics of different types of LSE, let us now pay attention to specific examples to highlight the science of improvement in resistance to surface dependent degradation processes like wear and corrosion due to LSE. In this connection, we will at first review the scope and current status of understanding about LSE of Fe-based or ferrous alloys and super-alloys. For the ease of discussion, we will address implication of LSE to enhance more common types of degradation in metals like corrosion, oxidation and wear. A summary of major studies carried out in the recent years in this direction is provided in Table 6.2 [27-45].

Corrosion Resistance

Parvathavarthini *et al.* [27] have attempted to eliminate the susceptibility to inter-granular corrosion of the cold worked and sensitized AISI 316 stainless steel by Laser Surface Melting (LSM). Similarly, LSM in nitrogen atmosphere or of nitrogen bearing steel is reported to improve the resistance to pitting corrosion of AISI 304 stainless steel [27,28]. In either case, this improvement may arise due to the presence of chromic oxide and nitrogen compound on the surface and consequent reconstruction of the passive layer or barrier to the electrolyte.

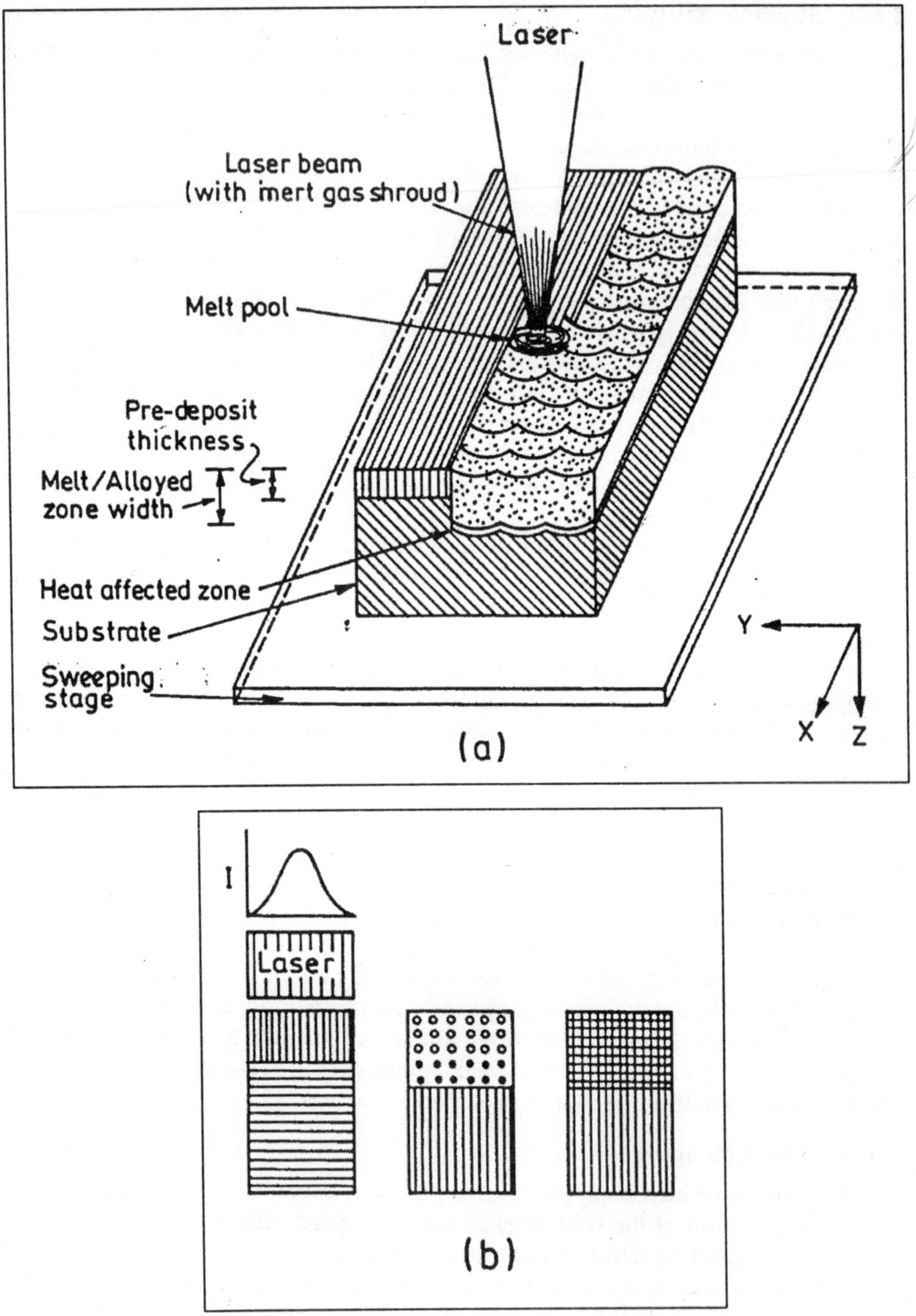

Figure 6.9(a): Schematic Hardware Set Up for Laser Surface Alloying (LSA), (b) The Processes of Heating/Melting, Intermixing and Solidification in LSA

Table 6.2: Selected Studies on Laser Surface Engineering of Ferrous Alloys

System	Year	LSE	Results	Reference
AISI 316 stainless steel	2001	LSM	LSM is effective in de-sensitization	Parvathavarthini *et al.* [27]
AISI 304 stainless steel	2001	LSM	Tendency for pitting decreases due to dissolution of nitrogen and formation of nitrides.	Conde *et al.* [28]
AISI 1040 (NiCoCrB) AISI 316L	2001	LSA	Corrosion resistance improves in mild steel but deteriorates in stainless steel. Both benefits are simultaneously impossible.	Kwok *et al.* [29]
17-4 PH stainless steel	2001	LSAn	Laser annealing produçes duplex microstructure that retards crack growth in quasi-cleavage fracture.	Tsay *et al.* [30]
G 10380 steel G 41400 martensitic AISI 316L austenitic	2000	LSS	Corrosion current decreases in martensitic steel. Laser peeing is more effective in reducing stress corrosion.	Peyre *et al.* [31]
UNS-S31603 austenitic stainless steel (Co,Ni,Mn,Cr,Mo)	2000	LSA	Resistance to erosion improves due to dispersed ceramic/intermetallic phase but pitting resistance decreases.	Kwok *et al.* [32]
AISI 304 stainless steel (Si)	2000	LSA	Si turns the matrix ferritic and segregates. Post LSA homogenization improves corrosion resistance.	Isshiki *et al.* [33]
AISI 304 stainless steel (Mo)	1999	LSA	Mo improves pitting and erosion corrosion resistance of 304-SS.	Dutta Majumdar and Manna [34]
Steel (Al)	2001	LSA	LSA with Al forms several aluminides that imparts good oxidation resistance at 600°C up to 200 h.	Pillai *et al.* [35]
Stainless steel	2000	LC	Thin oxide layers could be removed irrespective of chemical composition and history in a narrow energy band.	Psyllaki and Oltra [14]
Plain carbon steel (TiB_2)	2000	LSA	Complex oxide layers develop on steel surface exposed to 600-1000°C following parabolic growth rate.	Agarwal *et al.* [37]
S 31603 stainless steel (CrB_2,Cr_3C_2, SiC,TiC, WC,Cr_2O_3)	2001	LSA	Erosion resistance improves considerably for all carbides and borides except Cr_2O_3.	Cheng *et al.* [38]
Austenitic stainless steel (Nano-Zr)	2000	LSA	Hardness and wear/erosion resistance improves significantly due to dispersion of Zr-rich amorphous phase.	Wu and Hong [39]
Austempered ductile iron (Cr)	2001	LSA LSH	LSH (than LSA) is more effective in improving wear resistance and developing compressive residual stress.	Roy and. Manna [40]
AISI 1040 (TiB_2)	2000	LSA	LSA improves resistance to adhesive/ abrasive wear and reduces friction coefficient.	Agarwal and Dahotre [41]

Contd...

Table 6.2–Contd...

System	Year	LSE	Results	Reference
Mild steel (FeCr-TiC)	2000	LSR, SHS	LSR homogenizes the microstructure and improves wear resistance.	Tondu *et al.* [42]
Mild steel (Hadfield, Fe-Mn-C)	1999	LSC	LSC improves wear resistance and bulk mechanical properties of the clad.	Pelletier *et al.* [43]
Cr-Mo steel (Cr)	2000	LSA	LSA enhances oxidation resistance in 800-1000°C due to Cr_2O_3 rich scale.	Manna *et al.* [44]
Martensitic steel (Ni-alloy)	1999	LSC	LSC significantly increased hardness and erosion resistance	Zhang *et al.* [45]

Dutta-Majumdar and Manna [34] have investigated the effect of LSM and LSA of plasma spray deposited Mo on AISI 304 stainless steel (304-SS). Figure 6.10 shows the optimum conditions (shaded region) for the formation of a homogeneous microstructure and composition in this study for improvement in pitting corrosion

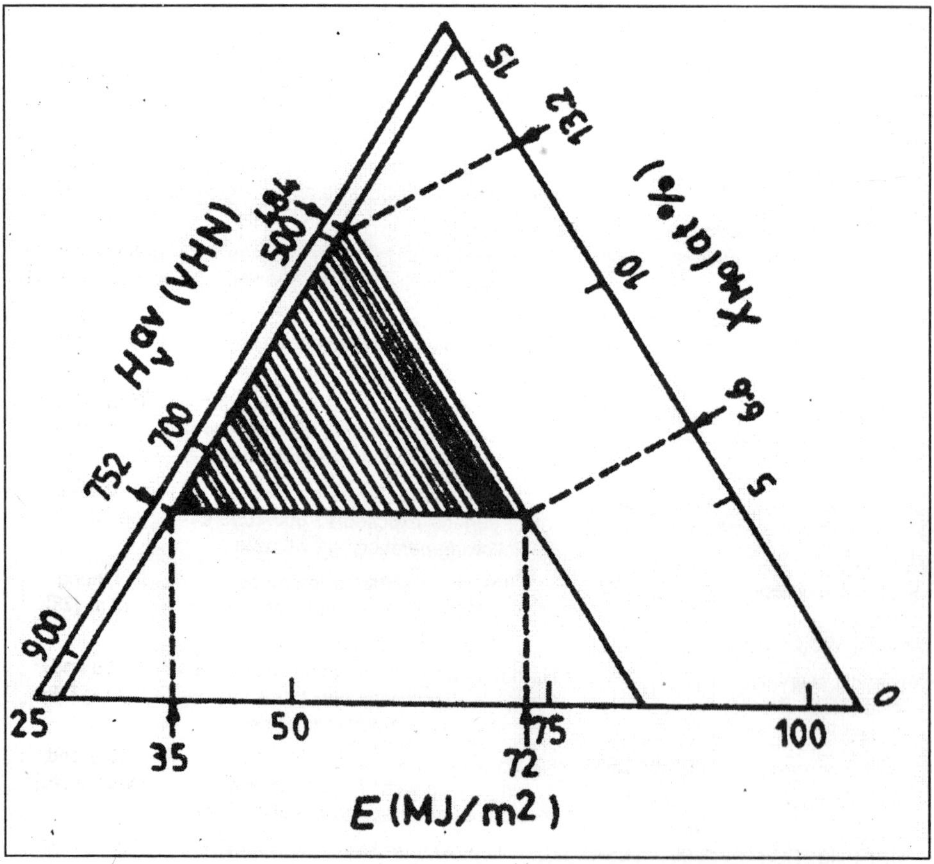

Figure 6.10: Process Optimization Diagram for Selecting the Necessary Energy Density (*E*) for Laser Surface Alloying of 304-SS with Mo to Achieve the Desired Composition (X_{Mo}) and Hardness (H_v^{av}) in the Alloyed Zone (Shaded region)

and mechanical property. Potentiodynamic anodic polarization tests of the substrate and laser surface alloyed samples (SS(Mo)) in a 3.56 wt. per cent NaCl solution (both in forward and reverse potential) showed that the critical potential for pit formation (E_{PP1}) and growth (E_{PP2}) have significantly (2-3 times) improved from 75 mV(SCE) in 304-SS to 550 mV(SCE) (Figure 6.11a). E_{PP2} has also been found to be nobler in as lased specimens than that in 304-SS. The poor pitting resistance of the plasma sprayed 304-SS samples (without laser remelting) was probably due to the presence of surface

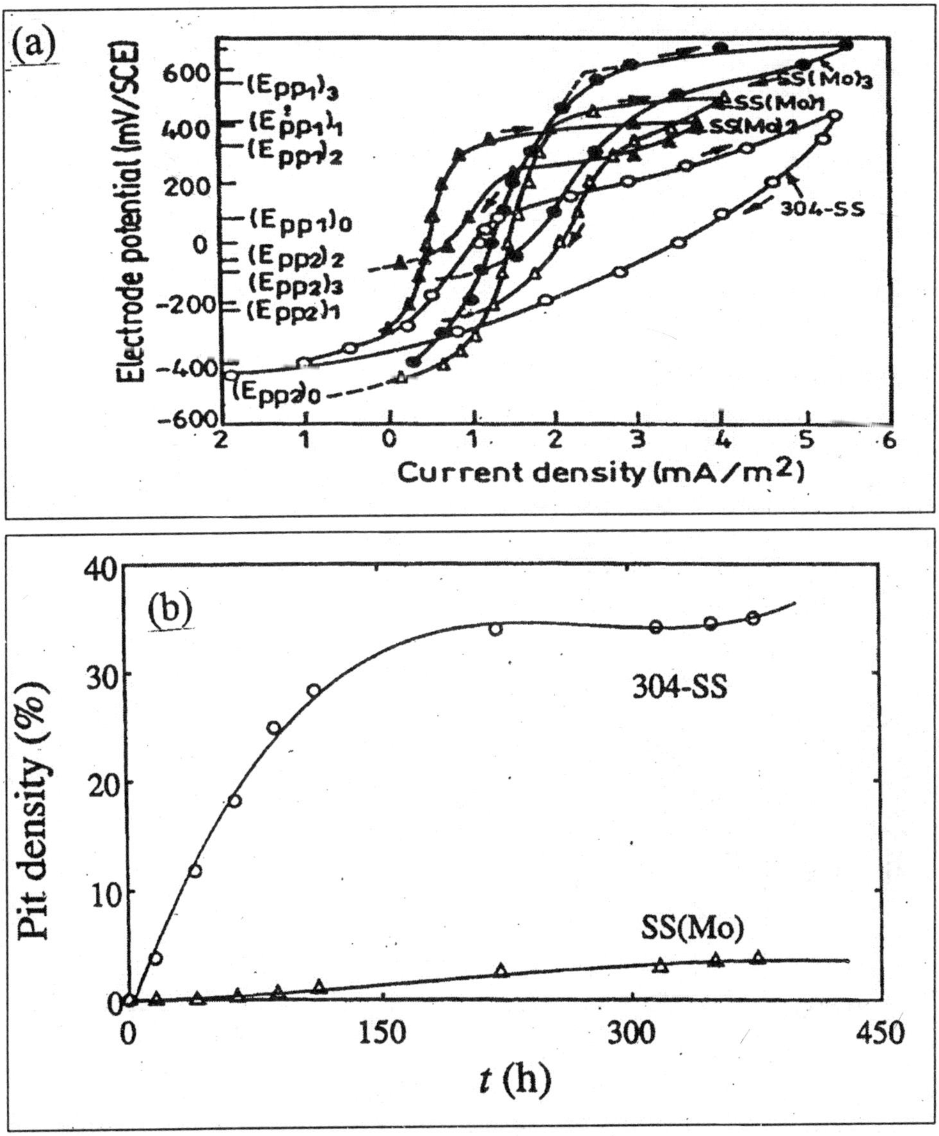

Figure 6.11: (a) Cyclic Potentiodynamic Polarization Behavior of 304-SS and Laser Surface Alloyed of 304-SS with Mo in a 3.56 wt. per cent NaCl Solution and (b) Variation of Pit Density in 304-SS and SS(Mo) as a Function of Time (*t*) in a 3.56 wt. per cent NaCl Solution

defects present in plasma deposited layer. Standard immersion test was conducted in a 3.56 wt. per cent NaCl solution to compare the effect of LSA on the pitting corrosion resistance. Figure 6.11b compares the kinetics of pit formation in terms of area fraction of pits determined by standard immersion test in a 3.56 wt. per cent NaCl solution as a function of time (t) between 304-SS and SS(Mo) lased with 1210 MW/m^2 power density and 31.7 mm/s scan speed for deposit thickness of 250 µm (corresponding to highest E_{PP1}). It is evident that both the extent and rate of pitting were significantly reduced in the SS(Mo) as compared to that in 304-SS. Furthermore, the process of pitting in 304-SS follows a sigmoidal nature marked by a substantially rapid initial stage than that of the later stage with no incubation time. In comparison, pits were noticed in SS(Mo) only after 50 h of immersion, and the number increases linearly following a much slower kinetics as compared to that in 304-SS. Continuous circulation of the samples at 750 rpm for 10 to 75 h in a medium containing 20 wt. per cent sand in 3.56 wt. per cent NaCl solution showed a significant decrease in the kinetics of erosion-corrosion loss in SS(Mo) than that in 304-SS (Figure 6.12). It was thus concluded that LSA is capable of imparting an excellent superficial micro hardness and resistance to corrosion and erosion-corrosion properties to 304-SS due to Mo both in solid solution and as precipitates.

Oxidation Resistance

Oxidation is another serious mode of surface degradation that gets aggravated under unabated counter ionic transport of cations and anions at elevated temperature. In the past, several attempts have been made to enhance resistance to oxidation by LSA, LSC and similar LSE techniques [35-37]. Manna *et al.* [44] attempted to enhance

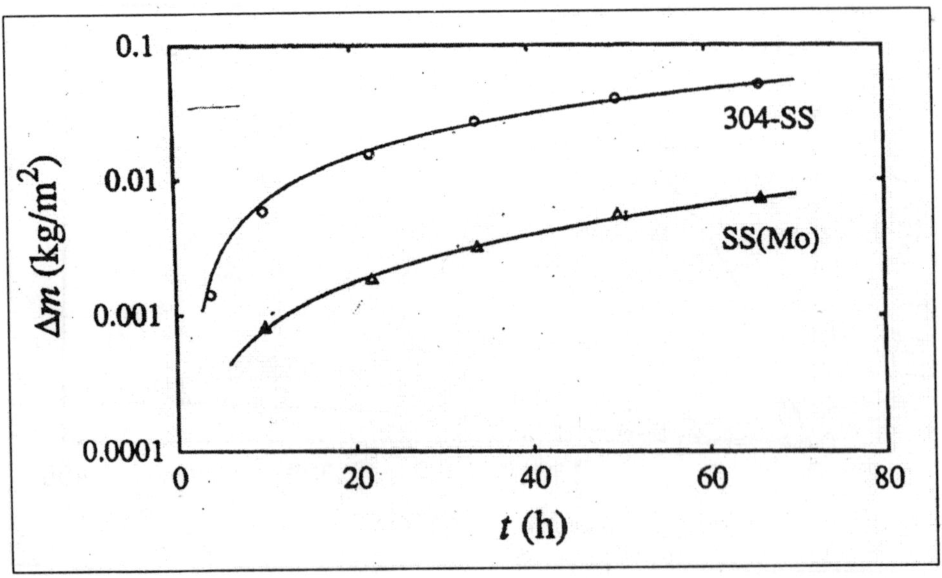

Figure 6.12: Comparison of Material Loss per Unit Area (Δm) Due to Erosion as a Function of t for 304-SS and Laser Surface Alloyed of 304-SS with Mo in a Mixture of 20 per cent Sand in 3.56 wt. per cent NaCl Solution. For details, please see [34].

the high temperature oxidation resistance (above 873 K) of 2.25Cr-1Mo ferritic steel by LSA with co-deposited Cr using a 6 kW continuous wave CO_2 laser. The main process variables chosen for optimizing the LSA routine were laser power (from 2 to 4 kW), scan speed of the sample-stage (150 to 400 mm/min) and powder feed rate (16 to 20 mg/s). Isothermal oxidation studies in air by thermo gravimetric analysis at 973 and 1073 K for up to 150 h revealed that LSA had significantly enhanced the oxidation resistance of ferritic steel during exposure to 973 and 1073 K (Figures 6.13a,b). Post oxidation micro structural analysis suggests that an adherent and continuous Cr_2O_3 layer is responsible for the improvement in oxidation resistance [44].

Wear Resistance

Attempt to improve wear resistance of steel by LSE seems to be more effective than the attempts to enhance the resistance to corrosion and oxidation. Cheng *et al.* [38] have added a mixture of WC-Cr_3C_2-SiC-TiC-CrB_2 and Cr_2O_3 to produce a metal matrix composite surface on stainless steel UNS-S31603. Following LSM, cavitation erosion resistance improved in all cases except for Cr_2O_3. Roy and Manna [40] have demonstrated that laser surface hardening (LSH), instead of LSA or LSM is more effective in enhancing hardness and wear resistance of unalloyed austempered ductile iron (ADI). Figures 6.14a,b show the typical martensitic microstructure developed by LSH and significant improvement in adhesive wear of laser hardened *vis-à-vis* as-received and laser surface melted ADI samples, respectively. Adhesive wear of austempered ductile iron consists of three distinct stages: the initial rapid, subsequent steady state and final accelerated (abrasive) wear. The improvement in wear resistance following LSH is attributed to the martensitic surface with residual compressive stress [40].

Agarwal and Dahotre [41] have reported a substantial improvement in resistance to adhesive/abrasive wear of steel following LSA with TiB_2. The surface composite layer containing about 69 vol. per cent TiB_2 particles recorded an elastic modulus of 477.3 GPa. Similar improvement in wear resistance of mild steel was reported by Tondu *et al.* [42] due to formation of a FeCr + TiC composite coating formed by laser assisted self propagating high temperature synthesis. Several other examples are cited in Table 6.2.

In order to establish LSE as a commercially viable technique, it is important that LSE is not contemplated only as a replacement for conventional techniques for routine and mundane applications. The lack of alacrity to adopt LSE is mainly due to the high installation cost and other economic factors involved. This initial antipathy could give way to enthusiastic consideration from the industry only when the novelty of the process is well established and cases that are amenable to LSE more than any other process are brought to notice. For this, continued efforts are needed both in fundamental research and industry level trials.

Specific Studies on LSE of Non-Ferrous Alloys

Surface dependent degradation by wear, oxidation and corrosion is as much a problem in nonferrous metals and alloys as that in ferrous alloys. Thus, LSE is widely applied to Al, Ti, Cu, Mg and other important nonferrous metals and alloys to extend

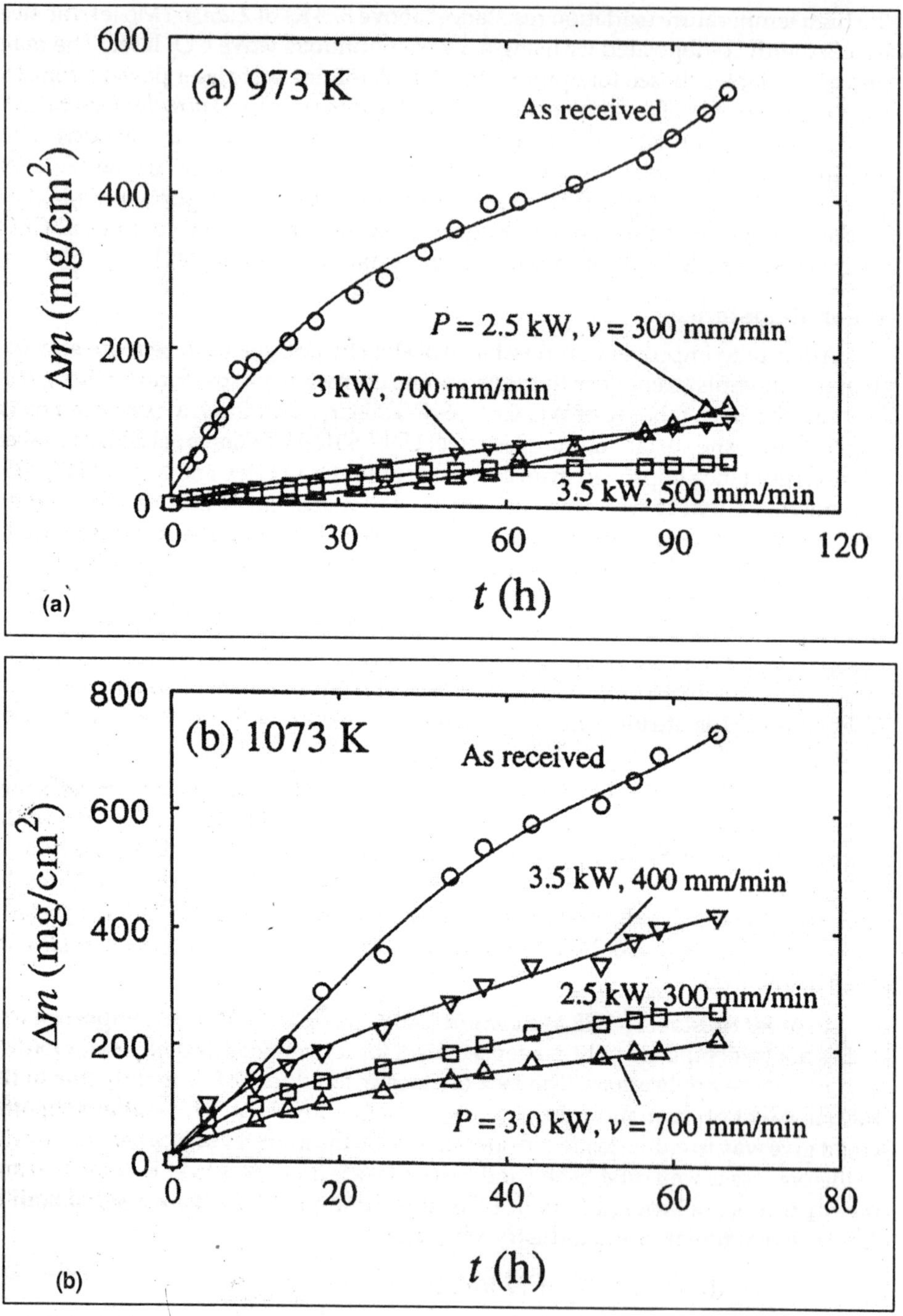

Figure 6.13: Kinetics of Isothermal Oxidation of 2.25Cr-1Mo Ferritic Steel Laser Surface Alloyed with Cr and Exposed to (a) 973 and (b) 1073 K in air, Respectively [44]

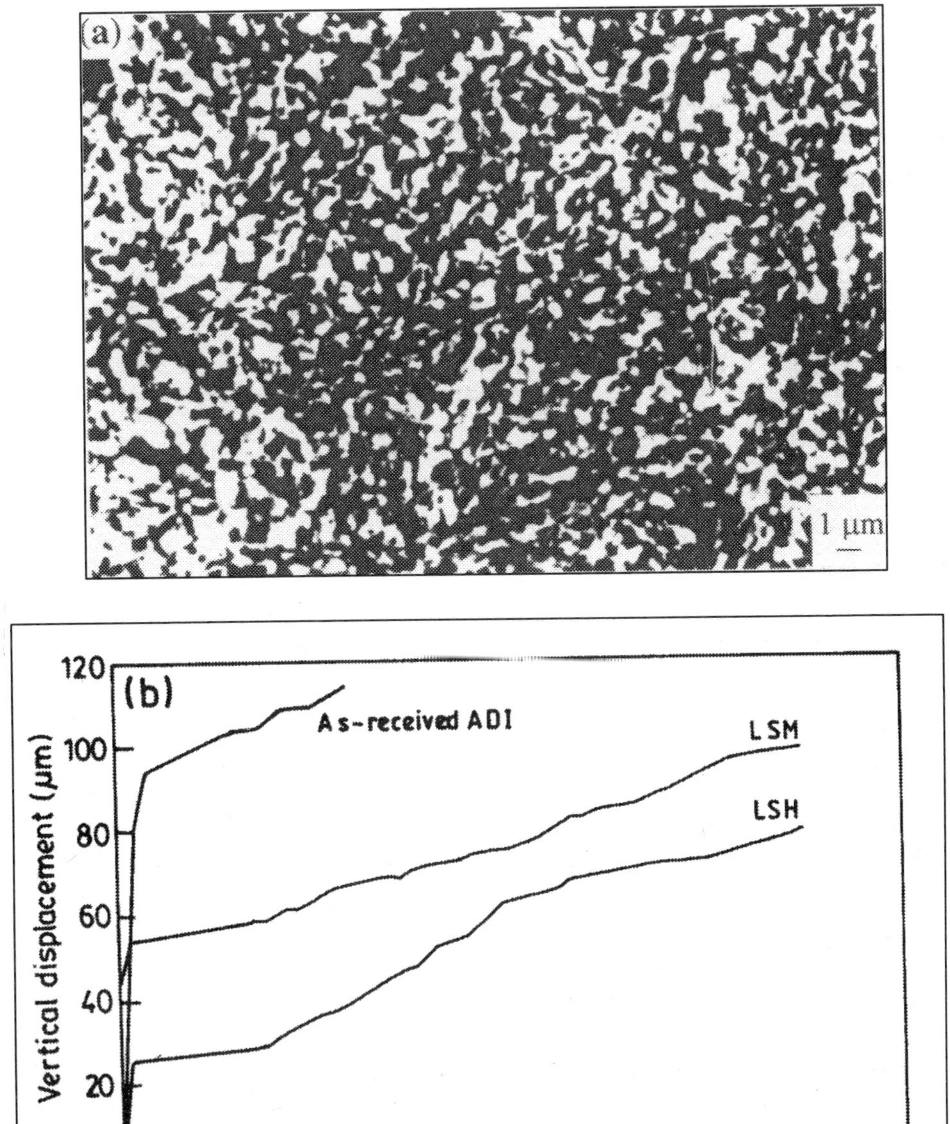

Figure 6.14: (a) A Predominantly Martensitic Microstructure of the Laser Hardened Sample Surface, and (b) Kinetics of Wear in Terms of Vertical Displacement of the Pin-head as a Function of Sliding Distance During Adhesive Wear Testing of Austempered Ductile Iron (ADI) Subjected to LSM and LSH with a Pin-on-Disc Machine in Dry Condition at 5 kg Load. For details, please see [40,64].

the service life of components subjected to severe conditions of wear, oxidation and corrosion. In this section, the scope and current status of understanding regarding the application of LSE to enhance surface dependent properties of non-ferrous alloys will be briefly reviewed. For a ready reference, Table 6.3 enlists a selected number of notable studies done in the recent time in this area [46-57].

Table 6.3: Selected Studies on Laser Surface Engineering of Non-ferrous Alloys

System	*Year*	*LSE*	*Results*	*Reference*
Al-6013 + SiC_p composite	1999	LSM	Rapid solidification of LSM refines the surface microstructure and improves the corrosion resistance.	Yue *et al.* [46]
6061 Al alloy (Al+TiC)	2001	LSC LSA	TiC+Al composite coating improves oxidation resistance in 200-600°C	Katipelli and Dahotre [47]
Al+SiC	1999	LSS	Both hardness and wear resistance improve by LSS due to residual compressive stress	Schnick *et al.* [48]
Al (Al_3Ti)	1999	LSC	Composite intermetallic layer on surface enhances wear resistance	Uenishi and Kobayashi [49]
Cu-Cr-Fe	2000	LSR	LSR improves hardness and wear resistance and reduces friction.	Geng *et al.* [50]
Cu (Cr)	1999	LSA	LSA enhances resistance to adhesive/ abrasive wear and erosion due to solid solution and dispersion hardening.	Dutta Majumdar and Manna [51]
Mg and its alloys				
Mg+SiC (Al-Si)	2001	LSC	LSC improves corrosion resistance. LSC parameters have strong influence.	Wang and Yue [52]
AZ91D and AM60B Mg alloys	2001	LSM	Uniform and refined microstructure improves corrosion resistance.	Dube *et al.* [53]
AZ91 Mg alloy	1999	LSM	Pulsed excimer laser irradiation improves corrosion resistance.	Schippman *et al.* [54]
MEZ, Mg alloy	2002	LSA	Laser parameters have influence on wear resistance	Dutta Majumdar *et al.* [55]
Ti (Si,Al,Si+Al)	2000	LSA	LSA with Si is more effective than Al/Si+Al to improve wear resistance.	Dutta Majumdar *et al.* [56]
Ti (Si,Al,Si+Al)	1999	LSA	LSA with Si/Si+Al is more effective than Al in enhancing cyclic oxidation resistance between ambience/750°C	Dutta Majumdar *et al.* [57]

Several attempts have been made to improve corrosion resistance of Al-based alloys by LSM [46], LSA [47] and LSC/LSR [49]. While grain refinement and micro structural homogeneity are responsible for better corrosion resistance after LSM, dispersion of inter-metallic and/or amorphous phases seems beneficial for corrosion resistance following LSC/LSR. Katipelli and Dahotre [47] have achieved considerable improvement in oxidation resistance of 6061 Al alloy by LSA with Al + TiC.

Addition of ceramic or inter-metallic particles to Al-alloys by LSC or LSA seems effective in improving wear resistance [48,49]. Usually, the alloyed or clad layer possesses a high hardness and consists of inter-metallic or amorphous phases [48,49]. Schnick *et al.* [48] induced residual compressive stress and achieved a significant improvement in wear resistance by laser surface shocking in HVOF spray coated SiC on Al.

Dutta-Majumdar and Manna [51] attempted to enhance the wear and erosion resistance of Cu by laser surface alloying with Cr (electrodeposited with 10 and 20 μm thickness, t_z). Total Cr content (X_{Cr}), Cr in the form of precipitates (f_{Cr}) and Cr in solid solution with Cu (Cu_{Cr}) were determined by energy dispersive spectrometry, optical microscope and X-ray diffraction technique, respectively. LSA extended the solid solubility of Cr in Cu as high as 4.5 at. per cent as compared to 1 at. per cent under equilibrium condition.

Figure 6.15a shows the variation of X_{Cr}, Cu_{Cr}, and f_{Cr} as a function of E for $t_z = 20$ μm. The higher the energy density (E), the smaller the X_{Cr}, Cu_{Cr}, and f_{Cr} in the alloyed zone. The micro hardness of the alloyed zone was found to improve significantly (as high as 225 VHN). Since hardness is related to Cr present in solid solution and dispersed as precipitates in the matrix, the variation of average micro hardness of the alloyed zone (H_v^{av}) as a function of the X_{Cr}, Cu_{Cr}, and w_{Cr} shows that hardness increases for all these micro structural factors, especially for Cu_{Cr} (Figure 6.15b). f_{Cr} was converted to the corresponding weight fraction (w_{Cr}) by the simple relation: $w_{Cr} = X_{Cr} - Cu_{Cr}$. Thus, suitable LSA parameters must be chosen to obtain the desired hardness.

Figure 6.16a shows the variation of scratch depth (z_{sc}) as a function of load (L) for pure Cu as well as laser surface alloyed Cu with Cr [Cu(Cr)] subjected to scratching with an oscillating steel ball in a computer controlled scratch tester. It is evident that the rate of increase of z_{sc} with both load and number of scratches is much higher in pure Cu than that in Cu(Cr). Figure 6.16b compares the kinetics of material loss (Δm) of Cu(Cr) lased with 1590 MW/m^2 and 0.08 s power and interaction time, respectively ($t_z = 20$ μm) with that of Cu as a function of time (t) under an accelerated erosive wear condition conducted both at room and high temperature in a slurry bath containing 20 wt. per cent sand. In addition to reducing the magnitude of Δm by over an order of magnitude, LSA has significantly decreased the kinetics of erosion loss in Cu(Cr) than that in Cu under comparable conditions. Though the extent of material loss increases with an increase in T for both pure Cu and Cu(Cr), the rate of erosion loss for Cu(Cr) is negligible as compared to a substantial change in Δm with T for pure Cu, especially beyond 370 K.

Uniform and refined surface microstructure by LSM seems quite effective in improving corrosion resistance in AZ91D/AM60B [53], AZ91 [54] and other commercial Mg-based alloys. Recently, Wang and Wue [52] have achieved significant improvement in corrosion resistance of SiC dispersed Mg by LSC with Al-12Si alloy layer. Similarly, Dutta Majumdar *et al.* [55] have been able to increase the hardness of a magnesium alloy (MEZ) by LSA with 76Al+24Mn and 55Al+45Mn. Figures 6.17a,b present the micro hardness profiles for laser surface alloyed MEZ specimens with (a) 76Al+24Mn and (b) 45Al+55Mn (lased under different processing conditions) as a

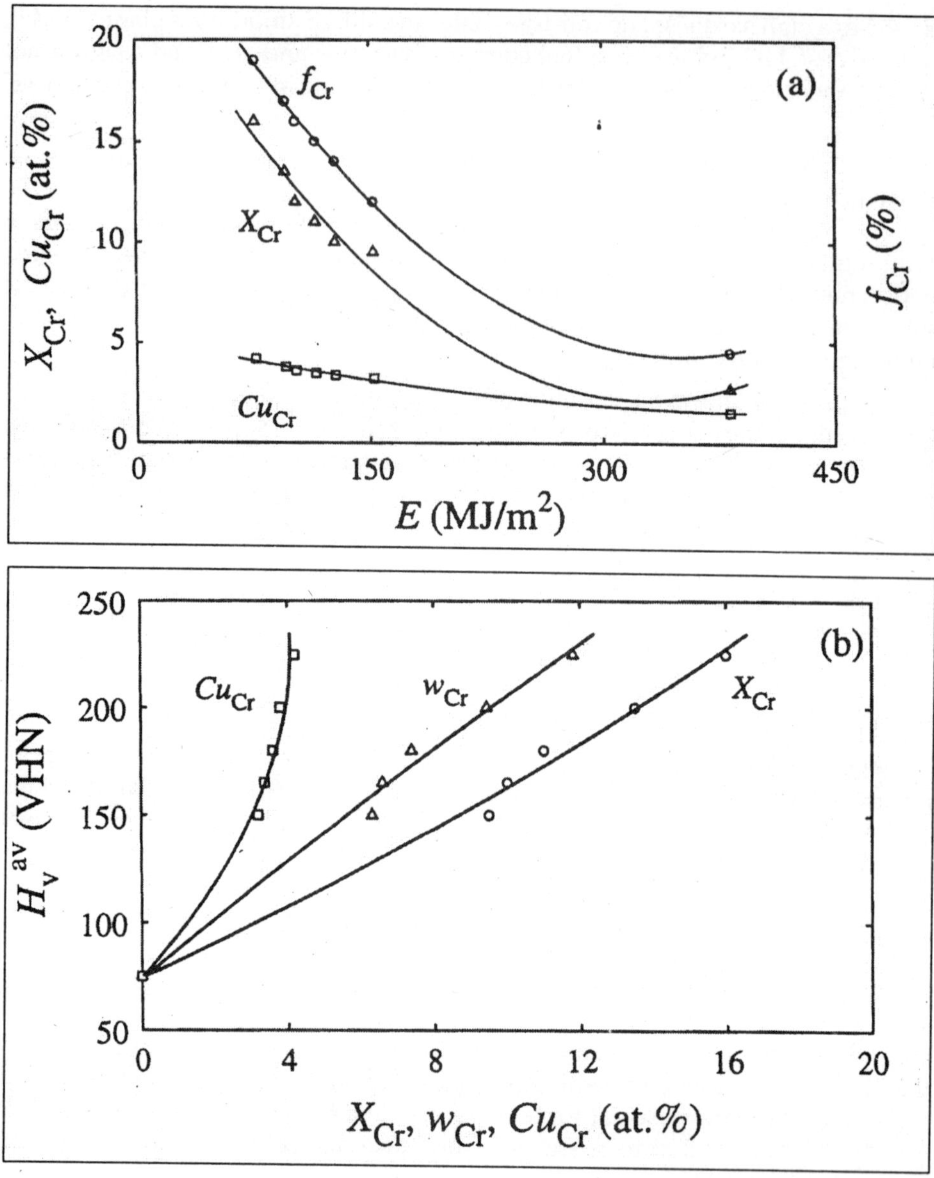

Figure 6.15: (a) Variation of Cr-Content (X_{Cr}), Dissolved Cr in Solid Solution (Cu_{Cr}) and Volume Fraction of Cr-precipitates (f_{Cr}) as a Function of Energy Density (E), and (b) Variation of H_v^{av} as a Function of X_{Cr}, Cu_{Cr} and f_{Cr}. For details, please see [51].

function of depth from the surface measured on the cross sectional plane. Micro hardness of the alloyed zone has increased as high as 350 VHN as compared to 35 VHN of the as-received one. In both the cases, micro hardness of the alloyed zone is the maximum near the surface and decreases with the distance from the surface,

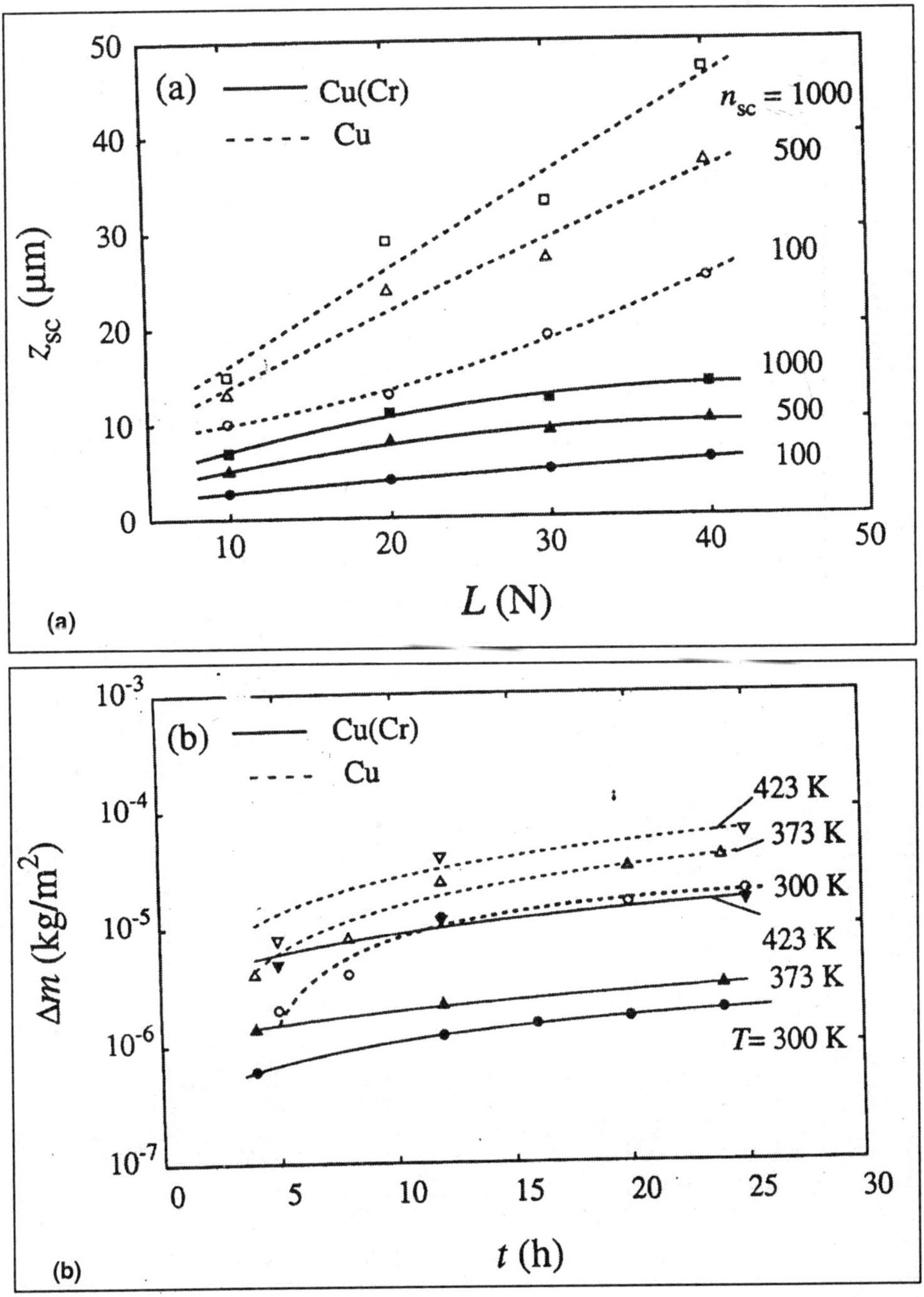

Figure 6.16: (a) Variation of Scratch Depth (z_{sc}) with Load (L) for Different Numbers of Oscillations (n_{sc}) for Pure Cu and Laser Alloyed Cu(Cr), and (b) Comparison Between Material Loss per Unit Area (Δm) Due to Erosion of Cu(Cr) and Cu as a Function of Time (t) at Different Temperature (T) Due to Erosion in Flowing SiO_2 Dispersed Oil Medium. For details please see [51].

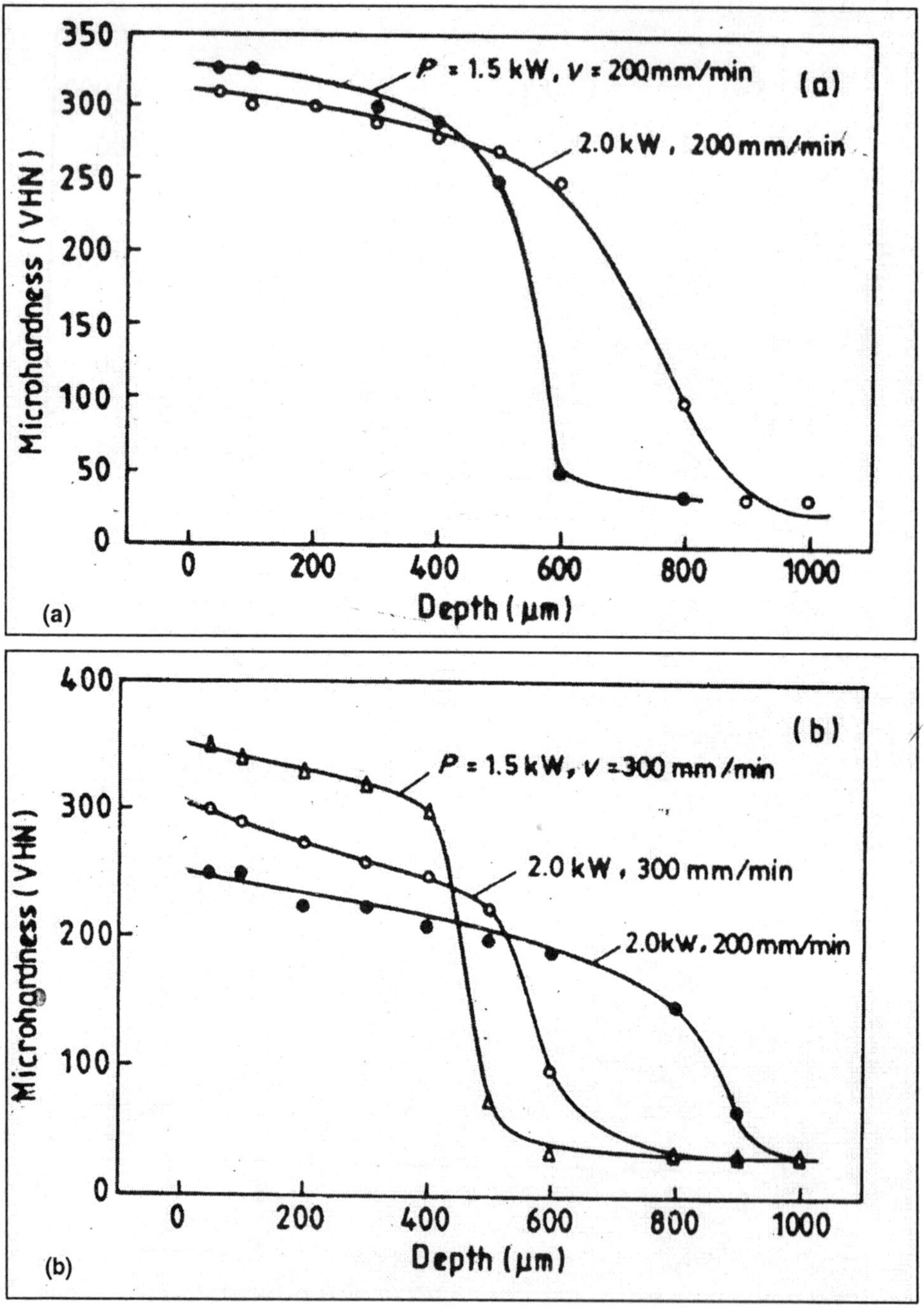

Figure 6.17: Microhardness Profiles as a Function of Depth from the Surface Measured on the Cross Sectional Plane in Laser Surface Alloyed MEZ with (a) 76Al+24Mn and (b) 45Al+55Mn (Processed under different lasing conditions as written in respective Figures)

perhaps due to the decrease in alloying content and volume fraction of inter-metallic phases with depth. Furthermore, micro-hardness decreases with increase in applied laser power.

LSA of Ti with Si, Al and Si+Al (with a ratio of 3:1 and 1:3, respectively) was conducted to improve the wear and high temperature oxidation resistance of Ti by LSA [56,57]. Oxidation studies conducted at 873–1023 K showed that LSA of Ti with Si and Si+Al significantly improved the isothermal oxidation resistance (Figure 6.18).

In addition to oxidation, LSA of Ti with Si or Si+Al was beneficial to enhance wear resistance was also studied. Figure 19b shows the variation of depth of scratching (z_w) with load (L) due to scratching of pure Ti, Ti(Si), Ti(Al) and Ti(3Si+Al) with a hardened steel ball. It may be noted that z_w varies linearly with L for all the cases. The effect of L on z_{sc} is more prominent at higher number of scratching ($n_{sc} \geq 1000$) than that at a lower value of the same (= 25). Under comparable conditions of scratching, Ti undergoes the most rapid wear loss followed by that in Ti(Al), Ti(3Si+Al) and Ti(Si). Ti(Si) undergoes the minimum wear loss. The improved wear resistance of laser surface alloyed Ti with Si was attributed to the formation of a hard Ti_5Si_3 precipitates in the alloyed zone [56].

LSE is useful also for enhancing functional properties and micro-components. Manna *et al.* [58-60] made an attempt to develop an appropriate material for neural stimulation electrode by LSA of Ti with Ir that can mimic the normal spatio-temporal pattern of neuronal activation by reversible charge transfer. The usual electrode made of iridium is expensive, brittle and not amenable to miniaturization by plastic deformation. On the other hand, titanium is cheaper, bio compatible and amenable to drawing/etching. Figure 6.19a shows the indigenous set up used for fabricating the electrode by LSA. Judicious combination of laser parameters, powder composition and post LSA etching were able to develop the desired microstructure (Figure 6.19b). Though charge density could not be measured due to exceedingly uneven surface intentionally developed by special etching (meant for increasing the surface area), the total charge was comparable to that of pure iridium in appropriate solution [60]. This study underlines the importance of micro structural evolution during the rapid cooling of LSA in deriving the desired property of the component.

Mathematical Modelling on LSE

Studies on mathematical modelling of LSE mostly focus on predicting the thermal profile and composition in the laser irradiated volume. A number of mathematical approaches in modelling different laser assisted surface modification and processing phenomena has been reviewed by Mazumdar [13]. While knowledge on thermal history and solute distribution is essential for predicting the properties, attempts have seldom been made to correlate the microstructure and property of the laser treated zone with the LSE parameters. Manna and Dutta Majumdar [61] developed a one dimensional heat transfer model based on explicit finite difference method to predict the thermal history (*i.e.* temperature profile, thermal gradient, cooling rate and solid-liquid interface velocity) and hence, the microstructure of the alloyed zone developed by laser surface alloying. Figure 6.20a shows the temperature (T) profile as a function of time (t) during LSA of AISI 304 stainless steel with 100 μm thick pre-

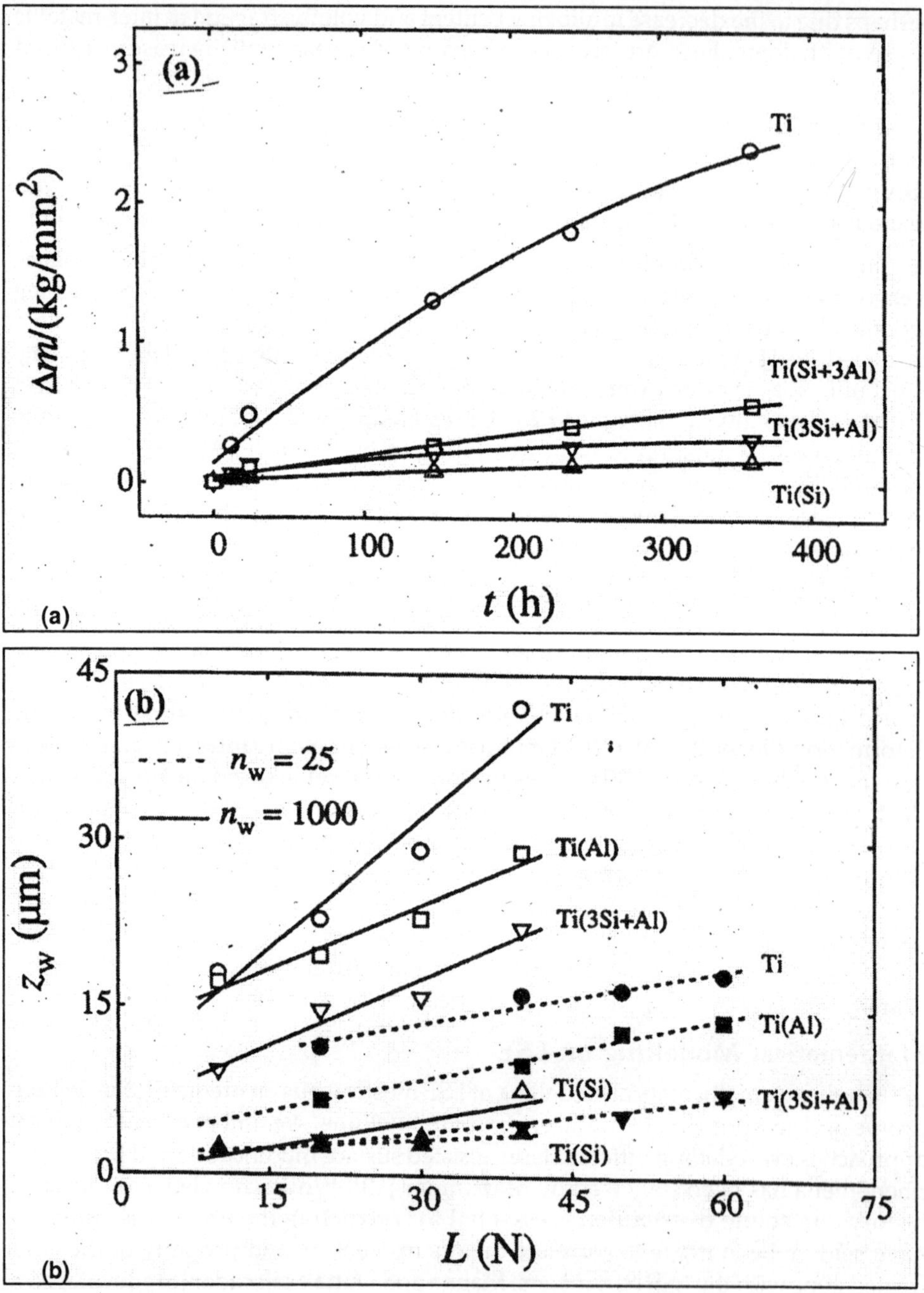

Figure 6.18: (a) Kinetics of Oxidation Expressed as Total Weight Gain per Unit Area (Δm) as a Function of Time (t), and (b) Kinetics of Wear in Scratch Test Expressed as Depth of Scratch (z_w) as a Function of Load (L) for as-Received and Laser Surface Alloyed Ti with Si, Al and Si+Al. For details about LSA, oxidation and wear test, please see [56,57].

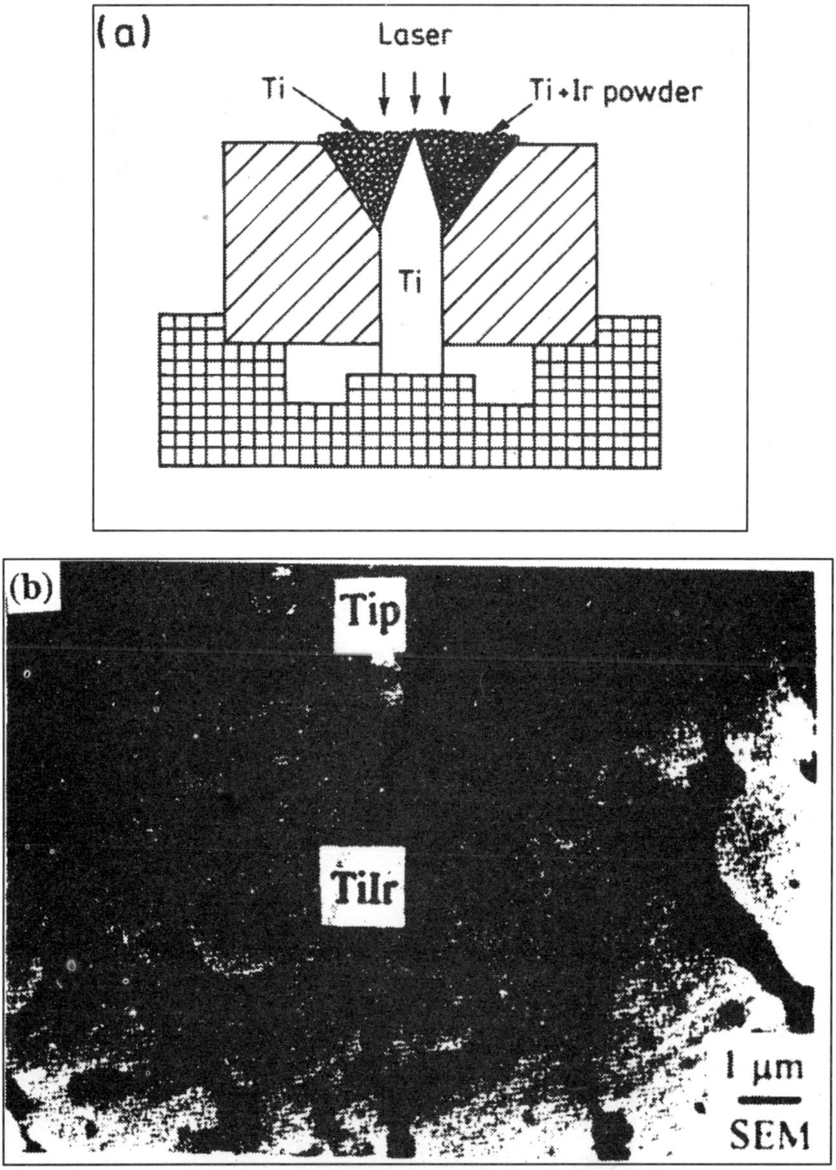

Figure 6.19: (a) Schematic Diagram Showing the Set Up Used for LSA of Ti with Ir (or Ti+Ir) by CO_2 Laser Pulse for Developing Neural Stimulation Electrode, and (b) Microstructure of the Alloyed Zone Showing a 3 Fold Increase in Surface Area Following Special Etching [58,59].

deposited Mo using a CW-CO_2 laser delivering 1800 MW/m^2 power irradiation with a 25 ms interaction time. The calculation considered the effect of intermixing between the bi-metallic layer after melting and temperature dependence of concerned material properties. The corresponding thermal quenching rate has been shown in

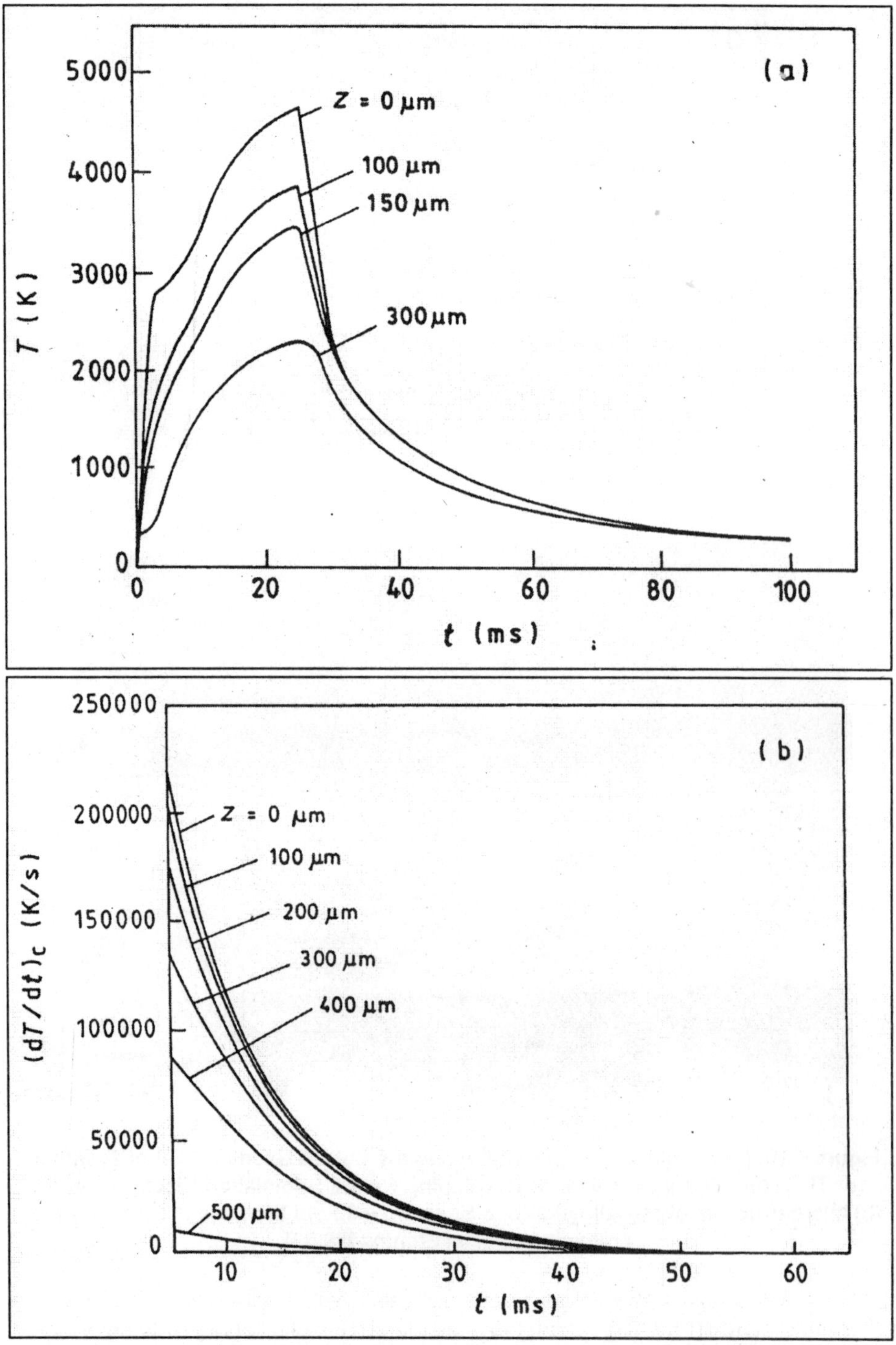

Figure 6.20: (a) Temperature (T) Distribution as a Function of Time (t) for SS(Mo) at Different Levels of Depth (z), and (b) Variation of Corresponding Cooling Rate $(dT/dt)_c$ as a Function of Time (t) at Different z During LSA of AISI 304 Stainless Steel with Mo. For details, please see [62].

Figure 6.20b. The model allows calculation of other thermal parameters like thermal gradient, solidification velocity, *etc*. that are useful in predicting the microstructure [62]. Similar model could be extended to two-dimensional radially symmetric heat transfer condition [63].

Roy and Manna [64] have used a simple analytical approach, based on the treatment of Ashby and Easterling [65], to predict the thermal profile in LSH of austempered ductile iron and explain the phenomenon of partial or complete 'localized melting' of graphite nodules without liquefaction of the surrounding ferritic matrix. Figure 6.21 reveals that a graphite nodule has undergone partial melting during LSH of austempered ductile iron with 650 W power and 60mm/s scan speed. Note that melting initiates at the graphite-matrix interface and assumes only a part of the nodule leaving the core unaffected. The extent or width of this incipient fusion primarily depends on the thermal cycle and concomitant carbon diffusion profile from the nodule into the matrix. The mathematical model to predict the thermal and composition profile, and hence, the width of annular molten region by Roy and Manna [64] could establish a correlation between the melt width (y_m) and laser parameters used for LSH. Figure 6.22 shows the variation of y_m as a function of depth (z) from the surface predicted for two predetermined conditions of laser surface hardening. It is evident that y_m decreases as z increases bearing a linear relationship between them. The experimental data (open symbols) are fairly close to the predicted trend of the model, particularly for higher laser power.

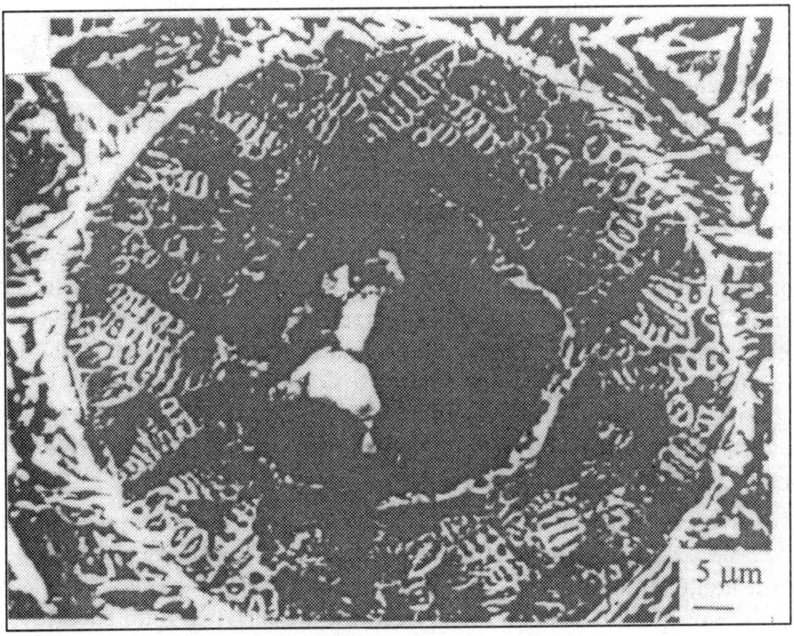

Figure 6.21: Partial Melting of a Graphite Nodule Embedded in the Ferritic/Martensitic Matrix Following LSH with 650 W Power and 60 mm/s Scan Speed. Note that melting initiated at the graphite-matrix interface and consumed nearly half the nodule from the circumference.

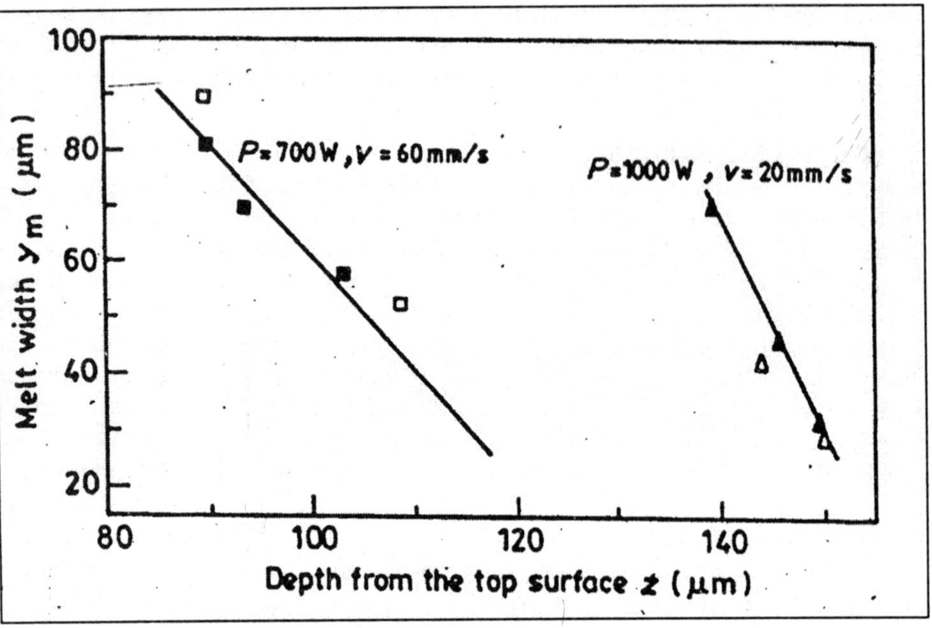

Figure 6.22: Variation of Predicted Melts Widths (y_m) Around Graphite Nodules as a Function of Depth (z) from the Surface. Open and solid symbols represent experimental and predicted data, respectively [64].

Laser Surface Amorphization: A Challenge

It is known that environmental degradation by corrosion and oxidation alone accounts for more than 30 per cent discard and loss of engineering components of all dimensions from bridges/railway-carriages to needles/bolts made of metals and alloys. While replacing a bolt is easy and inexpensive, a similar problem with an orthopaedic implant or turbine blade may cause serious trauma and endanger life. Corrosion preferentially initiates at defects and heterogeneity in a crystalline material. An amorphous state is devoid of both crystalline anisotropy and inter-crystalline defects and can retain more than equilibrium limit of solute content. Thus, amorphous microstructure is generally more resistant to corrosion and oxidation. However, an amorphous or glassy material is too brittle to use for a structure. Thus, a crystalline solid with an amorphous surface or over-layer is obviously superior in terms of resistance to any galvanic attack.

LSE with its inherent extreme heating/cooling rates, thermal gradient and re-solidification-velocity may enable retention of an amorphous or glassy surface on top of a metallic component. However, suppression of epitaxial nucleation and prevention of growth of crystallites from the underlying solid substrate must be insured to achieve this laser assisted surface amorphization. In this regard, Dutta Majumdar and Manna [66] have recently reviewed the theory, status and issues related to laser surface amorphization. It seems that LSC of a bulk amorphous alloy and not LSA would be the appropriate approach for achieving an amorphous surface on a bulk crystalline component. For practical applications as an industrial process, particularly for large-sized and complex-shaped components, the following difficulties and hurdles are likely to be encountered in an attempt to amorphize the surface:

- Epitaxial nucleation and growth of crystalline phases from the well compatible liquid-substrate interface as a competitive process to partition less amorphization.
- Re-crystallization or nucleation of crystalline phase at the overlap regions between adjacent laser tracks due to thermal activation.
- Possibility of diffusion-controlled partial crystallization from the melt or adjacent amorphous regions due to a relatively longer interaction time necessary for a CW-CO_2 laser assisted LSE process.
- Deviation from the intended melt composition due to a possible compositional dilution effect in the melt arising out of a higher interaction time and hence higher melt depth.

Thus, laser surface amorphization of bulk components is an example of the challenges ahead of the materials scientists and engineers that warrant immediate attention. At the moment, the present author is pursuing this problem in collaboration with University of Tennessee, Knoxville, USA through a DST-NSF research grant

Summary and Concluding Remarks

High power laser is a clean, contact-less and well-characterized source of directed energy that can perform several types of material processing with immense flexibility and novelty. Laser surface engineering (LSE) that concerns enhancing one of the surface dependent properties like hardness, friction, fatigue and resistance to wear and corrosion, is one of them. In this discourse, an attempt has been made to explain the history of evolution of LSE as a tool for modifying the surface microstructure and composition of engineering materials. The fundamentals as well as specific examples of research studies have been cited to underline the main issues involved. For further studies, the participants are encouraged to consult the relevant books [1,6-8] and review articles [4,5,67] exclusively dealing with LSE and its application.

Apart from structural applications, LSE could be equally useful in enhancing functional properties like magnetism, emission/absorption characteristics useful in sensors, actuators and microelectronic devices. These applications require monolithic or functionally/compositionally graded microstructures confined to smaller dimensions. The LSE processes described in this lecture are equally applicable to such functional applications, provided the correct laser parameters (wavelength, mode, energy, *etc.*) are selected for the given materials (semiconductor, polymer, *etc.*).

In India, the major LSE facilities exist in CAT-Indore, ARCI-Hyderabad and BARC-Mumbai. Since laser is an expensive facility to install and maintain, it is logical that exploratory investigations are carried out in collaborations with any of these organizations. In fact, most of our own results cited in this lecture were carried out in such collaborative effort (both in India and abroad), as we do not have the necessary facility at the IIT-Kharagpur. In any case, it is time that the materials community in India pays adequate attention to extend the research efforts in LSE from laboratory to the shop floor of industry. Particularly, immediate and justifiable opportunities exist in the automobile, bio-medical and machine tool sectors. Since this lecture was focused on LSE, areas like machining, joining and manufacturing were not addressed to. However, application of laser as a tool is already more successful in many of those areas. In this regard, the participants of this course may refer to a recent review on the status of application of laser for material processing in India by the present author [68].

Finally, the major outstanding issues concerning LSE are utilization of shorter wavelength for precision engineering, development of compositionally/functionally graded microstructures, elimination of structural defects and combining of dissimilar materials [68].

Acknowledgement

The authors are indebted to all their students and collaborators for their supports and help. Financial supports from various funding agencies like DST, N. Delhi; CSIR, N. Delhi; BRNS; Bombay and AICTE; N. Delhi at different stages of the work is gratefully acknowledged.

References

1. P. A. Molian, Surface Modification Technologies-An Engineers Guide, (ed. T. S. Sudarshan), Marcel Dekker Inc., New York, (1989), p. 421.
2. W. W. Duley, Laser Surface Treatment of Metals, (eds. C. W. Draper and P. Mazzoldi) NATO-ASI Series (E) No.: 115, Martinus Nijhoff Publishers, Boston (1986), p. 3.
3. B. L. Mordike, Materials Science and Technology, (eds. R. W. Cahn, P. Haasen, E. J. Kramer), vol.15, VCH, Weinheim (1993), p. 111.
4. C. W. Draper and J. M. Poate, Inter. Met. Rev., 30 (1985) 85-108.
5. W. Draper and C. A. Ewing, J. Mater. Sci., 19 (1984) 3815.
6. W. M. Steen (ed.), Laser Material Processing, Springer Verlag, N. York, (1991).
7. C. W. Draper, Laser and Electron Beam Processing of Materials, (eds. C. W. White and P. S. Peercy), Academic Press, New York, (1980), p. 721.
8. C. W. White and M. J. Aziz, Surface Alloying by Ion, Electron and Laser Beams, (eds. L. E. Rehn, S. T. Picraux and H. Wiedersich), ASM, Metals Park, Ohio, (1987), p. 19.
9. S. T. Picraux and D. M. Follstaedt, Laser-Solid Interactions and Transient Thermal Processing of Materials (eds. J. Narayan, W. L. Brown and R. A. Lemons), North-Holland, New York, (1983), p. 751.
10. J. H. Perepezko and W. J. Boettinger, Surface Alloying by Ion, Electron and Laser Beams, (eds. L. E. Rehn, S. T. Picraux and H. Wiedersich), ASM, Metals Park, Ohio, (1987), p. 51.

11. T. H. Maiman, *Nature*, 187 (1960) 493.
12. C. K. N. Patel, Physics Review, 136 A (1964) 1187; C. K. N. Patel, Applied Physics Letters, 7 (1965) 15.
13. J. Mazumdar, Lasers for Materials Processing, (ed. M. Bass), North Holland Pub. Co., New York, (1983), p. 113.
14. P. Psyllaki, R. Oltra, Mater. Sci. Eng. A, 282 (2000) 145-152.
15. Y. F. Lu, Y. P. Lee and M. S. Zhou, J. Appl. Phys., 83 (1998) 1677-1684.
16. A. Tsunemi, K. Hagiwara, N. Saito, K. Nagasaka, Y. Miyamoto, O. Suto, H. Tashiro, Applied Physics A, 63 (1996) 435-439.
17. Y. F. Lu and H. Qiu, J. Appl. Phys., 88 (2000) 1082-1087.
18. Y. F. Lu, W. D. Song, M. H. Hong, T. C. Chong and T. S. Low, Appl. Phys., A64 (1997) 573-578.
19. G. Seifert, M. Kaempfe, K. J. Berg and H. Graener, Appl. Phys., B71 (2000) 795-800.
20. P. Muggli, R. Brogle and C. Joshi, J. Optical Soc. America B, 12 (1995) 553-558.
21. L. Chefneux, Y. Renauld, B. Dauby, D. Bouquegneau, and M. Gerday, Ten Years of Laser Texturing in Cockerill Sambre, Processing of METEC Congress'94: 2nd European Continuous Casting Conference, 6th International Rolling Conference 1994, (Dusseldorf: Verein Deutscher Eisenhuttenleute, Germany), pp. 356-362.
22. I. Ursu, I. N. Michailesku, A. Popa, A. M. Prokhorov, V. P. Ageev, A. A. Gorbunov, and V. I. Konov, Journal of Applied Physics, 58 (1985) 3909-3913.
23. J. Liu, Lasers in Engineering 3 (1994) 67-72.
24. J. Singh and J. Mazumder, Metall. Trans., 18A (1987) 313.
25. J. De Damborenea and A. J. Vasquez, Journal of Materials Science, 28 (1993) 4775.
26. V. M. Weerasinghe and W. M. Steen, Applied Laser Tooling, Perez-Amour and Soares, eds., Martinus Nijhoff, Dordrecht, Holland, 1984.
27. N. Parvathavarthini, R. V. Subbarao, S. Kumar, R. K. Dayal, H. S. Khatak, Journal of Materials Engineering and Performance, 10 (2001) 5.
28. A. Conde, I. Garcia and J. J. Damborenea, Corr. Sci., 43 (2001) 817-828.
29. C. T. Kwok, F. T. Cheng and H. C. Man, Surf. Coat. Tech., 145 (2001) 206-214.
30. L. W. Tsay, T. Y. Yang and M. C. Young, Mater. Sci. Eng. A, 311 (2001) 64-73.
31. P. Peyre, C. Braham, J. Ledion, L. Berthe and R. Fabbro, J. Mater. Eng. Perform., 9 (2000) 656-662.
32. C. T. Kwok, F. T. Cheng and H. C. Man, Mater. Sci. Eng. A, 290 (2000) 74-88; and C. T. Kwok, F. T. Cheng and H. C. Man, Mater. Sci. Eng. A, 290 (2000) 55-73.
33. Y. Isshiki, J. Shi, H. Nakai and M. Hashimoto, Applied Physics, A70 (2000) 651-656; and Y. Isshiki, J. Shi, H. Nakai and M. Hashimoto, Applied Physics, A70 (2000) 395-402.

34. J. Dutta Majumdar and I. Manna, Mater. Sci. Eng. A, 267 (1999) 50-59.
35. S. R. Pillai, P. Shankar, R. V. Subba-Rao, N. B. Sivai and S. Kumaravel, Mater. Sci. Tech., 17 (2001) 1249-1252.
36. D. Raybould, M. Meola, R. Bye and S. K. Das, Mater. Sci. Eng., A241 (1998) 191-201.
37. A. Agarwal, L. R. Katipelli and N. B. Dahotre, Metal. Mater. Trans. A, 31 (2000) 461-473.
38. F. T. Cheng, C. T. Kwok and H. C. Man, Surf. Coat. Tech., 139 (2001) 14-24.
39. X. L. Wu and Y. S. Hong, Metal. Mater. Trans. A, 31(2000) 3123-3127.
40. A. Roy and I. Manna, Mater. Sci. Eng. A297 (2001) 85-93.
41. A. Agarwal and N. B. Dahotre, Wear, 240 (2000) 144-151; and A. Agarwal and N. B. Dahotre, Metal. and Mater. Trans. A, 31(2000) 401-408.
42. S. Tondu, T. Schnick, L. Pawlowski, B. Wielage, S. Steinhauser and L. Sabatier, Surf. Coat. Tech., 123 (2000) 247-251.
43. J. M. Pelletier, E. Sauger, Y. Gachon and A. B. Vannes, J. Mater. Sci., 34 (1999) 2955-2969.
44. I. Manna, K. Kondala Rao, S. K. Roy and K. G. Watkins, in: Surface Engineering in Materials Science I (conf. proc.), (eds. S. Seal, N. B. Dahotre, J. J. Moore and B. Mishra), TMS, Waarandale, PA 15086-7528, USA, (2000), pp. 367–376.
45. D. W. Zhang, T. C. Lei, J. G. Zhang and J. H. Ouyang, Surf. Coat. Tech., 115 (1999) 176-83.
46. T. M. Yue, Y. X. Wu and H. C. Man, Surf. Coat. Tech., 114 (1999) 13-18.
47. L. R. Katipelli and N. B. Dahotre, Mater. Sci. Tech., 17 (2001) 1061-1068.
48. T. Schnick, S. Tondu, P. Peyre, L. Pawlowski, S. Steinhauser, B. Wielage, U. Hofmann and E. Bartnicki, Journal of Thermal Spray Technology, 8 (1999) 296-300.
49. K. Uenishi and K. F. Kobayashi, Intermetallics, 7 (1999) 553-559.
50. H. R. Geng, Y. Liu, C. Z. Chen, M. H. Sun and Y. Q. Gao, Mater. Sci. Tech., 16 (2000) 564-567.
51. J. Dutta-Majumdar and I. Manna, Mater. Sci. Eng. A, 268 (1999) 227-235; and J. Dutta Majumdar and I. Manna, Mater. Sci. Eng. A, 268 (1999) 216-226.
52. A. H. Wang and T. M. Yue, Composites Science and Technology, 61 (2001) 1549-1554.
53. D. Dube, M. Fiset, A. Couture and I. Nakatsugawa, Mater. Sci. Eng. A, 299 (2001) 38-45.
54. D. Schippman, A. Weisheit and B. L. Mordike, Surf. Eng., 15 (1999) 23-26.
55. J. Dutta Majumdar, B. L. Mordike, R. Galun and I. Manna, Lasers in Engineering, 12 (2002) 171-190.

56. J. Dutta Majumdar, B. L. Mordike and I. Manna, Wear, 242 (2000) 18-27.
57. J. Dutta-Majumdar; A. Weisheit, B. L. Mordike, I. Manna, Mater. Sci. Eng. A, 266 (1999) 123-134.
58. I. Manna, W. M. Steen and K. G. Watkins, in: Surface Engineering in Materials Science I (conf. proc.), (eds. S. Seal, N. B. Dahotre, J. J. Moore and B. Mishra), TMS, Waarandale, PA 15086-7528, USA, (2000), pp. 377–384.
59. I. Manna, W. M. Steen and K. G. Watkins, Scripta Mater., 37 (1997) 561-568.
60. K. G. Watkins, W. M. Steen, I. Manna, D. F. Williams, S. Rhodes, P. Mazzoldi, S. L. Russo, M. G. S. Ferreira, J. T. Rito, T. M. Silva and A. M. P. Simoes, in: ICALEO'96–Laser Materials Processing and Surface Modification (conf. proc.), (eds. W. Duley, K. Shibata and R. Poprawe), Laser Inst. of America, USA, vol. 81A, (1996), p. 37-46.
61. I. Manna and J. Dutta Majumdar, Z. Metallkde., 86 (1995) 362-364.
62. J. Dutta Majumdar and I. Manna, Lasers in Engineering, 12 (2002) 171-190.
63. I. Manna, J. Dutta Majumdar and P. K. Das, in: International Conference on Advances in Physical Metallurgy (ICPM) (conf. proc.)- 1994, (eds. S. Banerjee and R. V. Ramanujan), Gordon and Breach Publishers, New York, (1995), p. 49-54.
64. A. Roy and I. Manna, Optics and Lasers in Engineering, 34 (2001) 369 383.
65. M. F. Ashby and K. E. Easterling, Acta. Metall., 32 (1984) 1935.
66. J. Dutta Majumdar and I. Manna, Sadhana–*Special Issue* 28 (2003) 495–562.
67. Surface Engineering–Special issue, vol. 21(2005).
68. I. Manna, Surface Engineering–Special issue, 21 (2005) 81.

Chapter 7

Surface Engineering with Low Pressure Plasmas

S. Mukherjee
FCIPT, Institute for Plasma Research,
B15-17/P, GIDC, Gandhinagar – 382 044, Gujarat, India
E-mail: mukherji@ipr.res.in

ABSTRACT

Modification of material surfaces, popularly known as surface engineering, has gained significance in the recent years. Low pressure plasmas play a significant role in such a surface modification. The present article discusses basic aspects of low pressure plasma, low pressure plasma sources used in surface engineering. Case studies related with plasma nitriding, plasma immersion ion implantation, plasma enhanced chemical and physical vapour deposition will be presented. Emerging areas of research related with the above fields are discussed.

Keywords: Surface engineering, Low pressure plasmas, Surface modification, Low pressure plasma sources, Vapour deposition.

Introduction

Modification of material surfaces has been very popular since a long time. The necessity for doing such a modification came into existence because of protection of the material surface from being exposed to the destructive environment it is interacting with. The destructive environment can induce physical as well as chemical changes on the material surface and sometimes even in the bulk material. Some examples of damage caused by interaction of a material surface with its environment are,

- Loosing lustre (*e.g.* brass door handle)
- Developing cracks (*e.g.* wood when exposed to atmosphere)

- Getting worn out (*e.g.* drill bits)
- Getting oxidised (*e.g.* rust)
- Getting corroded (*e.g.* pipes carrying crude oil)

Surface engineering or modification of a material surface protects the surface from the damage caused by the environment. Painting is one of the oldest forms of surface engineering, which is still practiced today. Plasmas play an important role in surface engineering. Plasmas can work at elevated pressures (*e.g.* plasma spray) or at low pressure (*e.g.* plasma based chemical vapour deposition). The present article discusses low pressure application of plasmas in surface engineering.

The following sections consist of basics of plasma, low pressure plasma sources used in surface engineering, basics of ion surface interaction, case studies related with plasma nitriding, plasma immersion ion implantation, plasma enhanced chemical and physical vapour deposition will be presented. Emerging areas of research related with the above fields are discussed.

Basics of Plasma

Plasma is defined as the 4th state of matter, distinct from the solid, liquid, and gaseous states. In this state, matter comprises of positive and negative charged particles, produced by ionization of atoms in a gas. Plasma is also an electrically conducting medium, though it is quasi neutral in nature, *i.e.*, the net charge in a given volume is nearly zero, and thus plasma medium prefers to remain free of external applied fields. Plasma can also be defined as a collection of charged particles, where the interaction between the charged particles control the dynamics and leads to collective effects like oscillations, waves, instabilities and self organization [1-3]. Figure 7.1 demonstrates the entire plasma parameter space.

Plasma constitutes of

- Free electrons whose energy depends on the translational speed or equivalent temperature in eV (1 eV = 11,600 K)
- Ions of molecules/atoms, which have lost one or more electrons
- Negative ions–electrons attach to the atoms of electronegative gases like oxygen
- Atoms/molecules in their ground or excited states
- Photons created by interactions between the plasma species

The uniqueness of the plasma state is obtained by the ability of the charged particles to respond to electromagnetic energy fields. Each specie may have independent energy distribution, not necessarily in equilibrium with other species. Thus on the application of an external field, which is usually done in the case of low pressure plasma applications in surface engineering, the charge particles respond and arrange themselves so that the net field in the plasma is minimised.

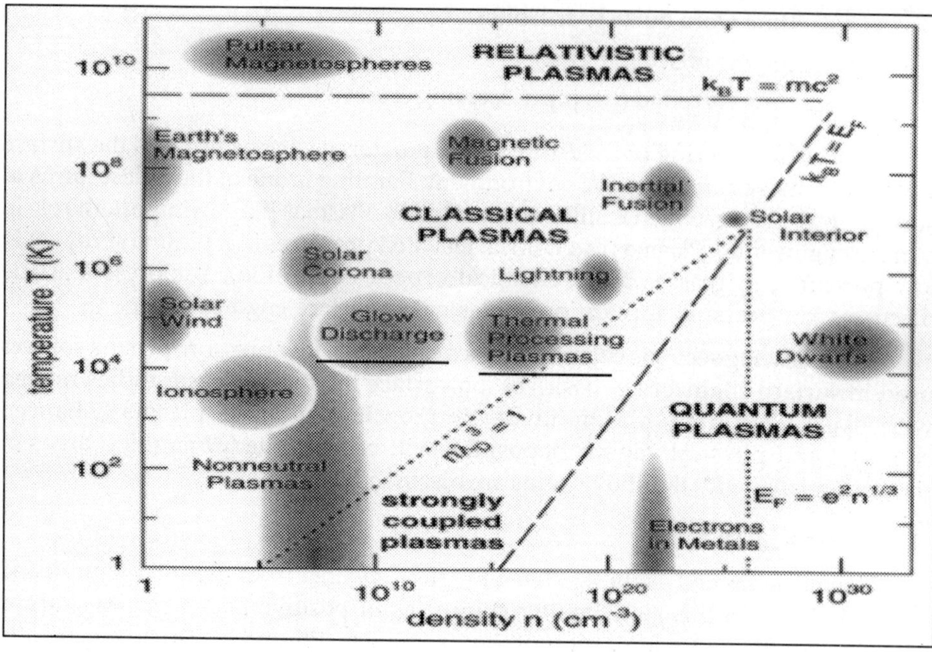

Figure 7.1: The Entire Plasma Parameter Space is Shown Above. Glow Discharge Plasmas and Thermal Processing Plasmas are used in plasma application (underlined). Surface Engineering is usually performed with Glow Discharge Plasmas.

Low pressure plasmas have electron temperature (T_e) much higher than ion (T_i) and neutral (T_n) temperature. Such plasmas are called non equilibrium plasmas. In the remaining text, the word plasma refers to non equilibrium plasma only. In such plasmas electrons move at very high velocities because of lower mass and high temperature, compared with ions and neutrals. For application in surface engineering, plasmas are usually formed inside a metallic vacuum chamber and hence the plasma is in contact with the chamber walls. Plasma develops a positive potential (~ few volts) positive with respect to the most positive biased electrode it is in contact with, which is usually the chamber wall. This positive potential is called the plasma potential.

When a substrate in contact with plasma is biased negative, the plasma shields this potential in a region where electric fields are present; this region is called the ion sheath region. If the potential difference in between the plasma and the substrate is ~ T_e, then the sheath is few Debye lengths in thickness. However for higher potential differences, Child's law or other laws govern the sheath thickness. A large amount of low pressure plasma surface engineering depends on how the sheath dynamics is controlled. The sheath dynamics determines, substrate placing, treatment uniformity, substrate bias power supply ratings and in some cases, even the plasma properties [3].

The ions enter the sheath at speed usually known as Bohm speed (also known as ion acoustic speed, ion sound velocity). The Bohm speed depends on the electron temperature and ion mass, and lie in between electron thermal and ion thermal speeds in plasmas. Inside the sheath there are ion-neutral charge exchange collisions at high pressure, which gives rise to a distribution of ion and neutral velocities at the substrate. Electron neutral collisions lead to ionization and excitation of neutrals. Such incidents usually take place in the sheath plasma boundary or in the bulk plasma [1,3].

The measurement of plasma properties, ion and electron density (usually they are equal for plasma comprising of a single type of ion) and electron temperature is needed, as it characterises the plasma and also determines the rate at which surface modification takes place. One of the most common diagnostic tools is the Langmuir probe. The Langmuir probe is a small metallic electrode, which is biased positive and negative with respect to the reference (usually the experimental chamber). For probe bias at plasma potential or above, in an ideal plasma, the probe draws a constant negative current known as the electron saturation current. This current has information of electron density of the plasma. When the probe is biased negative than the plasma potential, then only energetic electrons can reach the probe and thus the electron current reduces. When the probe is so much negatively biased that most of the electrons are repelled and the amount of electron current reaches the probe is exactly equal to the ion current collected, then the net current collected by the probe is zero. The potential at which this happens is known as the floating potential. An insulating substrate or an unbiased metallic substrate in contact with the plasma will be biased at floating potential. The difference between the plasma potential and the floating potential is proportional to $T_{e.}$ If the probe is biased more negative, then all the electrons are repelled, and the probe collects only ions. This current is known as ion saturation current. It needs to be remembered that this is the highest ion current that the plasma can supply under normal circumstances [3].

The plasmas used in surface engineering are usually multi species plasmas. Characterising such plasmas only with Langmuir probes and finding the ion densities are always not easy. In such cases, optical emission spectroscopy is used in conjunction with Langmuir probes. With the help of spectroscopy, the ion species in the plasma can be identified. Additionally, information of plasma density, T_e and T_i can also be obtained [4].

Low Pressure Plasma Sources Used in Surface Engineering

One of the simplest plasma sources but widely used for surface engineering is the glow discharge plasma source. The name originates from the glow formed around the cathode, which also fills the space between the anode and cathode. In a dc glow discharge, a potential difference is applied between the anode and the cathode. The ions from the plasma are accelerated by sheath electric field, and they impinge on the cathode. Secondary electrons are emitted from the cathode because of ion bombardment, which are accelerated towards the plasma. These electrons collide with the neutrals and causes excitation and ionisation, thereby creating plasma. This type of plasma source finds wide application in glow discharge plasma nitriding.

Even when the substrate (cathode) bias is pulsed, the plasma production mechanism remains unchanged. DC and pulsed glow discharges are formed between ~ 10^{-2} to few mbar [4].

When the source of secondary electrons, formed by ion bombardment on the cathode, is replaced by thermionically emitted electrons, then the operating pressure of the discharge can be further reduced. Such type of sources is used in hot filament assisted plasma enhanced chemical vapour deposition and in plasma immersion ion implantation. The sources can work at much lower pressures (10^{-5} to 10^{-3} mbar) [5,6].

Capacitively coupled rf discharges are very widely used in the field of plasma enhanced chemical vapour deposition. In this mode of plasma production, rf power (usually 13.56 MHz) is capacitively coupled with an electrode which is placed inside the vacuum chamber. The electrode behaves as an anode and as a cathode depending on the polarity of the pulse. Thus the electrode potential and the plasma potential oscillate at the driving frequency, and have an effect in terms of trapping of electrons. As the driving electrode has area less than the chamber area, it has a net dc negative bias. The substrates are usually placed on the driving electrode. Such a plasma source is very widely used in plasma enhanced chemical vapour deposition and also for plasma cleaning of surfaces, in plasma immersion ion implantation for biasing insulating/semi-conducting substrates, etc. This plasma source can work at lower pressures (10^{-4} to 10^{-2} mbar) [3].

In rf inductive coupled plasmas, rf (usually 13.56 MHz) is connected to an inductor placed inside the vacuum chamber. The inductor behaves as the primary of the transformer and the plasma as the secondary. In this inductive field, electrons get accelerated and collide with the neutrals thereby producing plasma. Rf inductive sources work between 10^{-5} to 10^{-2} mbar and are used in plasma enhanced chemical and physical vapour deposition, plasma immersion ion implantation, etc [3,7].

The highest plasma densities at low pressures are obtained with electron resonance plasma sources, popularly known as ECR plasma sources. In ECR source, microwave (2.45 GHz) is fed in the vacuum chamber through an insulating view port. A magnetic zone is created having strength of 875 Gauss. In this magnetic field, the electron cyclotron frequency is equal to the driving frequency and hence the electrons are trapped for an extended period. The electrons collide with the neutrals and make plasma. ECR sources can work at low pressures (~10^{-6} mbar) and produces a dense plasma (~10^{12} ions/cc). Such sources are widely used in semiconductor processing [3].

For generation of metallic ions the magnetron sputter sources and the cathodic arc sources are widely used. Magnetron sputter sources are available in a wide variety, planar circular (balanced and unbalanced), rectangular and cylindrical (post and inverted). A simplistic design comprises of arranging a suitable magnetic field in front of the target (the negatively biased electrode from which sputtering take place), so that ExB forces confine the electrons. The magnetic field is such that it magnetises the electrons keeping ions unmagnetized. The electrons move in a closed path and are continuously accelerated. The electrons ionise the gas. The ions from the ionisation region bombard the target and cause sputtering. The sputtered atoms may get ionised by the plasma and deposit on the substrate [8,9].

Cathodic arc sources generate metallic ions at a much faster rate compared with the magnetron. Cathodic arc sources can work at very low pressures and the arc is sustained by ionisation of the metallic ions emitted from the cathode. The source emits highly charged ions and also macroparticles. To prevent the macroparticles from interacting with the substrate various schemes for guiding the arc plasma has been tried. A higher deposition rate though reduces the deposition time, may not be always desirable as it enhances the surface roughness and sometimes leads to compressive stress being generated in the coating [10,11].

The above sources can generate singly and multiply charged ions, generates a large number of radicals, and in some cases even negative charged ions. However, from the viewpoint of surface engineering, it is not necessary that a single type of plasma source will be able to achieve a desired end product. Many a times a mixture of plasma sources is used and there are examples where using additional plasma sources are beneficial. It has been observed that in most of the cases, the substrate on which some modification is sought has to be biased negative, or it can be stated that a negative substrate bias enhances the surface modification process. The negative bias accelerates ions, which interacts with the substrate surface. The next section gives a brief introduction to the fundamentals of ion surface interaction.

Basics of Ion Surface Interaction

As discussed in earlier sections, the plasma boundary physics determine the characteristics of the ions impinging onto the surface of the substrate such as their flux and their energy and angular distributions. With these boundary conditions at the solid surface, solid state phenomena determine the effects of ion incorporation on the final properties of the modified surface and subsurface region. Generally, the surface modification results from an extremely complicated interplay of physical and chemical effects, which act on widely different time scales. An individual ion is slowed down to energies in the order of the solid state binding energies (few eV) within about 10^{-14} s. Associated collision cascades thermalize after about 10^{-12} s. During these phases, radiation defects are formed and partially annealed again. Subsequently, thermal diffusion of the implanted atoms and of the defects occurs at sufficiently high substrate temperature. The time constants of these diffusional processes may reach the typical durations of experiments or practical applications, *i.e.* 10^3 s. During all these processes, the stoichiometry and the structure of the subsurface region may be altered. Different kind of chemical reactions may occur such as the formation of new phases, which may be homogeneous or form precipitates. In particular during the initial collisional phase and the thermalization of the cascade, an intimate mixing of the subsurface atoms occurs, which again may be enhanced or even hindered by hot chemistry. All these phenomena are influenced by the presence of radiation defects [12-14].

The fluxes of ions are sufficiently low to exclude any interaction between the individual collision cascades, as far as the fast collisional phase ($t < 10^{-12}$ s) is concerned. Therefore, for a discussion of the basic phenomena, it is often sufficient to describe the phenomena associated with an individual incident ion. The interaction with the substrate atomic nuclei and the substrate electrons, being of statistical nature, results in significant scatter of the individual trajectories. Along each trajectory, an average energy loss per travelled distance is defined by the stopping power or the stopping cross section (S). Assuming that the interactions of the ion with the atomic nuclei (so-called "nuclear" or "elastic" interaction) and the electrons ("electronic" or "inelastic" interaction) act independently, the total stopping cross section is composed of a nuclear and an electronic fraction. Basically, the nuclear interaction is described by the elastic scattering of the ions at the substrate atoms in a screened Coulomb potential, resulting in simultaneous energy loss and angular scattering.

The above ideas form the base of semi-empirical tables, which list the electronic stopping over a wide range of ion energies, and for many ion-substrate combinations. The precise stopping data for all elemental combinations are also available from the computer program package SRIM (Stopping and Ranges of Ions in Matter) which can be downloaded via the internet [15].

It is further worthwhile to address the case of molecular ions, which are often present in low pressure plasma assisted processing. When entering the surface, the energetic molecule quickly splits into its atomic constituents, with the kinetic energy being distributed proportionally to their masses. Thus, a diatomic molecular ion of only one element, such as N_2^+, can be regarded as two incident atoms with half the energy. Consequently, the mean projected range reduces as the energy is reduced. The projected range can be empirically described for a wide variety of surfaces and impinging particles as 1 nm/keV.

A significant amount of substrate atoms can be sputtered [16] from the surface provided the collision cascade penetrates the surface and contains atoms of sufficiently high energy to overcome the surface binding energy, U_s. The surface binding energy is conventionally set equal to the enthalpy of sublimation, with values between about 2 eV and 8 eV for different materials. For the regime of linear cascades, Sigmund [17] has derived a simple expression for the sputtering yield Y_s, which denotes the number of sputtered atoms per incident ions.

During deposition experiments with substrate being biased very highly negative, there is simultaneous deposition and energetic incorporation. Such phenomena, the analytical treatment of which is very difficult, can be addressed by so-called 'dynamic' computer simulations. An example of such a process is ion beam assisted deposition. SRIM code is not capable of handling this type of situation. Growth of a coating under these conditions can be understood by the TRIDYN code [11,12,18,19]. TRIDYN is capable of treating a multi-component and inhomogeneous substrate material being subject to energetic or thermal fluxes of up to five different species. It needs to be remembered that both SRIM and TRIDYN assume that the substrate is amorphous in nature, and excludes any chemical reactions between the impinging ion and the substrate. However, most of the modifications performed by plasmas on the surface are of nanocrystalline in nature and hence these codes are applicable.

Case Studies

Plasma Nitriding

Plasma nitriding also popularly known as ion nitriding or Glow Discharge Plasma Nitriding (GDPN) is one of the most promising and commercially exploited applications of low pressure plasmas in surface engineering. Plasma nitriding improves the surface hardness, and enhances the wear resistance and corrosion resistance of the surface.

GDPN uses an abnormal glow discharge produced in a gas mixture of nitrogen and hydrogen. Abnormal glow discharges are discharges where the cathode of the discharge is fully covered by the glow. The substrate, which is also the cathode of the discharge is typically biased negative between 400-800 V, and draws a current of 1-5 mA/cm^2. The operating pressure is typically between 1-10 mbar. The ratio of nitrogen in the nitrogen-hydrogen gas mixture varies from 5-95 volume per cent [20]. The bias applied on the substrate may be dc or pulsed dc. However, pulsed dc is more favourable as it gives rise to the formation of more radicals and prevents transition to an arc (Figure 7.2).

The cross section of a GDPN treated iron alloy has a compound layer of few microns thickness which typically consists of Fe_3N followed by a diffusion layer of few tens of microns consisting of mainly Fe_4N. The thickness of the compound layer and the diffusion layer together is usually called as the case depth. To get reasonable

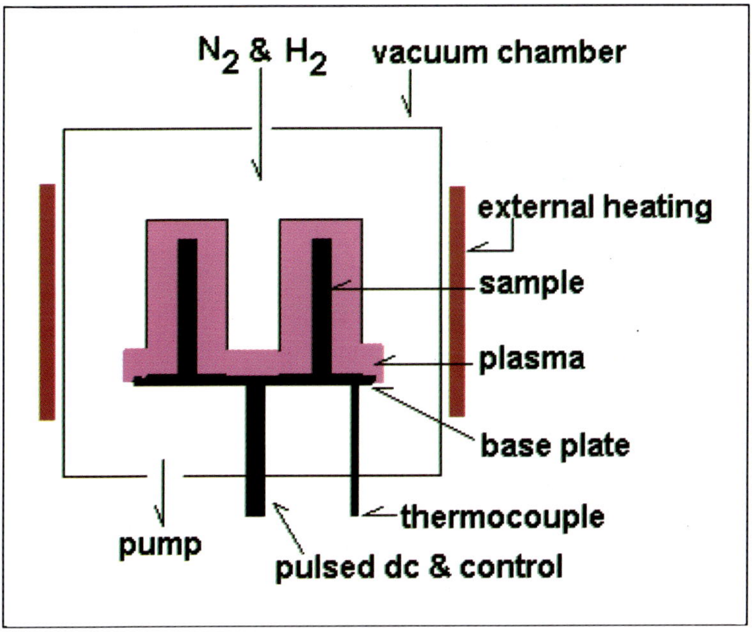

Figure 7.2: Schematic of a Plasma Nitriding Reactor. The plasma formed around the sample (substrate) envelops the sample uniformly.

case depths, the substrate temperature is typically kept between 400–560 C for iron alloys and between 700–900 C for titanium alloys. The compound layer, also known as white layer, is responsible for wear resistance improvement. The diffusion layer is mainly responsible for the improvement of the micro-hardness, fatigue life, etc. The compound layer thickness is usually controlled by the amount of hydrogen in the gas mixture; higher hydrogen gives a lower compound layer thickness. Figure 7.3 indicates how the different gas mixtures control the formation of white layer.

(a)

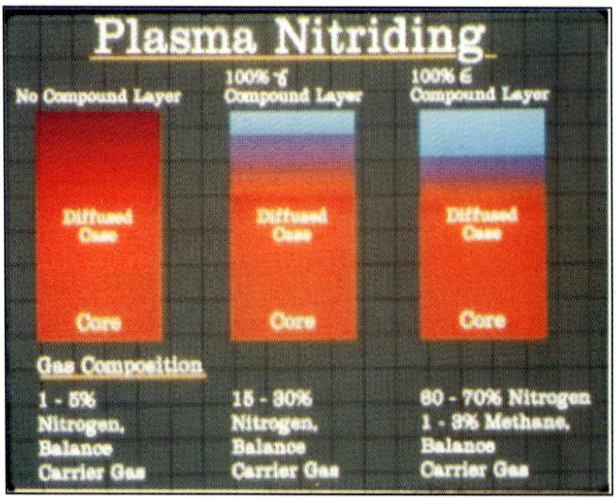

(b)

Figure 7.3: (a) Picture of a 50 kW Plasma Nitriding Reactor at FCIPT, Institute for Plasma Research, Gandhinagar, India; (b) The Dependence of the White Layer Formed after Plasma Nitriding on the Process Parameters is shown above. The parameters in which no white layer is formed have increased resistance to contact fatigue and increased fatigue strength. With increase in nitrogen percentage in the gas mixture, a gamma prime layer is formed which has good resistance to wear and deformation. With further increase in nitrogen percentage, an epsilon layer is formed which has increased corrosion resistance.

To increase the corrosion resistance of the nitrided surface a small amount is carbon is also introduced in the modified surface. The process is then referred to as plasma nitro-carburizing. A small amount of methane is used in gas mixture in that case along with nitrogen and hydrogen. For austenitic stainless steels, in the compound layer precipitation of CrN also takes place, which reduces the corrosion resistance. To avoid CrN precipitation, the treatment temperature is usually kept below 400 C.

GDPN derives its effectiveness by suitable manipulation of the glow discharge, working in conditions in which the glow covers the entire cathode surface. This is usually the condition of the abnormal glow discharge, in which case the discharge has positive dynamic impedance. The substrate behaves as the cathode and the chamber wall as the anode. Surrounding the substrate conformally, is a non-glowing region called as the ion sheath, after which there is the negative glow. In the ion sheath region, strong electric fields accelerate ions from the negative glow to the substrate. In GDPN, the substrate bias, background pressure and the substrate current are inter-related.

In abnormal glow conditions, the substrate surface is bombarded with a nitrogen ions, energetic nitrogen molecules and nitrogen hydrogen radicals [21,22]. Detailed calculation shows that the number of energetic nitrogen molecules bombarding the substrate is much more than the number of nitrogen ions. These energetic molecules also deposit more energy and momentum and thus contribute to net nitrogen flux, as well as heat the surface to facilitate diffusion of nitrogen. Additionally, hydrogen attaches itself with nitrogen forming NH* radicals, which liberates nitrogen in contact with the surface. All these findings indicate, that ions in an abnormal glow discharge plasma, plays a major role in generating the energetic neutral molecules and radicals, however in nitriding, ion's role is minimal [23,24].

A particular case study is presented where the base material is 13CrNi4 stainless steel. Plasma nitriding was performed on it for 40 hours to get an extended case depth of 250 microns. An XRD plot, showing the various phases formed on the surface is shown in Figure 7.4. It has been also shown that the nitrogen concentration and the hardness of the treated zone follows closely [Figure 7.5], which essentially shows that the amount of nitrogen incorporation is directly responsible for modification of material surfaces [25].

GDPN is suitable for the complete range of ferrous materials from cast irons and plain carbon steels to the high alloy content materials such as tool steels and stainless steels. The nitrided steels now dominate the range of steel currently being introduced. GDPN when compared to conventional (gas) nitriding offers more precise control of the nitrogen supply at the substrate surface and the ability to select either an epsilon or gamma prime layer or to prevent white-layer formation entirely. A comparison is presented in Table 7.1. Other advantages of plasma nitriding are:

- Improved control of case thickness
- Lower temperatures
- No environmental hazard
- Reduced energy consumption
- Ability to shield areas where nitriding is not desired by mechanical masking

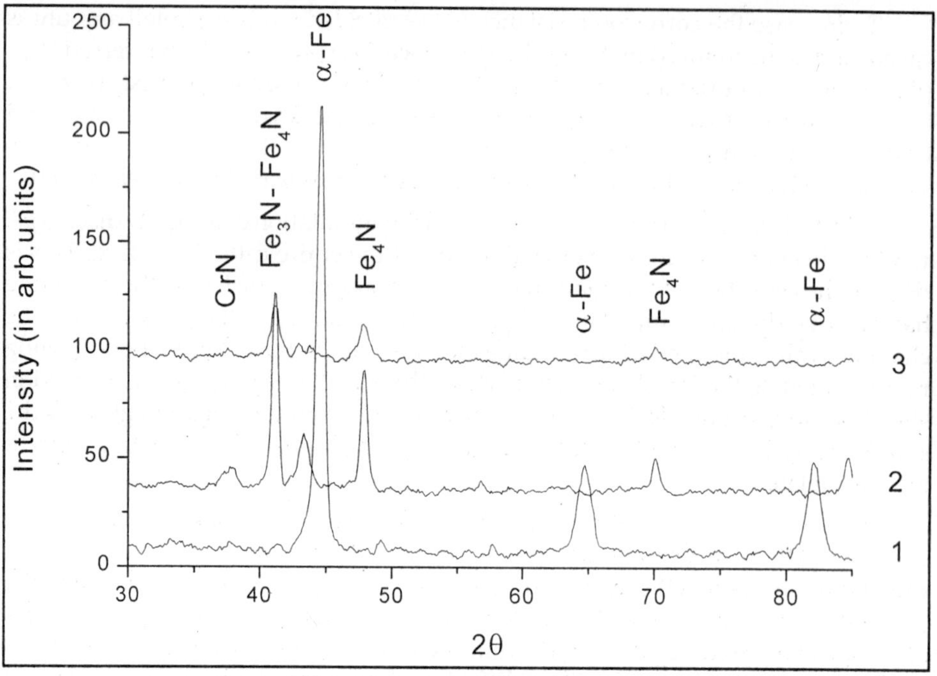

Figure 7.4: XRD Plots of 13CrNi4 Steel Treated with Different Gas Mixtures for 40 hours; (1) Untreated sample, (2) Treated with 35:65 nitrogen-hydrogen gas mixture, (3) Treated with 1:4 nitrogen-hydrogen gas mixture

Table 7.1: Comparison of Plasma Nitriding and Gas Nitriding

Criterion	*Plasma Nitriding*	*Gas Nitriding*
Minimal distortion	Yes	Yes, but not in all cases
Increase in mass	Negligible	Yes, minimal
Treatment temperature	340-565°C	520-540°C
Corrosion protection	Yes	Yes, but less
Increase of surface roughness	Negligible (can decrease)	Yes
Very good polish capability	Yes	Not in all cases
Nitriding hardness depth up till 1mm	Yes	Yes
Treatment duration at the same hardness depth	Low	Approx. 3 times as long
All iron-based materials can be treated	Yes	Not all cases (no high alloy steels)
Nitriding of bore holes is possible	Yes (in nearly all cases)	Yes
Covering of areas not to be nitrided	No problem	Difficult
Weldability	Yes	No
White layer on surface is free of pores	Yes	No
Exact specification of surface layer production is possible	Yes	No
Thin white layer possible	Yes	Yes, (with low hardness depth)
Thick white layer possible	Yes	Yes

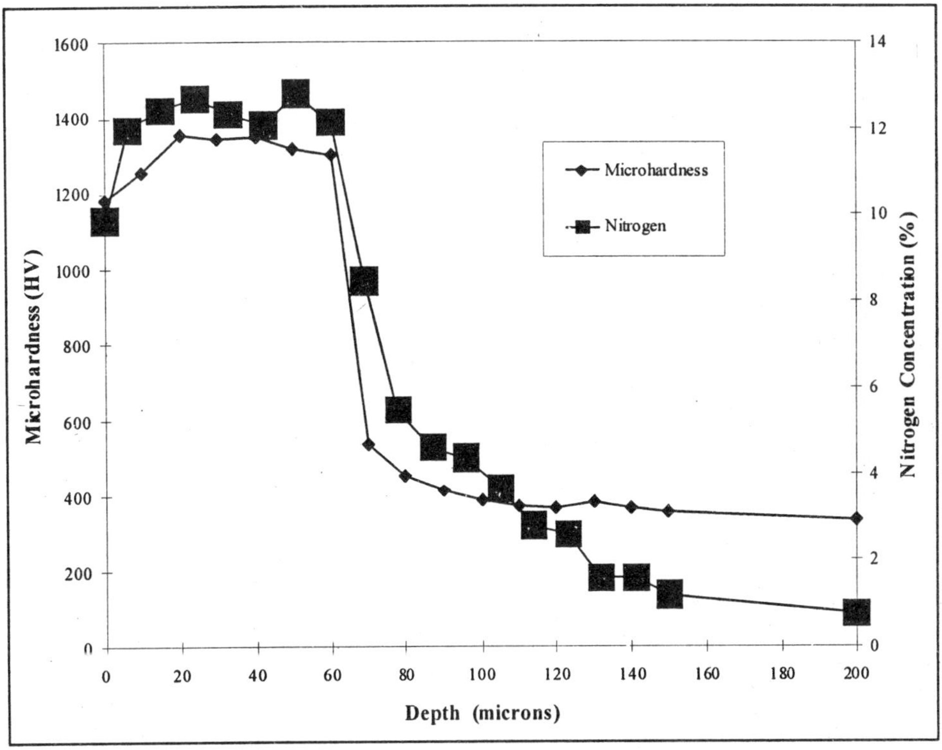

Figure 7.5: Cross-Sectional Microhardness and Nitrogen Concentration is Plotted as Function of Depth. The nitrogen profile closely matches with the hardness profile in the nitrided region.

Plasma Immersion Ion Implantation

The technology of plasma immersion ion implantation (PIII) originated from the necessity to implant ions uniformly on components of odd shapes and sizes. In conventional beam lime implantation, the ions are extracted from an ion source and accelerated to a substrate. The accelerated ion beam is rastered and the substrate is placed on a manipulator so that the incident fluence is made uniform. Inspite of such efforts the implanted dose is not uniform in conventional beam line implantation when performed on samples of odd shapes. Figure 7.6 describes the comparison between both the processes.

PIII, also popular as plasma source ion implantation, derives its effectiveness, by inserting substrates in plasma and biasing pulsed negative (Figure 7.7). The essential components of a PIII reactor are;

- High vacuum chamber with appropriate pumps
- Plasma source, usually made with RF inductive, ECR or filament assisted arc plasma sources (see section 3)
- Substrate holder
- Repetitive negative pulsed biased power supply

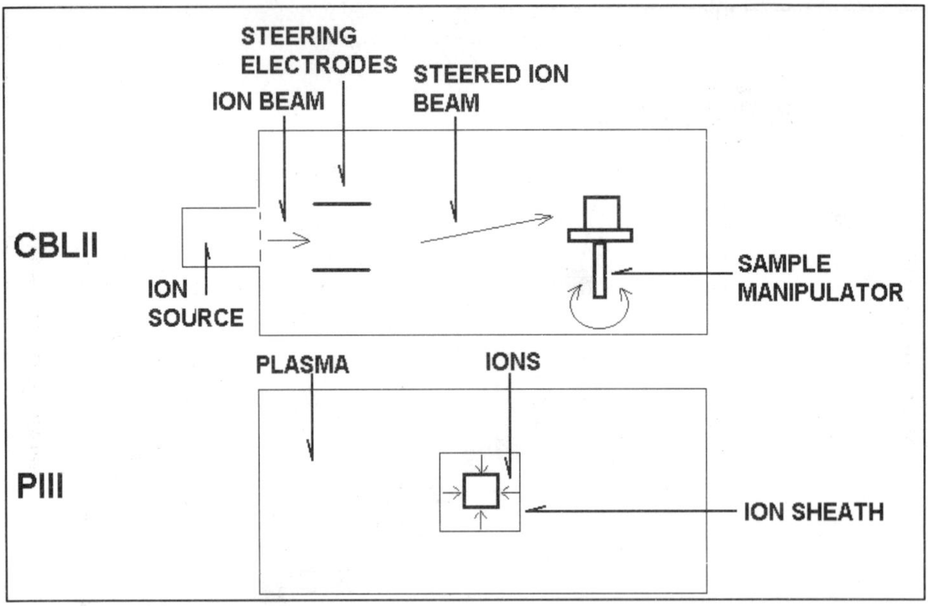

Figure 7.6: Schematic of CBLII and PIII

In PIII, the ion sheath surrounds the sample uniformly ensuring uniform implantation irrespective of the sample (substrate) shape. No need for sample manipulation or ion beam rastering.

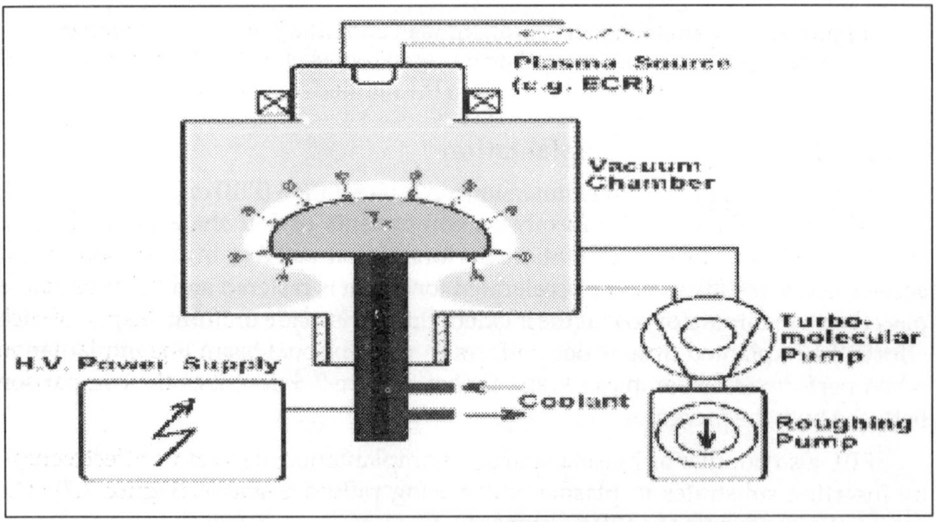

Figure 7.7: Typical Configuration of a PIII Device.

Plasma production with RF-ICP and thermionic electron assisted plasma production has also been performed. The choice of plasma source depends on the implanted dose per unit time, and size of substrate. The –ve pulsed HV power supply accelerates ions to the target during pulse on time. The off time is usually restricted by power supply hardware and for low temperature implantation experiments by the substrate material properties.

According to the theory of PIII, on the application of a large negative bias on the substrate [26], in the initial phase (for times less than an ion plasma period), ions remain stationary and electrons are repelled by the bias to create an ion matrix sheath (Figure 7.8). The potential profile in the ion matrix sheath is parabolic in nature. On longer time scales (for times ~ ion plasma period), ions begin to move towards the substrate. This phase is described by a quasi-static expansion of Child's law. The laws are described in detail in Refs. 26 and 27. Understanding of these laws is important in PIII, as they determine the spacing between substrates, the implanted ion dose, the power supply requirements and hence design parameters of the PIII reactor.

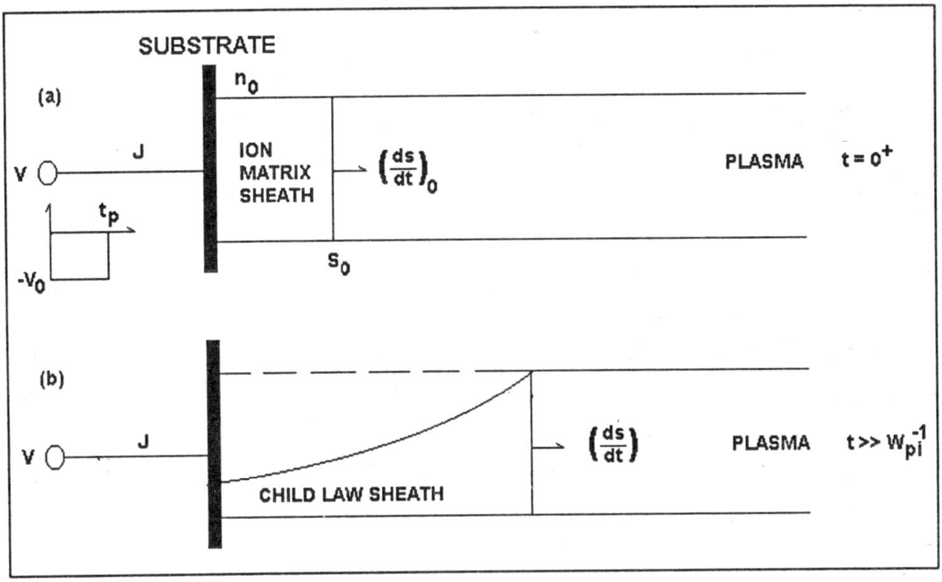

Figure 7.8: Schematic of the Sheath Dynamics during PIII

PIII derives its effectiveness by imparting kinetic energy to ions of the plasma and making them impinge on the substrate surface and is thus an ion dominated process. In PIII, the pressure is usually much lower than GDPN and so are the plasma densities. Thus the substrate bias does not ionise the background gas like in GDPN. This gives the freedom to vary the ion dose and bias independently, which is difficult in GDPN.

PIII, which was introduced as an alternate to beam line implantation, usually operates at room temperature. Thus the implanted ions have a Gaussian profile in the substrate with the mean at the range of the ion penetration. As the temperature is low the diffusion of implanted ions is negligible. This renders the surface a high hardness and also gives corrosion resistance and wear resistance [28].

PIII is also used as an intermediate step in diamond like carbon deposition on aluminium, where methane is used as the plasma forming gas. PIII finds it application for materials where there is a restriction on the maximum substrate temperature for surface modification.

PIII is also performed at high temperature when the implanted ions undergo diffusion inside the substrate surface after implantation. The ion bombardment to the substrate surface is applied at a much higher repetition rate, so as to increase the substrate temperature. This facilitates the diffusion of the implanted ions to depths significantly higher than their range of penetration. PIII has been successfully used to modify the surface properties of a wide variety of steels and titanium alloys. The major contribution from the PIII technique is in understanding the formation and properties of the expanded austenite phase, formed typically at processing temperatures just below 40°C in austenitic stainless steels (Figure 7.9). Expanded austenite phase has nitrogen present in solid solution in the austenitic stainless steel, without precipitation as nitrides. This gives stainless steel a higher surface hardness without compromising the corrosion resistance [3,7,29].

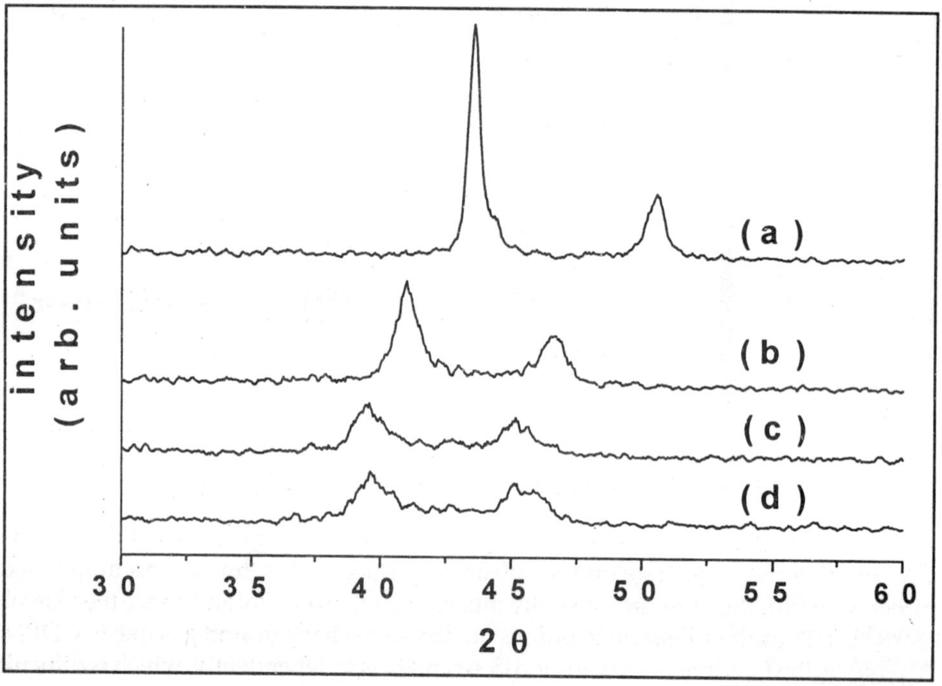

Figure 7.9: Glancing Angle XRD Patters for AISI316 under Different Substrate Bias (a) Untreated, (b) –20 kV, (c) –10 kV and (d) –1 kV, after Nitrogen PIII. A shift to lower 2θ without formation of CrN and Fe_4N indicates the formation of expanded austenite.

One of the emerging applications of PIII is in electronics industry in the doping of semiconductors. The formation of ultra-shallow p^+/n junctions by PIII has been successfully demonstrated. PIII has also been used for modification of oxidation behaviour of nitrogen and argon implanted Si wafers is reported [30]. Nitrogen and argon ions are implanted in Si at various energies at different implantation dose. Post implantation wet oxidation results show that argon implanted Si has higher oxide

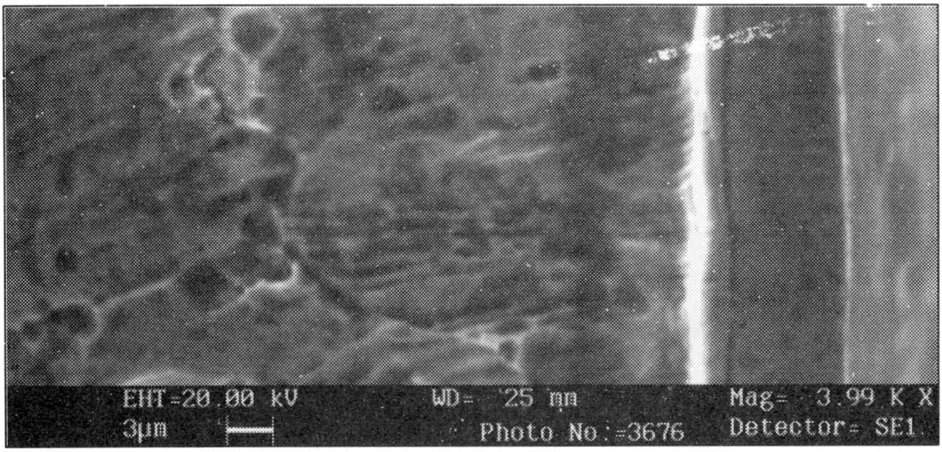

Figure 7.10: Cross-Sectional SEM Picture for AISI316 with Substrate Bias of –20 kV after Nitrogen PIII. The expanded austenite region has a different appearance than the untreated portion. The treated region has a thickness of ~ 8 microns

layer thickness with increase in the implanted ion dose. The results can be used to control the gate oxide layer thickness for memory and logic transistors.

While PIII can perform all that GDPN can do also, then it is natural to compare the results of both the processes. An experiment in that direction was also performed [31] in which substrates made of AISI316 stainless steel was treated in PIII and GDPN with same substrate bias parameters. The substrates were biased by pulses of magnitude of ~ –1 kV, repetition rates of 10 kHz and duty cycle of 50 per cent. This was done so as to keep the substrate temperature at 400 C. In PIII, the plasma was prepared using a filament assisted arc source working at 10^{-3} mbar, whereas in GDPN, it was a glow discharge working at few mbar.

The results of the comparison reveal that PIII is superior to GDPN as far the nitrogen incorporation inside the material is concerned. Both forms expanded austenite phase however the degree of expansion is more for PIII. One of the most remarkable results is that in GDPN, hydrogen is always needed for nitrogen incorporation whereas for PIII, hydrogen is not required; however presence of hydrogen is beneficial. This comparison also proves the superiority of PIII especially for those steels which require enhancement in surface hardness, but cannot be exposed to hydrogen as they suffer from hydrogen induced embrittlement (Figure 7.11).

Plasma Enhanced Chemical Vapour Deposition

Plasma enhanced chemical vapor deposition (PECVD) is one of the promising applications of low pressure plasmas. In PECVD, the plasma is usually formed by capacitive rf discharge, and the substrate on which the coating is to be formed is kept on the powered (live) electrode. However there are other examples of PECVD where the substrate may be kept on floating electrodes also.

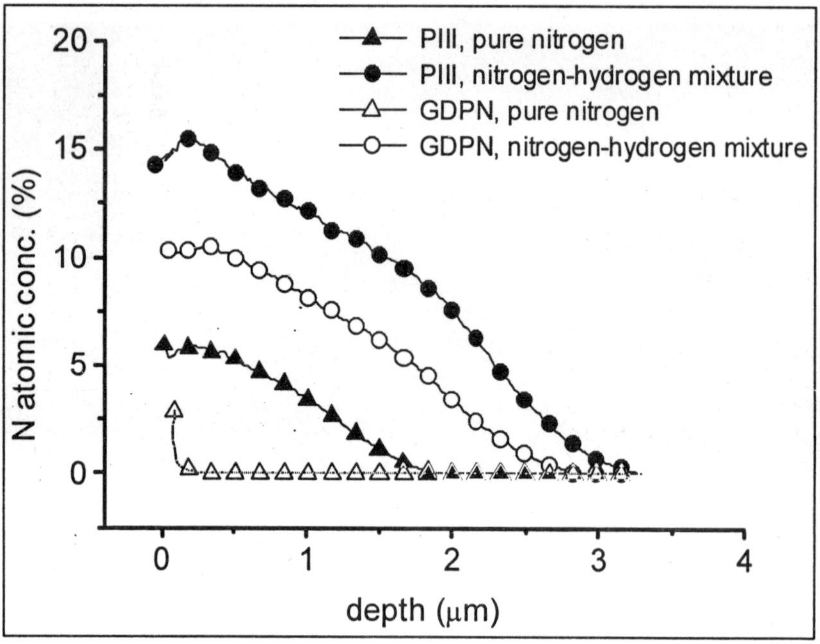

Figure 7.11: Nitrogen Concentration Profiles in AISI316, Measured with Glow Discharge Optical Emission Spectroscopy, after 2 hours PIII and GDPN Treatment with Nitrogen and Mixture Containing 20 per cent Nitrogen and 80 per cent Hydrogen as Working Gas

PECVD derives its effectiveness by using the plasma to fragment large molecules to smaller fractions and then suitably using the fragments to combine on the substrate surface to form the desired coating. For example, to get a TiN coating, $TiCl_4$ is used as the precursor, and this is ionized in the presence of nitrogen and argon. A large number of precursors used for PECVD are liquid at room temperature and hence they have to heated and entered inside the chamber as in gas phase. Table 7.2 contains a list of coatings and the precursors used to get such a coating.

Table 7.2: Precursors Needed for Various Coatings as Deposited by PECVD Technique

Coating	*Precursor*
SiO_x	SiH_4+H_2 or Hexamethyl Disiloxane (HMDSO)+O_2
TiN	$TiCl_4$+N_2
SiO_xN_y	Hexamethyl Disilazane (HMDSN)+O_2
Diamond Like Carbon (DLC)	CH_4+H_2 or C_2H_6+H_2

Silicon Oxy-nitride films were grown by PECVD technique on single crystal Si(100) and textured Si-solar cells, using a safe organic precursor, Hexamethyl-Disilazane. High quality thin (~20 A°) and thick (up to 2700 A°) films which with reduced carbon (<1.0 per cent) was formed. Using detailed XPS studies it was concluded that $SiO_{1.6\pm0.1}N_{0.3\pm0.05}$ type of coating with thickness of 870 Angstrom was formed which gave an increase of efficiency of 1 per cent of the solar cell [32].

PECVD in conjunction with PIII has been used to develop diamond like carbon films for aluminum based automobile components. In this hybrid process, initially aluminum components are sputtered in an argon environment to remove the natural oxide layer. Subsequently at a much lower pressure, methane+ hydrogen plasma is formed, and the components are biased with high negative voltage pulses as done in PIII. Thus carbon implantation takes place up to the depth as determined by the substrate bias. Subsequently the pressure is increased, substrate bias magnitude reduced, and diamond like carbon coatings are deposited on it. The advantage of combining both the techniques comes from the fact that the diamond like carbon coating formed on the surface becomes very strong adherent on the substrate. The implanted carbon gets bonded with the substrate material and also modifies the interface so that the coating is strong adherent. Such a hybrid process is also known as plasma immersion ion processing (PIIP) [33].

Using PECVD with a pulsed dc constricted anode plasma source, pinhole free SiO_x coating has been developed on Al. Such type of coatings is used to protect the Al coating present on headlight reflectors. The process is described in detail in Refs. 34. The process uses an inexpensive plasma source based on the self-confinement of plasma by choosing appropriate geometry of cathode and anode. The process uses an inexpensive plasma source based on the self-confinement of plasma by choosing appropriate geometry of cathode and anode.

SiO_x coatings deposited by PECVD techniques on Al also form an intermediate layer prior to adhesive bonding. It has been observed that adhesive bonding strength increases if such an intermediate step is followed [35]. Application potential of PECVD is enormous, as PECVD techniques are non line of sight processes, and though in some cases hazardous precursors are used, there are well documented methods for safe handling of such precursors. PECVD process has the ability to form simple coatings from complex molecules, but it needs to be remembered that PECVD is not always easy, as most of these coatings are a result of very complex gas-plasma phase reactions.

Plasma Enhanced Physical Vapor Deposition

Physical vapour deposition is one of the oldest deposition techniques based on physical evaporation (thermal or electron beam) and deposition at low pressures. It deals with the generation of metal atom from a source, which then deposits on a substrate. The transit path of the metal atom is usually collisionless; *i.e.*, the background pressure is usually low. In cases of reactive deposition the background neutrals are chosen such that the desired compound is formed on the substrate.

With the advancement of plasma techniques, magnetron sputtering and cathodic arc sources also played a major role in generation of the metallic species in the forms of ions and atoms. Auxiliary plasma is also employed very often to enhance the degree of ionization of the metallic species generated. Application of such deposition technique ranges from various products, some of which are;

- Anti-wear coatings on high speed steel and carbide Cutting Tools
- The reflective surface on Compact Disks
- Anti-reflective coating on Computer Displays
- Data Storage Media:
- Hard and Floppy Disks
- Digital Video Disks (DVD)
- Solar management coatings on Architectural Glass
- Decorative and anti-corrosive coatings in high end Bathroom Fittings
- Anti-reflective coatings on Aircraft Canopies
- Low friction coatings on Automotive Engine Parts
- Sensors on Aero-Engine Turbine Blades
- Decorative coatings on Car Wheels
- Interconnects on Computer Processor Chips

The above applications essentially indicate the huge spectrum plasma enhanced physical vapour deposition (PEPVD) offers. One of the most widely used PEPVD coatings is TiN coating, which is used for decorative, wear prevention, biocompatible coatings and in semiconductor applications.

TiN coatings have a wide variety of applications and have a golden yellow colour that makes it attractive for decorative applications, like in watchcases with golden colour. Usually in watchcases, there are 3 or more multilayered coatings. The innermost layer comprising either of Ni or Cr is done by electroplating. TiN coating (~ 1-2 micron) deposited by cathodic arc follows this. On top of this there is a thin gold coating.

Deposition of TiN is performed with magnetron sputter sources or cathodic arc. The factor, which decides which source to choose, is governed by:

- Deposition rate–cathodic arc usually has a much higher deposition rate than that of magnetron
- Whether it is the final step or not–as coatings developed by cathodic arc are more rough than that deposited by magnetron
- Deposition temperature–a high ion flux of cathodic arc increases the substrate temperature

The above factors have an influence on the overall coating performance and even on adhesion of the coating with the substrate. A typical PEPVD deposition system comprises of a cathodic arc/magnetron that generates Ti (ions and atoms). A background nitrogen plasma is also present (with auxiliary plasma source or ionized by the Ti plasma). The arrival rate of the titanium and nitrogen is adjusted so as to get the stoichiometric coating.

TiN coatings are hard coatings having hardness ~ 2500 HV. It is oxidation resistant in air up to 600 C. It is non-toxic and also has low friction against steels and ceramics. It has also been shown that the addition of Al and Si in TiN coatings during deposition improve not only the oxidation resistance of TiN, but also its adhesion behaviour [36]. Under certain conditions (addition of Al/Mo and Si during deposition such that there is a phase segregation) the coating is a composite of nanometer sized grains (popularly known as nano-composite coatings) with hardness ~ 7000 HV [37].

TiN coatings deposited by PAPVD techniques have a high compressive stress. A high compressive stress can delaminate the coating from the surface during use. Applying a high temperature during deposition can reduce compressive stress. Usually TiN coatings with high compressive stress have preferred direction as <111>. Using a negative pulsed substrate bias ~ 20 kV (similar to that used in PIII), it has been shown, that the compressive stress reduces, the coating colour changes to purple and <200> is the preferred direction [38,39]. The negative substrate bias also enhances the adhesion of the coating with the substrate. A combination of PAPVD and PIII technique is usually referred to as plasma immersion ion implantation assisted deposition (PIIIAD) [40] which is shown in Figure 7.12. Such a pulsed bias also determines the composition of coating when multiple metallic species are deposited. It has been observed that when high voltage pulse biasing of the substrate is used and the depositing ions are Ti and Al from a cathodic arc plasma source, the coating composition also depends on the bias magnitude. Al preferentially gets sputtered while deposition takes place [11, 40].

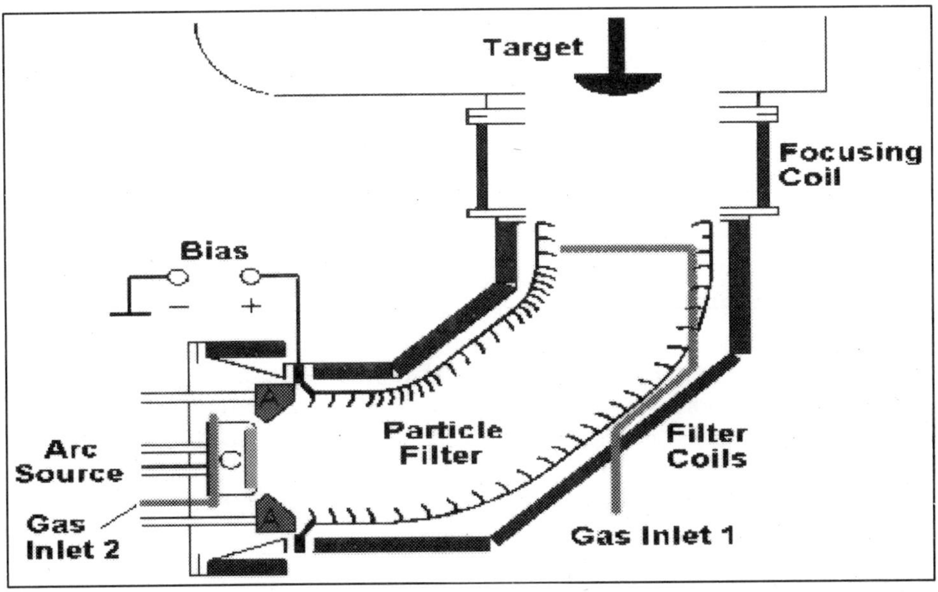

Figure 7.12: PIII Assisted Deposition (PIIIAD) by Means of a Filtered Cathodic Arc Device. Reactive deposition and implantation assisted deposition is possible. PIIIAD gives more control on the compressive stress at lower deposition temperatures.

PAPVD has been extended for developing various types of coating. One of the major limitations of this technique is its line of sight deposition. However, for substrates of relatively complex shapes, fixtures of 3-dimension rotation capability are developed which can make uniform deposition. And the major advantage of this technique is its simplicity in making even complex coatings.

Other Areas of Application of Low Pressure Plasmas in Surface Modification

Low pressure plasmas find wide application in other areas. The entire semiconductor industry uses such plasmas to modify the surface properties. Plasma etching is such as an example. Plasma etching [3] is anisotropic and hence does not give undercut as happens in chemical etching. Plasma etching uses capacitively coupled plasma and the Si wafer is placed on the live electrode. A mixture of argon and fluorine/chlorine is used for making the plasma and etching.

Plasma assisted cleaning is also such an example. In plasma cleaning the plasma is made from a mixture of CCl_4, argon and hydrogen in rf discharge. The substrate to be cleaned is first cleaned by other techniques (ultrasonic cleaning, etc), dried and then exposed to plasma. The ions of the plasma react with the surface and clean it. The plasma parameters must be chosen so that physical sputtering or chemical etching does not take place.

Plasma nitriding with rf inductive coupled plasmas is a recently developed technique for incorporation of nitrogen in steels at a much lower pressure than what is used in GDPN. In this process, the plasma potential is controlled separately by removing electrons by a positive biased yak immersed in the plasma [41]. The substrate is also biased negative (~ 1 kV) so as to accelerate ions. It is possible to get nitrided layers of few microns in thickness. The major advantage of this over GDPN is that this process does not need hydrogen for nitrogen incorporation.

Emerging Areas of Research

Apart from the above mentioned areas of surface engineering, there are emerging areas where basic research has been initiated. In this section, some such areas are identified.

Nitrogen Incorporation in Al Alloys

A large number of components are made from Al alloys, especially in the automobile, textile and aeroplane sectors. Al alloys usually have a lower melting point and also gets worn out much faster. Thus there is a need to enhance the surface hardness of such components. Al has a very high affinity for nitrogen, however incorporating nitrogen in Al alloys using low pressure plasma for wear and hardness enhancement has not been successful [42]. The penetration depth of nitrogen is less (Figure 7.13). This issue requires basic research.

Plasma Carburizing

Plasma carburizing is less popular than plasma nitriding as the technology is difficult even though there is a need for this. Plasma carburizing, in principle can be

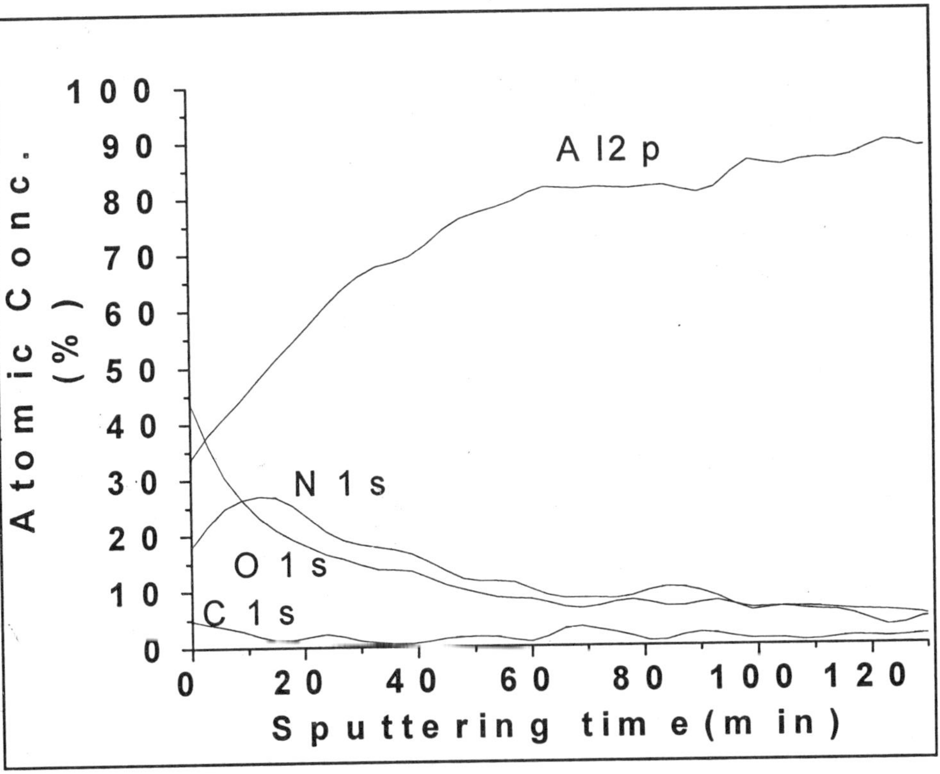

Figure 7.13: Depth Profile of Al2p, N1s, O1s and C1s for 10 keV Ion Energy and Dose 1×10^{18} ions/cm^2, as Measured with Auger Spectroscopy for PIII Done at Elevated Temperatures. Even at such a high dose there is no diffusion. Also residual oxygen in the chamber penetrates alongwith the nitrogen to similar depths.

performed by modifying a vacuum carburizing plant, provided high voltage connectivity to substrate is possible. By plasma carburizing it is possible to reduce the treatment time and hence also reduce the distortion of the substrate. A detailed understanding of the process and developments in instrumentation is needed for making effective plasma carburizing.

Retaining Weld-Ability after Nitrogen Incorporation

One of the problems associated with plasma nitriding is that after a component (substrate) is plasma nitrided it is difficult to weld anything on the component surface. Though plasma nitriding, like all surface engineering applications is intended to be the last technique, certain applications demand a post nitriding application. This is an issue that requires basic investigation.

Duplex Treatments

In a high erosive environment (*e.g.* silt erosion on hydroelectric power components) a single surface modification technique may not be the best solution. To achieve much better results there has to be a combination of a subsurface modification technique (like GDPN or PIII) along with a deposition technique (like PECVD or PEPVD or even high pressure plasma spray techniques). Such duplex treatment has been developed only for small components where GDPN is followed by TiN deposited by PAPVD. Basic understanding of the process and upscaling it for large components is needed.

Code Development including Diffusion along with Energetic Incorporation

It has been mentioned that both the TRIDYN and SRIM codes that are used to predict ion surface interaction have some major assumptions. The assumptions are;

- The substrate temperature is at 0 Kelvin
- The substrate is completely amorphous
- All thermal effects (like diffusion) are excluded
- All chemical reactions are excluded

However, in reality all the above factors are not fulfilled. Most of the substrates are crystalline and during plasma assisted surface modification there is rise in temperature. Most of the desired results are obtained because of very definitive chemical products formed on the surface. Thus a requirement for developing a code without the above assumptions is needed.

Nitriding with Plasmas in Electrolytes

Though not directly coming under the field of low pressure plasma surface modification, but this recent technique of nitriding using electrolytes has shown a lot of promises. In this technique, $NaNO_3$ is diluted in water and used as an electrolyte. In normal electrolysis, the anion of the electrolyte move towards the anode, take electrons and becomes a neutral and then a gas and gets liberated as bubbles. The cations move towards the cathode and become a neutral and gets deposited or liberated if it is a gas. In normal electrolysis, the voltage-current relationship is nearly linear. When the voltage is increased further, some of the gas molecules at the anode undergo a breakdown and the current reduces. On further increase in voltage a glow surrounds the sample uniformly and under these conditions nitriding is carried out. Altering the concentration of $NaNO_3$ can alter the value of this voltage, but in typical cases, like with 45 per cent dilution the voltage required is 175 V with a current density ~ 1 A/cm^2. It is observed that the glow is generally present around the electrode with lower surface area. Nitriding carried out using plasma electrolysis on low carbon 1020 steel show the formation of a hard compound layer (1200 HV) of thickness of 20 μ. Below that a diffusion layer of 300 μ is formed in time duration of 3 minutes. The treated surface exhibited excellent mechanical and corrosion resistance properties typically 30 times higher than untreated one (Figure 7.14).

Low Pressure Plasma Surface Engineering at FCIPT

FCIPT, a unit of Institute of Plasma Research is dedicated to industrial plasma applications and transfers the institute's knowledgebase in plasmas to benefit Indian

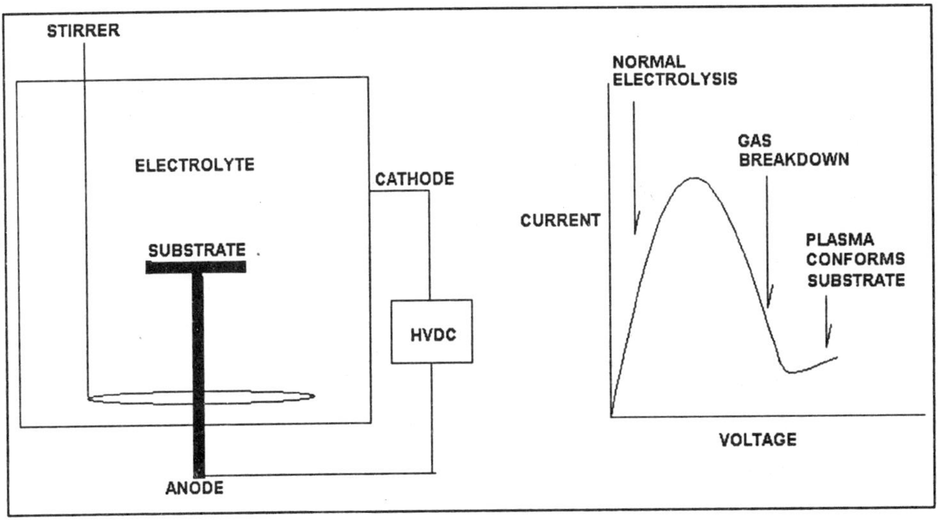

Figure 7.14: Details of Plasma Electrolysis

industries. Most of the low pressure plasma surface engineering described above is available at FCIPT and FCIPT holds patents in them (Figure 7.3a). FCIPT works on technology development and transfer, process development and equipment manufacturing. More details about FCIPT are available on www.plasmaindia.com.

Conclusions

The article discusses various aspects of application of low pressure plasmas in surface engineering. From the above description it is clear that this area is multidisciplinary in nature. To do surface engineering effectively, it is needed to develop a detailed understanding of vacuum, plasma production and control of its properties, substrate biasing techniques and placement of substrates on the substrate holder, and material properties of the substrate. It is often observed, *e.g.*, that a coating delaminates from the substrate during use, which is because of stress not being relieved before use. Thus to make surface engineering with low pressure plasmas a success, such a multidisciplinary detailed understanding is needed.

Acknowledgements

The author acknowledges the support extended by his colleagues at FCIPT, namely Prof. P.I. John, Dr. P.M. Raole, Dr. S.K. Nema and Ms. J. Alphonsa who have shared their experimental data with him and also made significant inputs to the article. The author also acknowledges the advice of Prof. W. Moeller of Institute for Ion Beam Physics and Materials Research, Germany, who was his host during his Alexander von Humboldt fellowship period, in the field of plasma immersion ion implantation assisted deposition.

References

1. F.F. Chen, "Introduction to Plasma Physics", Plenum Press, London

2. Y. Raizer, "Gas Discharge Physics", Springer-Verlag, New York

3. M.A. Lieberman and A.J. Lichtenberg, "Principles of Plasma Discharges and Materials Processing", John Wiley and Sons, New York

4. A. Ricard, "Basic Physics of Plasmas/Discharges: Production of Active Species" in Plasma-Surface Interactions and Processing of Materials, ed. O. Auciello, Kluwer Academic Publishers, Amsterdam

5. K.N. Leung, T.K. Samec and A. Lamm, Physics Letters, 51A, 1975, 490

6. J. Conrad, J. Radtke, R. Dodd, F. Worzala and R. Tran, J. Appl. Phys. 62, 1987, 4591

7. G. Collins, K. Short and J. Tendys, Surf. Coat. Technol. 93, 1997, 181.

8. B. Chapman, "Glow discharge processes", John Wiley and Sons, New York

9. R.F. Bunshah, "Deposition Technologies for Films and Coatings", Noyes Publications, Los Angeles

10. A. Anders, Surf. Coat. Technol. 93, 1997, 158

11. S. Mukherjee, F. Prokert, E. Richter, H. Reuther and W. Möller, Surf. Coat. Technol. 160, 2002, 93

12. W. Moeller, and S. Mukherjee, Current Science, 83, 2002, 237

13. M. Nastasi, J. Hirvonen and J. Mayer, "Ion Beam Processing: Fundamentals and Applications", Cambridge University Press, Cambridge

14. F. Nolfi, "Phase Transformation During Irradiation", Applied Science Publishers, London

15. J.F. Ziegler, "The Stopping and Ranges of Ions in Matter" (SRIM-2000), Computer software package. Can be downloaded via internet http://www.SRIM.org.

16. P. Sigmund, "Sputtering by Particle Bombardment", Springer-Verlag, Berlin.

17. P. Sigmund, Phys. Rev. 184, 1969, 383

18. W. Moeller, W. Eckstein, and J.P. Biersack, Comp. Phys. Comm. 51, 1988, 355.

19. W. Moeller and M. Posselt, *TRIDYN_FZR User Manual*, Forschungszentrum Rossendorf, Report FZR-317, 2001.

20. K.T. Rie, E. Menthe, A. Matthews, K. Legg and J. Chin, MRS Bulletin, Aug. 1996, 46

21. S. Mukherjee and P.I. John, Surf. Coat. Technol., 93, 1997, 188

22. L. Petitjean and A. Ricard, J. Phys. D. Appl. Phys., 17, 1984, 919

23. S. Mukherjee, Phy. Plasmas, 8, 2001, 364

24. A. Ricard, G. Henrion, M. Michel and M. Gantois, Pure and Appl. Chem., 60, 1988, 747

25. I. Alphonsa, A. Chainani, P.M. Raole, B. Ganguli and P.I. John, Surf. Coat. Technol., 150, 2002, 263

26. M.A. Lieberman, J. Appl. Phys., 66, 1989, 2926
27. S. Mukherjee and P.I. John, Pramana–Journal of Physics, 44, 1995, 55
28. S.M. Malik, K. Sridharan, R.P. Fetherson, A. Chen and J.R. Conrad, J. Vac. Sci. Technol. B 12, 1994, 843
29. S. Mukherjee P.M. Raole and P.I. John, Surf. Coat. Technol. 157, 2002, 111.
30. Rajkumar, Mukesh Kumar, P.J. George, S. Mukherjee and K.S. Chari, Surf. Coat. Technol. 156, 2002, 199
31. R. Guenzel, M. Betzl, I. Alphonsa, B. Ganguly, P.I. John and S. Mukherjee, Surf. Coat. Technol. 112, 1998, 307
32. A.Chainani, S.K. Nema, P. Kikani and P.I. John, J. Phys. D: Appl. Phys. 35, 2002, L44
33. K.C. Walter and M. Nastasi, Surf. Coat. Technol., 156, 2002, 306
34. S.K. Nema, P.M. Raole, S. Mukherjee, P. Kikani and P.I. John, "Plasma polymerization using a constricted anode plasma source", accepted in Surf. Coat. Technol.
35. J.C.S. Fernandes, M.G.S. Ferreira, D.B. Haddow, A. Goruppa, R. Short and D.G. Dixon, Surf. Coat. Technol., 154, 2002, 8
36. F. Vaz, L. Rebouta, M. Andritschky, M.F. da Silva and J.C. Soares, J. Mat. Proc. Technol. 92-93, 1999, 169
37. S. Veprek, J. Vac. Sci. Technol A17, 1999, 2401
38. M.M.M. Bilek, D.R. Mckenzie, R.N. Tarrant, S.H.M. Lim and D.G. McCulloch, Surf. Coat. Technol., 156, 2002, 136
39. S. Mukherjee, F. Prokert, E. Richter and W. Moeller, Thin Solid Films, 445, 2003, 48
40. S. Mukherjee, F. Prokert, E. Richter and W. Moeller, Surf. Coat. Technol., 196, 2005, 312
41. G. Collins, Surf. Coat. Technol. 93, 1997, 181
42. J. Chakraborty, S. Mukherjee, P.M. Raole, P.I. John, Mat. Sci. Engg. A 304-306, 2001, 910

Chapter 8

Thin Film Processing: An Overview

A. Subrahmanyam
Department of Physics,
Indian Institute of Technology, Chennai
E-mail: manu@iitm.ac.in,

ABSTRACT

In this paper, three important techniques: Electron beam evaporation, Laser Ablation and Magnetron Sputtering used for depositing thin films have been discussed. The physics of these techniques, the kinetics have been elaborated. These techniques are being employed for manufacturing functional thin films: hard multilayer coatings, multilayer optical coatings etc. The main idea of the paper is to introduce the three important techniques and to outline the basic principles.

Introduction

Thin film technology is pervasive in many applications, including microelectronics, optics, magnetic, hard and corrosion resistant coatings, micro-mechanics, etc. Progress in each of these areas depends upon the ability to selectively and controllably deposit thin films–thickness ranging from tens of angstroms to micrometers–with specified physical properties. This, in turn, requires control–often at the atomic level–of film microstructure and microchemistry. There are a vast number of deposition methods available and in use today. However, all methods have their specific limitations and involve compromises with respect to process specifics, substrate material limitations, expected film properties, and cost. This makes it difficult to select the best technique for any specific application. Three important techniques: Electron beam evaporation, Laser Ablation and Magnetron Sputtering are discussed in this article. The physics of the technique, the kinetics are elaborated.

Electron Beam Evaporation

The most common deposition technique for metals and metal-oxide film materials is electron-beam evaporation because the boiling temperatures of metal oxides are generally in excess of ~1200°C. The ebeam evaporation is now an industry standard and most of the thin films for optical applications are being coated by ebeam evaporators.

The basic principle of ebeam evaporation is that the high energetic electron beam strikes the material and transfers the energy into thermal energy; the material to be deposited melts and evaporates with this thermal energy. While the source is a composite material, it decompose into the basic constituents during evaporation. In the reactive electron beam evaporation, the basic evaporant reacts chemically with the active species present in the evaporation chamber forming a composite material onto the substrate. The electron beam heated sources differ from the resistance heated sources in two ways: (*i*) the heating energy is supplied to the evaporant by the kinetic energy of a high energetic electron beam, and (*ii*) the evaporant is contained in a water cooled cavity or hearth. Heating by electron beam allows attainment of temperatures limited only by radiation and conduction to the hearth. Evaporants contained in a water cooled hearth do not significantly react with the hearth, thus providing a nearly universal evaporant container.

The bent beam evaporation source, as is now used in thin film fabrication for electronics, optics and research, was first envisioned by Holland [1] and the modern 270° gun was developed in the early 1960s by Hugh Smith and Charles Hanks of Temescal Corporation [2]. In the late 1960s and 1970s electron beam guns were principally used for the deposition of aluminium metallization on semiconductor devices. Today these sources have been largely supplanted for semiconductors because of the switch to alloy metallization coupled with the development of high rate, easily automated planar magnetron sputtering sources. Magnetron sources are much better suited for alloy deposition and more stable in automated production applications. The use of lift-off metallization technology for gallium arsenide and other high performance devices is causing a resurgence of electron beam deposited semiconductor metallization. The development of the new generation of hard multilayer optical coatings has made electron beam evaporation the technology of choice in optics.

Evaporation Kinetics

The beam is formed in the electron gun (basically a tungsten filament powered), passes through the magnetic lens and is focused upon the evaporant. Application of an electron beam heated source is governed by three complex relationships.

1. The energy balance of the evaporant charge and the requirement for stable dissipation of the beam energy.
2. The complex distribution of the evaporant vapour flux from the evaporant surface caused by the pressure within this vapour and the resultant evaporant surface geometry.
3. The ionizing effect of the electron beam, as it passes through the evaporant vapour cloud, impacts the melt surface and is partially reflected from that surface.

For useful evaporation to occur, the evaporant surface must reach a temperature such that the surface vapour pressure is greater than 10^{-1} Torr. At very high evaporation rates, this pressure may reach 10 Torr. The electron beam is conventionally accelerated to 10 kV at a current of up to 1.5 A. This beam impacts an area of 0.25–1 cm^2 with energy of up to 60kW/cm^2. Stable evaporation requires that a thermal equilibrium exist in the evaporant and the energy dissipation must be stable. Energy is dissipated from the evaporant principally by conduction to the hearth, then by evaporant phase change heat and least by thermal radiation. A portion of the beam, often containing considerable energy, is also reflected from the evaporant surface [3]. A 10 kV electron beam upon impact with a surface gives up its energy essentially at the surface (within a small fraction of a millimetre). For an ideal evaporant (Ag, Al, Au, Cu) as the temperature is increased, evaporation will begin at a vapour pressure of about 10^{-4} Torr, increasing in rate with increasing energy input until the evaporant pressure over the beam impact point reaches the viscous flow range (about 10^{-1} Torr), and the mean free path is reduced to a fraction of a millimetre. At this point, added beam energy is principally absorbed by the evaporant cloud, limiting the further increase in evaporation rate. Because the temperature change in degrees K is relatively small from the point of first evaporation to instability, (for aluminium 1245 at 10^{-4} Torr to 1640 K at 10^{-1} Torr), heat losses by conduction to the hearth and radiation do not increase rapidly. As a result, the evaporation rate increases exponentially with increasing power until limited by the vapour density over the evaporant surface. The evaporation rate can be further increased at constant or increased power only by increasing the electron beam impact area, thereby reducing the power density.

Before evaporation begins, an operating electron beam source emits a small current of energetic electrons toward the substrates most likely reflected from the evaporant surface. In addition, the electron beam source emits a modest flux of soft x-rays. These emissions from the source, although small, can damage sensitive substrates/devices. Because similar or more intense ion and electron bombardment emanates from sputtering sources the most radiation sensitive devices must be coated using resistance heated sources. As the electron beam passes through the evaporant cloud, impacts the evaporant and is absorbed or reflected, several processes occur. Measurements of the ion current at 25 cm above an operating electron beam evaporation source [4] show that the evaporant is partially ionized by the electron beam. For aluminium the ionization is only 1 per cent but for silver and copper it may be much higher. The path of these ions in the evaporant during their first 1 to 2 cm of travel is influenced by the magnetic lens and the 10 kV beam accelerating field of the evaporation source. Some ions follow curved trajectories, impacting the high voltage feed-throughs and wires, sputtering material and causing arcs. The material sputtered from the feed-throughs is significant enough to destroy them over a period of years and to slightly contaminate the films. These feed-throughs and wires must be protected by a grounded metal shield.

Evaporation Sources

The electron beam heated source is made up of a power supply and evaporation source. Great care must be taken to assure that their design is sufficiently durable to survive continual arcing and attack by hot molten metal.

The filament and magnet power supplies are independent of the high voltage power supply and are unusual only in their exceptional immunity to electromagnetic noise. The SCR filament supply is normally controlled by the 0 to 9 V power voltage with the output to the gun coupled through a transformer with 20 kV of isolation. There are two feedback paths from the filament supply to the pass tube/switch, one for beam current regulation (if used) and a second to 'fold back' the filament power during arcs to assist arc recovery. The magnet power supply is of basic current regulated design with adjustable current limits, putting out ±2amps and crossing through zero smoothly. This supply is often interlocked with the filament to minimize the chance of the beam being directed outside the hearth.

The magnetic lens and the evaporant containing hearth are usually a totally integrated assembly. A magnetic field of a few hundred gauss is required to bend the electron beam from the electron gun through 270° and focus it upon the evaporant. The magnetic circuit consists of a pair of soft iron side plates with adjustable beam formers and two magnets, an electromagnet and a permanent magnet, in parallel. The field of the permanent magnet is chosen such that the beam is positioned within the hearth, and the parallel electromagnet is used to increase and decrease the main magnetic field for longitudinal beam position control. Like the cathode cavity, the magnetic lens is empirically designed to focus the beam to about 0.5 cm diameter at a nearly vertical angle of incidence upon the evaporant charge. The principal focusing elements of the magnetic lens are the beam formers [5] which create an adjustable sharp field concentration and gradient. This focus adjustability is of critical importance, for the evaporation of dielectrics. For lateral beam positioning, a small secondary magnetic circuit is often included, particularly in large hearth (over 20 cm^3) sources. An electromagnet with one pole near the electron gun and the other loosely coupled to the side plates performs this function. The required field is very low, and no detailed lens design is necessary. All the permanent and electromagnets in a evaporation source must be embedded within the water-cooled hearth or incorporate water cooling because of the high ambient temperature. Particular care must be taken in winding the electromagnets or they will be damaged by voltage spikes induced into them by high voltage arcs.

The core of an electron beam heated source is the hearth assembly, upon which the electron gun and magnetic lens are mounted. The water-cooled hearth, containing the evaporant, also serves as the mounting base for the entire gun. The gun hearth must make excellent electric contact to the vacuum chamber base because the ground leg of the high voltage flows through it. A good earth ground, that is connected to the power supply and the vacuum chamber by a large (5 cm) ground strap is often required to control RF arcs. As the container for the superheated evaporant, the hearth must, above all, have sufficient water cooling to be inert. It has been shown that for aluminium, a particularly reactive metal, evaporated from well-designed hearth, there is no detectable hearth-related contamination [6]. The conventional hearth water cooling consists of a series of narrow water passages that require flow of 15 litres/min. Flow velocity is the key aspect of the cooling, rather than water temperature, and great care must be taken to insure the actual flow rate is sufficient at all times. Because the water may carry away up to 90 per cent of the beam energy any interruption or reduction in flow during operation can destroy the hearth.

The important extension of the basic single hearth source is the multi hearth source. By fabricating the hearth rotary turntable with four or more independent hearth cavities and a water-distributing central axis, a heat increase in versatility can be achieved. Most films today are multiplayer films, and multi hearth sources commonly used, particularly for optics. These hearths are significantly larger and more complex to use and fabricate than single hearth sources.

In addition to containing the evaporant, the hearth must contain the electrons reflected from the evaporant. Because the energy in the reflected electron beam may contain 30 per cent of the beam energy, capturing electrons is of significant importance. This is usually done by mounting the permanent magnet on top of the hearth, behind a water-cooled copper shield, in the path of the reflected electrons. This creates a magnetic field concentration which bends these electrons into the shield over the magnet, capturing electrons.

Evaporation Characteristics

The depositing films using electron beam heated sources is much more sensitive to application lower than is the case for sputtering. The evaporant quantity required by an electron beam heated may be as little as 1 cm^3, yet the selection and conditioning of the evaporant are of utmost importance for successful film deposition.

The conditioning of the evaporant consists of a slow increase in power (or rate for subliming materials) over several minutes to one and one half to two times the level desired before opening the shutter to clean and degas the charge. Occasionally, during this process, a melting material will give off such large quantities of gas that the system pressure will rise above the maximum acceptable (4×10^{-5} Torr for example), and degas very slowly, even at high power. Severe gas evolution may also be a sign of that the evaporant is decomposing. Conditioning is completed by running the gun at maximum conditioning power, while sweeping the beam back and forth over the gun charge and watching for instability in the charge. With subliming materials, the amount of conditioning that occurs is necessarily low, and slow due to low gas mobility in solids.

The deposition rate is strongly influenced by the variable characteristics of the electron beam gun and the evaporability of the material [7,8]. For melting materials (those which melt before evaporating, Al, Au, Cu), the rate increases with increased power density (decreased spot size) in the melt, to the power dissipation limit. With semi-melting materials (those which melt only in the beam, sapphire), and subliming materials (chromium and silicon monoxide), which are not able to absorb the full power of the beam, the power and rate can be increased only by increasing the spot area. For these materials a fixed beam spot is similar to a swept one of equal area. The small spot, swept at 60 Hz, though to the eye a good approximation of a fixed large spot, is still only an approximation and is not fast enough to allow nearly as much power into silicon monoxide as does the fixed spot at 1 cm^2 area.

The melting materials deposit at higher rates from the large hearth due to the longer thermal path to the water cooling. Some subliming materials also exhibit increased rates in the large hearth, due perhaps to the larger charge permitting heating a greater volume to evaporation temperature. With all dielectrics, the beam power must be increased slowly (30 seconds to 1 minute) to red heat, allowing the electrical conductivity to increase to a point where the surface charge on the material can dissipate to ground.

The following are four widely applicable guidelines for successful deposition of thin films from an electron beam heated evaporation source.

1. Select a charge form with the largest possible volume to area ratio. Avoid trying to evaporate powdered or granular materials.
2. Use the largest hearth volume consistent with available evaporant charge and the desired film.
3. Use the largest beam spot area possible but still attain the required deposition rate.
4. Increase the spot size if increasing the beam power causes instability or film pinholding.

The limitations of ebeam evaporated films:

The applications such as imaging, displays, high ophthalmic anti reflection (AR) coatings, laser applications, etc. demand that these coatings should possess stable optical and mechanical/chemical properties and low light scatter. Unfortunately, e-beam deposited films grow with a columnar microstructure, meaning that there are sub-microscopic voids in the structures of the layers. This under-dense structure can and will absorb moisture and other gaseous-state materials, and in doing so, change in all of the above desired properties. Such films can have high scatter properties and sometimes high stress. Changes in optical properties with environmental exposure are evidenced by shifts in performance wavelength of filters and in reflection value for AR coatings. Mechanical degradation may take the form of loss of adhesion, strength, or hardness because the grain sizes are large and loosely bound. For these reasons, other, more energetic forms of film deposition and growth were developed. When more energy is made available, either in the form of the momentum of the arriving adatoms or surface and species activation, the adatoms have the mobility to find sites to nucleate on and to grow with a more compact and finer grain structure. High substrate temperatures assist with the surface mobility, but not necessarily with densification. Often surface energy barriers are present if atomic contamination is present on the surface or if the surface is otherwise chemically inert.

Laser Ablation

Film depositions by laser ablation have been widely employed for a variety of materials: high-T_c superconducting [1], ferroelectric [2], ferromagnetic[3] materials, etc. Film deposition by laser ablation is carried out by irradiation of the target by a focused laser beam. The laser beam ablates target materials from the target and materials are transferred to the substrates [4]. During laser ablation a luminous cloud can be seen along the normal to the target. This cloud is called a 'plume'. Various kinds of pulsed lasers are often employed for film deposition by laser ablation. Compared with conventional film deposition techniques, e.g thermal evaporation, molecular beam epitaxy (MBE) sputtering, organo-metallic chemical vapour deposition (OMCVD) etc, laser ablation has the following characteristics:

1. Materials with high melting-points can be deposited if the materials absorb the laser light.
2. Almost no contamination is present, unlike the situation often observed in films prepared using an evaporation heater or filament.
3. It is possible to prepare films in an oxidation environment with relatively high pressure because of the absence of a heater or filament in the deposition chamber.
4. The target composition is transferred to the film, leading to stoichiometric deposition.
5. A large number of droplets of submicron size are often seen on the surface of the deposited film.

The interaction between a high-power laser and the target materials was discussed by Ready[5]. Following his work the decomposition or ablation of materials by employing lasers was applied to the processing of a variety of materials, and then to the deposition of films by condensation of the decomposed or ablated particles. For laser ablation of inorganic materials Paek *et al.*, and Chun *et al.*, reported that the ablated materials from metals or non-metals are ejected explosively[6,7]. Srinivasan *et al.*, reported that organic or inorganic materials can be drilled with desired patterns by an excimer laser [8-10]. They called this phenomenon ablative photo-decomposition (APD). Features of APD are (i) a rapid removal of the reactant products, (ii) a flattened surface of the ablated material and (iii) a precise pattern.

A CO_2 laser with a high power output is not suitable for film preparation because of its cw-operation. Irradiation by a cw-operated laser induces a continuous melting of the target materials, resulting in segregation of the constituent materials in the target, and the film deposition having an off-stoichiometric composition. This deposition resembles the thermal evaporation in the melting process, leading to essentially no difference between the two cases. The velocity distribution of the ejected particles from the target is well described by the Maxwell-Boltzmann distribution in this case.

In contrast to the CO_2 laser, the excimer laser is the most popular for laser ablation because of its short wavelength and small pulse width. The wavelength of the ablation laser influences the absorption coefficient of the target material and the cross section of ambient gas excitation. The pulse width also seems to be a key factor in the ablation mechanism.

Two kinds of particle are often observed on films prepared by laser ablation. One is the droplets deposited by the laser ablation; the other is the outgrowth caused by the crystallization of films. The outgrowth problem occurs not only for laser ablation but also for other techniques. Droplet formation is the most important issue to be solved for the application of films prepared by laser ablation. The formation of droplets, or particulates, is greatly affected by the wavelength of the pulsed laser employed. Koren *et al.*, reported lower normal resistivities and higher critical current densities in $YBa_2Cy_3O_x$ (YBCO) superconducting films deposited by a shorter wavelength laser [11]. Their result also revealed that the surface morphology of the films is rough with large droplets when a 1064nm laser is used, whereas much smoother surfaces with fewer and smaller droplets are obtained with a UV laser (355 nm). Thus ablation wavelength is a crucial parameter for clarifying the mechanism. They ascribed the improvement to the fact that the ejected clusters are decomposed efficiently as a result of irradiation by a second laser beam with a shorter wavelength.

Based on these experimental results, various models (though tentative) can be proposed for the mechanism of the laser ablation:

1. The temperature increase caused by electronic excitation due to photo-absorption and successive lattice relaxation induces thermal evaporation.
2. In addition to the temperature increase, an explosive phenomenon takes place due to the laser irradiation.
3. As opposed to a temperature increase, bond breaking by electronic excitation due to the irradiation has a dominant role in laser ablation.
4. Coulomb explosion of positive ions takes place as a result of emission of the photoelectron.

Venkatesan *et al.*, investigated the velocity distribution by post-ablation ionization (PAI)[18]. Compared with the Maxwell-Boltzmann distribution, a velocity shift to the high-velocity side was observed, suggesting that high-energy particles are effective for the deposition of high-T_c superconducting thin films.

Recently, pulsed laser ablation has been classified into two categories: one is typical laser ablation for a congruent growth with a single target and the other is laser MBE for a layer-by-layer growth with multitargets [23]. For deposition of each layer from the respective target in laser MBE, the requirement for stoichiometric deposition is not crucial because each target consists of typically one metal component, *e.g.*, CuO, BaO etc. In that case the laser fluence can be reduced to a value just above the threshold value that brings about a thermal evaporation mode, resulting in deposition with a smooth surface morphology, almost without droplets. Very often the laser MBE mode is confused with the laser ablation mode.

The outstanding features of deposition by laser ablation lead us to deposition of multi-component oxides. This requires stoichiometric deposition, a highly oxidizing ambient, and a contribution from particles with moderate kinetic energy, for low temperature growth. Laser ablation is applicable also for ferroelectric thin films. PZT films are expected to be useful in non-volatile memory devices and were prepared by laser ablation by Otsubo *et al.* [2].

The Hardware

During deposition, the target is often rotated to eliminate changes in composition [1]. For the growth of multilayer structures, the target is often exchanged. Well designed apparatus gives films of high quality. Generally the target and substrate are arranged as a parallel configuration. The advantage of the parallel configuration is symmetric distribution of thickness on the substrate.

In order to guide the laser light into the deposition chamber, an incident window should always be transparent to the incoming laser light. If an ArF excimer laser is used for ablation, synthesized quartz is the preferred window material. Occasionally, the window material is turned opaque or luminous by the UV laser irradiation. To avoid a reduction of laser fluence onto the target, special care is required.

Introducing oxygen or an inert gas from the window is preferred for eliminating unintentional deposition on the window. In the preparation of the film by laser ablation it is important to control the plume in order to obtain high quality films. The optimum laser fluence for film preparation depends on the substrate-target spacing and the ambient gas pressure. The substrate-target spacing is, typically, around 2-5 cm, depending on the laboratory. In any case, the crucial point for laser ablation is how the plume interacts with the substrate.

No special treatment is required for the preparation of the substrate for deposition by laser ablation. At present, compared with the conventional film preparation techniques, the homogeneity of the film thickness is not good enough to allow the use of a large-area substrate. So small-area substrates are preferred for laser ablation. The substrate temperatures are often calibrated by a thermocouple attached to the substrate by an Ag paste. Radiation thermometer through an IR-transparent BaF_2 window can be used for the calibration of the substrate heater.

Target preparation is also important for film growth. The target morphology affects the surface morphology of the films [23]. A fresh target surface is preferred because after prolonged irradiation of the target its composition at the ablated surface often deviates from the original stoichiometric one, especially in ablation with a low fluence [24]. A high density target is also required for high quality film without a large amount of droplets. For example, the use of a melt-quenched amorphous target is useful for eliminating particle formation[25].

Optical emission, mass analysis, optical absorption and laser-induced fluorescence of the plume from the target have been measured by many groups. These real-time diagnoses are expected to clarify the mechanism of deposition and to monitor the ejected particles in order to obtain reproducible films. A kinetic study using real-time monitoring of gas species was reviewed by Chen *et al.* [26].

To eliminate droplet formation, various techniques have been proposed. In early experiments,a mechanical shutter, synchronously triggered with the laser through a delay circuit, was employed as a velocity filter for preparation of Ge or Si films[37]. Similarly, second laser irradiation and a pulsed supersonic oxygen fluid shutter were employed for this purpose [38]. A crossed fluxes technique is also valid [39]. This technique suppresses the incoming droplets by shadowing masks using two lasers and two targets. Our recent result on droplet formation revealed that the number of droplets strongly depends on the etching depth of the target per pulse caused by the laser ablation [40]. Accordingly, to eliminate droplet formation, a reduction of etched depth is required, leading to the use of a dense target, a pulsed laser with a low fluence, a short wavelength, and a short pulse width.

Safety Aspects

For laser ablation a pulsed laser such as an ArF or KrF excimer laser is often employed. These lasers are operated by the excitation of fluorine and inert gases. Fluorine gas is known to be hazardous to the human health, although it is not explosive. It can cause severe chemical and thermal burns. The gas attacks the nose, eyes, skin and the respiratory system. Exposure to high concentrations is usually fatal, respiratory damage and pulmonary edema being the cause of death. The threshold limit values (TLV) are as follows: the time weighted average concentration (TWA) is 1 ppm and the short time exposure limit (STEL) is 2 ppm for fluorine [31]. The fluorine gas cylinder is kept in the cylinder cabinet with a ventilation system. The gas detectors are to be installed in this cabinet and a ventilated room. The exhausted gas is trapped by an adsorbate installed in the evacuation system.

In order to avoid operator exposure to the laser, special care is required. A UV pulsed laser such as ArF (193 nm) is not visible but is hazardous to eyes and skin. In addition, it causes the formation of ozone from oxygen. Too much ozone is harmful. For this reason the ArF excimer laser beam is covered with a UV-cut plastic guide filled with nitrogen gas. If it is difficult to assemble the beam guide, the use of UV-cut goggles is recommended. The most hazardous laser is a visible-pulsed one, for instance, the second and third harmonics of the Nd:YAG laser. Visible pulsed lasers have a severely damaging effect on the human eye. In any case, confinement of the laser beam line into a waveguide is preferred.

Film deposition by laser ablation is undoubtedly an attractive technique for the preparation of a variety of materials. Disadvantages are droplet formation on the deposited film, the limitation on large area deposition, the high cost of the excimer laser, etc. The problem of droplet formation can be suppressed by a variety of techniques. For the moment, the laser ablation technique is of great importance in sophisticated material and device research in laboratories and the small-scale production of high cost-performance devices. In the near future an improved pulsed laser with a low cost will be available, leading to mass production.

Magnetron Sputtering

The first magnetron glow discharge device employing a cylindrical-post cathode with out-turned end flanges was communicated by Penning and Moubis to a scientific meeting in Holland in December of 1939 [1,2]. The device was used for sputtering copper, nickel and silver. The cathode spool was 23 cm long and 2.3 cm in diameter. The end flanges were 5 cm in diameter. At an argon pressure of 14 m Torr and a magnetic field of 350 G it operated at 550 V and 3 A (18 mA/cm^2) and had a voltage index value of 6.

From an historical point of view [3] there appear to have been two factors preventing Penning, and others, from quickly recognizing and developing the full sputtering potential of this early work: (i) the upheaval caused by the outbreak of war in the fall of 1939, and (ii) the small diameter of the apparatus which required the use of magnetic fields of 300-700 Gauss.

In the years following the war a 1939 US patent authored by Penning [4] became well known and his designs became the basis for a popular pressure (the Penning Ionization Gauge). Such 'magnetron ion gauges' are intended for use at very low gas pressure and are devices wherein a significantly large electric field permeates the geometry. This accounts for the name 'magnetron' which was borrowed from war-related microwave device development. Because sputtering does occur, even at very low pressure, his designs also became the basis for vacuum pumps of the ion getter type[5]. In 1961 Helmer and Jepson demonstrated that Penning low-pressure discharges produce sputter erosion patterns that defy simple explanation [6,7].

Approximately, thirty years elapsed between Penning's 1939 sputtering work and the surge in sputter magnetron development which occurred in the late 1960's and early 1970s and resulted in the recognition of three generic types of sputter magnetrons: (1) conical magnetrons [8,9]; (ii) cylindrical magnetrons [10-12]; and (iii) planar magnetrons [13-15]. In the same time period a little-known hemispheric magnetron was also studied [17,18].

The generic type used here is based on the new (*i.e.* not eroded) shape of the cathode surfaces. Within each type there may be substantial variations of design. In particular the planar magnetron designation includes devices in which the sputter erosion track is circular, square, rectangular, or oval (race-track like). In addition there may be a single erosion track or a nested series of tracks.

Figure 8.1 is a sketch of a simple planar magnetron whose sputter erosion track is quasi-rectangular. The magnetic field is produced by an assembly of permanent magnets such that the field lines emerge from, arch over, and re-enter the sputter target plate. The case which contains the magnet assembly is connected to ground potential and, in this sketch, functions as the anode of the discharge. An intense glow discharge forms in the tunnel defined by the field line arches.

By now the term 'sputter magnetron' has been applied to a bewildering array of devices; in fact, to almost any device which employs magnetic and electric fields in which the so-called Ex B drift currents form closed paths. However not all of these embody the end–containment principle employed by Penning. For present purposes the term 'sputter magnetron' is reserved for devices which operate in the positive space charge mode and satisfy the following 'Penning conditions'[10, 11]:

1. An annular, or annular-like, of space threaded by lines of magnetic field which, at either end, intersect surfaces at cathode potential;
2. A gaseous glow discharge maintained in the volume by the application of a negative voltage to the cathode surfaces. The dominant voltage drop occurs across positive ion sheaths which form and adhere to the cathode surfaces. (An alternate mode of operation has been observed [19,20].);
3. A magnetic field strength high enough to retain, or 'trap' the electrons, which are released from the cathode surfaces by ion bombardment, within

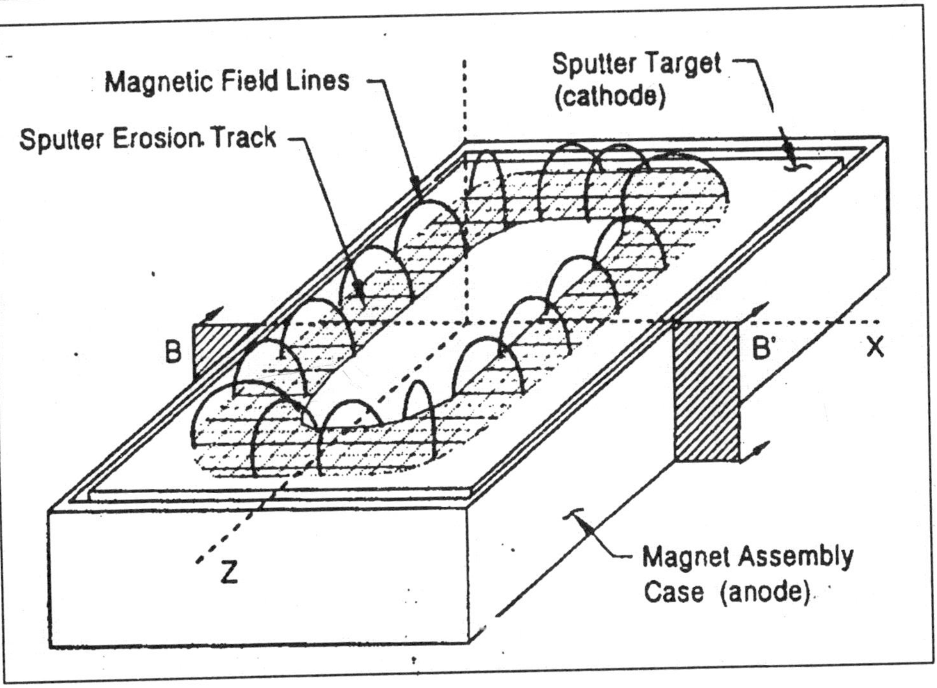

Figure 8.1: Sketch of a Generic Rectangular Planar Magnetron

the volume until a substantial fraction of their energy is lost to ionizing collisions with ambient gas molecules;

4. A geometry which allows a substantial fraction of the gaseous ions produced in the trap volume to be attracted to, and collected by, the cathode surfaces which delimit the volume. These ions are accelerated by the positive ion sheath at the cathode and cause sputter erosion of cathode material.

Present day magnetrons have characteristic dimensions which range from a few cm to more than 500 cm. They operate at voltages in the range 200-900, at pressures in the range 0.6-50 m Torr, with field strengths in the range 50-400 Gauss, and at power levels ranging from 1 watt to 120kW. At power densities in excess of 80 W/cm^2 copper has been self-sputtered [21]. In this case enough ions are formed from the sputtered copper atoms so that a stable discharge is maintained without the help of argon gas.

Properties of the Magnetron Discharge

In sputter magnetrons there is a substantial induced drift current which flows parallel to the cathode surface and is located adjacent to the cathode ion sheath. This self-closing flow of electrons is along the sputter erosion track[22] and is usually called 'the E x B drift current'. However the only large electric field in a magnetron discharge occurs in the cathode ion sheath which is typically very thin[23] and does not encompass, or directly cause, the drift current. The drift current is actually caused by: (i) reflection of γ-electrons from the ion sheath; (ii) the curvature of the magnetic field lines; and (iii) the gradient of the magnetic field strength.

The electron trap of a sputter magnetron can be quite efficient at extracting energy from a γ-electron. Sheridan and Goree[24] used a small circular planar magnetron, whose etch track was approximately 5 cm in diameter and 1 cm wide, to determine the trapping times in an argon discharge. At an applied voltage of 400, a tangential magnetic field of 250 Gauss, and an argon pressure of 1 Pa they found that the trapping time was 0.85 μs at a cathode current density of 15 mA/cm^2. Monte Carlo modeling of the system [25] revealed that the average γ-electron produced about 14 ions (out of a maximum possible number of 25 and a probable maximum number of 15[26]) before escaping from the trap. Good trapping ceased to occur when the tangential magnetic field was reduced to about 100 Gauss[27]. The speed of an electron of kinetic energy K (expressed in eV) is:

$$v = \sqrt{\frac{2e}{m}K} = 59\ \sqrt{K}\ \text{cm}/\mu\text{s}$$

where,

m is the electron mass and e is the unit electronic charge. Consequently an electron with an initial energy of 400 V would travel a distance of about 2/3 x 59 x 20 x 0.85=669 cm while losing energy in the trap. This is very much larger than the dimensions of the magnetron.

Rossnagel [28,29] and Escrivao *et al.* [30] have demonstrated that a considerable reduction of neutral gas density occurs in the trap region as the discharge current is raised. This should lead to trapping times which are an increasing function of cathode current density because it is primarily electron-neutral collisions which cause electrons to diffuse out of the trap volume. The measurements of Sheridan and Goree confirm this trend. Hoffman [31] using a novel form of a cylindrical-post magnetron, observed a related phenomenon which he called 'sputtering wind'.

The usual textbook description of the interaction of an electron with a magnetic field pictures the electron following a helical orbit, of perhaps varying diameter, which encloses a magnetic field line called the guiding centre [32-34,11]. Such a picture is not appropriate for describing γ-electron orbits in the magnetic trap at a magnetron cathode[25] and an alternative picture must be found. This will be discussed later.

The electron temperature in an argon magnetron plasma has been observed to be bi-modal [35] with a 7-8 e V hot component (whose source appears to be the cathode electron trap) and a cold component of about 0.5 eV. This is surprisingly similar to the bi-modal distributions which are observed in classical parallel-plate discharges [36].

The optical emissions from the trap region of a circular planar magnetron have been studied as a function of discharge current at a pressure of 12m Torr [37]. At high currents the emission from sputtered metal atoms can overwhelm that from neutral argon atoms [27]. Some impressive tomographic displays of the emission from the trap region of a circular planar magnetron of somewhat unusual design were published by Miyake *et al.* [38]. These 3D displays show that, as expected, the emission originate from the plasma of the electron trap which lies over the circular cathode erosion track.

The magnetic field strength of planar magnetrons usually falls rapidly with distance from the cathode surface and the radius of curvature of the field lines becomes relatively large. Consequently the rather low-energy electrons in the bulk of the plasma tend to gyrate about field-line guiding centres in helical orbits as described by the classical textbook pictures. The product of the radius of gyration, ρ, and the magnetic field strength, B, can be calculated from the following Larmor equation:

$$B\rho = \sqrt{\frac{2m}{e}k} \perp = 3.37 \sqrt{K} \perp \text{G cm} \tag{1}$$

where,

K is the component of the kinetic energy (expressed in eV) which is perpendicular to the field-line guiding centre. It is apparent that magnetic field strengths as low as 5 Gauss can profoundly affect the motion of the 'cold' electrons in the bulk plasma.

References to section on Electron Beam Evaporation

1. 1951 British patent 754 210.
2. 1969 US patents 3230 110.
3. Yamagishi K 1965 Proc. Ist Int.Conf. on Electron Ion Beam Science, ed Bakish (New York: Wiley) pp 245-63.
4. Graper E B 1970 Charged particle flux generated by an electron-beam deposition source J.Vac. Sci. Technol. 7 (1) 282-5.
5. Hanks C W 1969 US patent 3483 417.
6. Graper E B 1972 Deposition of aluminium from an electron beam source J. Vac. Sci. Technol. 9(1) 33-6.
7. Graper E B 1971 Evaporation characteristics of materials from an electron-beam gun J.Vac.Sci.Technol. 8(1) 333-7.
8. Graper E B 1987 Evaporation characteristics of materials from an electron-beam gun II J.Vac.Sci.Technol. A 5(4) 2718-23.

References to section on Laser Ablation

1. Dijkkamp D, Venkatesan T, Wu X D, Shaheen S A, Jisrawi N, Min-Lee Y H, McLean W L and Croft M 1987 Appl.Phys.Lett. 51 619.
2. Otsubo S, Maeda T, Minamikawa T, Yonezawa Y, Morimoto A and Shimizu T 1990 Japan, J.Appl.Phys. 29 L133.
3. Kidoh H, Morimoto A and Shimizu T 1991 Appl.Phys.Lett. 59 237.

4. Venkatesan T, Wu X D, Inam A and Wachtman J B 1988 Appl.Phys.Lett 52 1193.
5. Ready J F 1963 Appl.Phys.Lett 3 11.
6. Paek U C and Gagliano F C 1972 IEEE J. Quantum Electron. QE-8 112.
7. Chun U C and Rose K 1970 J.Appl.Phys. 41 614.
8. Srinivasan R and Leigh W J 1982 J.Am.Chem. Soc. 104 6784.
9. Srinivasan R 1982 Polymer 23 1863.
10. Srinivasan R and Mayne-Banton V 1982 Appl.Phys.Lett 41 576.
11. Koren G, Gupta A, Baseman R J, Lutwyche M I and Laibowitz R B 1989 Appl.Phys.Lett 55 2450.
12. Otsubo S, Minamikawa T, Yonezawa Y, Morimoto A and Shimizu T 1990 Jpn J.Appl.Phys. 29 L73.
13. Eryu O, Murakami K and Masuda K 1989 Appl.Phys.Lett. 54 2716.
14. Eryu O, Murakami K, Masudu K 1989 Appl.Phys.Lett. 54 2716.
15. Tabata H, Murata O, Kawai T and Kawai S 1990 Appl.Phys.Lett. 56 1576.
16. Venkatesan T, Wu X D, Muenchausen R and Pique A 1992 MRS Bull. (February) p 54.
17. Kwok H S 1992 Thin Solid Films 218 277.
18. Wu X D, Muenchausen R E, Foltyn S, Estler R C, Dey R C, Garcia A R, Nagar N S, England P, Ramesh R, Hwang D M, Ravi T S, Chang CC, Xi X X, Li Q and Inam A 1990 Appl.Phys.Lett. 57 523.
19. Kidoh H, Ogawa T, Morimoto A and Shimizu T 1991 Appl.Phys.Lett. 58 2910.
20. Chrisey D B, Horwitz J S and Grabowski K S 1990 Mat.Res.Soc. Symp.Proc.191 25.
21. Kawasaki M, Gong, J, Nantoh M, Hasegawa T, Kitazawa K, Kumagi M, Hirai K, Horiguchi K, Yoshimoto M and Koinuma H 1993 Jpn.J.Appl.Phys 32 1612.
22. Otsubo S, Minamikawa T, Yonezawa Y, Maeda T, Morimoto A and Shimizu T 1989 Jpn.J.Appl.Phys. 28 2211.
23. Van de Riet E, Nillesen C J C and Dieleman J 1993 J.Appl.Phys. 74 2008.
24. Geohegan D B, Mashburn D N, Culbertson R J, Pennycook S J, Budai J D, Valiga R E, Sales B C, Lowndes D H, Boatner L A, Sonder E, Eres D, Christen D K and Christie W H 1988 J.Mat. Res.5 2075.
25. Agostinelli E, Bohandy J, green W J, Kim B F, Adrian F J and Moorjani K 1990 J.Mat.Res.5 2075.
26. Chen C H, Philips R C and Morrison P W 1992 Thin Solid Films 218 291.
27. Lubben D, Barnett S A, Suzuki K, Gorbatkin S and Greene J E 1985 J.Vac.Sci.Technol. B 3 968.

28. Eryu O, Yamaoka K, Murakami K, Masuda K 1991 Proc. Int. Conf. on Solid state Devices and Materials Extended abstracts p 438.

29. Strikovsky M D, Klyuenkov E B and Gaponov S V, Schubert J and Copetti C A 1993 Appl.Phys.Lett. 63 1146.

30. Yonezawa Y, Segawa K, Katayama S, Minamikawa T, Morimoto A and Shimizu T 1994 Jpn.J.Appl.Phys.33 L1178.

31. Threshold Limit Values and Biological Exposure Indices for 1986-1987 American Conference of Government Industrial Hygienist (ACGIH).

References to section on Magnetron Sputtering

1. Penning F M 1936 Physica III 9 873.
2. Penning F M and Moubis J H A 1940 K.Ned.Akad.Weten. 43 41.
3. Penfold A S 1989 Thin Solid Films 171 99.
4. Penning F M 1939 US Patent 2146 025.
5. Jepson R L 1959 Le Vide 80 80.
6. Helmer J C and Jepson R L 1961 Proc. IRE 49 1920.
7. Vossen J L and Kern W (eds) 1978 Thin Film Processes (New York: Academic).
8. Clark P J 1971 US Patent 3616450 (Filed 1968).
9. Fraser D B 1978 See chapter II-3 of reference [7].
10. Penfold A S and Thornton J S 1975 US Patent 3884 793 (Filed 1971).
11. Thornton J A and Penfold A S 1978 Thin Film Processes (New York: Academic).
12. Kirov K I, Ivanov N.A, Atanasova E D and Minchev G M 1976 Vacuum 26 237.
13. Chapin J S 1979 US Patent 4166018 (Filed 1974).
14. Chapin J S 1974 Research/Development Magazine (Jan) p 37.
15. Waits R K 1978 Thin Film Processes (New York: Academic) ch II-4.
16. Kasaev I G and Pashkova VV 1959 Sov. Phys-Tech Phys. 4 254.
17. Mullaly J R 1969 RPt-1310 Dow Chemical Co. Rocky Flats Div.
18. Mullaly J R 1970, Research/Development magazine (Feb) p 40.
19. Wasa K and Hayakawa s 1992 Handbook of Sputter Deposition Technology (Park Ridge, NJ: Noyes Publications.
20. Hayakawa S and Wasa K 1965 J.Phys. Soc.Japan 20 1692.
21. Kukla R, Krug T, Ludwig R and Wilmes K 1990 Vacuum 41 1958.
22. Rossnagel S M and Kaufman H R 1987 J.Vac.Sci.Technol.A 5 88.
23. Rossnagel S M and Kaufman H R 1987 J.Vac.Sci.Technol A 5 2276.
24. Sheridan T E and Goree J 1989 J.Vac.Sci.Technol.A 7 1014.
25. Sheridan T E, Goeckner M J and Goree J 1990, J.Vac. Sci.Technol.A 8 30.

26. Christophorou L G 1971 Atomic and Molecular Radiation Physics (London: Wiley-Interscience) ch 2.

27. Miranda J E, Goeckner M J, Goree J and Sheridan T E 1990 J.Vac.Sci.Technol. A8 1627.

28. Rossnagel S M 1988 J.Vac.Sci Technol. A 6 1821.

29. Rossnagel S M 1990 Handbook of Plasma Processing Technology ed S M Rossnagel, J J Cuomo and W D Westwood (Park Ridge, NJ: Noyes Publications) ch 6.

30. Escrivao M L, Moutinho A M C and Maniera M J P 1994 J.Vac.Sci. Technol. A 12 723.

31. Hoffman D W 1985 J.Vac.Sci.Technol. A 3 561.

32. Nicholson D R Introduction to Plasma Theory 1983 (New York: Wiley) ch 2.

33. Seshadri S R Fundamentals of Plasma Physics 1973 (New York:Elsevier) ch3.

34. Thornton J A 1982 Deposition Technologies for films and Coatings ed R F Bunshaw (Park Ridge, NJ: Noyes Publications) ch 2.

35. Sheridan T E, Goeckner M J and Goree J 1991 J.Vac.Sci.Technol. A 9 688.

36. Godyak V A, Piejak R B and Alexandrovitch B M 1991 Phys.Rev.Lett.68 40.

37. Rossnagel S M and Saenger K L 1989 J.Vac.Sci.Technol. A 7 968.

38. Miyake S, Shimura N and Makabe T 1992 J.Vac.Sci.Technol. A 10 1135.

39. Cormia R L 1975 US Patent 4 046 659.

Chapter 9

Micro Arc Oxidation Vs Hard Anodizing: Process Features and Coating Properties

L. Rama Krishna
International Advanced Research Centre for Powder Metallurgy and New Materials, (ARCI), Balapur (P.O), Hyderabad – 500 005, India
Email: lingamaneni2000@yahoo.com, info@arci.res.in, arcittm@yahoo.co.in

ABSTRACT

The author here examines the development of various surface modification techniques for aluminium alloys, such as hard anodized coatings, micro arc oxidation coatings and detonation spray coatings etc. Points out that the Micro Arc Oxidation (MAO), which is relatively a new process, has an edge over the other processes in the sense that the process requirements, coating deposition procedure and the properties of the coatings obtained through MAO are radically different from the conventional anodizing processes. In recognition of its wide-ranging advantages, the MAO coating process is gaining worldwide momentum and attracting greater research attention as well as industrial interest. Highlights the achievements made by the ARCI in this direction and points out that it has taken the lead in addressing various issues relevant to commercialization of the MAO technology for the benefit of the Indian industry. In order to facilitate MAO coating on large components and the subsequent technology transfer, an industrial version of the MAO set-up with mass production capabilities, as well as improved design and added safety features is successfully accomplished at ARCI. Finally discusses possible further directions for R&D efforts.

Keywords: *Aluminium alloys, Anodizing, Ceramic coatings, MAO coatings, Plasma electrolytic oxidation, Hard-anodized coatings, Detonation spray coatings, Corrosion resistance, Wear resistance.*

Introduction to Anodizing

Aluminium alloys, which are strategically and structurally important materials due to their high specific strength, suffer from moderately poor tribological properties. Surface engineering of this class of alloys has consequently been a subject of significant research effort in order to permit their satisfactory use in wear-prone conditions and in this connection numerous technological advances have been made in this field. As a result, surface modified wear-resistant aluminium alloys have been finding increasing application in today's industries.

Among the various surfacing options that exist to treat aluminium alloys, the process of anodizing or controlled oxidation of aluminium and aluminium alloys is more than seven decades old. The primary intent of anodizing aluminium and aluminium alloy parts is to protect the highly reactive surface against corrosion in aqueous environments, such as humid air and seawater. Since the anodic coating can be produced in a variety of colors, anodized parts are used in architectural applications. Furthermore, because the anodizing process produces a hard ceramic coating, many times harder than that of the substrate from which it is formed, anodic coatings are also used to protect aluminium parts from abrasion, especially sand abrasion.

Traditional anodizing is an electrochemical oxidation process. The part to be anodized is connected to the positive terminal of a DC power source and a non-reactive metal, such as stainless steel, is connected to the negative terminal. The aluminium part, which is the anode, and the stainless steel cathode are immersed in an electrolytic bath and a DC voltage is applied across them. The potential difference is of the order of 20-100 V, and the current densities are in the range 0.01-0.1 A/cm^2. The electrolytic baths comprise aqueous solutions of chromic acid, orthophosphoric acid, sulfuric acid, oxalic acid, or combinations thereof. As the electrolytic baths have appreciable resistivity, and because the anodizing process itself is exothermic, the temperature of the electrolytic bath increases greatly during anodizing. Since the anodizing process is quite sensitive to temperature, the bath temperature is controlled rather closely by heat exchanger or refrigeration equipment.

Despite many decades of experience and the expensive equipment employed by the traditional anodizing plants, the acid-bath-based DC anodizing process has severe limitations:

- By the very nature of the low-voltage DC power employed, the anodic coating is quite porous–often the volume per cent of pores is as much as 50 per cent
- The electrolytic baths are made up of extremely low-pH acidic electrolytes, and thus the process does not meet many of today's environmental regulations
- The traditional process, for reasons that are not fully understood, cannot be used for anodizing aluminium alloys containing high concentrations of Cu and Si. Thus, many aerospace and automotive parts cannot be satisfactorily anodized, if at all.

- The present process, while appropriate for a limited range of wrought-aluminium alloys, cannot be used for anodizing other reactive metals, such as Ti, Zr, Mg etc., and inter metallic compounds and metal-matrix composites. Thus, most of the promising aluminium based advanced alloys and composite materials cannot be protected by the traditional anodizing process.
- Above all, the hardness of even the so-called hard anodic coatings is far below the hardness of alpha-alumina, the principal component of the anodic coating. Accordingly, the traditional process cannot rationalize the full strength potential of the anodic layer. Indeed, the other potentially beneficial properties of aluminium oxide, such as the high thermal and electrical resistivities and the high dielectric breakdown strength, are not even addressed.

This state of affairs is primarily due to the porosity of the coating produced by the traditional acid-based electrolytic processes at low power levels, and to a certain extent the poor bonding between the aluminium-alloy substrate and the anodic layer.

A relatively new process called Micro Arc Oxidation (MAO), also differently known as Plasma Electrolytic Oxidation (PEO), Spark Anodizing (SA) and Micro Discharge Oxidation (MDO), has emerged as a unique technique to produce dense, hard, thick ceramic oxide coatings on different Al, Ti, Mg and Zr alloys. Essentially, the MAO process combines electrochemical oxidation with a high-voltage spark discharge treatment and provides numerous benefits in comparison to the conventional anodizing process.

MAO: A Unique and Innovative Process

Like the traditional processes, the MAO process is also an electrochemical process but there the similarity ends. The MAO process is radically different from the traditional anodizing process in many respects:

- The process employs an alkaline electrolyte whose composition is extremely critical to the coating deposition rate and the coating properties of the anodic film that is formed. The pH of the electrolyte is in the range of 8- 12 and thus is environmentally sound.
- The process employs AC at high voltage and high current. Because of the high voltage, a micro plasma surrounds the electrodes and the oxygen ions produced in the plasma diffuse through the anodic film into the aluminium substrate to react and form a more anodic film.
- The high voltage and high current enable the production of anodic films of the same thickness as that in the traditional process but only a fraction of the time is required in the latter.
- As the voltages are higher than the breakdown voltage of the film formed, open channels are not necessary for sustaining the process and hence dense, thick layers of nonporous film can be readily formed.
- Since the process employs AC power, its productivity is higher.

- The temperature of the electrolytic bath need not be precisely maintained. Successful coatings can be obtained even if the temperature excursions are as much as 10-20 degree C, further simplifying the process.
- Significantly, unlike the traditional anodizing process, the MAO process can successfully anodize aluminium alloy parts of any composition. Even more importantly, a variety of ceramic composite coatings, such as Al_2O_3-SiO_2, Al_2O_3-MgO, Al_2O_3-CaO etc., can also be produced by the MAO process.
- The MAO process is also suited for hard-coating the inside surface of a part, for example cylindrical, conical, or spherical hollow parts. Many commercially available coating processes, such as Chemical Vapor Deposition (CVD), Physical Vapor Deposition (PVD), Plasma-Enhanced Chemical Vapor Deposition (PECVD), sputtering, thermal spraying, etc., are unable to coat the inside surface of a long part.

Due to the fact that the MAO process produces a thick, well-bonded ceramic coating on a variety of reactive light metal alloys, it can be used for a broad range of applications. The primary applications are driven by the need to replace heavier metallic alloys, or the more expensive composite materials required by the aerospace and automotive industries, with light metals (*e.g.*, Al, Ti, Mg and their alloys) that are coated by the process for enhanced surface properties. Other applications in the chemical, mechanical, thermal, electrical, and electronic industries also exist. Specific features of the MAO coating technique that make it suitable for applications in certain key industry segments are exemplified below:

- The ceramic coatings produced by the MAO route can resist both aqueous environments and moderately high temperatures with demonstrated resistance to strong acids and bases. Thus, the technique can be used in the chemical and food processing industries also.
- The hardness of the film produced by the MAO technique being more than 1400 kg/mm^2, it can be used to resist mechanical sliding, abrasive, and erosive wear. In addition, the coefficient of friction is low, thereby making it appropriate for use in marginally lubricated systems.
- The thermal conductivity of the anodic film is much less than that of metals. As a consequence, anodized parts can be used to maintain uniform temperature distribution, and to resist thermal shock.
- The dielectric breakdown strength of the MAO film is comparable to that of a-Al_2O_3, and hence the coating can be used as an insulating film on electrical and electronic components.
- Additionally the MAO process is also well suited for hard-coating interior surfaces, recesses, blind holes, threaded sections, and so on.

From the above information, it is apparent that the MAO process is unique and innovative in the sense that the process requirements, coating deposition procedure and the properties of the coatings obtained through MAO are radically different from the conventional anodizing processes. In recognition of its wide-ranging advantages, the MAO coating process is gaining worldwide momentum and attracting greater research attention as well as industrial interest.

Background of MAO Process

Micro arc sparking phenomenon in anodizing was initially observed by scientists in the former Soviet Union, particularly in the current Ukraine and Russia, about 40 years ago. Further investigation led to the micro arc oxidation technology that was initially developed for military purposes. It has been claimed that this technology was used to prepare submarine parts in the former Soviet Union. Later on, research activities pertaining to the basic understanding about the MAO process were also initiated in few other countries like USA, Europe and Israel. More recently, researchers from China, Japan and Australia have also reported studies conducted to explore different aspects of the micro arc oxidation technology. To date, there are about 100 scientific publications dealing with micro arc oxidation technology that are distributed in a wide variety of journals and conference proceedings. In addition, there exist more than 50 patents pertaining to the MAO technology and the apparatus employed for carrying out the process. Now widely acknowledged as a fascinating technology, micro arc oxidation is attracting an increasing attention from both academic institutions and industries.

In India, International Advanced Research Centre for Powder Metallurgy and New Materials (ARCI), consistent with its mandate to develop surface-engineering related technologies and eventually transfer them to the Indian industry, has taken the lead in addressing various issues relevant to commercialization of the MAO technology for the benefit of the Indian industry. Towards this objective, collaboration between India and Moldova was initiated in 1997-98 and led to in-house fabrication of a bench-scale MAO unit necessary for generating technical know how as well as a process knowledge base. Since then, dedicated research activities have resulted in multidirectional achievements that have been responsible for substantial improvement in global status of the technology, prompting patent filing both in India and abroad. Details on various aspects of the MAO technology are presented in the following sections.

Basic Elements of MAO Technology

Micro arc oxidation technology basically consists of a special power source (Figure 9.1) and an alkaline electrolyte, which is generally proprietary in nature. Unlike conventional anodizing, MAO has to utilize a high voltage power supply ranging from 150 to 1,000 V in either AC or DC. An AC supply with asymmetric anodic and cathodic potential peak waveforms is usually preferred for achieving higher productivity in combination with the superior quality of the coatings. The artificial waveforms are sometimes misinterpreted as the semiconductor characteristic of aluminium oxide, which is actually a dielectric material. Current density can typically be as high as 100 amperes per square foot.

Electrolytes are typically composed of silicates, aluminates, meta-phosphates, borates and hydroxides with some additives such as carboxylic acids, vanadates, permanganates, polymers and dispersants, etc. Tanks made up of acrylic (Figure 9.2)

Figure 9.1: MAO Set Up at ARCI

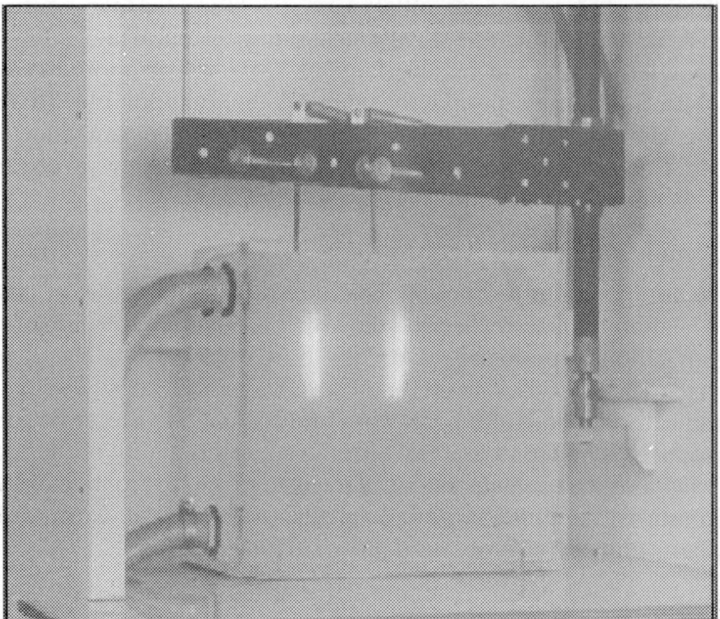

Figure 9.2: A Close View of Reaction Chamber during Coating Deposition

or stainless steel or mild steel are usually employed as the reaction chamber. However, in the case of metallic materials, the tank has to be provided with a proper lining to avoid the mild corrosive action of the alkaline electrolyte. The temperature of the electrolyte is generally controlled within the range of 10 to 60°C. Accordingly, a high capacity chiller is required to maintain the temperature range since the electrical power used in the MAO process generates a large quantity of heat.

Mechanism of Coating Formation and Typical Coating Characteristics

Despite several investigations carried out so far by numerous investigators, the coating formation mechanisms in the MAO process have not been fully understood. A few studies do exist which attempt to unravel the coating formation mechanisms. At the beginning of the process, a thin insulating oxide film is quickly formed on the surface of aluminium substrate in a suitable electrolytic solution. When the voltage is further increased above a certain critical value, the oxide film exhibits dielectric breakdown at discrete locations accompanied by spark discharge.

Three main steps lead to MAO coating formation. In the first step, a number of discrete discharge channels are formed in the oxide layer as a result of loss in its dielectric stability in a region of low conductivity. The material in the channel is heated up by generated electron avalanches up to temperatures of 10^4K. Due to the strong electric field, the anionic components are drawn into the channel. Concurrently, owing to the high temperature, aluminium and alloying elements are melted out of the substrate, enter the channel and get oxidized. Subsequently, the oxidized Al is ejected from the channel onto the coating surface in contact with the electrolyte, thereby increasing the coating thickness in that location. In the last step, the discharge channel gets cooled and the reaction products are deposited on to its walls. The above process repeats itself at a number of discrete locations over the entire coating surface, leading to overall increase in the coating thickness.

The very high cooling rate experienced by the oxide layer getting ejected out of the channel due to its immediate contact with the electrolyte promotes the formation of γ phase during solidification of alumina droplets. However, the low thermal conductivity of alumina, causes the underlying layers of the coating to get heated and, thus, further transformation of the initially formed γ-Al_2O_3 to the much harder α-Al_2O_3 occurs. Consequently, the proportion of α phase increases with increasing depth from the coating surface towards the coating–substrate interface. Transmission Electron Microscopy (TEM) investigation of MAO coatings has revealed that the coating close to the interface exhibits a nano scale polycrystalline microstructure.

A review of the literature clearly indicates that earlier studies have not addressed the interrelationship between the kinetics of the MAO coating formation and variation in the number and size of the discharge channels with reference to the treatment time. Hence, it is essential to investigate in detail, the mechanisms underlying the formation of the MAO coatings. A 7075 Al-alloy having the nominal composition 1.29 per cent Cu, 5.3 per cent Zn, 2.11 per cent Mg, 0.21 per cent Fe, 0.18 per cent Si, 0.15 per cent Mn, 0.12 per cent Cr was used as the substrate and the coating investigated in detail to gain greater insight into the associated mechanism.

Figure 9.3 illustrates the influence of MAO treatment time on the thickness of the coatings deposited and the data reveals that the kinetics of coating formation is linear. The linear rate constant is about 1.45 µm/minute, equal to the average coating thickness deposited per minute of the treatment time.

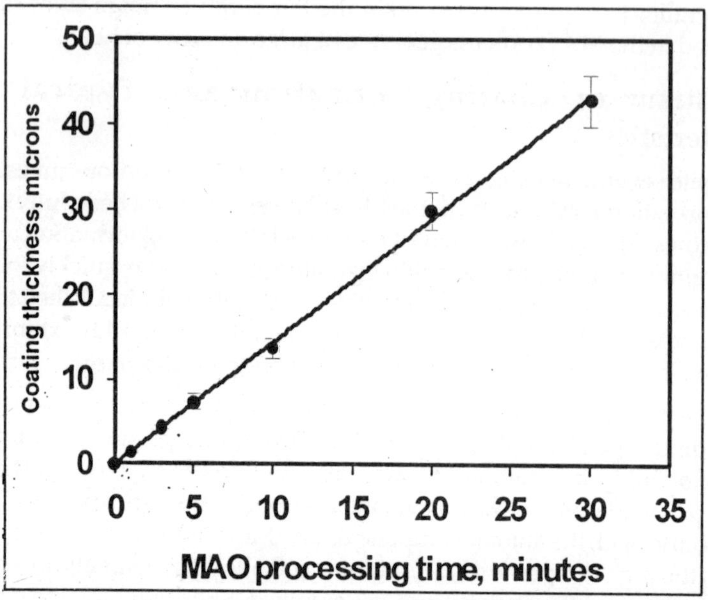

Figure 9.3: Influence of MAO Treatment Time on Coating Kinetics

The variation in surface roughness (Ra) as a function of coating time is illustrated in Figure 9.4. It is observed that the surface roughness increases linearly with increase in coating thickness. In other words, thinner coatings exhibited lower surface roughness and the thicker coatings exhibited higher surface roughness.

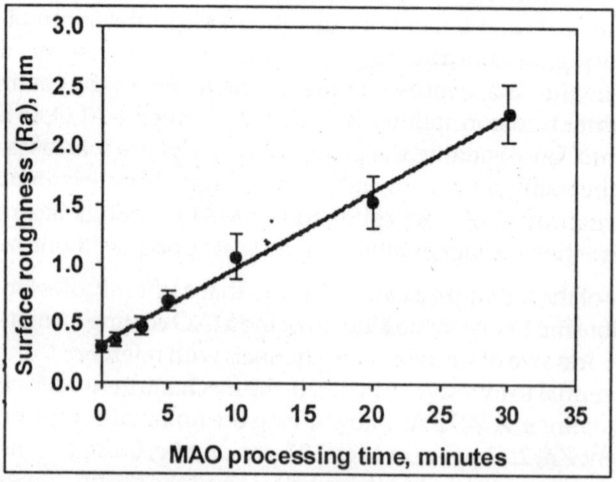

Figure 9.4: Influence of MAO Treatment Time on Coating Surface Roughness

The variation in micro hardness exhibited by a 90mm thick MAO coating obtained is presented in Figure 9.5. It is clear that the MAO coatings typically exhibit a micro-hardness gradient across the coating thickness, with maximum hardness close to the interface and the lowest hardness at the surface of the coating. Figure 9.6 illustrates

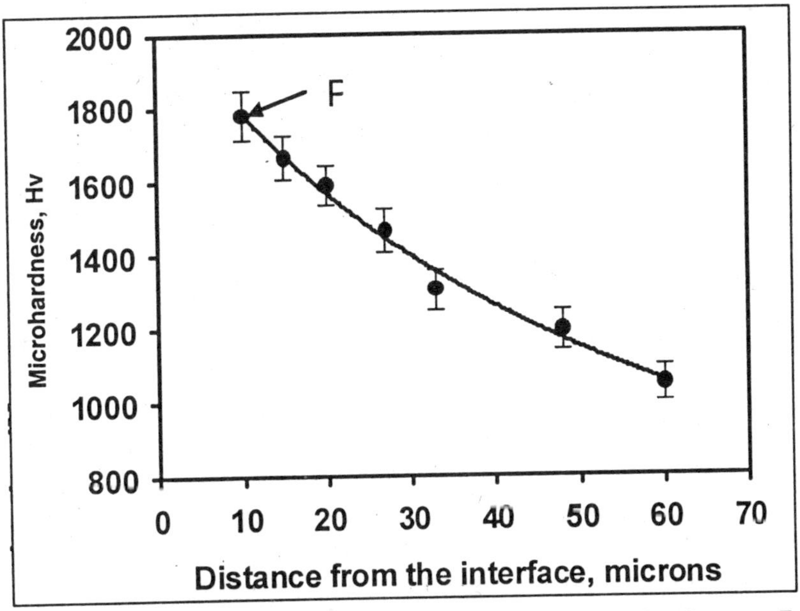

Figure 9.5: Variation in Micro-hardness of a 90 μm Thick MAO Coating as a Function of Distance from the Substrate-Coating Interface

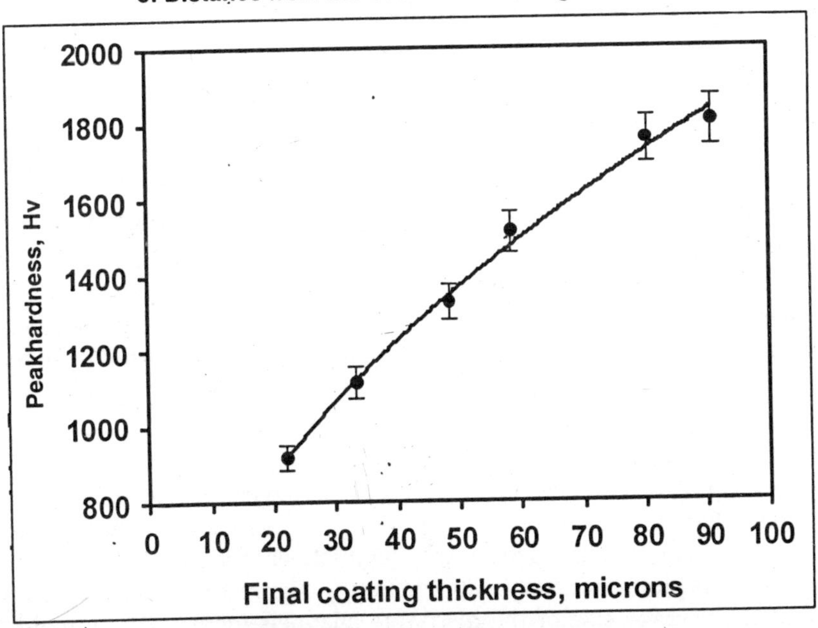

Figure 9.6: Influence of Final Coating Thickness on Peak Hardness Measured at 10 μm Distance from the Interface

the variation in micro hardness values measured at a distance of 10 μm from the substrate-coating interface (termed as peak hardness and indicated as point 'P' in Figure 9.5) as a function of the final coating thickness. It is to be noted that each data point in Figure 9.6 is from different coatings of varying thickness. Figure 9.6 suggests that the peak hardness increases substantially with increased final coating thickness and a similar trend is consistently observed in MAO coatings.

Figures 9.7(a-f) illustrates the XRD spectra of the as deposited surface layers of MAO coatings processed for 1, 3, 5, 10, 20 and 30 minutes respectively. The theoretically calculated maximum penetration depth of X-rays (Cu-kα radiation) is about 7 μm in 100 per cent dense and defect free alumina. The coating thickness measured on 1, 3, 5 minutes processed samples were less than 7 μm. Thus, in the case of Figure 9.7(a-c), the characteristic diffraction peaks of aluminium are very strong due to substrate effects and should be excluded from analysis. Keeping the above aspect in mind, it can be concluded from Figure 9.7(a-f) that the surface layers of the MAO coatings are comprised of mainly γ-Al_2O_3 phase with a small proportion of α-Al_2O_3 also present. In addition, Figure 9.7(d-f) exhibit identical spectra thereby indicating that the phases present in the near surface regions of the MAO coatings are identical irrespective of coating time (10 to 30 minutes).

Figures 9.8 (a-f) illustrate the surface features of the coated samples that were MAO treated for different periods of time at a magnification of 500X. High magnification SEM micrographs of the coated surfaces (*i.e.*, 2000X) are also presented in Figures 9.9 (a-f). The above micrographs clearly indicate the presence of discharge channels appearing as dark circular spots distributed all over the surface of the coatings. It is also apparent that the number of such channels decreases with increasing treatment time while the discharge channel diameter increases.

It can be noted from Figure 9.9 that the coatings are typically characterized by the presence of molten regions that were rapidly solidified around the circular discharge channels. The thickest coating in the present study (Figure 9.9f) typically represents a pancake kind of structure wherein the centre of each pancake is a discharge channel through which the molten alumina flowed out of the channel and rapidly solidified leaving the sharp and distinctly visible boundaries delineating each pancake.

Figure 9.9f also illustrate another interesting feature that the growth of the coating takes place by the individual contribution of different discharge channels formed as the treatment time increases rather than by means of few selective discharge channels in action for enormously long duration throughout the process.

For measuring the diameter of the discharge channels, the micrographs obtained at 2000X magnification were utilized. In particular, the diameter of five largest channels was measured after magnifying the selected channel area individually as illustrated in the inset of Figure 9.9(e). The average value of five such channel diameters for each sample corresponding to a particular treatment time was calculated and this average value is plotted against treatment time in Figure 9.10. It is clear from Figure 9.10 that the average size of the discharge channels increased gradually as the treatment time increased from 1 to 30 minutes.

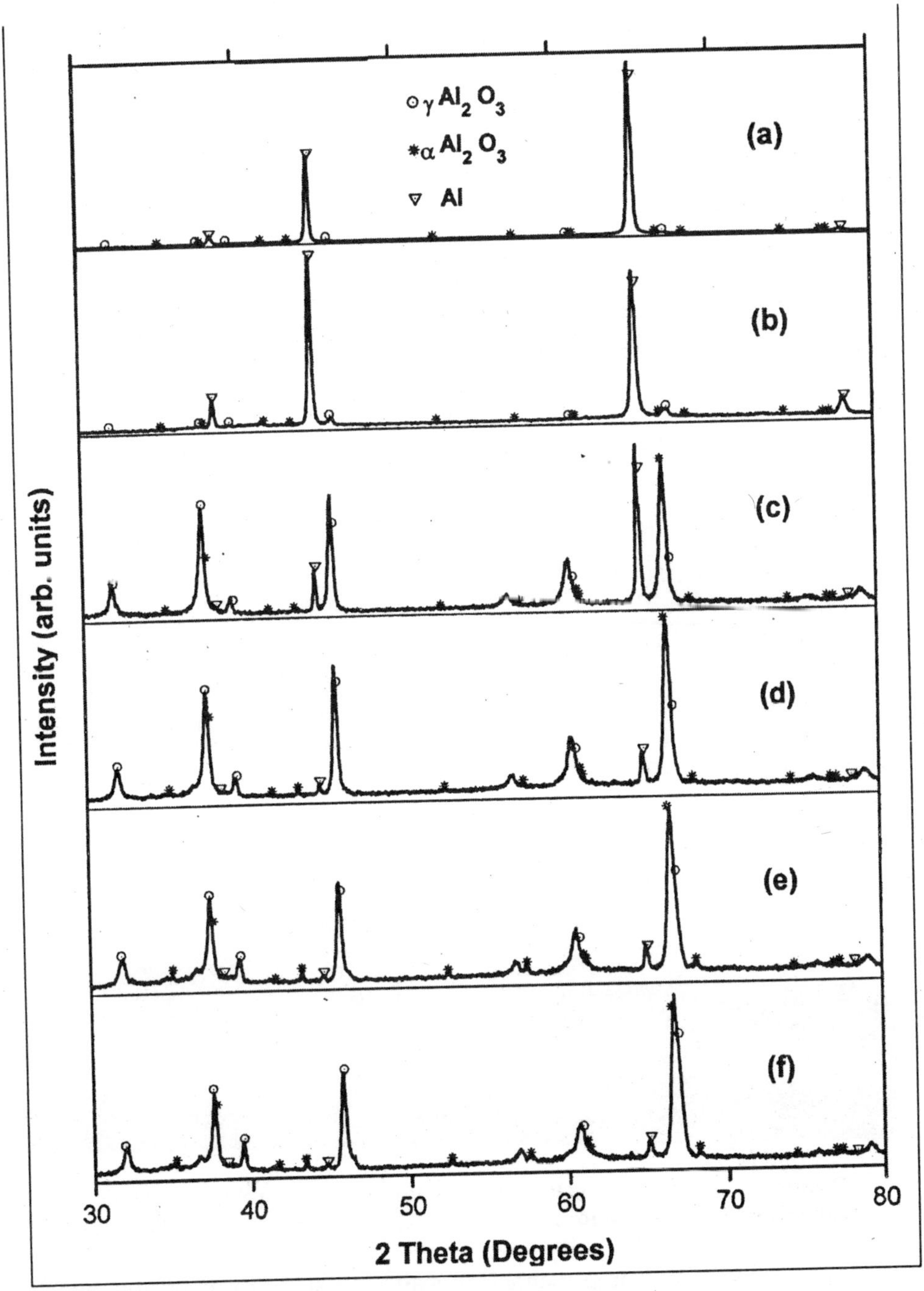

Figure 9.7: XRD Spectra of MAO Coatings Processed for (a) 1, (b) 3, (c) 5, (d) 10, (e) 20 and (f) 30 minutes

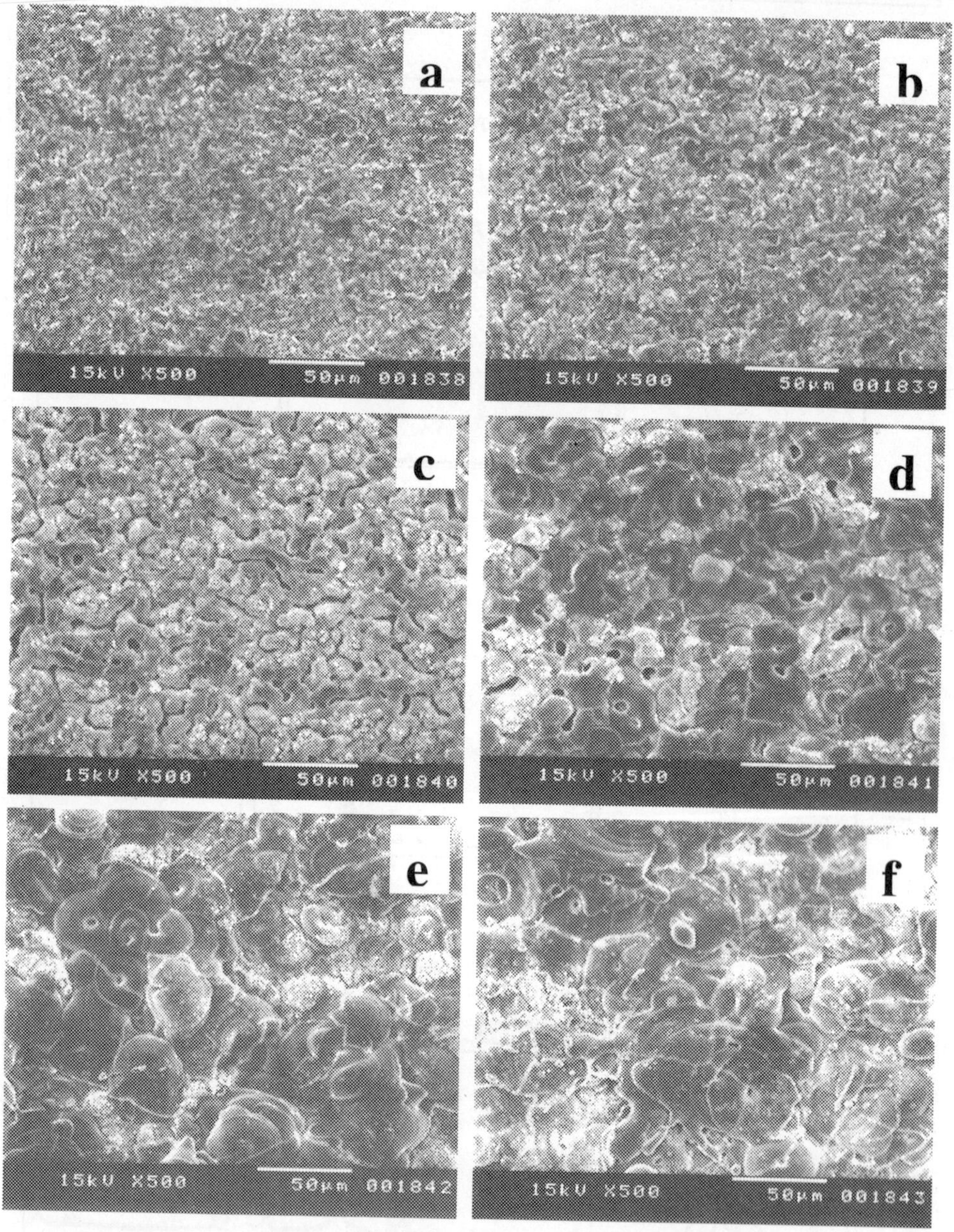

Figure 9.8: Secondary Electron Images of MAO Coating Surfaces at 500X Magnification Processed for (a) 1, (b) 3, (c) 5, (d) 10, (e) 20 and (f) 30 minutes

The number of discharge channels was also measured by analyzing the SEM micrographs obtained at 200X magnification which were then magnified further 200 times using Adobe Photo Deluxe software. Each channel was then marked and total number of channels was counted using the software. The density of the discharge

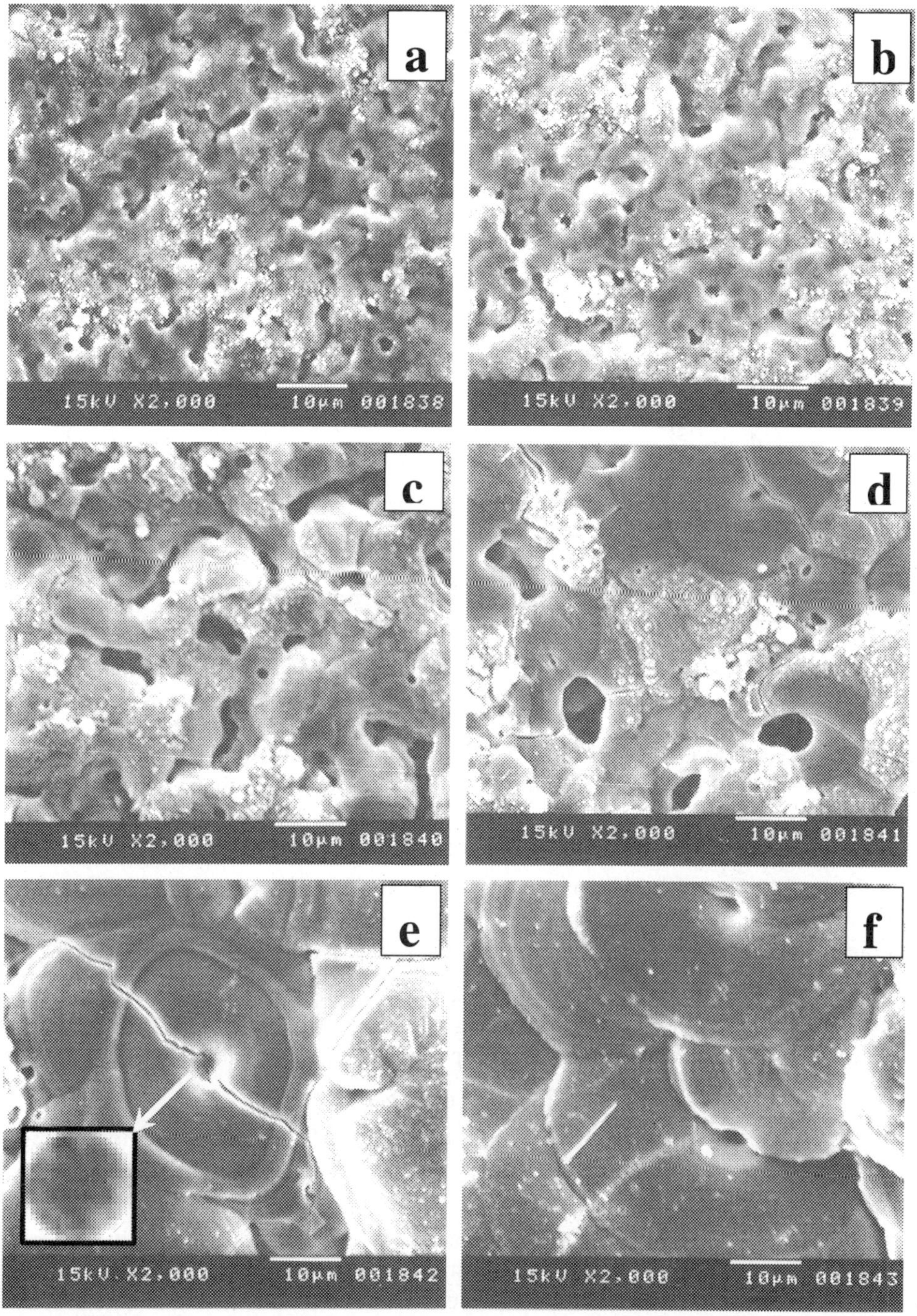

Figure 9.9: Secondary Electron Images of MAO Coating Surfaces at 2000X Magnification, Processed for (a) 1, (b) 3, (c) 5, (d) 10, (e) 20 and (f) 30 minutes

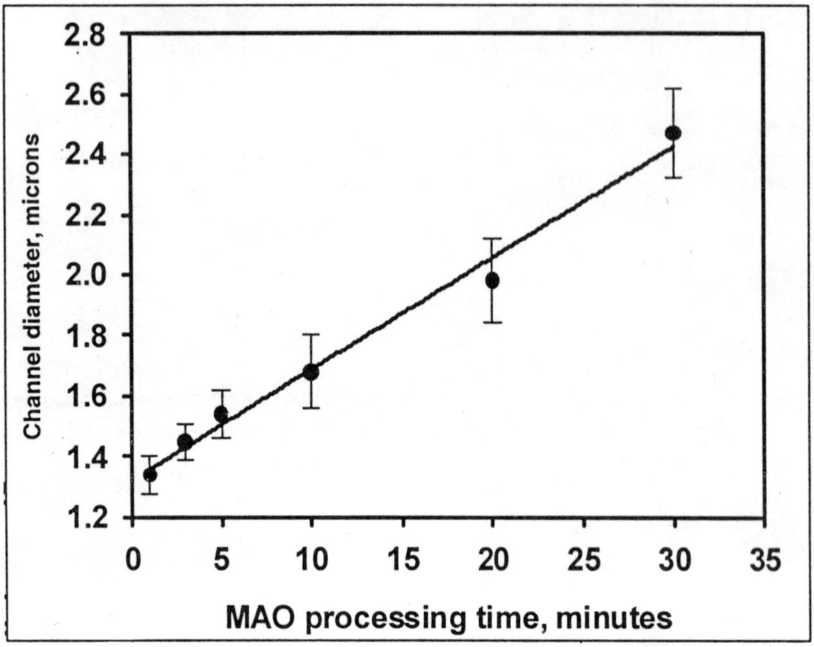

Figure 9.10: Variation in Diameter of the Micro Arc Discharge Channels as a Function of MAO Processing Time

channels *i.e.*, the average number of channels per square meter area, was then calculated. In Figure 9.11, the variation of the density of discharge channel as a function of coating time is presented. It is clear that the discharge channel density decreases exponentially with increasing coating time.

Figure 9.12 illustrate the variation in the pancake diameters measured on coatings obtained through MAO processing for 5, 10, 20 and 30 minutes. It is clear that the average diameter of the pancake increases linearly as a function of coating time.

Prior to a discussion of the possible MAO coating mechanism based on the above-presented results, it is important to clarify that the coating mechanism should explain the following features observed experimentally.

1. A linear coating deposition rate (Figure 9.3)
2. A linear increase in surface roughness of the MAO coating with increasing time of the coating or equivalently with increasing coating thickness (Figure 9.4).
3. The observed micro hardness profile which is characterized by a peak hardness at the substrate-coating interface and further a continuous decrease in hardness (from the peak value) with increasing distance from the substrate-coating interface (Figure 9.5).
4. The increase in peak micro hardness value (near the interface) with increasing coating thickness (Figure 9.6).

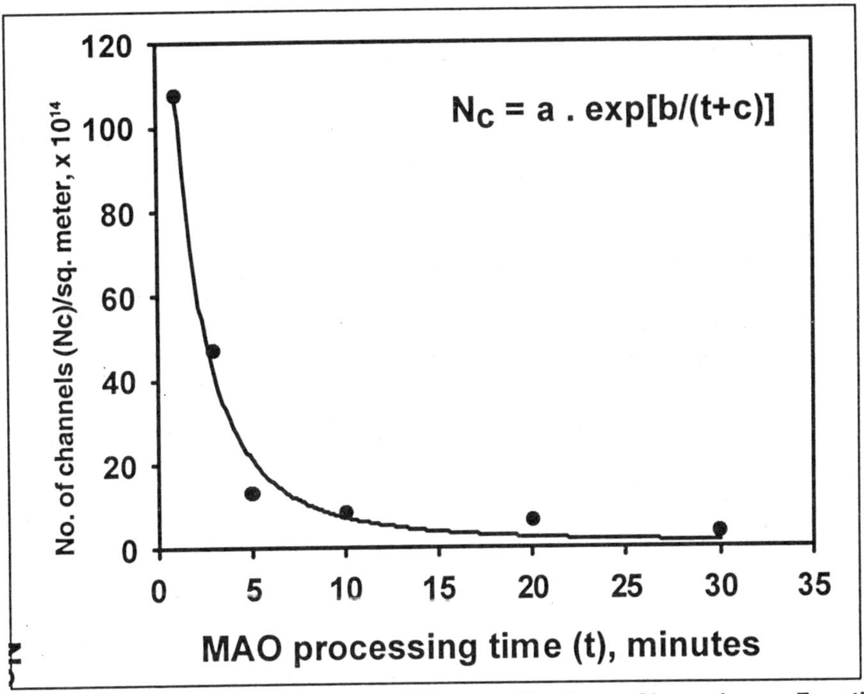

Figure 9.11: Variation in the Density of Microarc Discharge Channels as a Function of MAO Processing Time, Nc: Number of channels per sq. meter area

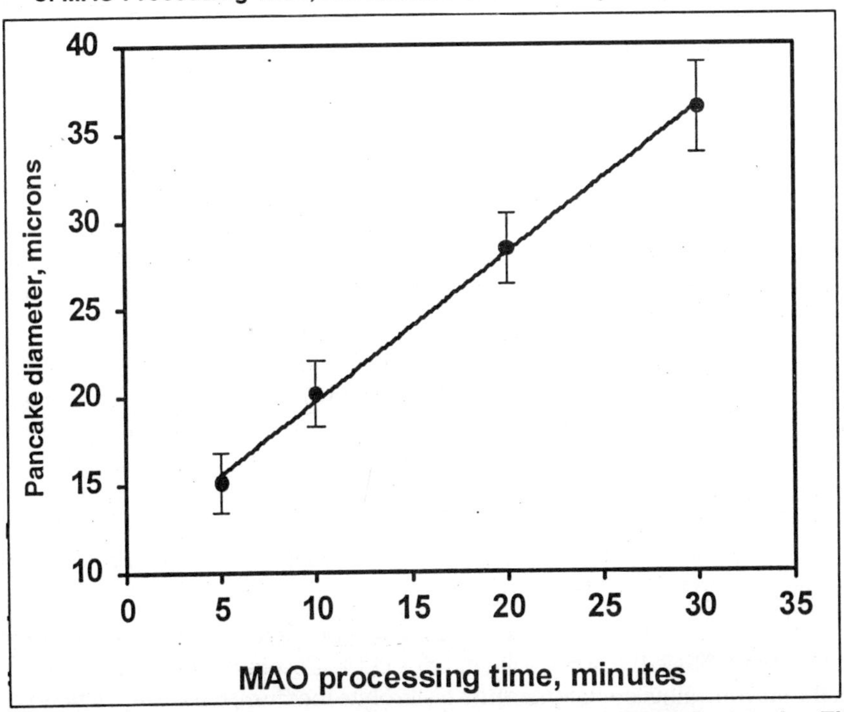

Figure 9.12: Variation in Pancake Diameter as a Function of MAO Processing Time

Based on the experimental observations reported above, a phenomenological mechanism for MAO coating formation has been proposed. In the MAO process, the observed linear deposition kinetics can be simply related to Faraday's laws of electrolysis. However, the micrographs illustrated in Figures 9.8–9.9 clearly indicate that the coating deposition occurs via the discharge channels which, at any time during the coating process, connect the coating-substrate interface all the way with the coating top surface which is in intimate contact with the electrolyte. It is also clear, as indicated by the visual observation of continuous movement of the sparks on the coating surface and also by the SEM micrographs, that the discharge channels have finite life. Thus, the discharge channels are continuously formed and closed all through the coating process. In fact, this accounts for linear kinetics in spite of the fact that the coating thickness is as high as 100 μm.

The experimental data (Figure 9.3) indicates that the coating thickness increases linearly with coating time and this in turn implies a constant volume rate of coating deposition since coating area is a constant. Now each discharge channel, during its life time, deposits material on the coating surface equivalent to the volume of the pancake surrounding it. The volume of the pancake (V_p) is given as,

$$V_p = (\pi/4)\, d_p^2.(d_c/2) = (\pi/8)\, d_p^2\, d_c \tag{1}$$

In eq.1, d_p is the pancake diameter and d_c is the channel diameter. A reasonable assumption that the pancake thickness should equal half the channel diameter (since material flows in all directions out of the channel) has been made. At any instance of time during the coating process, there are N_c number of channels and these channels will deposit a material of volume equal to $N_c.V_p$. Thus, a linear volume rate of the coating deposition implies that $N_c.V_p$ should be nearly a constant at any time during the coating process. Using eqn.1 for V_p, the above postulate implies that the product of the terms $N_c.d_p^2.d_c$ should be a constant independent of coating time. A perusal of Figures 9.10 to 9.12 which indicate the variation of d_c, N_c and d_p respectively with coating time indicates that the product $N_c.d_p^2.d_c$ has a fairly constant values with 10-15 per cent variation. Thus, as a first approximation, the experimental data (with regard to N_c, d_c and d_p) is consistent with the experimentally observed linear kinetics. In physical terms, the decrease in the number of channels with increasing coating time is compensated by the increase in the channel diameter and the concomitant increase in pancake diameter with increasing coating time.

As discussed above, the pancake diameter and thickness (of the order of channel diameter) increases with increasing coating time. Since the coating is formed with pancakes (discs of thickness of the order of channel diameter) as the basic building unit, it should be obvious that the surface roughness of the coating should be of the order of channel diameter as well. The present experiments substantiate the above inference as can be noted by comparing Figures 9.4 and 9.10. Thus, the linear increase in coating roughness is directly related to the fact that the discharge channel diameter also increases linearly with coating time.

The rapid solidification of alumina promotes the formation of meta-stable γ-Al_2O_3 phase. Since the surface layers of the MAO coatings are always in contact with the surrounding low temperature electrolyte medium, the surface layers contain γ-Al_2O_3 phase. According to this explanation, the phase contents of the surface layers should comprise predominantly γ phase irrespective of the coating thickness. The identical XRD patterns with predominantly γ-phase obtained for the coatings wherein the X-ray penetration is totally within the coating as shown in Figures 9.7(d-f) are in clear support of the above explanation.

However, the low thermal conductivity of alumina, causes the underlying layers of the coatings to remain hotter (since the heat generated from the channel cannot be easily dissipated) and the temperature is sufficiently high to cause the transformation of originally formed γ to α Al_2O_3. Thus, it is expected that the proportion of α-Al_2O_3 will continuously increase towards the coating-substrate interface. Such a phase gradient is responsible for the observed micro hardness variation across the coating thickness with peak (maximum) hardness close to the interface since α-Al_2O_3 is harder than γ-Al_2O_3. During the progress of MAO coating deposition, the new discharge channels that are formed heal the pores and cracks formed earlier thereby resulting in the dense inner portion of the coatings free from cracks.

However, for the successful deposition of high quality coatings, one needs to understand the influence of different process variables. Keeping this in the mind, the following section is dedicated for the discussion on the influence of different operating parameters.

Influence of Process Variables on Coating Deposition Kinetics

It is apparent from Figures 9.5 and 9.6 that the properties of the MAO coatings are considerably influenced by the final coating thickness. Hence, it is interesting to understand how the coating deposition rate is influenced by different MAO process variables. In order to exemplify the same, a 7075 Al-alloy has been used as the substrate material for the coating deposition studies.

Figure 9.13 illustrates the variation in coating kinetics with time during coating formation at two different current densities. It is evident that the coating kinetics is linear at both the current densities. It is also clear that an increase in the current density from 0.1 to 0.3 A/cm^2 increases the coating deposition rate. The linear rate constant calculated is 0.5 and 1.5 at the current densities of 0.1 and 0.3 A/cm^2, respectively. Nevertheless, it is to be noted that this linear rate constant is equal to the average coating thickness built up per minute. Interestingly, identically thick coatings synthesized on the same aluminium alloy by means of different processing times at different current densities are found to exhibit identical properties like surface roughness, micro hardness and its distribution across the coating cross-section. From this, it is quite apparent that the current density primarily controls the coating deposition kinetics; however, the coating properties depend on the final coating thickness alone.

The influence of electrolyte temperature on MAO coating kinetics studied at two different electrolyte temperatures at a constant current density of 0.3 A/cm^2 is illustrated in Figure 9.14. It is clear that the temperature of the electrolyte during the MAO process only marginally influences the coating kinetics. This observation is of significant practical importance, since it suggests that it is not essential to maintain the electrolyte temperatures close to 0°C as is the case with anodizing.

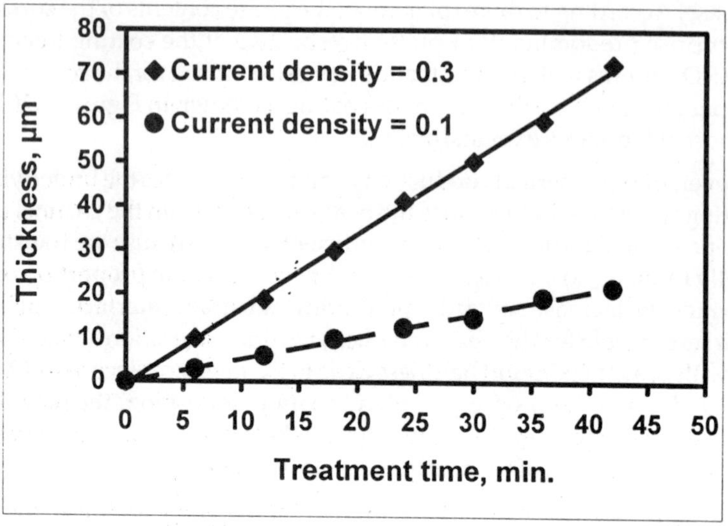

Figure 9.13: Influence of Current Density on MAO Coating Kinetics

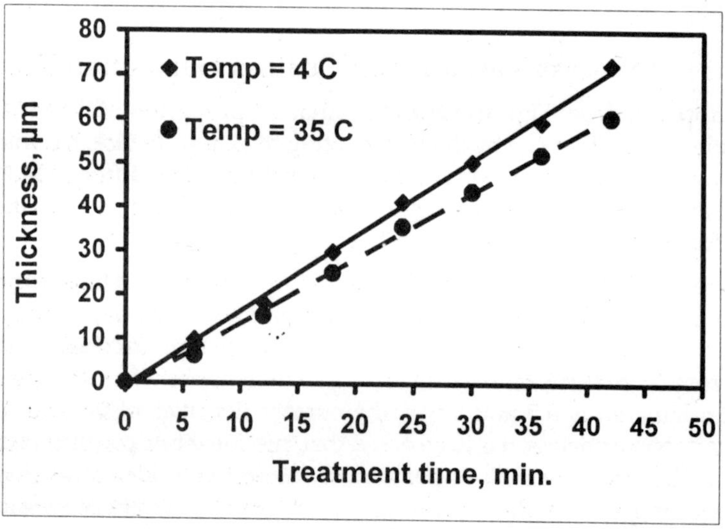

Figure 9.14: Influence of Electrolyte Temperature on MAO Coating Kinetics

Figure 9.15 depicts the influence of inter-electrode distance (distance measured from centre to centre of the work piece) on the final coating thickness formed in 60 minutes at a constant current density of 0.2 A/cm^2 and at an electrolyte temperature of 35°C. It is evident that, although the inter-electrode distance is doubled from 60 mm to 120 mm, the MAO coating growth rate (mm/hour) is not influenced significantly.

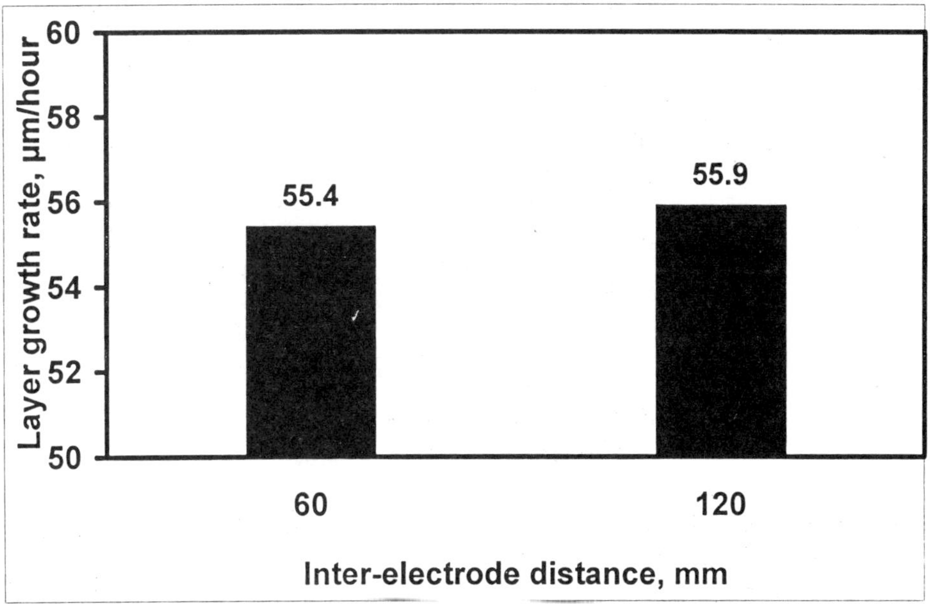

Figure 9.15: Influence of Inter-electrode Distance on MAO Layer Growth Rate

Tribological Performance of MAO Coatings

To realize the full potential of MAO coatings, the tribological performance of these coatings was comprehensively studied under three diverse wear modes namely abrasion, erosion and sliding. In addition, under selective wear modes, the MAO coating performance has also been separately compared against a) un-coated *i.e.*, bare substrate material of identical Al-alloy on which MAO coating has been synthesized, b) hard-anodized coatings and c) detonation sprayed Al_2O_3 and cold iso-statically pressed and sintered Al_2O_3.

Abrasion Resistance of MAO Coatings

Tribological performance of the 7075 Al-alloy in MAO coated and un-coated conditions was evaluated by conducting a dry sand rubber wheel abrasive wear test (ASTM G65). In order to find out the steady state wear loss, the abrasion tests were conducted in discrete steps, weight loss in each step was measured and converted to volume loss and reported. The abrasive wear test was performed at a normal load of 50N and with a step size of 200 wheel rotations each. The steady state volume loss of MAO coated and bare 7075 Al-alloy is illustrated in Figure 9.16.

From Figure 9.16, it is clear that the MAO coating exhibits extremely good abrasion wear resistance, which is almost 70 times higher than the bare substrate material under identical test conditions. It was also to be noted that, during the abrasion test, after every 200-disc rotations, the maximum depth of scar measured at the centre of the scar was very different for the two materials. In the case of bare substrate, this depth was more than 1 millimeter while in the case of MAO coating it was of the order

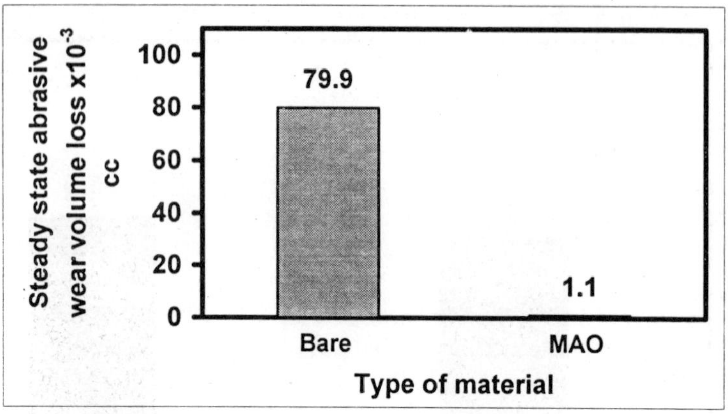

Figure 9.16: Steady State Abrasive Wear Performance of MAO Coating, 7075 Al-alloy

of few microns only. This clearly illustrates the superiority of MAO coatings in resisting material removal due to abrasive action.

Comparison with Hard-Anodized Layers

The primary objectives behind the comparison of MAO coatings with hard-anodized coatings were:

1. To identify and select an Al-alloy that is widely used by the industry in the hard-anodized condition for wear-resistant applications
2. Synthesize MAO and hard-anodized coatings on the above alloy with identical coating surface roughness and coating thickness to facilitate comparison
3. Comparative property evaluation of the above two coatings in comparison with the bare substrate in terms of microstructure, micro-hardness, metallurgical phase distribution and the tribological performance under different wear modes

Amongst the available aluminium alloys, 6000 series is widely used in hard-anodized condition for several industrial applications that require mild wear and corrosion resistant properties. In addition, hard anodizing of 6000 series aluminium alloys is well established. Hence, MAO and hard-anodized coatings were successfully deposited on a 6061 Al-alloy (1.1 per cent Mg, 0.7 per cent Si, 0.6 per cent Fe, 0.1 per cent Cu, 0.15 per cent Mn, 0.25 per cent Cr, 0.2 per cent Zn 0.1 per cent Ti and balance Al) and their coating characteristics as well as tribological performance has been compared.

The average coating thickness of hard-anodized coating was found to be about 50 μm and its surface in the as coated condition had an Ra value of 2.26 μm. MAO coatings, which were slightly thicker to start with, were polished down to 50 μm and it was also ensured that the relatively rougher MAO coated surfaces, which had an

Ra value as high as 3.56 μm, were also smoothened such that the Ra values in the two cases were almost identical.

Figure 9.17 illustrates the micro-hardness measured for the two above coatings in comparison with the substrate material. It is to be noted that the micro-hardness measured for MAO coatings is the value obtained at 10 μm distance from the substrate-coating interface. The variation in the micro-hardness of MAO coatings as a function of distance from the substrate-coating interface has already been discussed in detail in an earlier section. It is clear that, although both coatings yield a substantial increase in surface micro-hardness, the increase in surface micro-hardness due to MAO coating is more than 3-fold higher than that achieved by hard anodizing. The prime reason behind the large difference in micro-hardness, despite the fact that both the coating techniques involve the oxidation of the substrate material, is the difference in the kind of phases formed. X-ray diffraction studies have revealed that the hard-anodized coatings typically contain amorphous alumina layers while the MAO coatings are constituted entirely of the crystalline alumina phases.

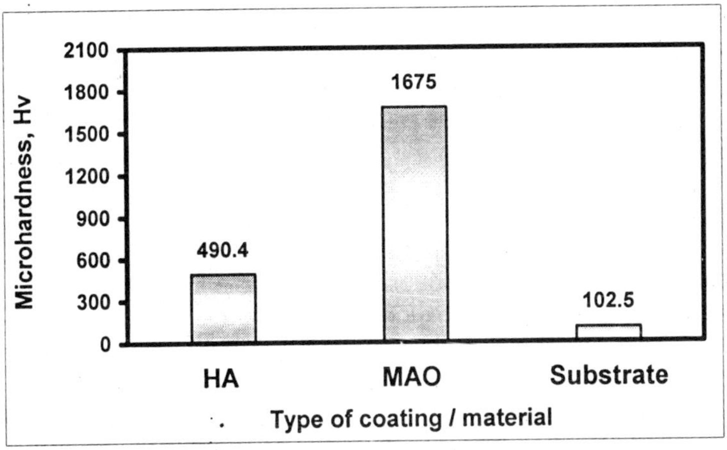

Figure 9.17: Micro-hardness Values of Hard-anodized (HA), MAO Coatings and Substrate Material

An SEM examination of the surfaces of these coatings revealed that the hard anodized coatings contain pin-hole porosity that arises due to the presence of unsealed pores generated during the coating deposition process while, the MAO coating does not exhibit such porosity. However, interestingly, the cross-section of neither of these coatings exhibited the presence of porosity.

Figure 9.18 illustrates the abrasive wear performance of the coatings in comparison with that of bare substrate at a nominal load of 5N with silica as the abrasive medium. It is evident that both the coatings improve the abrasion wear performance of the 6061 Al-alloy. However, there was found to be a major difference in the life of the two coatings as reflected by the total number of disc rotations that these coatings could withstand before their complete removal from the substrate.

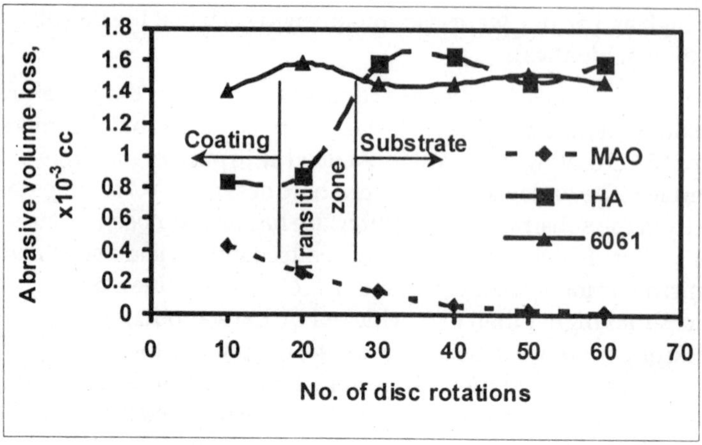

Figure 9.18: Abrasive Wear Performance of MAO and Hard-Anodized Coatings in Comparison with Bare 6061 Al-alloy

After 30 rotations, the hard-anodized coating was completely removed and the specimen began to exhibit wear loss similar to that of the bare substrate. On the other hand, the MAO coating could withstand 60 disc rotations without any evidence of substrate appearance. Interestingly, the wear losses in case of MAO coating was seen to exhibit a continuously decreasing trend as illustrated in Figure 9.20. The reason for such a trend is, as explained earlier, the increasing a-Al_2O_3 content from the surface to the interior of the coating. This clearly illustrates that the MAO coatings offer outstanding wear resistance compared to hard-anodized coatings and the bare substrate. It was also interesting to note that, even when the nominal load was increased by a factor of 10, *i.e.*, at 50N load, the MAO coatings could withstand 1600 rotations.

Figure 9.19 illustrates the solid-particle (silica) erosion wear volume loss per unit weight of erodent exhibited by MAO coatings in comparison with hard-anodized and bare 6061 Al-alloy substrate evaluated at an erodent impact angle of 30° and the particle velocity of 60m/s. It is clearly evident that the MAO coating and 6061 Al-alloy substrate exhibit a nearly identical steady state erosion rate. In the case of hard-anodized coatings, just after an initial erosion time of 5 minutes, the coating was found to be completely removed at the centre of the erosion scar, clearly exposing the substrate material. Gradually, the coating was completely removed from the entire area of erosion scar and eventually the erosion rate of the coated specimen matched with that of the bare 6061 Al-alloy after about 35 minutes. In general, the erosion resistance of ductile/soft materials like aluminium is much better than that of ceramic materials due to the differences in the modes of fracture. Despite this, the MAO coating's erosion resistance is found to be identical to that of 6061 Al-alloy which, in turn, is superior to that the hard-anodized coatings.

In the above discussion, it has been elucidated that the MAO coatings offer superior abrasion and erosion wear resistance than that of hard anodized coatings.

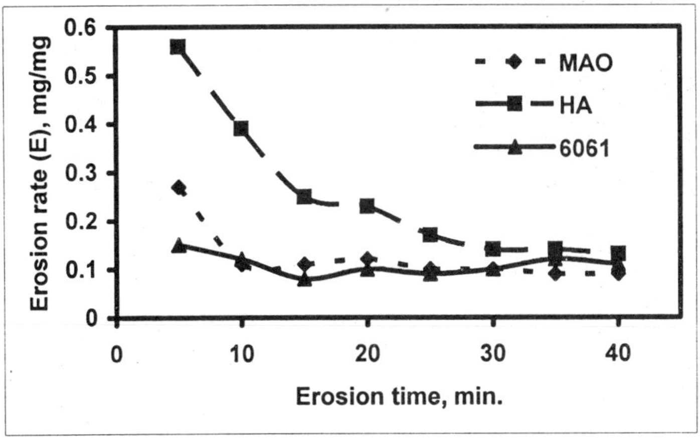

Figure 9.19: Erosion Wear Behavior of MAO, Hard-Anodized and Bare 6061 Al-alloy Substrate

Comparison of Detonation Sprayed and Bulk Alumina

It is also meaningful to compare the properties of MAO coatings with those achieved by another popular protective alumina coating technique, namely Detonation Spray Coating (DSC), as well as with those of bulk alumina. Accordingly, the DSC coated Al_2O_3 samples and hot isostatically pressed and sintered bulk alumina specimens were obtained and fully characterized along with MAO coated 7075 alloy.

The average surface roughness (Ra) measured on the bulk a-Al_2O_3 samples was 0.82 ± 0.1 µm, while DSC and MAO coatings deposited on 7075 Al-alloy exhibited Ra values of 3.1 ± 0.9 µm and 4.8 ± 0.4 µm, respectively. The average coating thickness measured using an eddy current coating thickness gauge was 110mm and 100mm for DSC sprayed Al_2O_3 and MAO coatings, respectively. The average micro-hardness exhibited by the bulk Al_2O_3 specimen was 1567±52 HV and that of detonation sprayed Al_2O_3 coating is 1295±37 HV. In the case of MAO coatings, the identical hardness gradient across the coating thickness as illustrated in Figure 9.5 has been observed.

SEM micrographs of the detonation sprayed Al_2O_3 and MAO coating cross sections are presented in Figures 9.20 (a-b). Both DSC-Al_2O_3 and MAO coatings are found to exhibit uniform coating thickness, and a clean and wavy interface indicating that the coating appears to be well bonded with the substrate. However, the MAO coating exhibited the presence of two distinct regions *viz.*, an inner dense region and an outer granular region. More importantly, the thickness of the outer granular region was less than 5-10 per cent of the total coating thickness.

The XRD spectrums of detonation sprayed Al_2O_3 coating and MAO coatings are shown in Figure 9.21(a-b). Both coatings exhibit an identical combination of γ-Al_2O_3 and α-Al_2O_3 phases in the coating. However, the relative contents of the two phases were found to differ in the two coatings, with the MAO coating having comparatively higher α-Al_2O_3 phase with the increased distance from the coating surface while no such phase gradient was found in the case of detonation sprayed coatings.

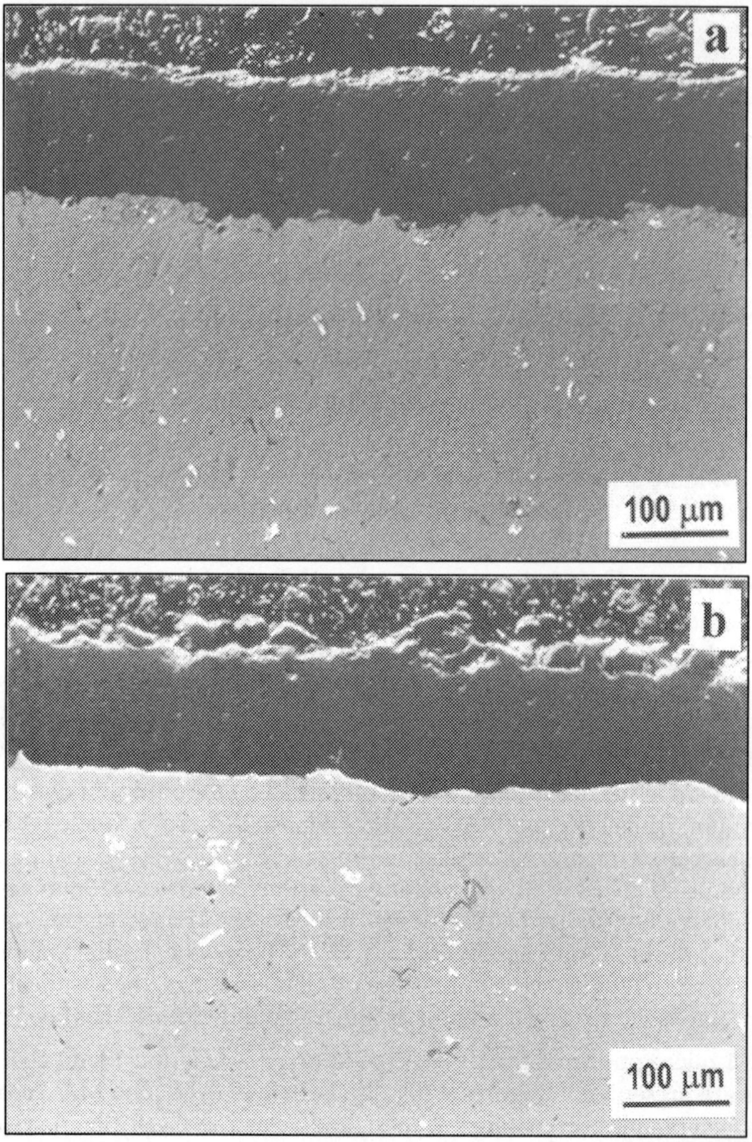

Figure 9.20: SEM Micrographs of (a) DSC Alumina and (b) MAO Coatings

Figure 9.22 illustrates the abrasive wear performance of MAO and DSC coatings in comparison with bulk Al_2O_3 material. It is clearly evident that bulk Al_2O_3 exhibits steady state abrasive volume loss right from the beginning while the DSC and MAO coatings are seen to reach a near steady state only fter completion of nearly 800 disc rotations. However, in the case of MAO coating, even after 800 disc rotations, the coating volume loss continues to exhibit a decreasing trend, and after about 1400 revolutions it is found to approach the bulk Al_2O_3 value.

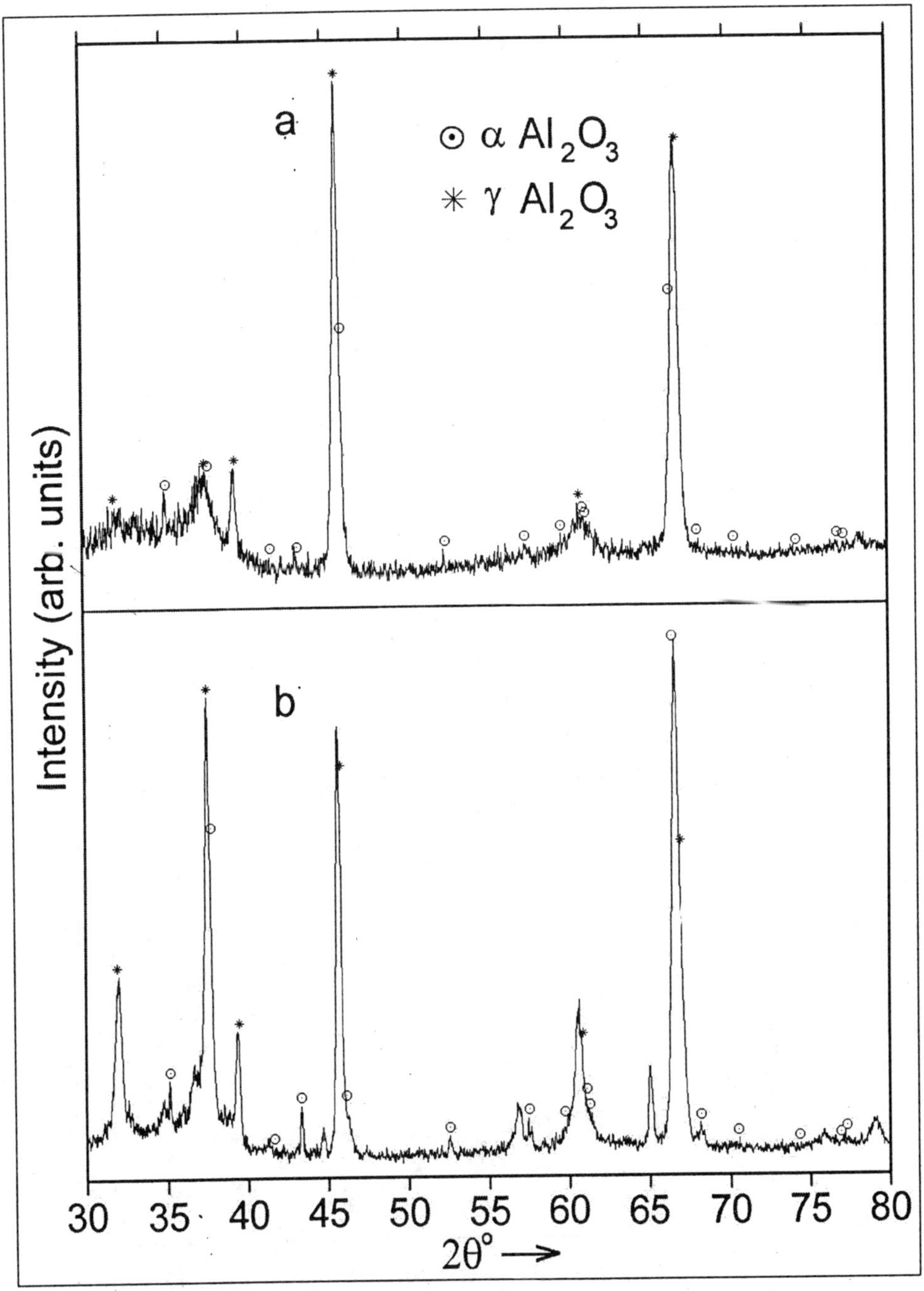

Figure 9.21: XRD Spectra of (a) Detonation Sprayed Alumina and (b) MAO Coatings

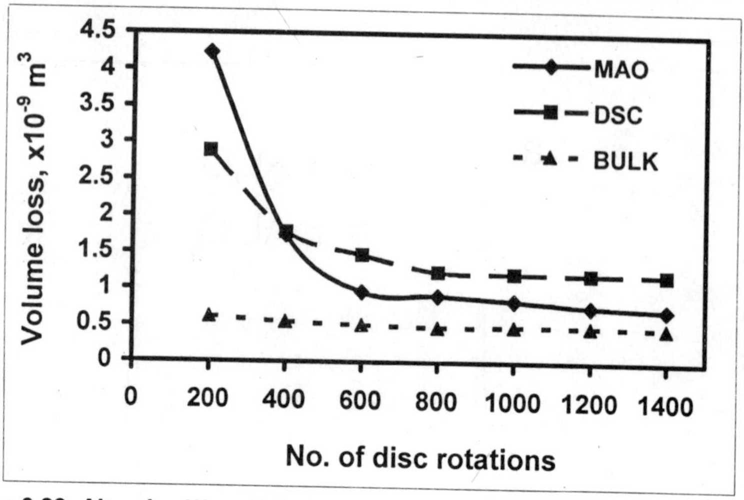

Figure 9.22: Abrasive Wear Behavior of MAO Coatings in Comparison with Detonation Sprayed Alumina Coating and Bulk Alumina

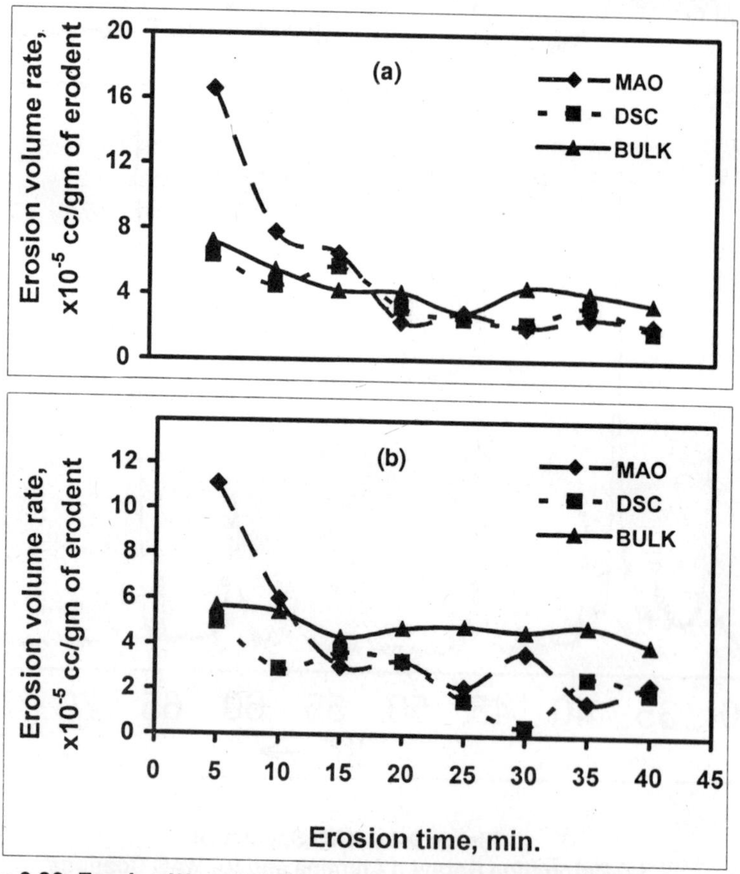

Figure 9.23: Erosion Wear Behavior of MAO and DSC Alumina Coatings in Comparison with Bulk Alumina at (a) 30° and (b) 90° Impact Angles

The erosive wear volume loss per kilogram of erodent at two different impact angles (30° and 90°) for the above specimens are shown in Figure 9.23. Surprisingly, bulk Al_2O_3 exhibits higher volume loss than the other two coatings at both the erodent impacting angles. Nevertheless, at both the angles, DSC and MAO coatings are found to exhibit nearly identical steady state volume loss.

Figure 9.24 illustrates the sliding wear performance of all the three materials under investigation. Bulk Al_2O_3 exhibits the lowest sliding wear volume loss, while the MAO coating exhibiting a marginally lower steady state volume loss than the detonation sprayed Al_2O_3 coating. However, the MAO coatings exhibit higher volume loss in the initial stages of wear followed by a dramatic decrease before finally reaching the near steady state in all the three wear modes. This can be attributed to the top 5-10 μm thick surface layer of the MAO coating being granular and irregular while the inner regions are dense and defect free (Figure 9.22b).

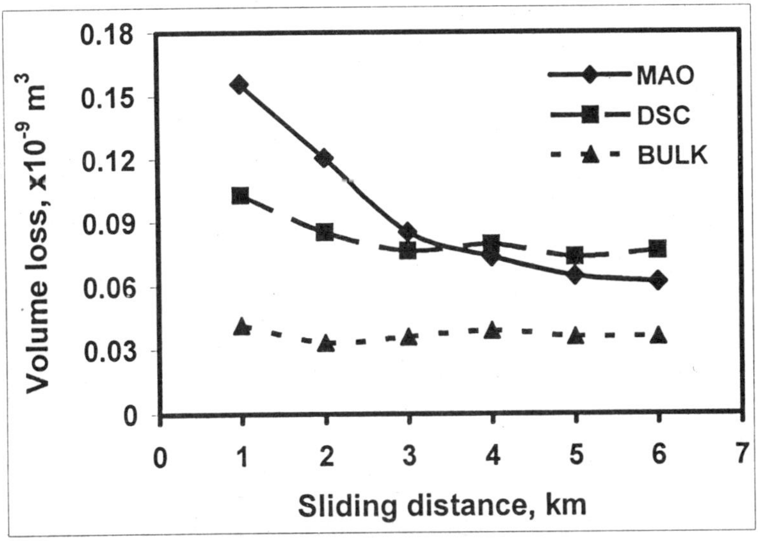

Figure 9.24: Sliding Wear Behavior of MAO and DSC Coatings in Comparison with Bulk Alumina

As expected, being more homogeneous, bulk Al_2O_3 exhibits the lowest steady state abrasive and sliding wear volume loss than the other two coatings. However, under erosion conditions, the steady state volume loss is surprisingly high. SEM analysis of the eroded surfaces as shown in Figures 9.25(a-c) revealed that the excessive grain ejection of bulk Al_2O_3 (Figure 9.25a) is responsible for its higher erosion rate. Interestingly, in the case of MAO coatings, wear is confined to very few, well-distributed regions which are unclosed discharge channels (Figure 9.25b) present in the coating. In the case of the detonation sprayed Al_2O_3 coating (Figure 9.25c), the material removal is predominantly due to inter granular spallation with some of the not fully molten particles resulting during spraying being easily knocked out.

Based on the above results, it is amply evident that the overall tribological performance of the MAO coatings compares well with that of the well-established DSC coatings as well as bulk alumina. However, as illustrated in Figure 9.27b, the discharge channels appear to influence the wear characteristics of the MAO coatings. This particular observation suggests the necessity to control the process parameters during MAO coating deposition process in such a way that all or most of the discharge channels are closed before actually stopping the coating deposition.

Figure 9.25: Eroded Surfaces of (a) Bulk Alumina, (b) MAO Coating and (c) DSC Alumina Coating

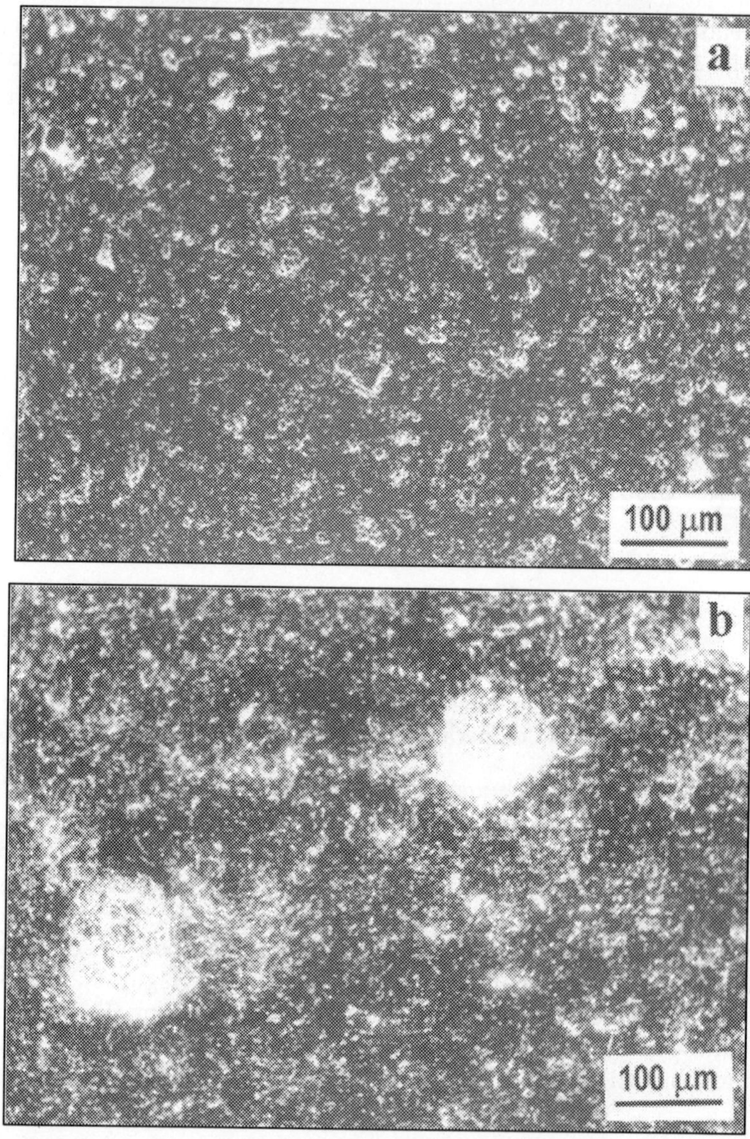

Contd...

Figure 9.25–Contd...

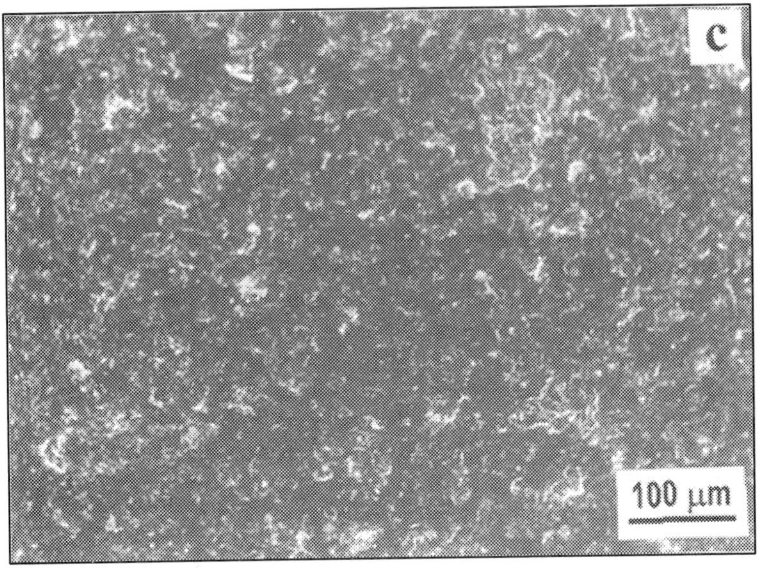

Process Advantages and Shortcomings

In summary, the micro arc oxidation technology is found to be unique in several respects. However, like any other technology, it has its distinct advantages and some shortcomings as listed below:

Advantages

Pre and post treatments are not strictly necessary for micro arc oxidation, except water rinsing which is carried out to remove traces of electrolyte from the surface of the component. This substantially simplifies the production process and reduces environmental concerns that are often associated with pre and post treatment solutions. This technology can be applied to anodizing a variety of materials, such as aluminium, magnesium, titanium, and zinc. In particular, alloys difficult to anodize by the conventional anodizing process, such as high copper containing alloys (typically 2000 series), high silicon containing die castings (typically A380), and magnesium alloys can also be treated. Micro arc oxidation has excellent throwing power and is capable of producing very thick coatings on substrate materials. For instance, more than 300-mm thick oxide coatings have been reported. Electrolytes used in micro arc oxidation are environmental friendly, non-corrosive, cheap, and easy to replenish and maintain. Micro arc oxidation has no "burning" problem, which is frequently encountered in most conventional anodizing process. The intrinsic colors of the ceramic coatings produced by micro arc oxidation are also attractive and can be explored for decorative applications. Micro arc oxidation technology results in less loss in the fatigue strength of an anodized material than the conventional anodizing process. These distinguishing features of micro arc oxidation strongly attract and encourage one to widely explore its potential applications.

Shortcomings

As any other technology, micro arc oxidation has its limitations. It requires high voltage (up to 1,000 V) and capacity even larger than 1 MW of power source to run a micro arc oxidation process in very large scale integrated production. As a consequence, the capital investment required is typically high. Owing to the generation of large quantity of heat in a micro arc oxidation process, a high capacity chiller is required to cool the electrolyte. In addition, racking tools and materials are of great concern due to the high voltage used.

Despite the above listed shortcomings, the micro arc oxidation process has been the subject of increasing interest in the academic and industry segments owing to its ability to yield high quality deposits that are capable of working at relatively high stress levels due to their superior abrasion, erosion and sliding wear resistance.

Applications

As a logical follow up to the success achieved at ARCI in depositing the high quality MAO layers, application development has been taken up in a big way as a prelude to technology commercialization. The MAO coatings have been successfully synthesized on the following industrial components:

- Al-Zn alloy rollers of textile industry
- Al-Cu-Si alloy hydraulic door lifting cylinders of aircraft industry
- Al-Mg-Si alloy filament bobbins and scanning machine guide rails
- Al-Mn alloy separator sheets for di-electric applications
- Al-Mg-Si alloy drilled plates for PCB applications
- Al-Si cast alloy components for automobile applications
- Al-Si-Mg cast alloy cones of textile industry
- Al-Si-Mg cast components for petroleum extraction in sea water

Influence of Substrate Chemistry

In order to evaluate the role of different alloying elements added to aluminium, 10 different Al substrates were selected based on their relevance and widespread utility in various industry segments. The MAO coating deposition was carried out on each of the 10 selected alloys under identical conditions. The results obtained have revealed that all the substrate materials investigated are eminently coatable without any pretreatment. The coating kinetics were, however, found to be considerably influenced by the type and quantity of alloying elements added to the base Al alloy. Of all the alloying elements, the presence of Si reduced the coating deposition rate most notably, with increasing Si content leading to a substantial reduction in the coating deposition rate as illustrated in Figure 9.26. Silicon being a passive element compared to Al, it is plausible that the coating deposition rate is adversely influenced by the presence of Si in the base alloy. Except the Al-Si alloys, all other systems, including commercial purity Al were found to exhibit linear coating deposition rates. The surface roughness of MAO coatings was found to be a function of final coating thickness alone and was not influenced by the alloy chemistry.

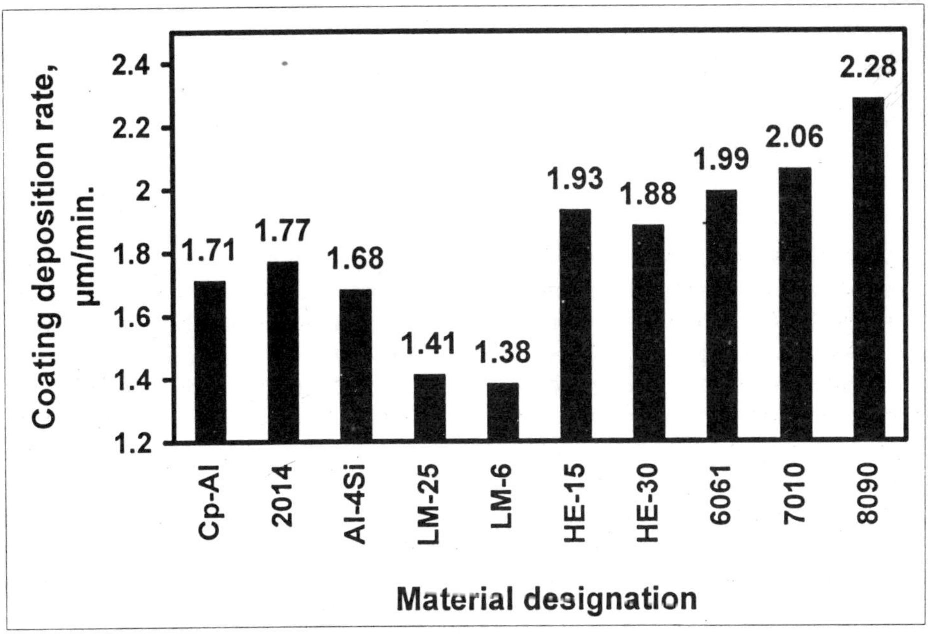

Figure 9.26: Average Coating Deposition Rate on Different Substrate Materials Investigated

SEM investigations also revealed that the MAO technology developed at ARCI is capable of depositing dense and uniform coatings, except in the case of high silicon (>7 per cent) containing alloys. With increasing silicon content from 4 per cent to 12 per cent, the porosity content also increased considerably. Independent of substrate chemistry, the γ- and α-Al_2O_3 phases were found to be the predominant phases in the MAO coatings. However, in the case of coatings deposited on the Al alloys (*i.e.*, except coatings deposited on commercial purity aluminium), the X-ray diffraction studies revealed the presence of additional diffraction peaks. In the case of Al-Si alloys, these additional peaks were found to correspond to alumino-silicate (mullite) phase. The phase content of mullite increased with increasing Si content in the Al alloy. On the other hand, Al-Li alloys exhibited the formation of Al_4Li_9 during coating deposition. In the case of all other alloys (those comprising of Cu, Mg and Zn as alloying elements), the low intensity of the extra peaks precluded the identification of the corresponding phases. The multi-point energy dispersive spectroscopy (EDS) analysis confirmed the presence of alloying elements like Cu, Mg and Zn in the coatings as well.

A typical characteristic of MAO coatings is the observed phase gradient across the coating thickness, with higher a-Al_2O_3 proportion towards the substrate-coating interface and higher γ-Al_2O_3 proportion towards the surface, thus leading to a micro-hardness gradient across the coating thickness. All the coatings deposited in the present study exhibited a similar trend, with the highest hardness (peakhardness) being noted close to the interface and thereafter gradually decreasing towards the surface. However, there were considerable differences in peak hardness values of MAO coatings deposited on different substrates. For example, coatings on 6061 Al-alloy exhibited a peak hardness value of 1900HV while LM6 exhibited a peak hardness value close to 1300HV only. The steady state abrasion, erosion and sliding wear volume loss of MAO coatings deposited on different coatings was observed to correlate well with the peak hardness thereby suggesting that the peak hardness is the critical parameter governing the tribological performance of MAO coatings irrespective of the substrate chemistry.

A multiple linear regression (MLR) model was developed to quantify the influence of each alloying element on coating deposition kinetics and peak hardness of MAO coatings. By comparing the predicted and experimental values of kinetics and peak hardness, excellent predictability of the MLR model was confirmed. The outcome of MLR analysis revealed that elements like Mg, Li, Fe and Cu when added to Al substrate enhance the MAO coating deposition rate, while Zn and Si decrease the coating deposition rate. In contrast, none of the alloying elements were found to increase the peak hardness value of the MAO coatings. The presence of Fe in the Al substrate was observed to dramatically reduce the peak hardness value while the effective decrease in peak hardness values due to all other alloying elements was only marginal. The above results can be useful in formulating guidelines for choosing appropriate alloy chemistry for any given application.

Technology Scale Up

In order to facilitate MAO coating on large components and the subsequent technology transfer, an industrial version of the MAO set-up with mass production capabilities, as well as improved design and added safety features is successfully accomplished at ARCI, the equipment is as illustrated in Figure 9.27.

The scaled up unit has been designed as per the industry requirements. It is possible to accommodate up to 1800 sq. cm. area per batch (or equivalently 6000 sq.cm. per component) with an upper limit of coating thickness in the range of 125-150 microns). Furthermore, several process monitoring facilities and the alarm facilities were built-in for easier process monitoring and for enhancing the operational safety. By utilizing the scaled up version, several industry concerns regarding the simultaneous coating on large number of parts, rate of aluminium dissolution in the electrolyte, coatability on complex shapes and internal surfaces, coatability of Al-Si cast alloy components, ability to color the coated surfaces, issues related to masking, tensile and fatigue properties of coated components, process economics etc., have already been effectively addressed.

Present Status of MAO Technology

In India, ARCI is the only institution where the task of developing the MAO technology has been initiated. The achievements at ARCI have been multidirectional. ARCI has successfully developed a huge reservoir of knowledge base generated through careful experimental investigations. The efforts have also led to the synthesis of globally superior coatings in terms of coating density and excellent mechanical and tribological properties. As an outcome of the above success, patent applications

(a) Control panel

(b) Reaction Chamber

Figure 9.27: MAO Scale Up Version Equipped with Advanced Process Monitoring Facilities

have been already filed in India and U.S.A., both the patents are expected to be granted shortly.

The industry has exhibited considerable interest in the MAO technology and in conducting the field trails on coated components for achieving service life enhancement. Towards this objective, the MAO coatings have been successfully deposited on several industry components and initial trails have yielded very encouraging results. The corrosion resistance of MAO coatings is found to be much superior than the anodic coatings and stainless steels. The technology transfer on a non-exclusive basis to different Indian industries has already been initiated; two technology transfer agreements were already signed and two more technology transfers are expected in a month's time.

Possible Directions for R&D Efforts

- Synthesis of suitable coatings on Ti and its alloys through MAO technique for applications demanding superior corrosion resistance and bio-compatibility
- Synthesis of sea-water corrosion/erosion resistant coatings on Mg and its alloys for possible applications in corrosive and wear environments.

References

1. G.Sundararajan and L.Rama Krishna Surf. Coat. Technol., 167 (2003) 269.
2. L.Rama Krishna, K.R.C.Somaraju and G.Sundararajan, Surf. Coat. Technol., 163-164 (2003) 484.
3. L. Rama Krishna, A. Sudha Purnima and G. Sundararajan, *A Comparative Study of Tribological Behavior of Micro Arc Oxidation and Hard Anodized Coatings*, Wear (in press).
4. L. Rama Krishna, A. Sudha Purnima, Nitin P. Wasekar and G. Sundararjan, "*Kinetics and Properties of Micro Arc Oxidation Coatings Deposited on Commercial Al Alloys*", Metallurgical and Materials Transactions A (in press).
5. P.I. Butyagin, Ye.V. Khokhryakov, A.I. Mamaev, Materials Letters, 57 (2003) 1748.
6. Yong Han, Seong-Hyeon Hong, Kewei Xu, Surf. Coat. Technol., 168 (2003) 249.
7. Yang Guangliang, Lu Xianyi, Bai Yizhen, Cui Haifeng, Jin Zengsun, J. of Alloys and Compounds, 345 (2003) 196.
8. W. Xue, C. Wang, Y. Li, Z. Deng, R. Chen, T. Zhang, Materials Letters 56 (2002) 737.
9. A.L. Yerokhin, A. Leyland, A. Matthews, Applied Surface Science 200 (2002) 172.
10. A.L.Yerokhin, A.A.Voevodin, V.V.Lyubimov, J.Zabinski and M.Donley, Surf. Coat. Technol., 110 (1998) 140.
11. X.Nie, A.Leyland, H.W.Song, A.L.Yerokhin, S.J.Dowey and A.Matthews, Surf. Coat. Technol., 116-119 (1999) 1055.

12. S.V.Gnedenkov, O.A.Khrisanfova, A.G.Zavidnaya, S.L.Sinebrukhov, A.N.Kovryanov, T.M.Scorobogatova and P.S.Gordienko, Surf. Coat. Technol., 123 (2000) 24.
13. Y.K.Wang, L.Sheng, R.Z.Xiong and B.S.Li, Surf. Eng., 15 (2) (1999) 112.
14. A.A.Petrosijants, V.N.Malyshev, V.A. Fedorov, and G.A.Markov, Friction Wear, 2 (1985) 350 (in Russian).
15. S.N.Bulychev and V.P.Alekhin, Testing of Materials under Continuous Pressure Indentation, Mashinostroenije, Moscow, 1990. (in Russian).
16. A.L.Yerokhin, X.Nie, A.Leyland, A.Matthews and S.J.Dowey, Surf. Coat. Technol., 122 (1999) 73.
17. A.A.Voevodin, A.L.Yerokhin, V.V.Lyubimov, M.Donley and J.S.Zabinski, Surf. Coat. Technol., 86-87 (1996) 516.
18. Y.K.Wang, L.Sheng, R.Z.Xiong and B.S.Li, Surf. Eng., 15 (2) (1999) 109.
19. W.Xue, Z.Deng, Y.Lai and R.Chen, J. Am. Ceram. Soc., 81 (5) (1998) 1365.
20. A.L.Yerokhin, V.V.Lyubimov and R.V.Ashitkov, Ceramic International, 24 (1998) 1.
21. R. McPherson, J. Mater. Sci., 8 (1973) 851.
22. Jun Tian, Zhuangzi Luo, Shangkui Qi and Xiaojun Sun, Surf. Coat. Technol., 154 (2002) 1.
23. X.Nie, E.I.Meletis, J.C.Jiang, A.Leyland, A.L.Yerokhin and A.Matthews, Surf. Coat. Technol., 149 (2002) 245.

Chapter 10

Nanostructured Multilayer Superlattice Coatings and Nanocomposite Coatings

Harish C. Barshilia
Surface Engineering Division, National Aerospace Laboratories,
Post Bag No. 1779, Bangalore – 560 01, India
E-mail: harish@css.nal.res.in, harishbarshilia@yahoo.co.in

ABSTRACT

Current research problems in the surface coatings technology include development of various thin coatings with exotic properties as part of an effort to modify the surfaces of a variety of engineering materials at lower cost. Among these coatings, multilayer/superlattice coatings have generated a lot interest in the scientific community because of their exotic properties and probable technological applications in diverse fields. This article outlines the latest developments in the field of multilayer coatings. Multilayer coatings of ceramic/ceramic materials are discussed in detail. Characterization techniques such as–X-ray diffraction, transmission electron microscopy, nanoindentation hardness tester, atomic force microscopy, low-angle X-ray reflectivity, potentiodynamic polarization, etc. have been used to discuss the structural and the mechanical properties of the multilayer coatings. We also discuss the thermal stability of the multilayer superlattice coatings.

Keywords: Multilayer superlattice coatings, Nanostructured coatings, Magnetron sputtering, Structure and mechanical properties, Thermal stability, Corrosion resistance.

Introduction

Nanomaterials are materials possessing grain sizes on the order of a billionth of a meter. They manifest extremely fascinating and useful properties, which can be exploited for a variety of structural and non-structural applications. Nanometer sized grains contain thousands of tens-of-thousands of atoms as compared to billions or trillions of atoms in the grains of conventional materials. The size of grains markedly affects the mechanical properties of the nanomaterials and the hardness of a material increases with a decrease in the grain size. This concept of reducing grain size can be used for designing new superhard materials (*i.e.*, hardness > 40 GPa).

In recent years a number of superhard coatings have been developed. These coatings can be divided into (i) intrinsic, such as diamond and cubic boron nitride (c-BN) and (ii) extrinsic, whose mechanical properties are determined by their microstructure. Diamond and c-BN are prominent intrinsic superhard coatings. Diamond film, however, reacts with oxygen and ferrous materials at higher temperatures. Cubic boron nitride is widely viewed as an ideal material for cutting tools. However, synthesizing c-BN coating has proved to be difficult. Of late, in search of extrinsic superhard coatings a number of multilayer superlattice coatings and nanocomposite coatings have been developed.

Since the publication of a classical paper on "attempt to design a strong solid" by Koehler in 1970, a lot of research has taken place in this field. Theoretically, Koehler has demonstrated that by using alternate ultra-thin layers of materials with high and low elastic constants, new materials with superior mechanical properties can be produced. Not only mechanical properties, the electrical, optical and magnetic properties can also be tailored by layering of different materials. In recent years, this concept has led to the development of so-called multilayer coatings or superlattice coatings. Multilayers represent what might be called materials engineering on atomic scales, with structures made up of layers only few atomic monolayers thick. Artificial multilayers are structures prepared by sequentially depositing two (or more) materials on a suitable substrate. Let us suppose that we have two substances, called A and B. The nature and thickness of the constituent materials (t_A and t_B) greatly affect the properties of the multilayers. The bilayer thickness, *i.e.*, $t_A + t_B$ is commonly known as modulation wavelength (Λ, also known as bilayer period). Depending upon the modulation wavelength generally we have two types of structures, namely: laminar (Λ >> atomic scale) and superlattice (Λ ~ atomic scale). Current interest in the multilayer coatings is mainly confined to superlattice coatings wherein the thicknesses of the constituent materials are of the order of few angstroms. The general interest in the multilayer coatings has been motivated both by the potential for discovering of new and yet unknown physical properties as well as by possible technological applications in diverse field. The common examples of multilayer superlattice coatings are: TiN/NbN, TiN/VN, TiAlN/CrN, etc., to name a few.

Most of the transition metal nitride superlattice coatings exhibit extremely high hardness. The multilayer films, in general, have better adhesion than single layer coatings as the stresses are relieved at the interfaces. Also, multilayers will have lesser defects than the single-layer films as the thicknesses of the individual layers are on atomic scales and dislocation generation mechanisms such as Frank-Read dislocations cannot be operative. The most important feature of the superlattice coatings is that the properties can be tailored depending upon the choice, and layer thickness of the constituent materials and the deposition parameters. This opens innumerous possibilities to design new materials with desired properties.

In 1987, Helmersson *et al.* reported that single-crystal transition metal nitride superlattice films, deposited on MgO (100) substrates by reactive magnetron sputtering, showed hardness 2-3 times those of the homogeneous nitrides. Maximum hardness values were 5600 kg/mm^2 for TiN/VN with a bilayer repeat period of 5.2 nm. In addition, these multilayers showed a higher wear resistance than homogeneous material. This has opened up a wide range of opportunities for the deposition of "tailor-made" superhard superlattice coatings, wherein a variety of combinations can be used. Recent literature data report the fabrication of a number of ceramic/ceramic multilayer systems. The most technologically important systems are: TiN/VN, TiN/NbN, TiN/ZrN, TiN/CrN, TiN/(V_xNb_{1-x})N, TiN/AlN, TiAlN/CrN, TiN/TaN, CrN/NbN, TiN/TiCN, AlN/ZrN and WC/TiN. Ceramic/ceramic multilayer coatings are of current interest because of their exceptionally superior mechanical properties. The properties of these coatings can be tailored by judicious control of the process parameters.

Unlike other superhard coatings (*e.g.*, diamond and c-BN) the ceramic multilayer coatings are stable even at higher working temperatures. In principle, these coatings can be deposited on any of the important engineering substrates with improved adhesion as wear resistance coatings. Each of the ceramic multilayer coatings has its advantages. For example, TiN/CrN coating has high potential as a tribological coating material. The TiN/CrN multilayer has a unique combination of high fracture resistance and high wear resistance which makes this coating a very interesting choice for a number of tool applications where high tensile stresses are generated in the coating. A multilayer coating of TiN/TaN is the best choice for applications where abrasive wear is the predominant wear mechanism. The CrN/NbN superlattice coatings are wear resistance and corrosion resistance. Similarly, a TiN/AlN multilayer structure greatly improves the oxidation resistance and increases the temperature range of possible applications.

Deposition Methodology

In the last twenty years many techniques have evolved which are capable of precision multilayer fabrication. Broadly speaking there are two deposition routes, namely: (i) electrodeposition and (ii) physical vapor deposition (PVD). Here we will briefly discuss about the PVD techniques, with an emphasis to sputtering.

The physical vapor deposition processes are versatile techniques to fabricate multilayer coatings. The important techniques in this category are: evaporation, molecular beam epitaxy (MBE) and sputtering. The evaporation (and also e-beam evaporation) is limited only for low melting point metals. Refractory materials cannot be deposited by this technique. The MBE process is very expansive and is used for depositing low melting point materials for research applications. The sputtering processes are the most widely used processes for multilayer deposition. The sputtering can be from many different types of sources (RF or DC, planar or cylindrical magnetron, triode or diode) operating at moderately high pressures 0.1 to 1 Pa, or can be ion-beam sputtering in which target is bombarded by an ion beam, generated independently. A typical multi-source RF/DC magnetron sputtering system for the deposition of multilayer coatings is shown in Figure 10.1 (Barshilia and Rajam, 2002). The sputtering system consists of 4 sputtering guns (3" diameter). In order to get varying thicknesses of TiN and NbN layers (in the case of TiN/NbN multilayers, for example), 6 mm thick high purity Ti (99.95 per cent) and Nb (99.99 per cent) targets are sputtered for different durations in high purity Ar (99.999 per cent) and N_2 (99.999 per cent) plasma. Typically, TiN/NbN multilayers are deposited under a base pressure of approximately 2.0×10^{-6} mbar and a total Ar+N_2 gas pressure of approximately 4.0×10^{-3} mbar. The flow rates of N_2 (1.5 sccm) and Ar (17 sccm) are controlled separately by mass flow controllers. A DC substrate bias (V_B) of–200 V is applied to improve the mechanical properties of the coatings. Typical power densities are 5.4 and 2.26 W/cm^2 for Ti and Nb targets, respectively. Under these conditions the growth rates are approximately 2 Å/sec each for TiN and NbN. The coatings are deposited at a substrate temperature of 400°C. Multilayer coatings are obtained by a computer operated substrate rotation assembly, which controls the dwell time of the substrate underneath each target very precisely. The total thickness of the films is approximately 2.0 μm. Substrates are cleaned in ultrasonic bath using acetone, ethyl alcohol and trichloroethylene. Further cleaning is done by *in situ* Ar^+ ion bombardment, wherein a DC bias of–850 V is applied to the substrate for 30 minutes prior to the film deposition. Sputter cleaning of the targets is done for 15 minutes prior to the deposition. A 0.5 μm thick metallic (*e.g.*, Ti) interlayer is incorporated between the substrate and the film for improved adhesion.

Characterization of Multilayers

The characterization of multilayers is essential in developing an understanding of the link between deposition conditions and properties of the coatings. Different techniques, namely: X-ray diffraction (XRD), electron diffraction, transmission electron microscopy (TEM), nanoindentation hardness tester (NHT), atomic force microscopy (AFM), etc. have been used by different workers to characterize the multilayer coatings. In the following section we discuss various characterization techniques. We have developed a variety of multilayer superlattice coatings. These coatings have been characterized extensively using a number of techniques. We will take TiN/NbN as an example to present our results related to structural properties, mechanical properties, corrosion properties, wear properties, etc. Finally, we will take up TiAlN/CrN multilayers as an example to demonstrate oxidation behavior of the multilayer coatings.

X-ray Diffraction

X-ray diffraction is an essential and universal tool for the structural characterization of the multilayer coatings. It is a nondestructive technique and provides structural information on the atomic scale. Numerous researchers have used this technique to investigate the lattice spacing and the texture (*i.e.*, preferred orientation of crystallites) of the superlattice films. The interpretation of XRD data provides most of the information needed to fully characterize the composition modulation, including L and the interplanar spacing. A superlattice film shows the Bragg peaks, each surrounded by a set of satellites with a spacing related to the

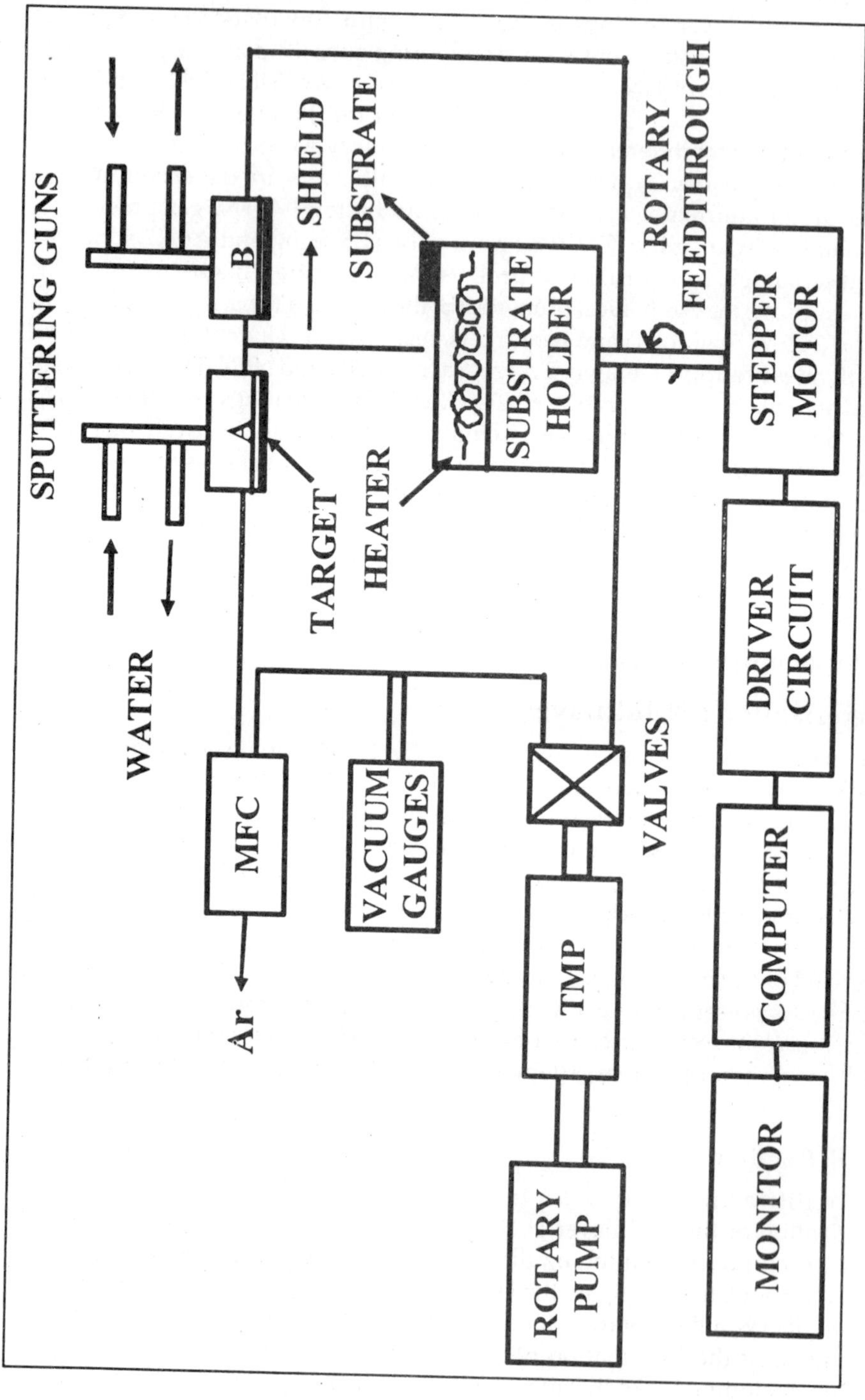

Figure 10.1: Schematic Diagram of a Multi-Target Sputtering System (After Barshilia and Rajam, 2002)

modulation wavelength (Fullerton *et al.*, 1992). Information may be obtained from the diffraction scan as follows: (i) The crystal structure from the positions of the Bragg peaks (or their absence). (ii) Preferred orientation in a polycrystal from the relative intensities of the Bragg peaks. (iii) The modulation wavelength from the positions of the satellites. (iv) The film thickness or structural coherence in the direction normal to the substrate from the width of Bragg peaks. (v) The variation in the modulation wavelength from the width of the satellite peaks.

The XRD data of TiN/NbN superlattices deposited at different modulation wavelengths are shown in Figure 10.2 (Barshilia and Rajam, 2004). Superlattice structure was clearly seen for the films deposited at 106 Å ≥ Λ ≥ 30 Å. Samples did not show prominent positive satellites, which was attributed to larger lattice spacing and higher X-ray scattering factor for NbN than TiN. At very low modulation wavelengths (Λ < 30 Å) only a broad principal reflection was observed and satellite reflections were absent, indicating that a long-range crystalline order was no longer

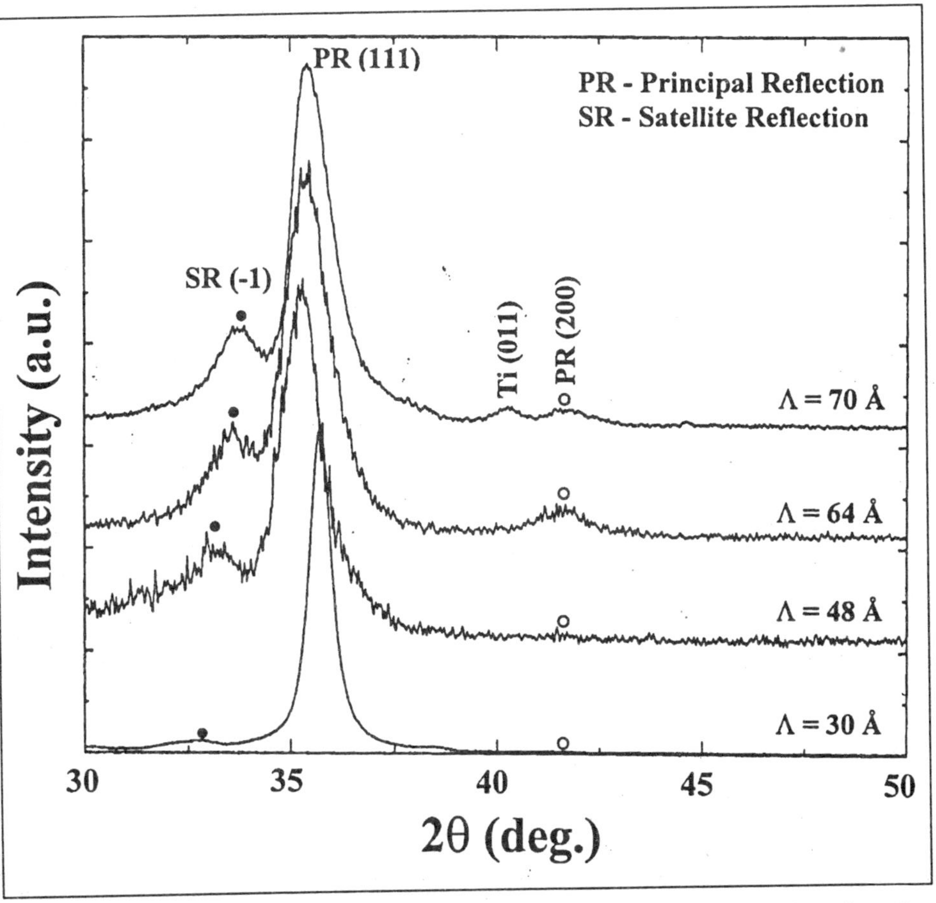

Figure 10.2: XRD Data of TiN/NbN Multilayers with Different Modulation Wavelengths. The spectra show (111) principal reflection, which is flanked by first-order negative satellite reflection. (After Barshilia and Rajam, 2004).

maintained. Multilayer coatings with 106 Å ≥ Λ ≥ 48 Å showed broader XRD peaks than TiN and NbN, which means that multilayers (approximately 110 Å) had smaller grain size than the single layer TiN (approximately 150 Å) and NbN (approximately 200 Å) coatings. The full-width-at-half-maximum (FWHM) of the (111) peak decreased with an increase in the modulation wavelength, however, its position remained unaffected. The first-order negative satellite reflection moved closer to the principal reflection with an increase in Λ. The intensity ratio of first-order negative satellite to the principal Bragg peak (I_{-1}/I_B) is maximum at Λ = 64 Å. It decreased markedly at low modulation wavelengths and at very low Λ the satellite reflections were completely absent. This signifies that there is a limited range of modulation wavelength, wherein superlattice formation was possible.

Transmission Electron Microscopy

The cross-sectional transmission electron microscopy (XTEM) is an ideal tool for studying ultra-thin layers of the multilayer coatings. As the thicknesses of the individual layers in the multilayer coatings are on the atomic scales, examination of the cross-section of the coatings becomes impossible by scanning electron microscopy. The examination of the cross-section of multilayer coatings under TEM requires special sample preparation techniques. A commonly used technique is based on sandwiching of the two multilayer samples. The sandwiched samples are cut into thin wafers. The wafers are thinned down mechanically up to a level of 100 μm thickness. Subsequent thinning is done in an ion-milling machine. Ion-milled samples are directly examined in TEM. The same sample can also be used for electron diffraction studies. The XTEM image provides observation on individual layer thickness along with the interfacial defects (*e.g.*, misfit dislocations and voids). The diffraction pattern shows satellite reflections around the main Bragg reflection, which can be used to calculate modulation wavelength. One added advantage of electron diffraction over X-ray diffraction is its higher sensitivity. Further, TEM yields direct information about the local defects, such as dislocations, twins, layer relaxation, uniformity of layer thickness and interface roughness not easily obtained from XRD. One of the most direct applications of XTEM is to study unevenness in the layers.

Figure 10.3 shows the cross-sectional view of the TiN/NbN multilayers (Barshilia *et al.*, 2006). The micrograph shows alternating layers of TiN and NbN with almost equal layer thicknesses (approximately 3.8 nm). The XTEM micrograph showed that the layers are well resolved and indicated that ion-induced mixing did not cause substantial layer intermixing. The HRTEM studies showed that the layers are crystalline with no amorphous regions (data not shown).

Nanoindentation Hardness Tester

The mechanical properties of ceramic multilayers are of interest because multilayers exhibit very high strength and hardness. For micron and sub-micron thick multilayer films the hardness becomes meaningful only if the influence of substrate material can be eliminated. Thus indentation depth should not exceed about one-tenth of the total coating thickness. The applied loads in the order of 0.5-50 mN are desirable if the indentation depths are to remain in the nanometer range.

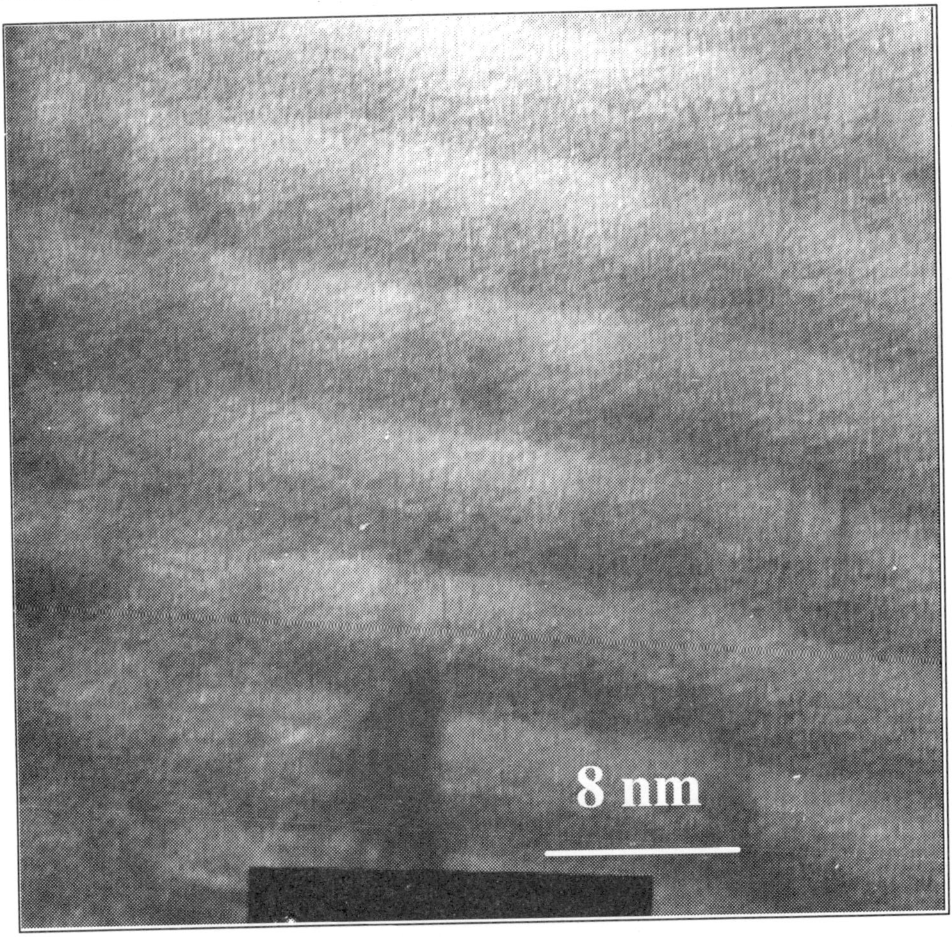

Figure 10.3: Cross-Sectional TEM Images of TiN/NbN Multilayers Showing Distinct Layers of TiN and NbN (After Barshilia *et al.*, 2006)

Nanoindentation hardness tester (NHT) has become widely accepted as the logical successor to micro-indentation, the later being limited by the resolution of an optical microscope as this is used to determine the imprint diagonal and thus the hardness of the tested material. Furthermore, micro-indentation limits the maximum depth for the indentation, meaning the depths of less than 1-2 μm make indentation too small for size measurement by conventional light microscopes (Barshilia and Rajam, 2004). Of late, various nanoindentation techniques have been developed. The principle here lies on the continuous measurement of force and displacement as an indenter, of known geometry, is pressed into the sample material. With force and displacement resolutions in the range 5 μN and 0.3 nm respectively, it is possible to produce load displacement curves representative of the material response in terms of hardness and elastic modulus.

Load vs. displacement curves measured for single layer TiN, single layer NbN, TiN/NbN multilayer coatings and tool steel substrate at 5 mN load are shown in Figure 10.4 (Barshilia and Rajam, 2004). From this plot the hardness was calculated using Oliver and Pharr method (Oliver and Pharr, 1992). The area formed by the loading and the unloading curves, defined as plastic deformation work, can be used to assess the resistance of plastic deformation and the wear resistance of the coatings. The resistance of plastic deformation is inversely proportional to the plastic deformation work. TiN/NbN multilayer coatings showed the smallest plastic deformation work and the largest resistance to plastic deformation as compared to the single layer TiN and NbN coatings. For the softest film (*i.e.*, NbN), the maximum indentation depth was 120 nm, which was much less than 1/10th of the coating thickness, thus eliminating the effect of substrate on the hardness measurements. Tool steel substrate showed a maximum indentation depth of 160 nm. For the hardest film (*i.e.*, TiN/NbN), however, the maximum indentation depth was only 85 nm.

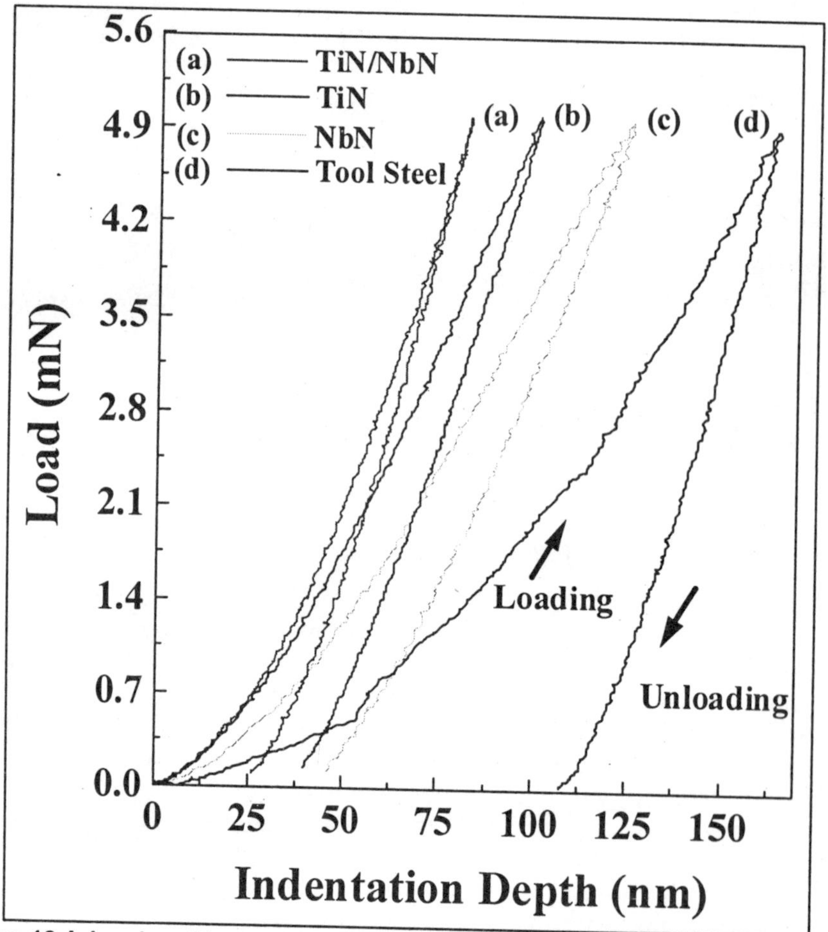

Figure 10.4: Load vs. Displacement Curves for the TiN/NbN Multilayer, Single Layer TiN Coating, Single Layer NbN Coating and Tool-Steel Substrate. For multilayer the modulation wavelength was about 8 nm. (After Barshilia and Rajam, 2004).

The variation of nanoindentation hardness of TiN/NbN multilayers with modulation wavelength is shown in Figure 10.5 (Barshilia and Rajam, 2004). The peak hardness occurred at Λ = 48 Å and the maximum hardness obtained was approximately 4000 kg/mm^2, which is near to that of a superhard material. The hardness values of TiN, NbN coatings and tool steel substrate were approximately 2000, 1400 and 1100 kg/mm^2, respectively. The maximum hardness obtained was much higher than the value of the rule-of-mixture, which for TiN-NbN is approximately 1700 kg/mm^2. For the multilayer coatings pronounced variations in the strength and hardness as a function of modulation wavelength are commonly observed. For example, TiN/NbN multilayers exhibit hardness varying from 3600-5000 kg/mm^2 depending on the deposition process and the growth conditions. Apart

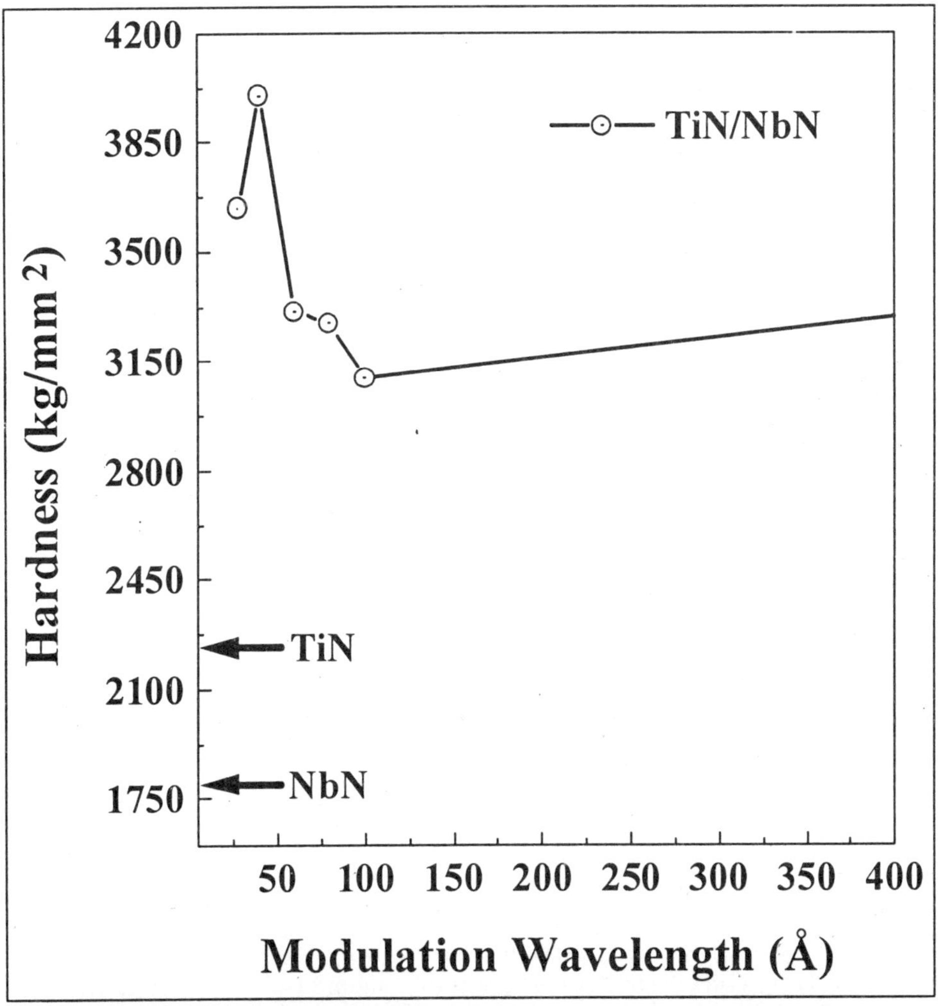

Figure 10.5: Variation of Nanoindentation Hardness of TiN/NbN Multilayer Coatings with Modulation Wavelength. Also, shown are the hardness values of single layer TiN and single layer NbN coatings. (After Barshilia and Rajam, 2004).

from the modulation wavelength, hardness of the coatings was also dependent on the substrate bias. A maximum hardness of 4000 kg/mm^2 (Λ = 48 Å) was achieved at V_B = –200 V. The hardness decreased from 4000 to 3300 kg/mm^2 as the bias was decreased from –200 to –150 V. At very high substrate bias (*e.g.*, $V_B \geq -225$ V) the hardness again decreased, presumably because of ion beam intermixing effects at the interfaces and interface roughening. This observation is consistent with the earlier reports on effect of bias on the mechanical properties of TiN/NbN superlattices. The enhanced hardness can be explained by ion-induced densification below –200 V and deterioration of the superlattice structure caused by interdiffusion and interface roughness above –200 V.

A variety of mechanisms have been put forward to explain the hardness enhancement in the multilayer superlattice coatings. These include suppression of dislocations (Koehler strengthening), grain refinement (Hall-Petch), coherency strain hardening, dislocation line energy effects and interfacial crack deflection (Chu and Barnett, 1995). In general, enhancement in the hardness results from the resistance to dislocation glide across the interfaces, which is proportional to the difference in the layer shear moduli. The observed enhancement in the hardness in TiN/NbN multilayer samples at low modulation wavelength (40-108 Å) is primarily believed to be due to the interfaces. At low modulation wavelengths, the interfaces between the layers act as pinning sites for dislocations. As the thicknesses of TiN and NbN layers are too small, the dislocation generation (such as Frank Read) cannot occur inside the layers. Even if the dislocations are generated in the layers they propagate towards the interfaces. As the interfacial energies are quite high, further movement of the dislocations is prevented and hence pile-up of the dislocations takes place near the interfaces. Furthermore, each interface serves as crack-tip deflectors. This leads to a substantial increase in the system hardness relative to that of the homogeneous materials. However, the same arguments are not true at very low modulation wavelengths, where the multilayer behaves like an alloy. At large modulation wavelengths, softening occurs as a result of dislocation motions within the individual layers. It must be mentioned that the maximum enhancement in the hardness of a superlattice coating also depends on the microstructure and stress level of the coating, irrespective of the modulation wavelength.

Atomic Force Microscopy

The study of microstructure, composition and surface morphology at relatively high resolution is needed to understand processes occurring at material surfaces, such as growth of the films, adhesion phenomena and breakdown and failure of coatings. The introduction of scanning force microscopy (SFM) has provided one of the few types of imaging instruments capable of making quantitative measurements at such small scales with nanometric precision and three-dimensional imaging capabilities. One of the very important techniques among the SFM is atomic force microscopy (AFM) which is capable of measuring the interaction forces and mechanical properties of solid surfaces on a subnanometric level. Since its invention in 1985, AFM has been used to study attractive van der Waal forces, repulsive contact forces, lateral friction forces and magnetic forces (Binning *et al.*, 1986). One interesting application of AFM is atomic scale imaging of surface topography via force interactions, with resolutions comparable to that obtained by scanning tunneling microscopy (STM). Unlike STM, however, AFM is not restricted to study electrically conducting materials. One of the very important applications of AFM is to visualize nanoindentation imprints in order to characterize the material response in terms of pile-up or sink-in and to investigate the surface morphology before and after indentation. This gives information about the plastic and elastic behavior of the material. A 3-dimensional AFM image of a typical TiN/NbN multilayer coating is shown in Figure 10.6 (Barshilia and Rajam, 2004). These films had an average roughness of ~5 nm.

X-ray Reflectivity

The refractive index in materials for X-rays of wavelength around 1 Å is slightly less than unity and thus total external reflection occurs at very low incidence angles. In the region just above the critical angle, the X-ray wave penetrates successively deeper into specimen as the angle is increased. If one or more thin films of different electron density to the substrate are present, interference oscillations are observed in

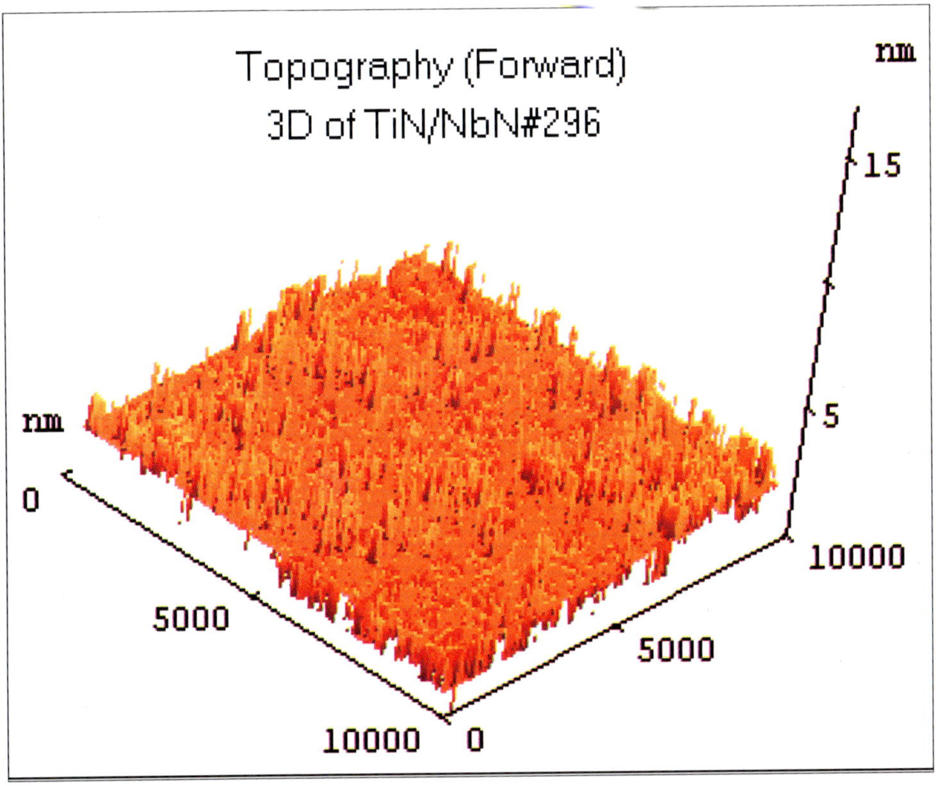

Figure 10.6: Three-Dimensional AFM Morphology of a TiN/NbN Multilayer Coating on Tool Steel Substrate (After Barshilia and Rajam, 2004)

the reflectivity profile which contains information on the film thickness, electron density, and interface and surface roughness. Figure 10.7 shows the typical X-ray reflectivity scans of about 1.5-μm thick TiN/NbN superlattices with different L values (Barshilia *et al.*, 2006). For all the coatings second-order satellite reflections are seen in the reflectivity data. The intensities of higher order satellite reflections can be taken as an approximate of the abruptness of the change in the composition between the two constituent layers (*i.e.* TiN and NbN). Since the low-angle superlattice peaks results from the electron density modulation that can be represented by a Fourier

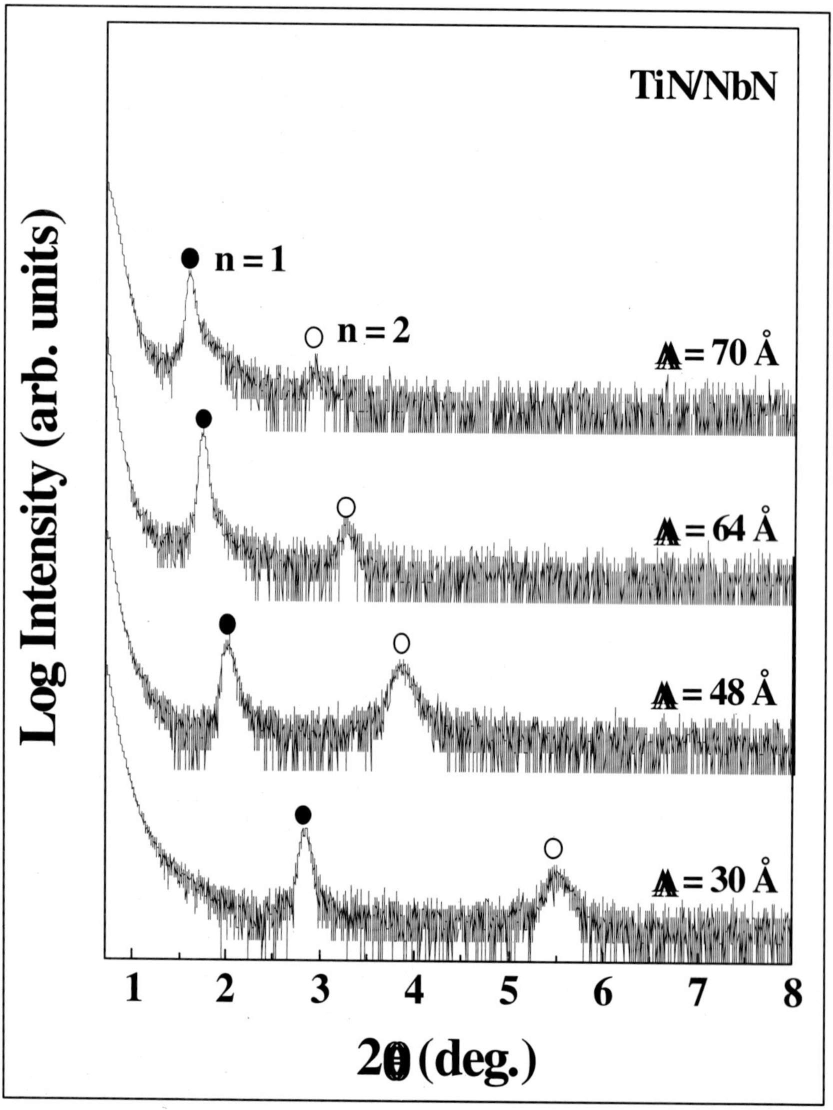

Figure 10.7: Low-Angle X-Ray Reflectivity Patterns of TiN/NbN Multilayer Coatings Deposited at Various Modulation Wavelengths (After Barshilia *et al.,* 2006)

expansion, characteristic interface sharpness can be estimated by $\Lambda/4n_{max}$, where n_{max} is the highest order of the observed low-angle X-ray reflectivity peaks and Λ is the modulation wavelength. The narrow peak widths and higher intensities indicate a well-defined superlattice structure and good interface flatness.

Potentiodynamic Polarization/Corrosion Testing

Physical vapor deposition (PVD) coatings, especially those of transition metal nitrides, contain high defect density and a columnar microstructure. The columnar structure can allow micropores/pinholes to run through the coating thickness, via which corrosive media may attack the coating/substrate interface. Thus, although, transition metal nitride coatings are chemically inert, the micropores/pinholes present in these coatings affect the corrosion resistance of the coated system, apart from other properties. In order to improve the corrosion protection afforded by the PVD coatings it is of great importance to inhibit the columnar growth and the porosity. In addition to impurities on the substrate surface, the substrate-state (roughness, scratches, etc.) and a large grain size can also induce porosity in the coating. Smaller grain size can be achieved by the judicious control of deposition parameters (*e.g.*, ion bombardment during deposition).

Various methods have been used to improve the corrosion resistance of transition metal nitride coatings. These include increased coating thickness, alloying of nitrides, incorporation of interlayers, intermediate plasma etching during deposition and multilayer deposition. Layered deposition is expected to minimize the columnar structure of the PVD coatings (Barshilia *et al.*, 2004). In the multilayer coatings re-nucleation associated with successive deposition of sub-layers lowers the grain size and prevents the growth of pores and defects all through the coating. A reduced grain size and a large number of interfaces in a multilayer result in dense and homogeneous microstructure, which lead to improved or novel coating properties. Furthermore, multilayer coatings have lower residual stresses mainly because of stress relaxation at the interfaces, thus inhibiting the crack propagation as a result of chemical attack. Potentiodynamic polarization technique is generally used to study the corrosion behavior of the PVD coatings.

The corrosion potential and corrosion current density obtained from the Tafel plots for tool steel substrate, single layer NbN, single layer TiN and multilayer TiN/NbN (Λ = 5.6 nm) coatings in 0.5 M HCl solution are presented in Figure 10.8 (Barshilia et al., 2004). The thickness of the coating was ~1.5 µm for all the samples. After coating the substrate with ~1.5 µm thick TiN/NbN multilayer, E_{corr} increased from–0.460 to–0.340 V, whereas, I_{corr} decreased from 240 to 0.9 µA/cm^2. Single layer TiN and NbN coatings exhibited almost identical corrosion behavior with corrosion currents of 1.2 and 1.6 µA/cm^2, respectively. These results conclusively demonstrate the positive effect of layering for improving the corrosion resistance of transition metal nitride coatings.

Thermal Stability of Multilayer Coatings

Apart from high hardness, the ceramic multilayer coatings also exhibit high strength, and wear resistance. Therefore, these coatings have great potential as

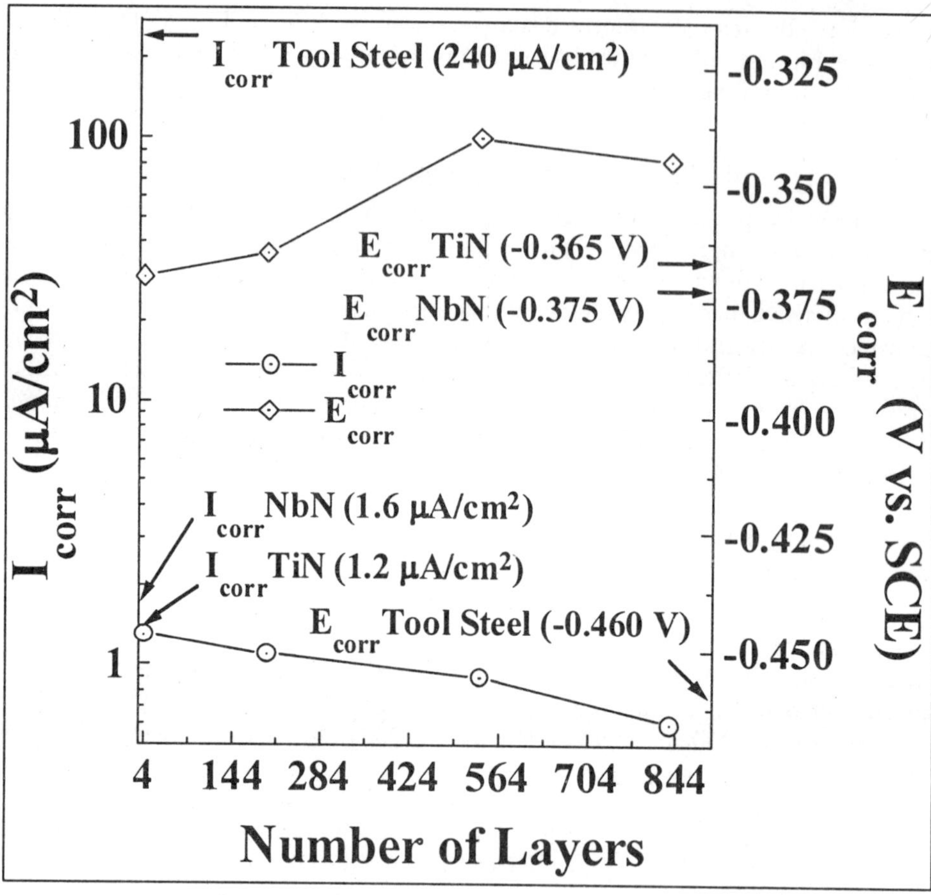

Figure 10.8: Variations of I_{corr} and E_{corr} of TiN/NbN Multilayers with Total Number of Layers in 0.5 M HCl Solution. Also shown are the values for tool steel substrate, single layer TiN and single layer NbN coatings. (After Barshilia *et al.*, 2004).

protective coatings on cutting tools and other mechanical components. For cutting tool applications one not only needs to prepare hard and tough coatings but their thermal stability is also very important. Tools protected by the hard coatings often operate at elevated temperatures in air. In the following section we will discuss thermal stability of TiAlN/CrN multilayers coatings.

Figure 10.9 shows high-angle XRD spectra of as-deposited TiAlN film and films heated up to 800°C for 30 minutes (Barshilia *et al.*, 2005). The spectra did not change up to 700°C, however the position of (111) reflection shifted to higher 2q values with an increase in the annealing temperature. This shift is likely being due to stress relaxation as a result of annealing. Compressive stresses are commonly observed in the bias sputtered transition metal nitride films because of defects created by Ar^+ ion bombardment. At 800°C, the film exhibited strong peaks of various oxides of Al, Ti and Cr along with a weak (111) reflection of TiAlN. No evidence of AlN phase

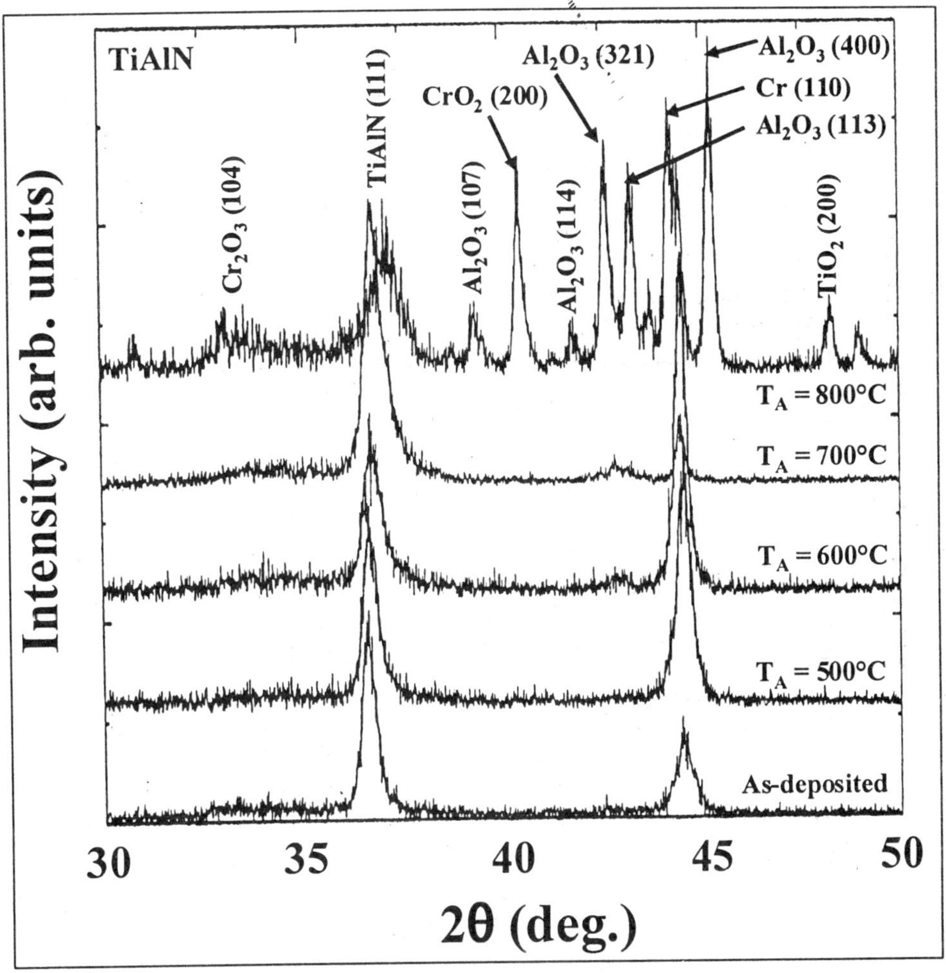

Figure 10.9: XRD Patterns of as-Deposited TiAlN Film and Films Heated at 500, 600, 700 and 800°C (After Barshilia *et al.,* 2005)

formation was observed at 800°C. These results indicated that TiAlN films heated under present set of conditions were stable up to 700°C and got oxidized at 800°C. Multilayer films of TiAlN/CrN (Λ = 56 Å) were also heated under identical conditions at 500, 600, 700, 800 and 900°C. As compared to TiAlN films intense oxide peaks were observed only at 900°C for multilayer films, although weak peaks of Al_2O_3 started to appear even at 800°C (see Figure 10.10). It must be mentioned that a weak unresolved satellite peak was observed for as-deposited multilayer film and it disappeared even after heating at 500°C. This shows that upon heating there was some interdiffusion between TiAlN and CrN layers. These results demonstrate that the onset of oxidation of TiAlN/CrN multilayers is higher than TiAlN and CrN, which was confirmed by characterizing the heat-treated coatings using Raman spectroscopy (Barshilia and Rajam, 2004).

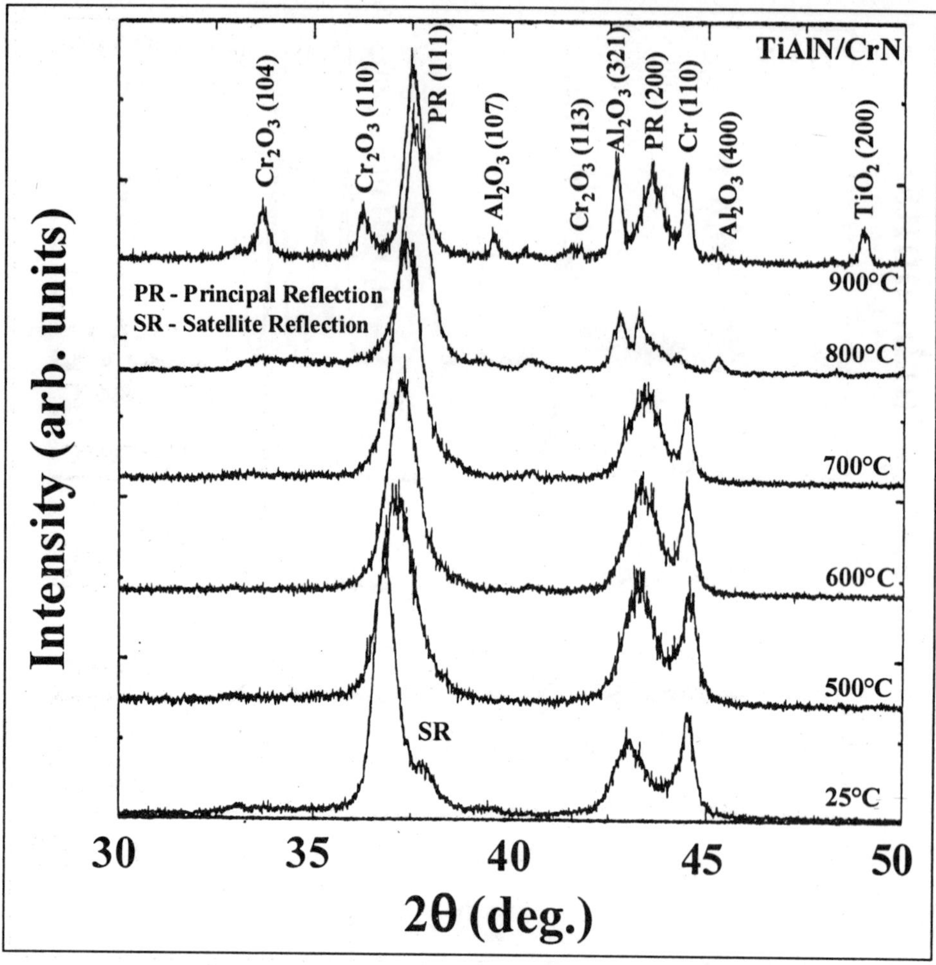

Figure 10.10: XRD Patterns of as-Deposited TiAlN/CrN Multilayer Film and Films Heated at 500, 600, 700, 800 and 900°C. The modulation wavelength was 56 Å. (After Barshilia *et al.*, 2005).

In order to check the suitability of TiAlN/CrN multilayer coatings for high temperature applications, hardness of the films was measured after heat treatment. In Figure 10.11 we plot the variations of hardness of TiAlN films and TiAlN/CrN multilayer films as a function of annealing temperature (Barshilia *et al.*, 2005). The hardness of both TiAlN films and TiAlN/CrN multilayers did not change up to 500°C. The hardness decreased for $T_A > 500°C$ and this decrease in hardness was more for single layer TiAlN films than multilayer films of TiAlN/CrN. For example, the hardness decreased to 2200 kg/mm^2 for TiAlN films after heating the films at 700°C for 30 minutes. This drastic fall in the hardness of TiAlN film at higher temperature could be attributed to the formation of soft and amorphous Al_2O_3 phase. The indentation measurements could not be performed for the film heated at 800°C as

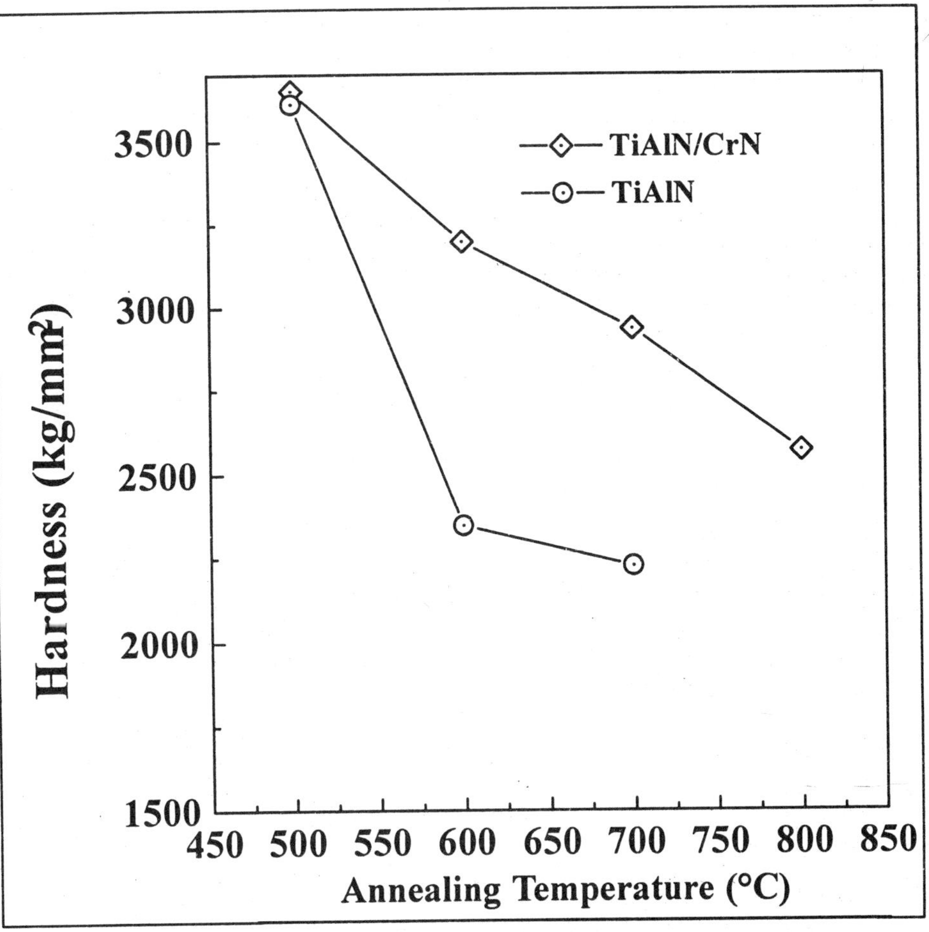

Figure 10.11: Variations of Nanoindentation Hardness Values of TiAlN/CrN and TiAlN Films with Annealing Temperature (After Barshilia *et al.*, 2005)

TiAlN films were completely oxidized at this temperature. TiAlN/CrN multilayers retained hardness as high as 2600 kg/mm^2 even after heating up to 800°C. In the case of multilayer films in addition to oxide layer formation the hardness also falls because of interdiffusion between TiAlN and CrN layers.

Applications and Outlook

The multilayers possess a variety of exotic properties depending upon the constituent materials. The multilayers are futuristic materials and an upcoming field. A lot of basic research is required in order to fully understand the properties of these materials which will ultimately lead to their possible technological applications in diverse areas. The main technological applications of metallic multilayers are in the field of: magnetoresistive heads, magneto-optical devices, giant magnetoresistance devices, mirrors for soft X-rays and neutron beams. The ceramic multilayers can be used as wear resistance coatings on a variety of engineering components such as–cutting tools and in high precision machining. Another area where these coatings can be used is for punching and forming tools. Here the idea is to increase the lifetime of the formimg tool and improve the surface quality of the final formed parts. These coatings can also be used in automotive segment such as fuel injection train. Another application of PVD deposited multilayer coatings is in shafts of the turbo compressor. Future applications may include high-performance car parts. This will certainly have spin-offs for the more normal car production. The continuous drive for improved energy efficiency, weight reduction, noise reduction and volume reduction are all favorable conditions for increased use of low-friction PVD deposited multilayer superlattice coatings. It is early as yet, however, to identify the main future applications of multilayer superlattice coatings, especially given the wide range of special properties.

Acknowledgements

We thank Director, NAL (CSIR) for giving permission to publish these results. Dr. (Mrs.) K. S. Rajam, Head, Surface Engineering Division, NAL is thanked for constant support and advice. Dr. D. V. Sridhara Rao, DMRL, Dr. Sujeet Chaudhary, IIT Delhi, Dr. Anjana Jain, NAL are thanked for doing TEM, X-ray reflectivity and XRD measurements, respectively. Technical support provided by all staff members of SED is greatly acknowledged.

References

1. Barshilia H. C. and Rajam K. S., 2002. Characterization of Cu/Ni multilayer coatings by nanoindentation and atomic force microscopy, Surf. Coat. Technol. 155, pp. 195.
2. Barshilia H. C. and Rajam K. S., 2004. Structure and properties of reactive DC magnetron sputtered TiN/NbN hard superlattices, Surf. Coat. Technol. 183, pp. 174.
3. Barshilia H. C., Rajam K. S. and Sridhara Rao D. V., 2006. Microstructural characterization of nanolayered TiN/NbN multilayer thin films by cross-sectional transmission electron microscopy, Surf. Coat. Technol. 200, pp. 4586.
4. Barshilia H. C. and Rajam K. S., 2004. Nanoindentation and atomic force microscopy measurements on reactively sputtered TiN coatings, Bull. Mater. Sci. 27, pp. 35.
5. Barshilia H. C., Jain A., Rajam K. S., Gopinadhan K. and Chaudhary S., 2006. A comparative study on the structure and properties of nanolayered TiN/NbN and TiAlN/TiN multilayer coatings prepared by reactive DC magnetron sputtering. Thin Solid Films 503, pp. 158.
6. Barshilia H. C., Prakash M. S., Aithu Poojari and Rajam K. S., 2004. Corrosion behavior of nanolayered TiN/NbN multilayer coatings prepared by reactive DC magnetron sputtering process, Thin Solid Films 460, pp. 133.

7. Barshilia H. C., Prakash M. S., Jain A. and Rajam K. S., 2005. Structure, hardness and thermal stability of TiAlN and nanolayered TiAlN/CrN multilayer films, Vacuum 77, pp. 169.

8. Barshilia H. C. and Rajam K. S., 2004. Raman spectroscopy studies on the thermal stability of TiN, CrN, TiAlN coatings and nanolayered TiN/CrN, TiAlN/CrN multilayer coatings, J. Mater. Res. 19, pp. 3196.

9. Binning G., Quate C. F. and Ch. Gerber, 1986. Atomic force microscope, Phys. Rev. Lett. 56, pp. 930.

10. Chu X. and Barnett S. A., 1995. Model of superlattice yield stress and hardness enhancements, J. Appl. Phys. 77, pp. 4403.

11. Fullerton E. E., Schuller I. K., Vanderstraeten H. and Bruynseraede Y., 1992. Structural refinement of superlattices from X-ray diffraction, Phys. Rev. B 45, pp. 9292.

12. Helmersson U., Todorova S., Barnett S. A., Sundgren J.–E., Markert L. C. and Greene J. E., 1987. Growth of single crystal TiN/VN strained layer superlattices with extremely high mechanical hardness, J. Appl. Phys. 62, pp. 481.

13. Koehler J. S., 1970. Attempt to design a strong solid, Phys. Rev. B 2, pp. 547.

14. Oliver W. C. and Pharr G. M., 1992. An improved technique for determining hardness and elastic modulus using load and displacement sensing indentation experiments, J. Mater. Res. 7, pp. 1564.

Chapter 11

Electro-Spark Coatings

K.R.C. Soma Raju

International Advanced Research Centre for Powder Metallurgy and New Materials (ARCI), Hyderabad – 500 005

Email: info@arci.res.in

ABSTRACT

Discusses in detail various aspects of electro-spark coating (ESC) technology. After a short historical introduction, examines its process fundamentals and mechanism of coating formation. Discussing coating materials and work pieces, states that the versatility of the ESC technique is a significant advantage from a commercial standpoint–nearly all metals and cermets that can be melted in an arc can be deposited on metallic substrates. Further, dwells upon the characteristics of ESC coatings–like microstructure, coating continuity and tribological performance. Regarding its applications, it is pointed out that the ESC process, due to its operational simplicity, low capital cost and negligible coating cost is increasingly used not only in high technology areas such as nuclear and aerospace but also in virtually every commercial industry. In view of its promising features, ARCI has successfully indigenized the ESC technology in collaboration with the Experimental Plant of Moldovia and manufactured 4 models with a capability to deposit the coating with different thicknesses.

Keywords: *Electro-spark coatings, Corrosion resistance, Surface modification, Arc welding, Surface erosion, Alloys, Substrate materials, Coating materials.*

Introduction

The continuous quest for enhanced durability and performance of engineering components has led to the development of various exotic materials in the last few decades. However, further improvements in material properties through alloy development alone are yielding diminishing returns. Furthermore, it is widely recognized that most engineering components, especially those subjected to different modes of wear during normal operation, fail predominantly due to surface degradation phenomena. Consequently, it is now acknowledged that alterations of the surface of engineering parts could lead to substantial improvement in properties such as hardness, wear resistance, low and high temperature strengths and corrosion resistance. This has lead to the introduction of many hard facing and or coating techniques such as welding, flame/induction hardening, nitriding, thermal spraying and vapour based coatings such as CVD, PVD and diamond and diamond like carbon coatings. These are now generically referred to as surface engineering/ modification techniques. Each of the above mentioned techniques have certain merits as well as demerits and none of them is universally applicable. Hence, it is the responsibility of the designer to choose one that is most appropriate for any given application. The Electro-spark coating technology is one of the many surfacing options that the modern-day designer has access to. Various aspects of this technology are discussed in detail in the sections that follow.

The phenomenon of hardening of metals due to electric sparking was reported as far back as in 1920s. However, efforts to develop the process for practical utility began only after another couple of decades and one of the earliest spark toughening equipment was produced in Russia in the early 1950s. Shortly thereafter, commercial systems also became available in the western world, and these are now popularly called electro-spark coating/deposition (ESC/ESD), electro-spark alloying (ESA), pulsed electrode surfacing/deposition (PES/PED) and spark hardening etc. Though Rawdon and Lazarenko are said to be the pioneers of this technology for their works conducted between 1920-1930, knowledge of the later's effort was restricted to the erstwhile U.S.S.R. One of the first landmark studies on the subject of electro-spark coatings in the open literature can be attributed to Welsh, who showed that the surface hardness could be influenced by the medium in which electro spark coating is performed. Welsh and his co-workers also proceeded to show that the ESC process with WC or WC-TiC electrodes could deposit highly wear resistant layers. However, available literature on the ESC technique as a whole is not very voluminous even to day, with an overwhelming majority of the reports emanating from countries belonging to the erstwhile Soviet Union and not widely read due to the language barrier. Furthermore, most of these studies have focused either on improving the efficiency of the coating unit through parametric control or on evaluating the quality of deposits achieved using 'new' electrode materials. A good review of the electro-spark coating deposition process has been authored by Johnson and Sheldon, who have dealt with a detailed description of the ESC process as well as its advantages and limitations and discussed the various parameters influencing coating quality.

The ESC process, due to its operational simplicity, low capital cost and negligible coating cost is increasingly used not only in high technology areas such as nuclear and aerospace but also in virtually every commercial industry. The ESC process is well-suited to deposit a wide range of coatings on diverse component materials and, in recent times, has also been evaluated as a means of increasing cutting tool life in machining and wood working operations. In view of its promising features, ARCI has successfully indigenized the ESC technology in collaboration with the Experimental Plant of Moldovia and manufactured 4 models with a capability to deposit the coating with different thicknesses. One of such ESC models developed is presented in the Figure 11.1.

Figure 11.1: Electro-Spark Coating Unit Developed at ARCI

Process Fundamentals

The role of surface modification in creation of tailored systems with a unique combination of surface and bulk properties is now well acknowledged. The Electro-Spark Coating (ESC) process is one such surfacing technique, which permits application of metallurgically bonded coatings, with a low heat input to the substrate, at the ambient temperature. This reduces the heat-affected zone (HAZ) and minimizes the changes in the properties of the substrate material. This coating methodology also offers certain unique advantages over other coating technique such as thermal spray, PVD/CVD and laser assisted surface modification.

The ESC technique essentially utilizes short-duration, high-current electrical pulses to melt material from a consumable electrode and deposit it on the substrate. Since there is very little total heat input because of the low duty cycle, the chance of distortion of the components is negligible. This ensures that there are no thermal stresses or metallurgical changes in areas other than the fusion zone. In principle, any electrically conductive material available in electrode form can be coated on any conducting substrate by the ESC process and this makes the ESC technology extremely versatile.

The heart of the ESC unit is a generator containing a bank of capacitors which releases stored energy in controllable bursts upon contact between the positive and negative poles of the generator. A consumable electrode made of the material to be deposited is connected to the positive pole (anode) while the work piece is connected to the negative pole (cathode). When there is contact between the electrode and the work piece, sparking occurs raising the temperature instantaneously to as high as 10,000°C and provides an intense heat source sufficient to melt and partially vaporize both the electrode as well as substrate in a very short time. Simultaneously, part of the electrode gets deposited on the job forming an alloyed layer with the molten metal of the substrate. The material transfer from the electrode to the job is rapid and self-quenching is extremely quick, since the bulk substrate material remains at near-ambient temperature and acts as a heat sink. This results in an extremely fine-grained coating of high density, hardness and strength. As the coating is metallurgically bonded to the substrate, its adhesion to the substrate is excellent and therefore the coating is spall resistant. The ESC process has often been referred to as a pulsed micro-welding process and is subject to some of the same effects of welding parameters as the conventional arc welding process. However, significant differences also exist– unlike arc welding, the electrode is in light continuous contact with the substrate in the ESC process and continuous electrode motion with respect to the substrate material is essential to prevent welding of the electrode to the part being coated.

The major difference between welding and an electro-spark coating lies in the pulsing nature and the stand off distance *i.e.* arc length in the two cases. The current discharge should be of short duration, typically less than 4 mille second, for the material transfer to satisfactorily occur. A condition known as "arcing" occurs when the electrical discharge is of low intensity and of longer duration. With all capacitor discharge circuits of ESC, the current discharges across the point of contact rather than across a gap as in welding. The above difference can be schematically represented in the Figure 11.2. As the spark durations in ESC are usually a few µ secs, or about three orders of magnitude less than in other pulsed welding processes, heat is generated in ≈1 per cent of the duty cycle and dissipates in ≈99 per cent. This allows the work piece to remain at near ambient temperature thereby eliminating or reducing the heat affected zones, metallurgical changes in the substrates and dimensional changes or warpage. In cases where prolonged sparking on thin sections is envisaged, the work piece can also be cooled with the help of fan or heat sinks.

The cooling rate of the deposit is estimated to be 10^5 to 10^6 °C/sec. The extremely rapid solidification of the tiny weld volume results in a nano-structured deposit that may be amorphous for some materials, exhibiting superior wear and corrosion properties. The Hall-Petch effect is evident in many deposits. A good example of this is Stellite6™ which ordinarily exhibits a hardness of about 40 Rockwell C, but an ESC deposit of this material can yield a hardness as high as 60 Rockwell C.

The typical air gap between electrodes is of the order of few microns usually 5-10µm. Hence, sparking preferentially occurs at the peaks of the surface profiles of the electrode and substrate that first contact with each other. As the coating thickness grows, high points on the substrate are observed to build more at the expense of surrounding areas. The smoothest deposited coatings are reported to be ≅ 2.5 μR_a and

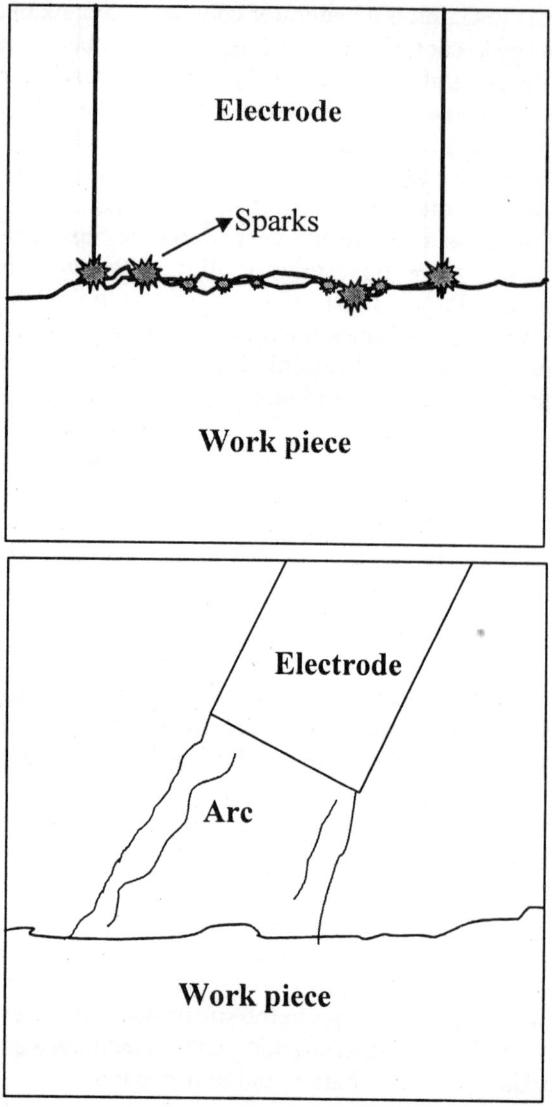

Figure 11.2: Schematic Representation of the Spark and an Arc that Occurs in ESC and Welding Techniques Respectively

are obtained with lower spark energies and in thinner deposits. The roughness increases with increase in spark energy and thickness, with some studies also reporting coated surfaces with R_a values as high as 30 μm.

In recent times, a method known as "TRESS", Thermo-reactive Electro-spark Surface Strengthening has also been introduced to increase the process efficiency and is being practiced increasingly, especially in CIS countries. In this process, electric discharge of the spark pulse initiates a chemical exothermic reaction between the electrode components. The additional heat that is generated from the chemical reaction of synthesis prior to the formation of final products on the substrate contributes to increased thickness and continuity of coating. The process of coating formation by this technique is also less energy consuming as compared to the basic technology of electro-spark coating due to the energy contributed by the exothermic reaction.

Mechanism of Coating Formation

The actual mechanism of material transfer by electro-spark coating method is subject to speculation and is rather complex. It is usually assumed that a gas bubble forms about (around) the spark discharge and persists for a time longer than the spark itself. Metal melted due to the high temperature of the spark is transferred from the electrode to the work piece surface via the expanding gas bubble. It involves not only transfer of material from the anode to cathode but also reaction of the molten material with oxygen and nitrogen from the surrounding air. This has been clearly established by sparking a Titanium alloy by a Ti or TiC electrode in air, which results in coatings with certain quantity of TiN. Generally, maximum material transfer is achieved by moving the electrode over the work piece and by making the consumable electrode the anode and the work piece as the cathode for the discharge circuit. This is similar to reverse polarity of the welding circuit. The highest temperature generation in the reverse polarity is at the anode, since the consumable electrode is fixed to the anode, its tip melts at a faster rate than the cathode and gets sprayed over the hot/molten spot of the work piece, thereby forming the alloyed layer of both work piece and electrode materials. Another reason for providing relative motion between the electrode and the work piece is the need to continually fracture the adhesive junctions forming as the electrical discharges occur and the molten metal deposits and solidifies. It has also been observed that the coating takes place not only in the liquid phase as explained above but also in the solid form. Based on the density of the electrode, certain portion of the electrode gets separated from the parent body and deposits in the molten pool over the surface of the work piece. This is clearly visible in the single splat studies conducted (Figure 11.3a). This mechanism is seen to prevail regardless of the spark energy level and duration of the pulses. The often observed variation in hardness in the coated film can also be attributed to this dual material transfer mechanism.

From the microstructure shown in Figure 11.3b, it is apparent that the coating is constituted of two distinctly different phases: One containing a nano-structured gray coloured material formed due to rapid solidification which corresponds to material transfer in molten state and the second phase distributed along the coated layers which appears as bright white spots after etching. These white spots are due to material transfer in solid form during the coating.

Though ESC coatings of the order of few mm have been reported for some materials, 25 to 100 μm is the most practical coating thickness that can be achieved from hard carbide materials. Researchers have observed that tough, lower thickness coatings perform better than brittle, thicker coatings. The difference has been attributed to the thermal stresses that are generated due to the repeated sparking and fracturing in the thicker coatings. Most practical coatings of hard and brittle materials such as

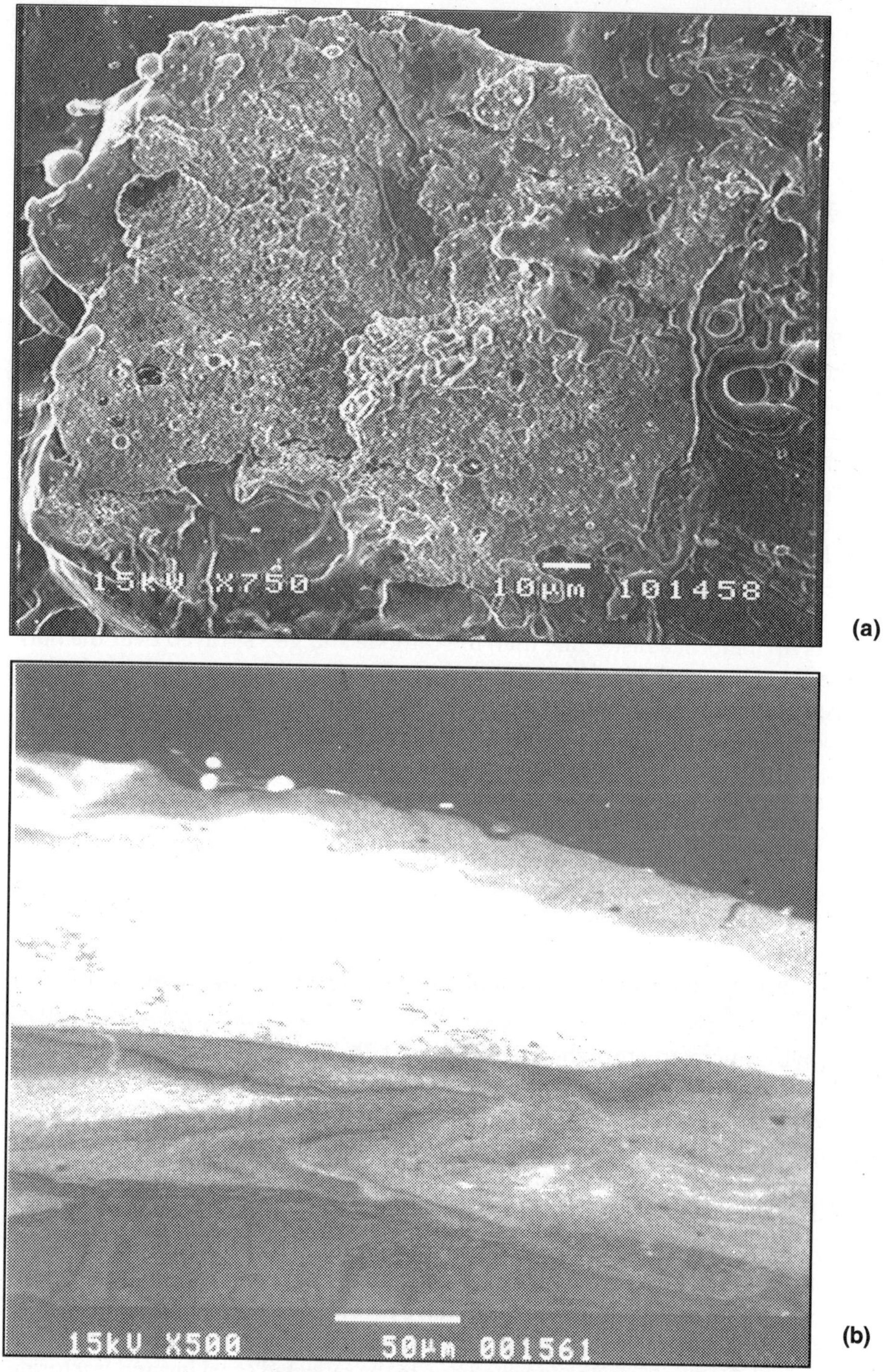

Figure 11.3: Micrograph Showing (a) Solid Material Transfer and (b) Resulting Microstructure

refractory carbides are 50–100 μm thick, whereas ductile materials can easily be deposited to thicknesses greater than 150 microns. The smoothest as deposited coatings can be obtained with lower spark energies and are thin. Whereas, thicker coatings are obtained with higher spark energies but surface roughness is high. However, in both cases the surface roughness increases with increasing coating thickness. Another limiting factor is that, for every spark energy, there is a certain limiting thickness beyond which further increase in thickness is not possible and even erosion of the already deposited material takes place.

The initial stages of the ESC can be related to the electro erosion process. This is clearly evident from the weight gain data for few parameters wherein the very first layer exhibits "negative weight gain". This need not be the same for all electrode work piece combinations that could be coated with ESC. The next stage involves the more complex physical phenomena where polar transfer of the anode material to the cathode surface and interaction of the materials of the electrodes (anode and cathode) occur under the conditions of high temperature and pressure (from the expanding gases present in between electrodes). Structural and phase transformations, diffusion processes and thermal stresses accompany this step.

The typical character of the relations between cathode weight gain, anode erosion, and time of electro-spark alloying with the use of various discharge energies is shown in Figure 11.4. Curve 1 is typically representative of any electrode-work piece combination depending on the atmospheric conditions and the coating application pressure etc. Curve 2 is usually obtained for mild regime of spark energies (E<1J) and curve 3 for harsh regimes (E>4J). Of course, these regimes may change based on the electronic circuitry adopted but the trends are similar for different spark energies. The dynamics of the formation of surface layers on the cathode is characterized by the fact that the rate of transport of the anode material to the cathode, which is maximum in the initial moments of the process, decreases with subsequent treatment. As a result for certain values of the discharge energy and treatment conditions, transport is replaced by erosion of the layer already applied.

Electro-spark coatings are usually evaluated by the amount of erosion of anode and the amount of mass transported to the cathode. This characteristic of mass transport determines the duration of the process or rate of growth of the coating layer. Limitation of the cathode weight gain and formation of the so-called limiting layer (maximum coating thickness) at certain specific treatment times are characteristic. The investigations of the process under different atmospheric conditions such as inert and vacuum with different electrode combinations have led to the following conclusions and presumed the limiting layer to be a result of one or more of the following reasons:

1. Chemical interaction of the electrode materials with elements of the inter-electrode medium during electro-spark alloying in air and the formation of solid solutions, oxides and nitrides that prevent interaction of the anode material newly arriving at the cathode with the already applied coating leading to embrittlement and destruction of the formed layer.

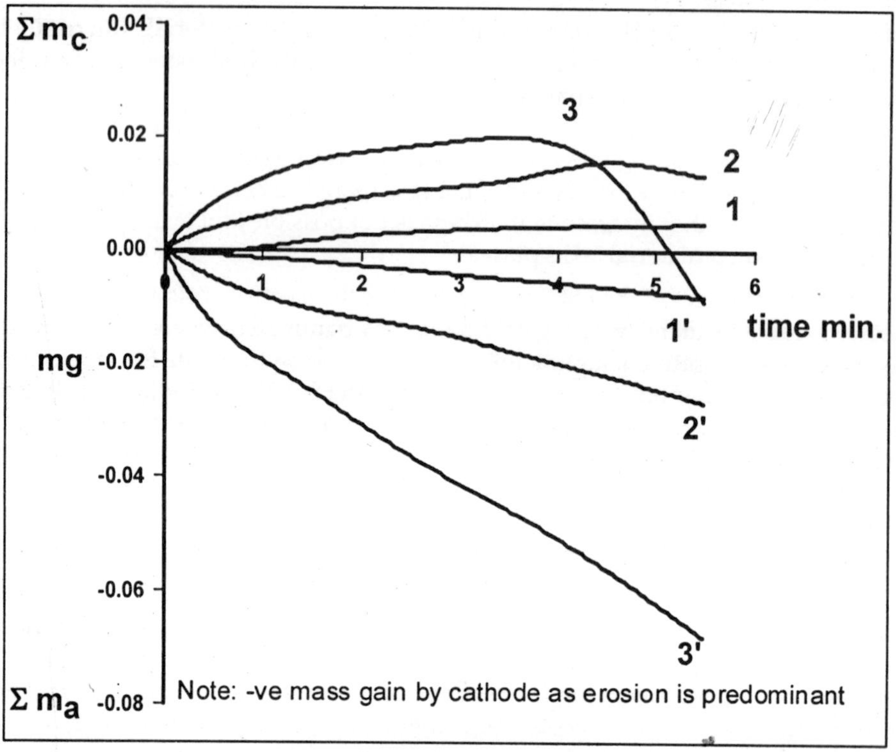

Figure 11.4: Typical Relation Between the Cathode Weight Gain (1,2 and 3), Anode Weight Loss (1', 2' and 3') and Duration of Electro-Spark Coating in Mild and Harsh Regimes

2. However, under inert atmospheric conditions when brittle oxides and nitrides do not form, the limitation of the thickness of the layer nevertheless exists. The inter-electrode medium is not the main influencing factor in such a case, although it can substantially influence mass transport.
3. Destruction of the formed layer begins due to the combined effect of the preceding stress-strain state and pulsed thermal stresses created by the discharges occurring at the given moment. The total stresses occurring in the layer adjacent to the surface of the zone of action of the discharge can exceed the ultimate strength of the material of the formed layer.
4. Long action of electrical discharges and the mechanical impact of the vibrator as the metal ceases to absorb energy (its possibilities for deformation are exhausted) the density of dislocations approaches the critical and the substructure of the hardened layer is characterized by maximum degree of micro-dislocations.
5. During long treatment, a progressive increase of density of dislocations occurs in the walls of the cells causing a change in the orientation of adjacent

sub grains. Once the dislocation density reaches a certain critical value, micro cracks occur on the cell boundaries. The cracks develop under the effect of stresses, which promotes destruction of the surface layer.

Coating Materials and Work Pieces

The versatility of the ESC technique is a significant advantage from a commercial standpoint–nearly all metals and cermets that can be melted in an arc can be deposited on metallic substrates. More recently, ultra-hard ceramic coatings have also been deposited. Some materials, such as Iron aluminum alloys that cannot be welded without cracking at compositions above 10 wt per cent aluminum, can be deposited by the ESC route without any cracks even up to 35 per cent aluminum. Rapid solidification of the deposit is believed to be the principal reason for this advantage. On the other hand, some electrically conductive materials can not be coated (ex. Bismuth telluride, chromium silicide etc). Presumably, these materials tend to vaporize or decompose in the arc without significant transfer of any molten material. The wide range of coating and substrate materials that can be addressed by the process is depicted in the Table 11.1. Graphite, which does not have a molten phase at atmospheric pressure, does not transfer from the electrode but can thinly carburize the strong carbide forming materials such as titanium and zirconium.

Table 11.1: Electrode and Substrate Materials Used for ESC Coatings

Coating Materials			*Substrate Materials*
For Wear Resistance	*For Corrosion Resistance*	*For Built Up and Other*	
Hard carbides of W, Ti, Cr, Ta, Mo, Hf, Zr, Nb, V etc.	Stainless steels, special alloys (hastalloys, inconels etc.), Fe, Ni, and Ti aluminides.	Nickel base and Cobalt base super alloys.	High and low alloys steels, stainless steels, Ni and Co- alloys, Al, Ti-alloys, Cu, Zr-alloys, refractory metals (W, Mo, Ta, Nb, Nb Hf and Re).
		Noble metals (Au, Ag, Pt, Ir, Pb, Rh).	
Hardfacing alloys (Stellites, Triballoys, Colmonoys etc.) Borides of (Ti, Zr, Ta)	Fe-Cr-Al-Y, Ni-Cr-Al-Y, Co-Cr-Al-Y.	Refractory metals and alloys (W, Mo, Ta, Re, Nb, Hf).	
Inter-metallics and cermets		Other alloys (Fe, Ni, Cr, Co, Al, Ti, Zr, Cu, Zn, V, Sn.)	

Influence of Process Variables

The quality of electro-spark coatings is dependent significantly on the process parameters. The key parameters can be classified into four categories : electrical (Voltage, pulse current amplitude, capacitance, pulse duration, frequency, inductance, packet of pulses etc.); substrate related (material, surface finish, cleanliness and geometry etc); environmental (cover gas composition, flow rate and flow geometry etc); and electrode related (material composition, density, geometry contact force, number of passes, overlap of passes and linear and rotational speeds in case of automation of the process). Automation of the process helps in reproducibility of the coating quality for repetitive jobs or adoption of ESC technology for regular production.

Though the overall process is identical, researchers from all over the globe have often used different circuits in their ESC units. The earliest equipment, some of which are still marketed today, consisted of a simple resistance-capacitance circuit wherein the spark frequency was controlled by the mechanical frequency of the vibrator only. Most modern equipment in contrast uses solid state devices, which enable the user to control the spark characteristics independent of electrode motion (vibrator or electrode holder). These are usually thyristor-fired circuits wherein the spark duration, its frequency and spark energy can easily be controlled. Higher voltages from 60 to 250 V, lower capacitance up to 600 µf are employed for generation of coatings. Process conditions employing a voltage of 50 to 200 V and a capacitance of 10 to 60 µf have been found to yield good results. The typical spark durations are 4 to 60 µsec with the spark energy of 1 joule max. and frequency of 0.1 KHz to 4 KHz, though much higher spark energies of up to 10 Joules have also been reported. The latest systems use MOSFETs for faster switching. Our circuit, wherein user safety was given utmost importance, utilizes capacitance as high as 80000 µf but lower voltage 35V for attaining the coatings at frequencies ranging from 50 Hz to 2000Hz.

Though spark pulses are created and controlled electronically, the electrode requires to move continuously to prevent welding/sticking of the consumable electrode to the job. This helps in making and breaking the contacts. Many means are used for achieving this, and among them, mechanical vibrator using solenoid and rotary motion using single or multiple electrodes are the most important.

Characteristics of ESC Coatings

Microstructure

The mode of transfer and properties of the electrode influence the efficiency of material transfer from the electrode to the substrate. The electrode mass transfer mechanisms can be classified into two categories. ESC coatings, when applied in dissociable gas like air, promote a molten globular mass transfer as the ionized gases form plasma of high thermal conductivity. In this mode of transfer, droplets form and are accelerated towards the substrate by the plasma jet. The droplets impact the substrate to yield the typical "splash" appearance on the surface of coatings as shown in Figure 11.5. On the other hand, with an inert gas like argon that produces lower thermal conductivity plasma, a "spray transfer" mechanism is more likely. This produces a fine matte appearance since the material is sprayed as a fine spray. In general, materials with a lower melting point and low heat capacities appear to transfer more efficiently. For example, Cr3C2-15NiCr was found to yield a transfer efficiency of 77 per cent in air and 66 per cent in argon, whereas higher melting WC-25TIC-6Co shows a lower transfer efficiency of 53 per cent in air and 57 per cent in argon.

Most of the ESC coatings are associated with micro cracks, which form due to the rapid quenching rates that result. Though these cracks apparently do not hamper the tribological performance, however care must be taken in the case of applications that

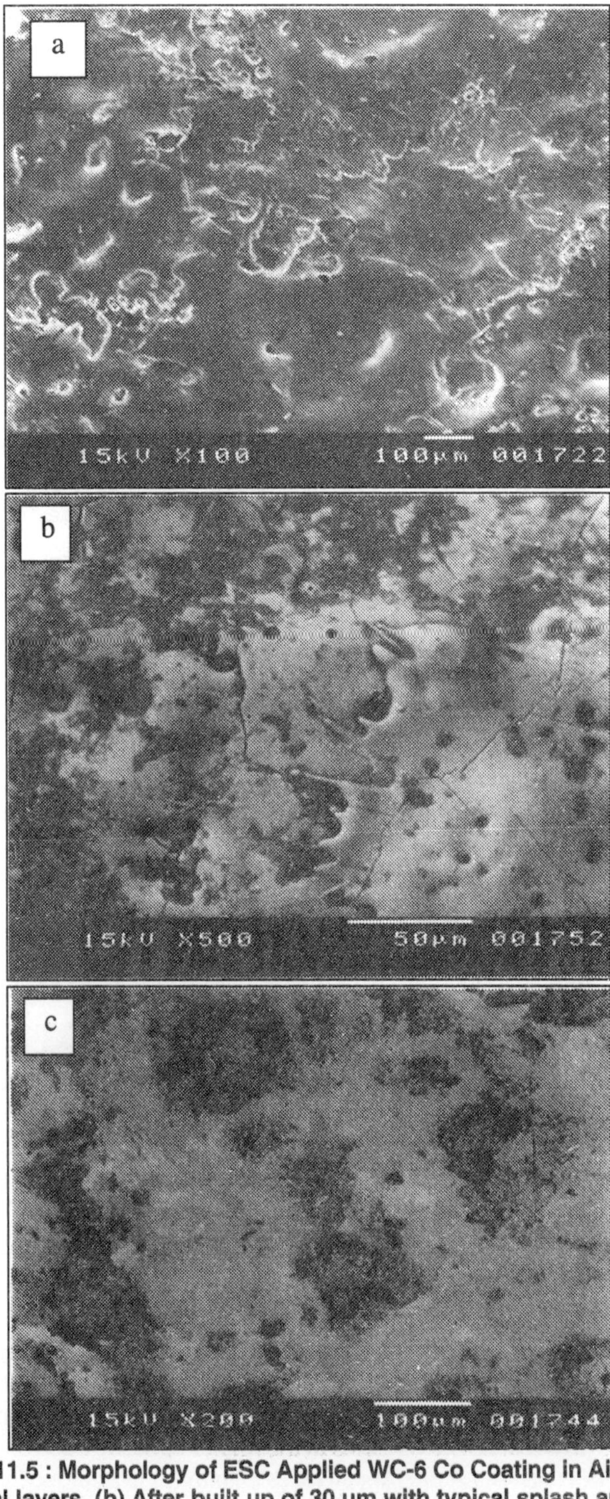

Figure 11.5 : Morphology of ESC Applied WC-6 Co Coating in Air on Mild Steel (a) Initial layers, (b) After built up of 30 µm with typical splash appearance and (c) After polishing

are sensitive to fatigue failure before employing the ESC coatings. The cracks that appear in the as-coated condition are usually observed to disappear in polishing, suggesting that the cracks are of shallow depth. Some of the investigations have even presented coatings with very high crack density and opined that the stress relief cracking, or crazing that sometimes occurs in hard facing, is not objectionable. This is especially true when the cracks are perpendicular to the surface and do not propagate parallel to the surface. In this context, it is also pertinent to note that chromium carbide coating specifications for sodium cooled nuclear reactor components, that are regularly coated using the ESC process, now specify a certain limit on the average crack density, that is acceptable in the coating.

Microstructures of various coatings that were obtained using different electrode and work piece material combinations are presented in Figures 11.6 and 11.7. A strong continuous interface explains the formation of a metallurgical bond between the coating and the substrate.

The coating thickness deposited per layer of the electro-spark coating depends significantly on the spark energy employed. Although there exists a definite limiting thickness, there is no marked difference in the microstructure of the coatings with respect to the spark energy employed. The microstructures shown in Figure 11.8 are of the same coating deposited to the same thickness level of about 10 microns employing different levels of spark energy. It has also been determined that the HAZ is negligible, the coatings are very dense and the porosity is low. The porosity shown is characteristic to ESC coatings and is due to the gas bubbles that form in the molten pool in micro level which can become further reduced in inert or vacuum atmospheres.

The electro-spark coatings microstructure also confirms dilution of the coating by the substrate material in a "transition zone" in the coating layer adjacent to the substrate. This is clearly seen when knoop/vickers indentations are taken as shown in Figure 11.9. The transition zone exhibits properties intermediate to the interface and surface. Absence of any variation in hardness of the substrate from the interface to the core of substrate also indicates negligible HAZ (heat affected zone).

Figure 11.10 depicts SEM micrograph of the cross section of a WC-6Co coated mild steel sample and elemental distribution of W, C, Fe and Co in the same cross section illustrated in the form of EPMA maps. The coating is seen to be dense, adherent and free from any defects such as cracks. Elemental EPMA maps presented in the Figure 11.10 show the coating to be W rich and the substrate to be Fe rich as expected. The EDAX analysis shows the presence of W up to 50 wt per cent and a relatively high concentration of Fe up to 40 wt per cent in the coating. The EPMA maps of different elements present in the coating and substrate supplement this. This provides ample evidence that Fe has migrated into WC-6Co layer. This occurs during melting and solidification of the localized melt pool of coating layer and the substrate under sparking.

The reason for high Fe content in the coating is the large difference between the melting points of the WC-6Co and the substrate (Fe). The lower melting point material in the pair (Fe in this case) forms a relatively larger molten volume as compared to that of the electrode material, which results in the higher concentration of the substrate

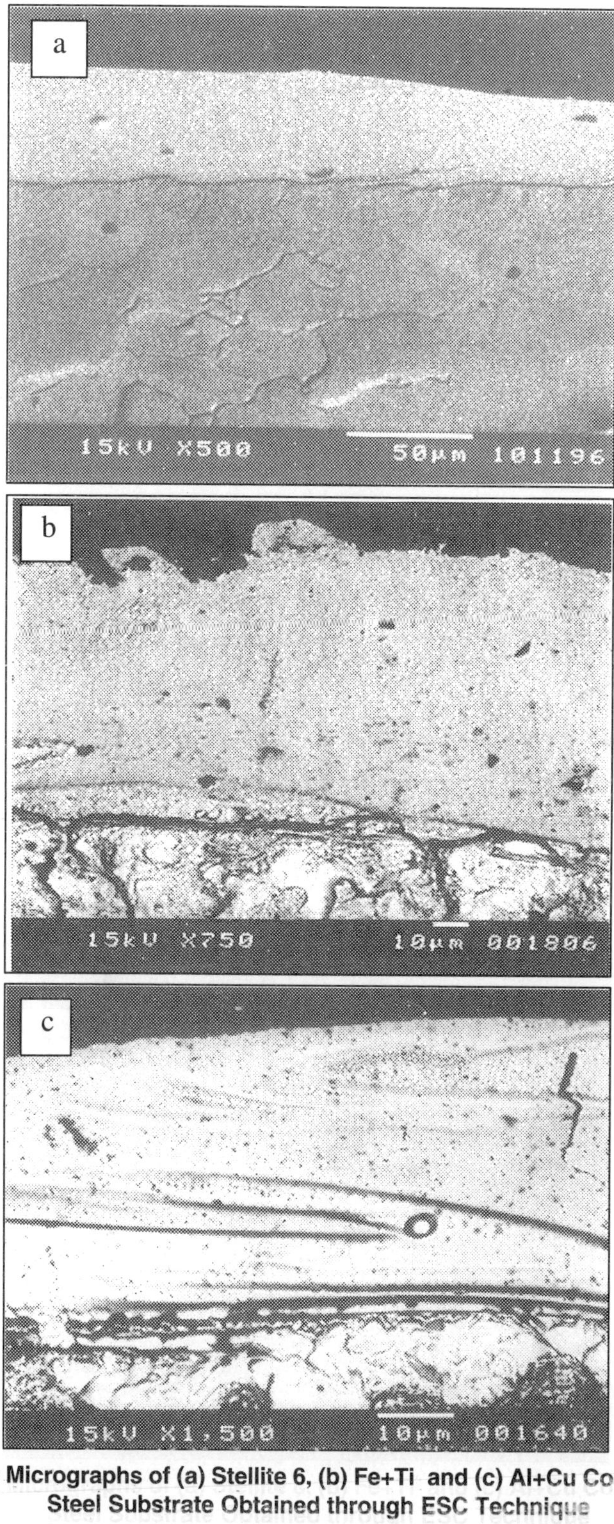

Figure 11.6: Micrographs of (a) Stellite 6, (b) Fe+Ti and (c) Al+Cu Coatings on Mild Steel Substrate Obtained through ESC Technique

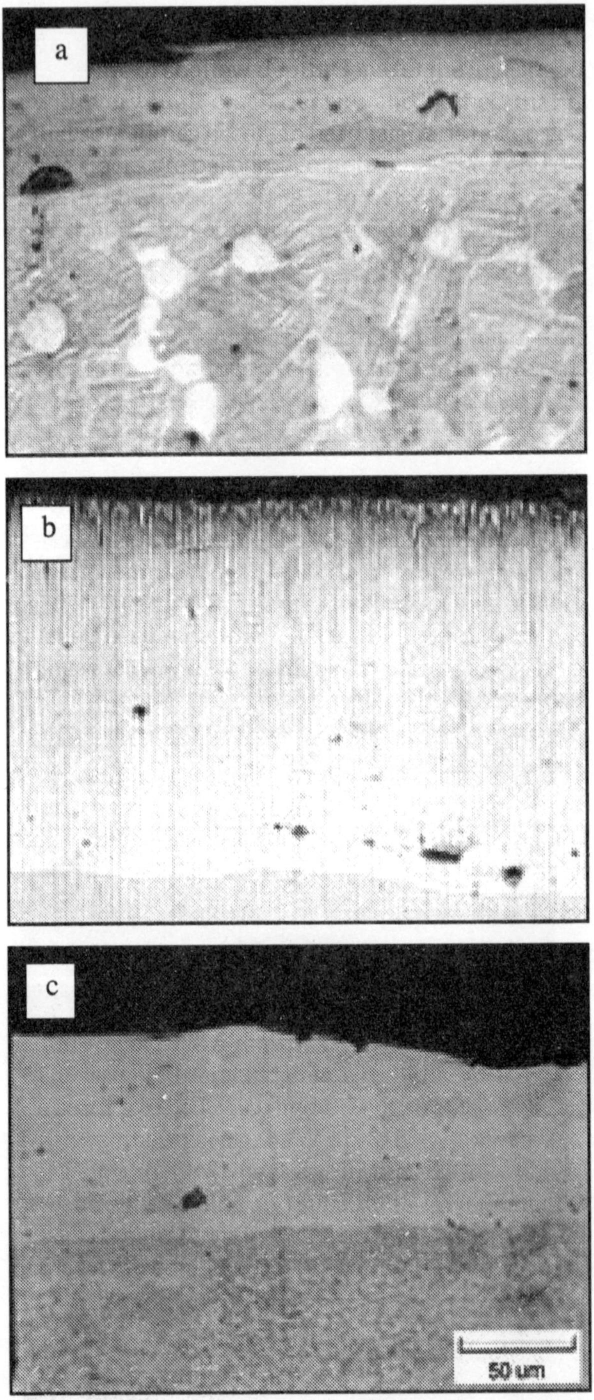

Figure 11.7: Microstructure of (a) TiC on Ti, (b) Inconel on D6 and (c) Alternate Layers of Stellite and Monel on Steel

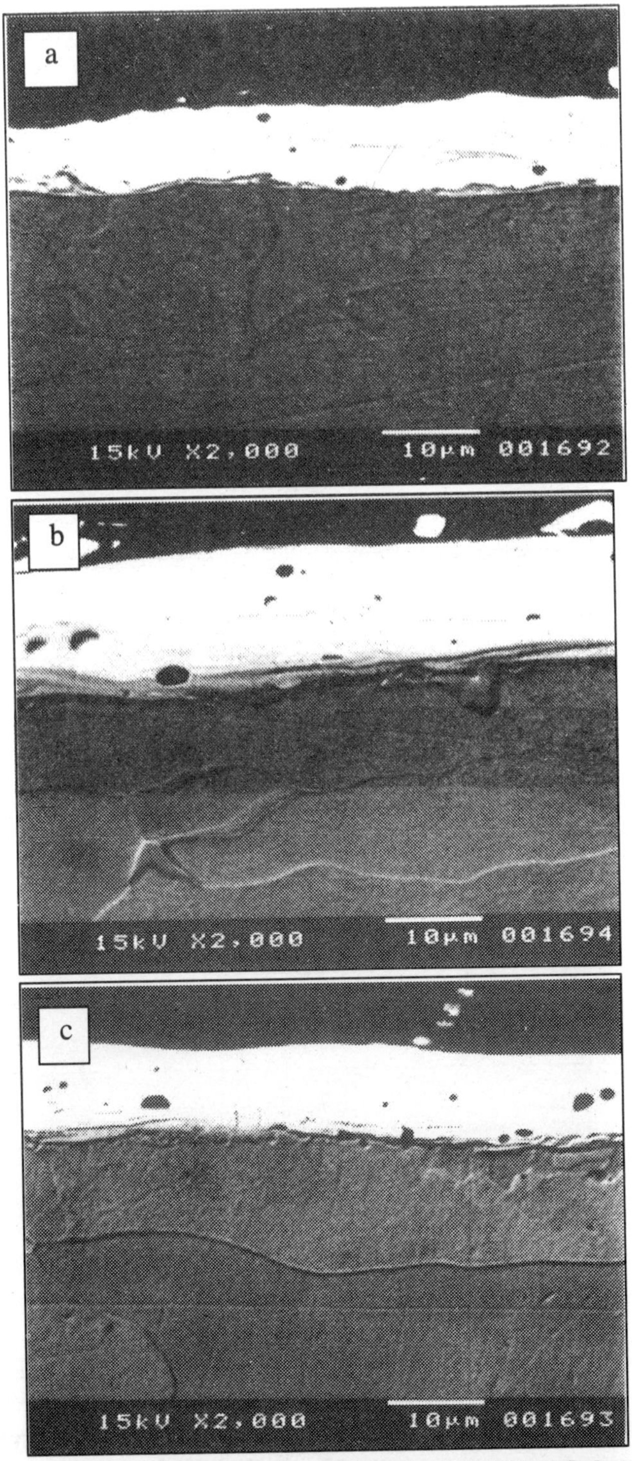

Figure 11.8: ESC Applied WC-6Co Coatings on Mild Steel Substrate at (a) 1.5 J, (b) 2.5 J and (c) 4.5 J Spark Energy Respectively

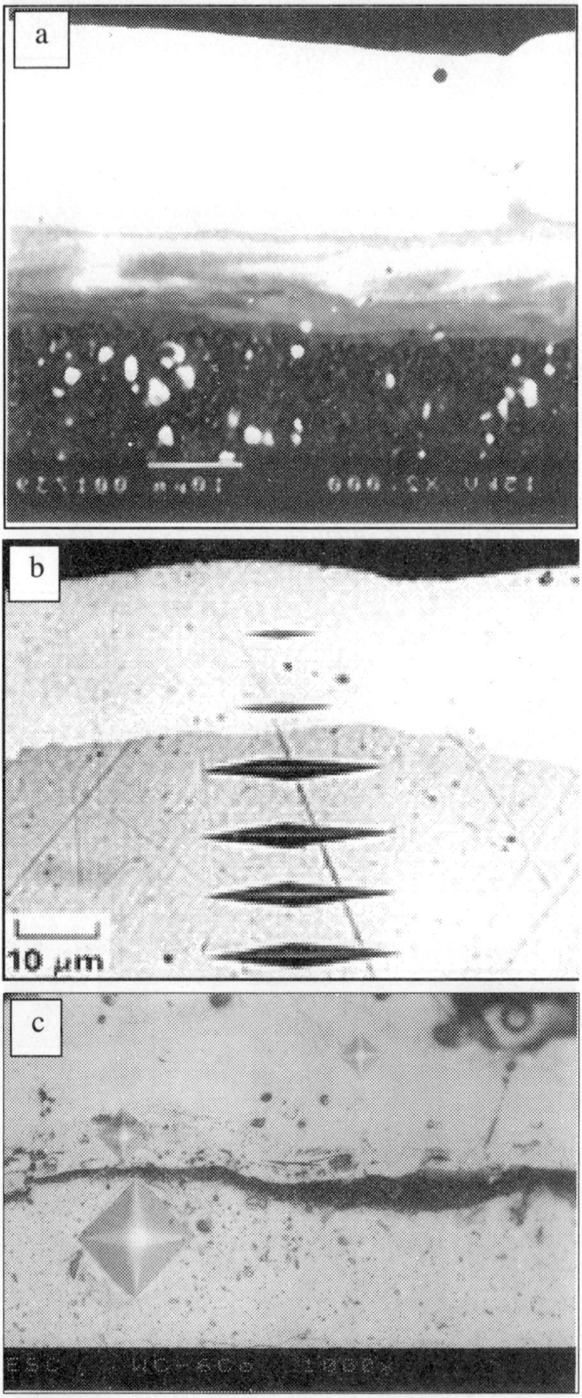

Figure 11.9: Micrograph of (a) WC-6Co on HSS, (b) Cr3C2-NiCr on Mild Steel and (c) WC-6Co on Mild Steel

Figure 11. 10: Microstructure of (1) WC-6Co EPMA elemental maps of (2a) W, (2b) C, (2c) Fe and (2d) Co Applied on Mild Steel

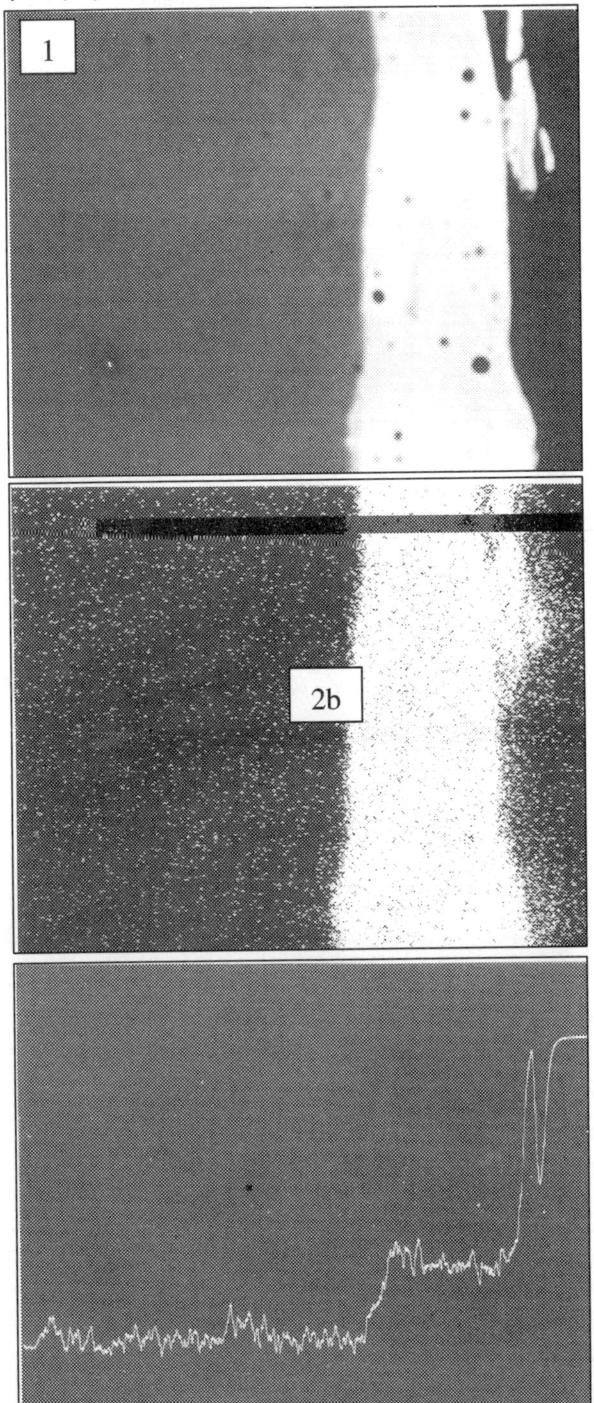

Contd...

Figure 11. 10–Contd...

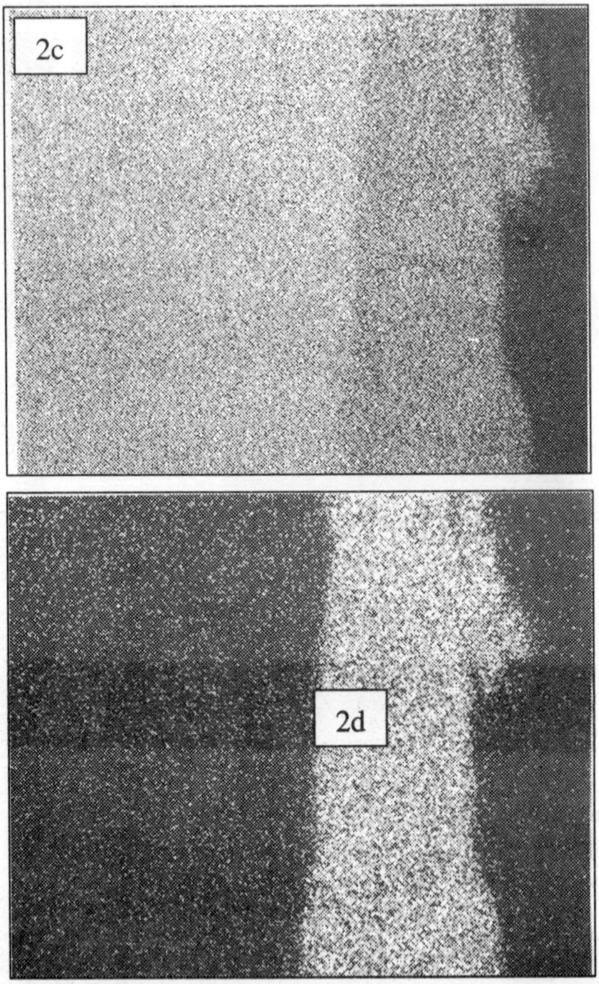

elements in the coating. In contrast, when lower melting point material like Silver is applied on a mild steel substrate, the coating is found to be silver without any dilution from the substrate material.

The spalling resistance of the coatings can be evaluated by a simple bend test using a flat steel strip of 5 mm thickness and a width of 25 mm coated to a uniform thickness of 100 microns on both sides. Two specimens, one coated by ESC and the other thermal sprayed, with a chromium carbide–nickel chrome coating being applied in both cases are comparatively evaluated. The metallurgically bonded electro-spark coatings show no delamination whereas the thermal sprayed mechanically bonded coating delaminated from both faces of the substrate in a severe 180º bend test using a 13 mm diameter fulcrum.

Although the basic equipment permits manual operation conveniently, a variety of automated electrode motions have also been used, *viz.* Vibratory, oscillatory and rotary motions, to increase the efficiency of the process. Mechanization by adopting, for example, a rotating tool electrode holder and a job manipulator has proved to enhance productivity. A variety of electrode holders have been developed providing capability to hold either a single electrode or multiple electrodes as depicted in the Figure 11.11. More recently, robots have also been employed (see Figure 11.12). Though

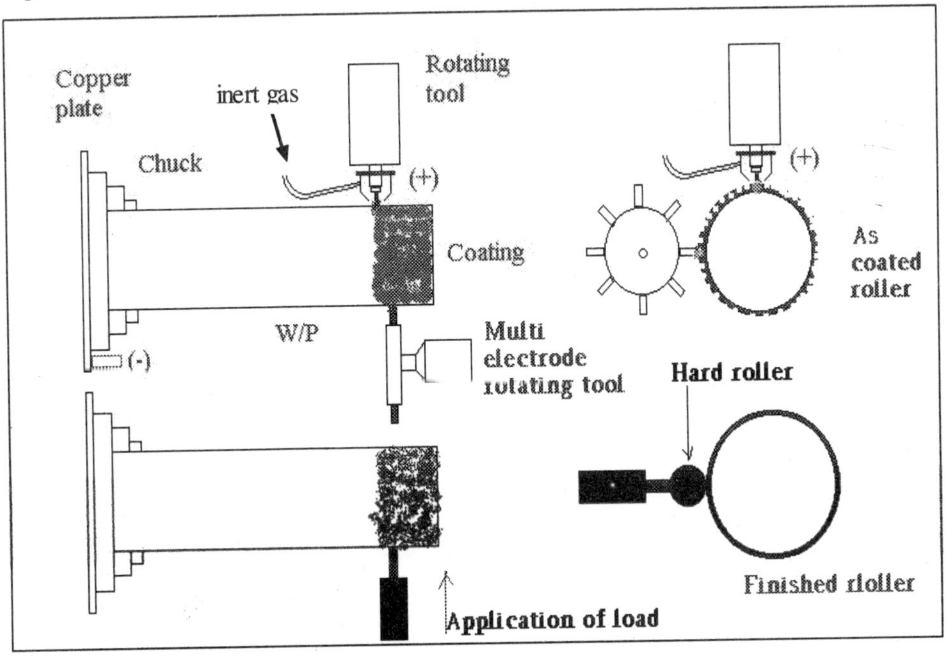

Figure 11.11: Schematic Representation of Mechanization Process Adopted for ESC Coatings

Figure 11.12: Robot In Use for ESC Applications

the coatings are applied using automation, they are associated with certain roughness and discontinuity as explained in the next section. However, these can be reduced by the use of a hard roller by cold deforming of the applied coating as depicted above. The typical coating rates that are reported vary from 1 to 20cm^2/min for a 25 micron coating thickness, depending on the electronic circuit, process parameters and the electrode-work piece combinations adopted. Researchers have tried to further increase the deposition rate, but increasing the spark energy and frequency beyond a certain point results in poorer deposits and become indistinguishable from arc welding process. This is considered to be due to the unavailability of sufficient time for solidification of the deposit resulting from the molten material transferred during the previous spark prior to the onset of the next burst/sparking.

Coating Continuity

Figure 11.13 shows the topographical features of the top surface of an as coated WC-6Co deposit on a steel substrate. The surface is irregular with several islands of globular shape which is characteristic of the electro-spark coating technique when the process is carried out in air. The ESC coatings are non-uniform in coating thickness and continuity, because of two reasons. The first reason is that since it is a micro welding technology, the material transfer is not uniform. The plasma channel that forms in the inter electrode gap while expanding under the intense heat pushes the molten metal forming crater-like features on the surface. As the coating is applied manually, there is a certain inevitable non-uniformity associated with the varying speed of movement and overlap that ultimately affects the coating uniformity. This is clearly evident from the above presented topography which was obtained from a 2 mm x 2 mm surface of WC-6Co HSS sample where 101 profiles were obtained over a 2 mm distance with each profile separated by 20 microns. Figure 11.14 presents the dynamically varying nature of roughness profiles at different locations from the same topographical evaluation.

Some of the micro-hardness values of the ESC coatings reported by various investigators obtained with various electrode compositions and atmospheric combinations are presented below in Table 11.2.

Generally, XRD spectra of the coated substrates reflect peaks characteristic of both the substrate material and material transferred from the electrode. In addition (see Figure 11.15) spectral lines from reaction products of the electrode and substrate materials and of possible products of reaction with the gaseous environment are also often observed. Some of the coatings have also been reported to be amorphous. When WC-6Co electrode is sparked on the mild steel substrate phases like γ Fe, Fe_6W_6C, WC_{1-x} and tungsten oxide are reported. The same electrode, when sparked on the Ti substrate, yields phases like Ti, TiN, $WTiC_{1-x}$ and $TiC_{0.7}N_{0.3}$. The presence of tungsten oxide and TiN phases in the substrates shows the strong influence of the atmosphere over the phases that are found in the coating. Sometimes formation of much complex compound phases can also result depending on the electrode-substrate combinations employed. This is also the reason for higher hardness levels being reported in the coating compared with that of the powders of the same composition applied by thermal sprayed coatings.

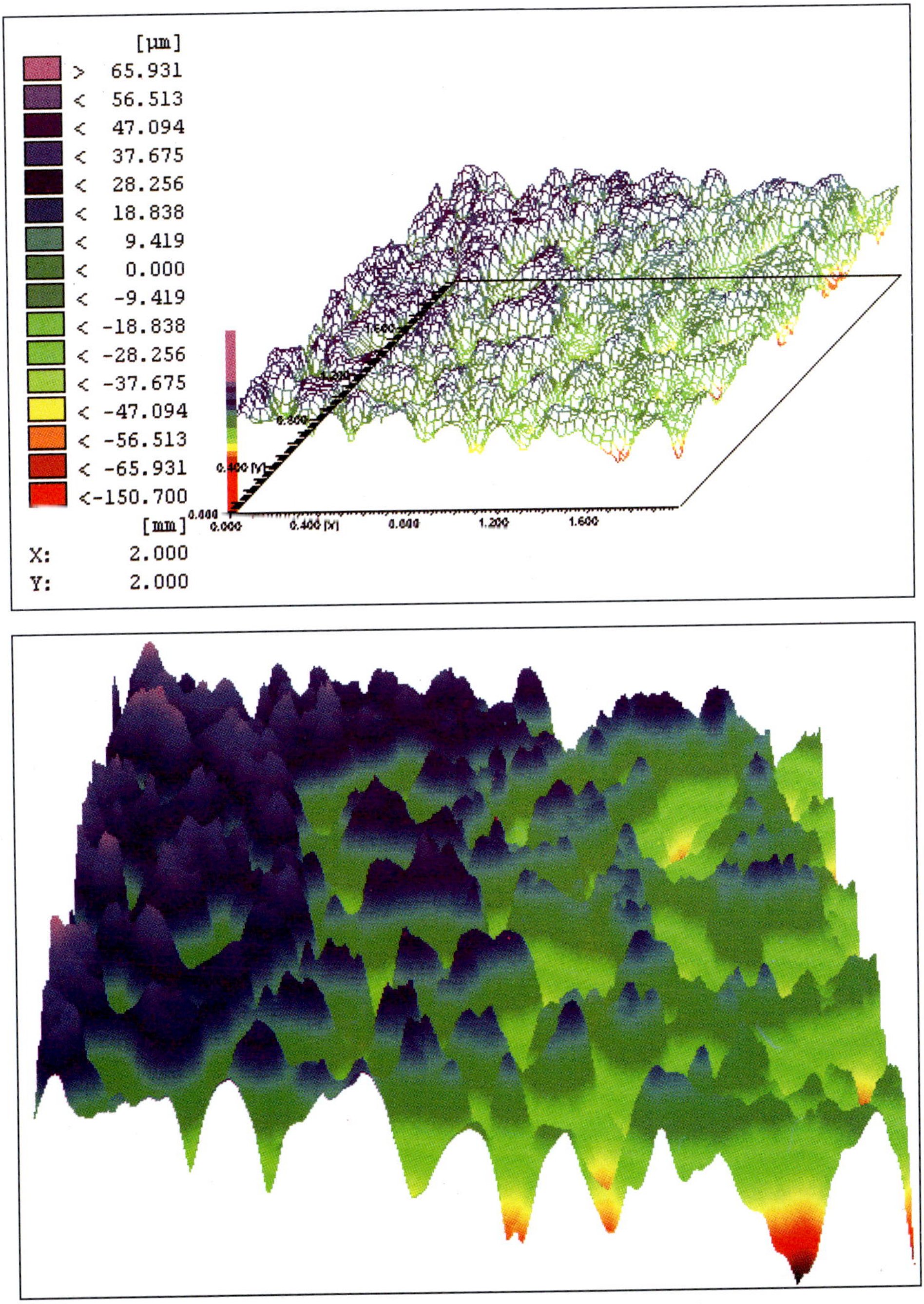

Figure 11.13: Topography of the ESC Coatings as Obtained by Perthometer Showing the Typical Peaks and Valleys Formed Due to the Sparks and Resulting Roughness

Figure 11.14: ESC Coating Profiles as Obtained by Topography Evaluation Unit (Perthometer) Showing the Dynamically Varying Nature of the Coating Thickness and Roughness

Profile 1 of 101

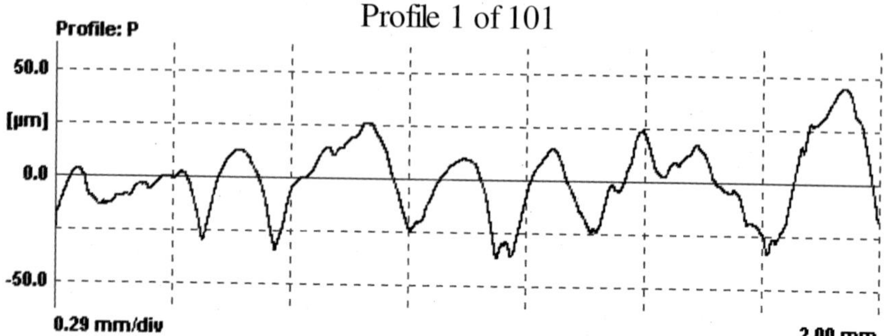

LC-R (GS)	0.25	mm
Ra	6.57	µm
Rmax	35.26	µm
Rz	28.32	µm
Rq	8.20	µm
Rp	11.90	µm
Rt	39.27	µm

Profile 49 of 101

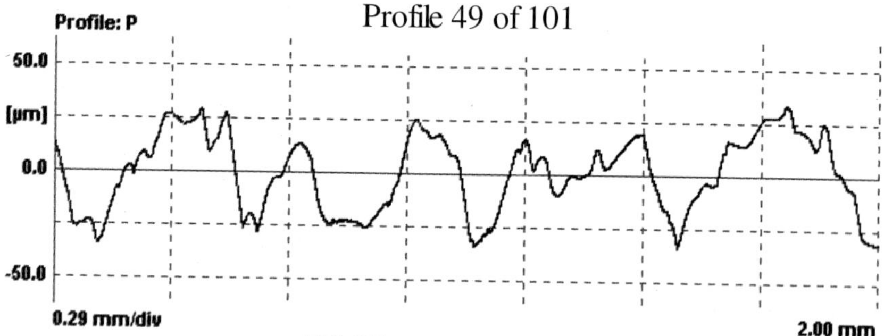

LC-R (GS)	0.25	mm
Ra	6.32	µm
Rmax	41.71	µm
Rz	31.63	µm
Rq	7.93	µm
Rp	15.58	µm
Rt	41.71	µm

Profile 50 of 101

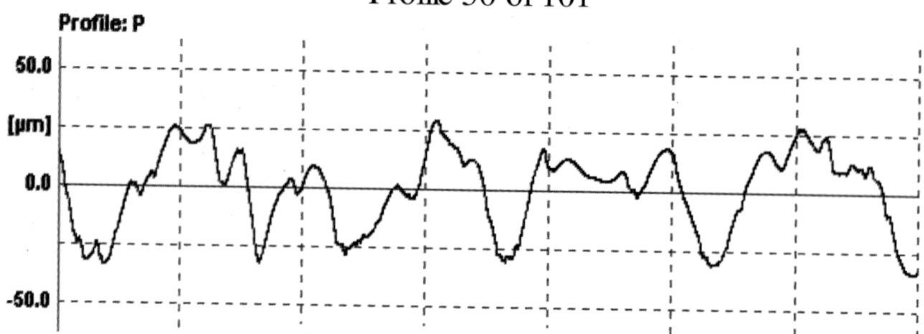

LC-R (GS)	0.25	mm
Ra	6.18	µm
Rmax	40.70	µm
Rz	28.95	µm
Rq	7.76	µm
Rp	13.57	µm
Rt	40.70	µm

Contd...

Figure 11.14–Contd...

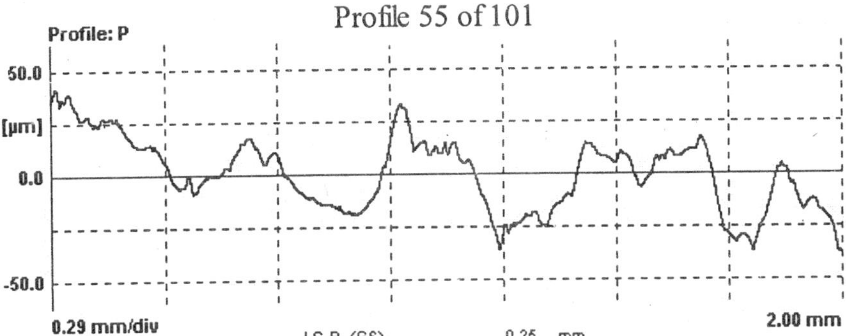

LC-R (GS)	0.25	mm
Ra	4.11	µm
Rmax	34.58	µm
Rz	21.78	µm
Rq	5.47	µm
Rp	11.75	µm
Rt	34.58	µm

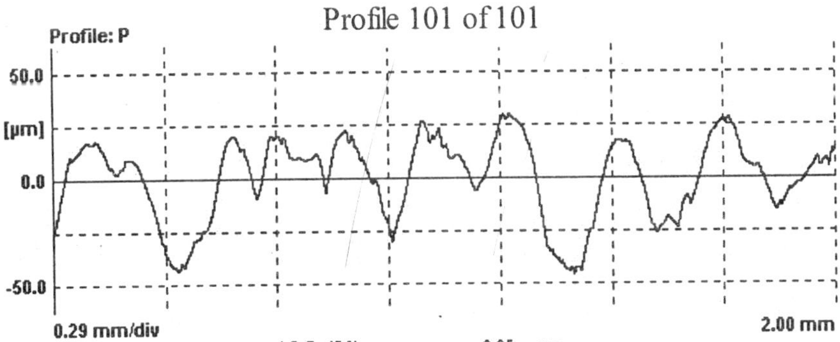

LC-R (GS)	0.25	mm
Ra	7.74	µm
Rmax	39.83	µm
Rz	31.56	µm
Rq	9.04	µm
Rp	14.16	µm
Rt	39.83	µm

Tribological Performance

The performance of the ESC coatings is observed to increase with the increase in the applied normal load. Investigators have found that ESC applied chromium carbide exhibited a lower friction coefficient of 0.25 when applied a second layer of Triballoy 700. When 316 SS samples coated with chromium carbide–Ni by Dgun (M/s Union Carbide) and ESC process were tested on, cylinder on flat sliding wear tester, it was observed that at low contact stress both coatings exhibited low wear, however, as contact stress increased wear rate of detonation gun coating increased rapidly while that of the ESC coating remained nearly unaffected (Figure 11.16). The investigators have related the better performance to that of the structure and bonding of the coatings. ESC coatings are much finer grained and more homogeneous with about 50 per cent higher hardness than the Dgun coating. ESC coating is a metallurgical bonding where as Dgun coating is mechanically fused coating. ESC applied chromium carbide

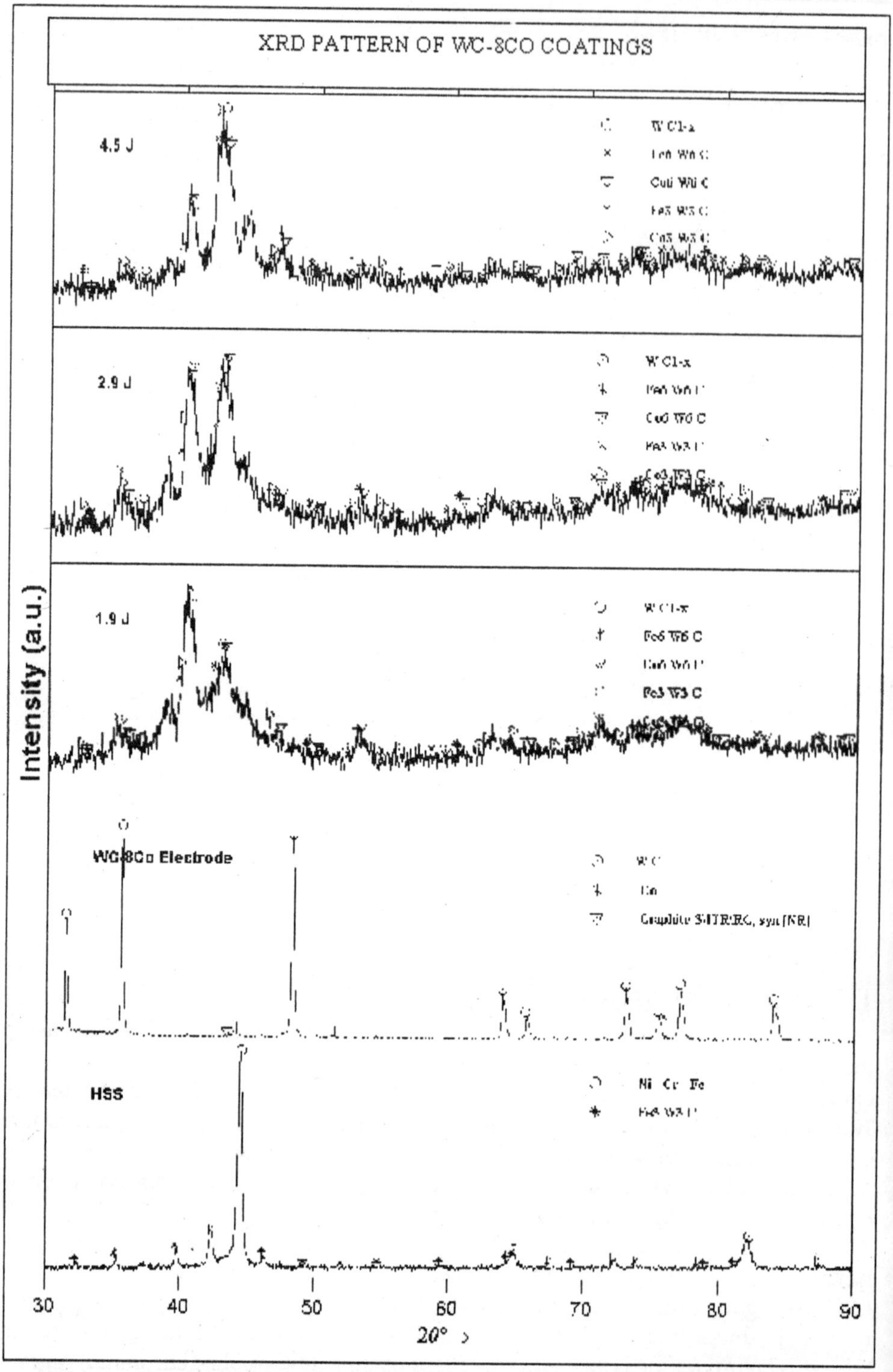

Figure 11.15: XRD Pattern of M2 Grade HSS Substrate, WC-8CO Electrode, and Coatings Generated Using 1.9, 2.9 and 4.5 J Spark Energies

nickel coatings were also found to be exceptionally wear resistant in the demanding environments of liquid metals.

Table 11.2: Micro-hardness Values of ESC Coatings Obtained with Different Coating Materials

Sl.No.	*Electrode Composition/Environment*	*Micro-hardness*
1.	WC-6Co/Air on M.S	900-1300 Hv
2.	WC-15TiC-6Co/Air on M.S	1000-1400 Hv
3.	Cr_2C_3-15Ni/Air on M.S	2000 knoop
4.	Stellite6/Air on M.S	700 Hv
5.	Stellite 12E/Air on M.S	450 Hv
6.	Mo,Ti (C,N)/Air on M.S	900 Hv
7.	Fe-TiC/Air on M.S	450 Hv
8.	Zr on En 31/Air on M.S	850-900
9.	B_4C/Air on M.S	1050 Hv
10.	Titanium carbide+NiCr and FeCr/Air on M.S	8-15.5 Gpa (800-1565 Hv)
11.	Titanium carbo nitride+AlN+NiCr and FeCr/Air on M.S	7-14.75 Gpa (700-1490 Hv)
12.	Titanium carbo nitride+Titanium diboride+NiCr and FeCr/Air on M.S	6.85-14.57 Gpa (700-1490 Hv)
13	Titanium carbo nitride+Zirconium nitride+NiCr and FeCr/Air on M.S	6.78-11.68 Gpa (690-1180 Hv)
14	Al on steel in N_2 environment to get–Fe_3Al and Fe_xAl_y and NH_3 environment to get–Fe_3Al, Fe_2N, Fe_3N and AlN	592Hv 774Hv
15	AlN-TiB_2/Air on M.S	2000Hv
16.	TiC-TiN-20 per cent Co, on Ni die steel	650-1400 HV

Similar trend was observed even with bare and WC-6Co and WC-15TiC-6Co coated HSS specimens. The ESC coatings have performed exceptionally well at higher contact stress in pin on disk test conducted with 5.54 m/sec sliding velocity and.35 to 1.76 Mpa contact stress conditions. The friction coefficients were ranged from 0.55 to 0.65 for WC-6Co and WC-15TiC-6Co coatings applied on HSS substrates with approximately same thickness (25 microns) and roughness levels (2-5 micron Ra). In all the three conditions HSS wear rate has gradually increased with the sliding distance where as the coated specimens maintained a steady state. When observed this under relative wear mode it was observed that the coatings performance increased many folds with increasing stress for the conditions studied.

The following are some of the results of the performance evaluation studies conducted employing ESC coatings on different cutting tools.

HSS Lathe Bits

Cutting tests have shown that ESC coated HSS cutting tools greatly reduce wear and significantly increase tool life (Figure 11.17). Under specific cutting conditions (rake angle = 15°, clearance angle =6°, cutting speed=170ft/min and feed=0.011"/

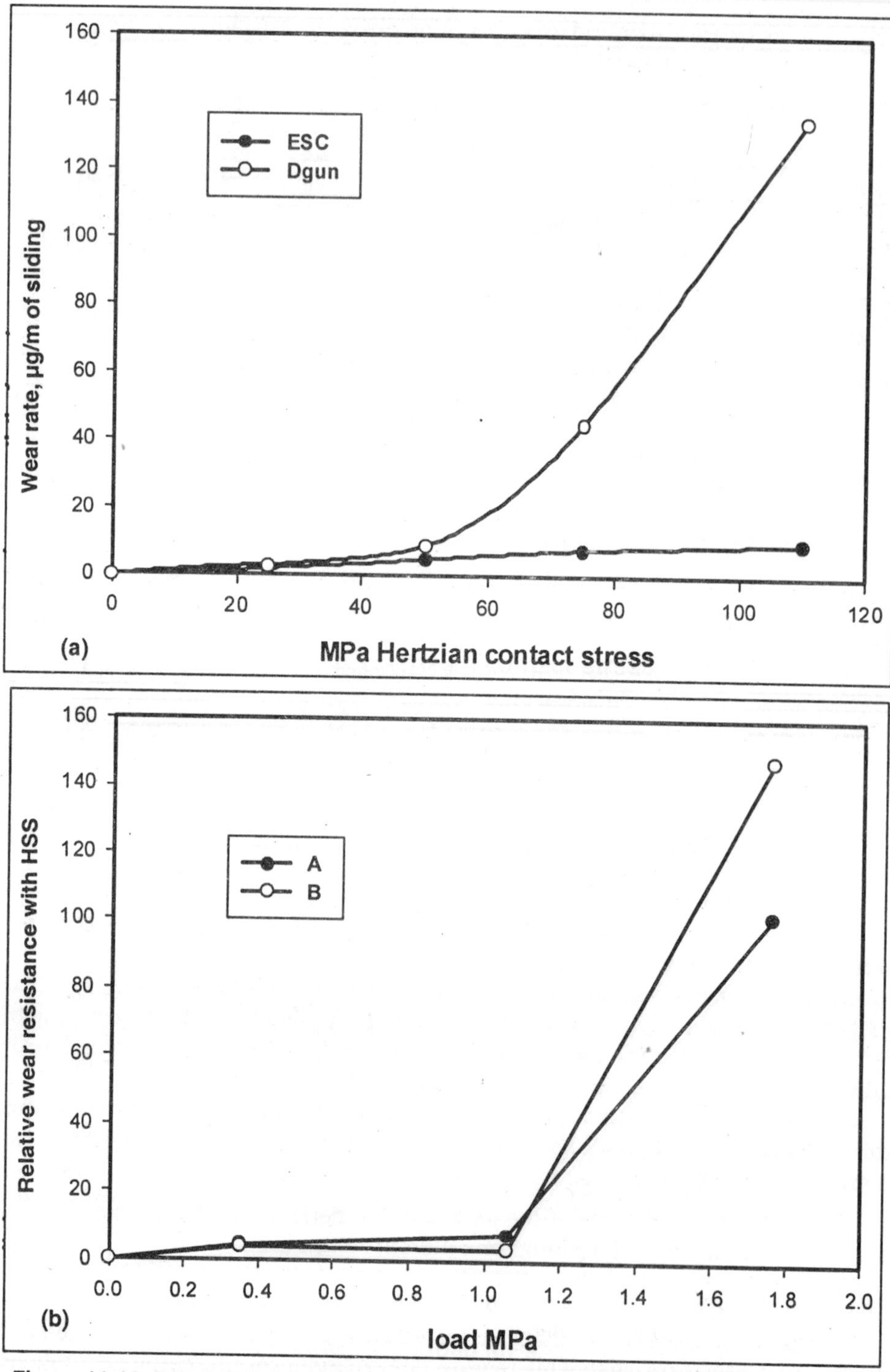

Figure 11.16: Effect of Contact Stress: on the Wear Behavior of (a) Dgun and ESC Chromium Carbide Coatings and (b) Relative Wear Resistance of (A) WC-6Co, (B) WC-15TiC-6Co Coatings

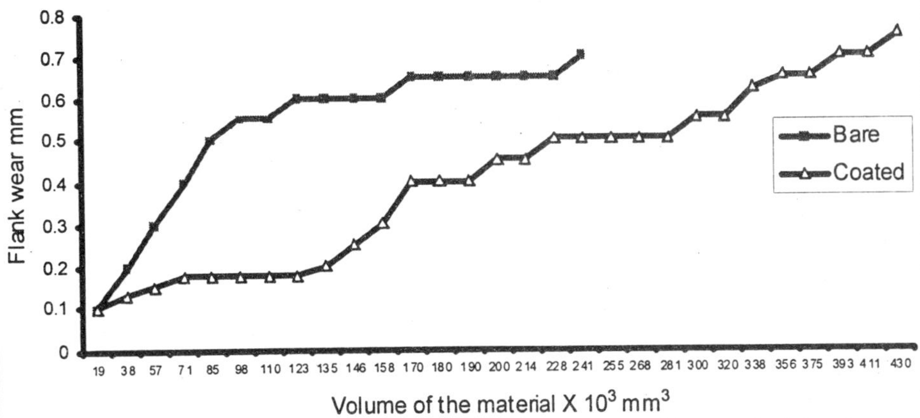

Cutting conditions: Bare and WC-6Co coated HSS tool (at 1.9J spark energy for 15 mm thickness), 41.78 m/min speed, feed 0.21 mm/rev and 0.5 mm depth of cut

Figure 11.17: Flank Wear Vs Volume of the Material Removed

rev), ESC coating of WC-6Co applied only on the tool face were found to increase the tool life to the same extent as commercially produced TiN PVD coated HSS tool when applied side to side and at least twice as good when applied front to back. Results obtained with other electrode compositions are summarized below in Table 11.3.

Table 11.3: Cutting Tool Conditions and Respective Tool Life Reported

Sl.No.	*Composition*	*Increased Tool Life*	*Machining Conditions*
1.	69 per cent WC, 20 per cent TiC, 4 per cent TaC and 6 per cent Ni	10 times	rake angle = 15°, clearance angle =6°, cutting speed=170ft/min and feed=0.011"/rev
2.	71.5 per cent WC, 12 per cent TiC, 10.5 per cent TaC and 6.5 per cent Co	8 times	Do
3.	TiB_2 and unknown binder	2.75 times	Do
4.	94 per cent WC and 6 per cent Co	2 times	8°, 8°, 6°, 6°, 10° and 10° with appropriate nose radius

HSS End Mills

A 16 mm HSS end mill in bare and WC-6Co coated condition was employed for performance evaluation runs. The machining conditions used were: 540 RPM, 0.04 mm/rev feed, 1mm depth of cut and 100 mm length of cut/pass. The increase in tool life was assessed to be 2 fold. Coatings were applied at two pulse durations for assessing its efficacy on enhancing the tool life for the same current amplitude. The investigations showed that the larger pulse duration further augmented the performance of the cutting tool under the test conditions employed (Figure 11.18).

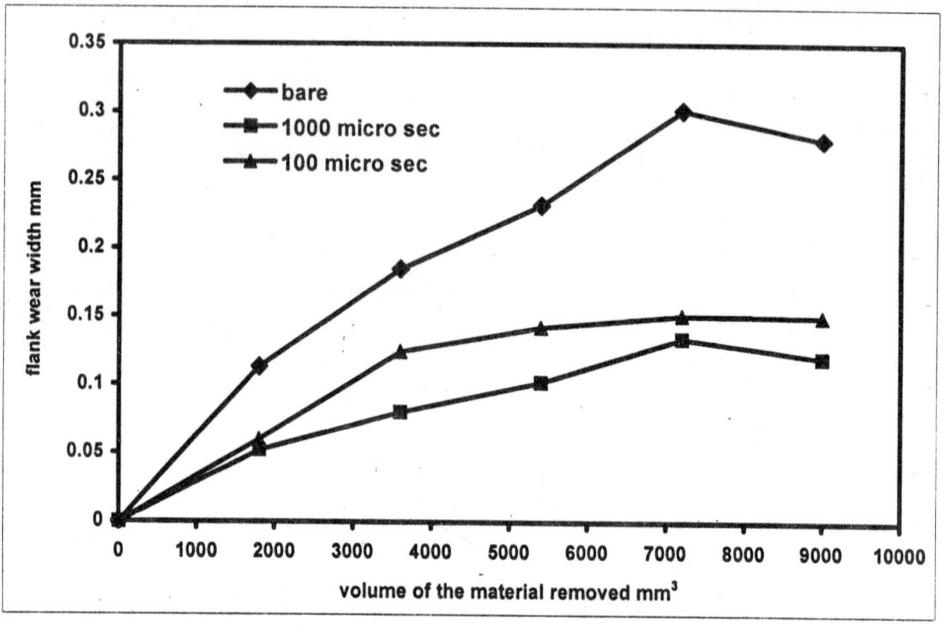

Figure 11.18: Flank Wear Vs Volume of the Material Removed

HSS Twist Drills

WC-6Co coating deposited by ESC route have also been found to significantly enhance the performance of M2 grade HSS twist drills by 3 to 6 times depending on the machining conditions adopted (Figure 11.19). The specific cutting energy for drilling was also seen to be marginally reduced. Investigation of the resulting hole dimensions indicated that the deviations observed were well within the acceptable limits (Figure 11.20). The mean average roughness of the holes produced by bare and ESC coated drills was almost identical.

Variants

Mould Mendor or Micro Spot Welder

This comes under resistance welding family. Resistance welding differs from other welding programs as there are no fluxes involved and the joints that are of lap type. The amount of heat generated depends on the amplitude of the current, resistance of current conducting path and the time of current flow.

H = heat generated = $I\,V\,t$ where $V = IR$

Or $H = I^2 R\,t$

H = heat in joules, I= current rms. amps, R = resistance in Ohms, t = time of current flow in sec.

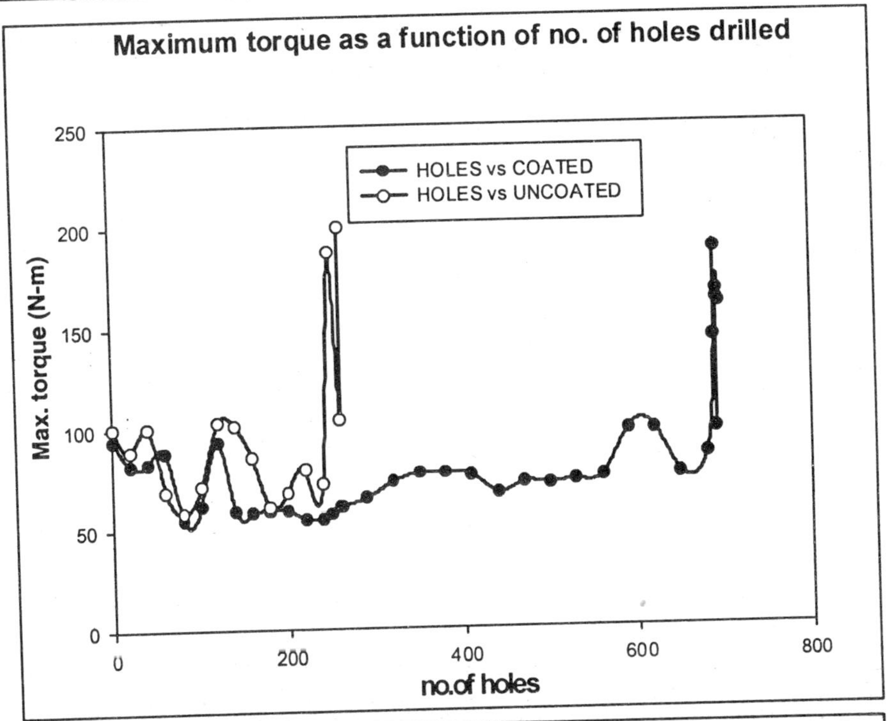

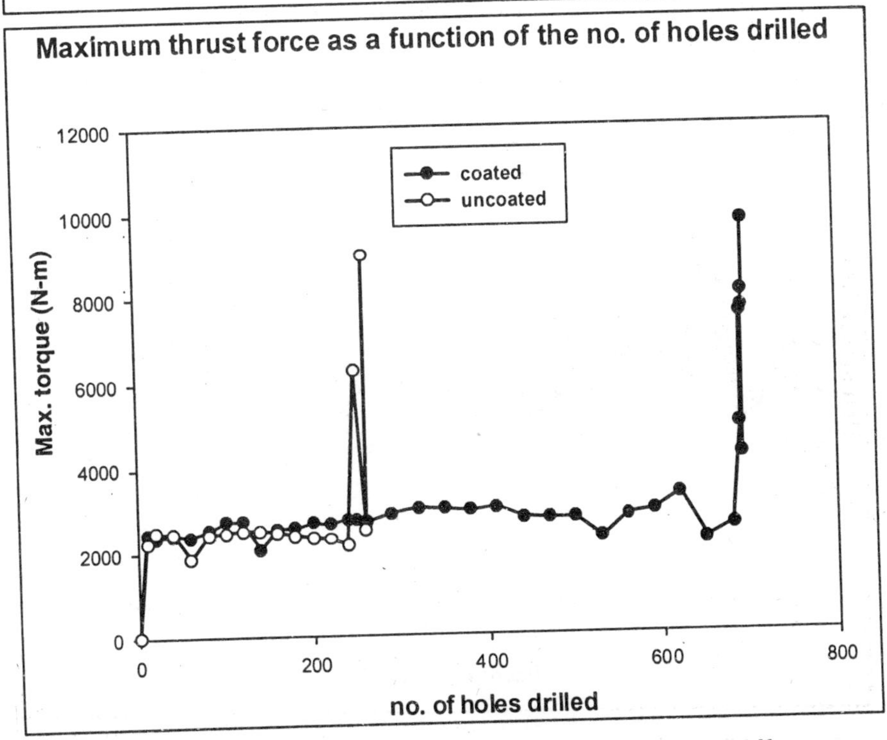

Figure 11.19: Torque and Thrust as a Function of Tool Life

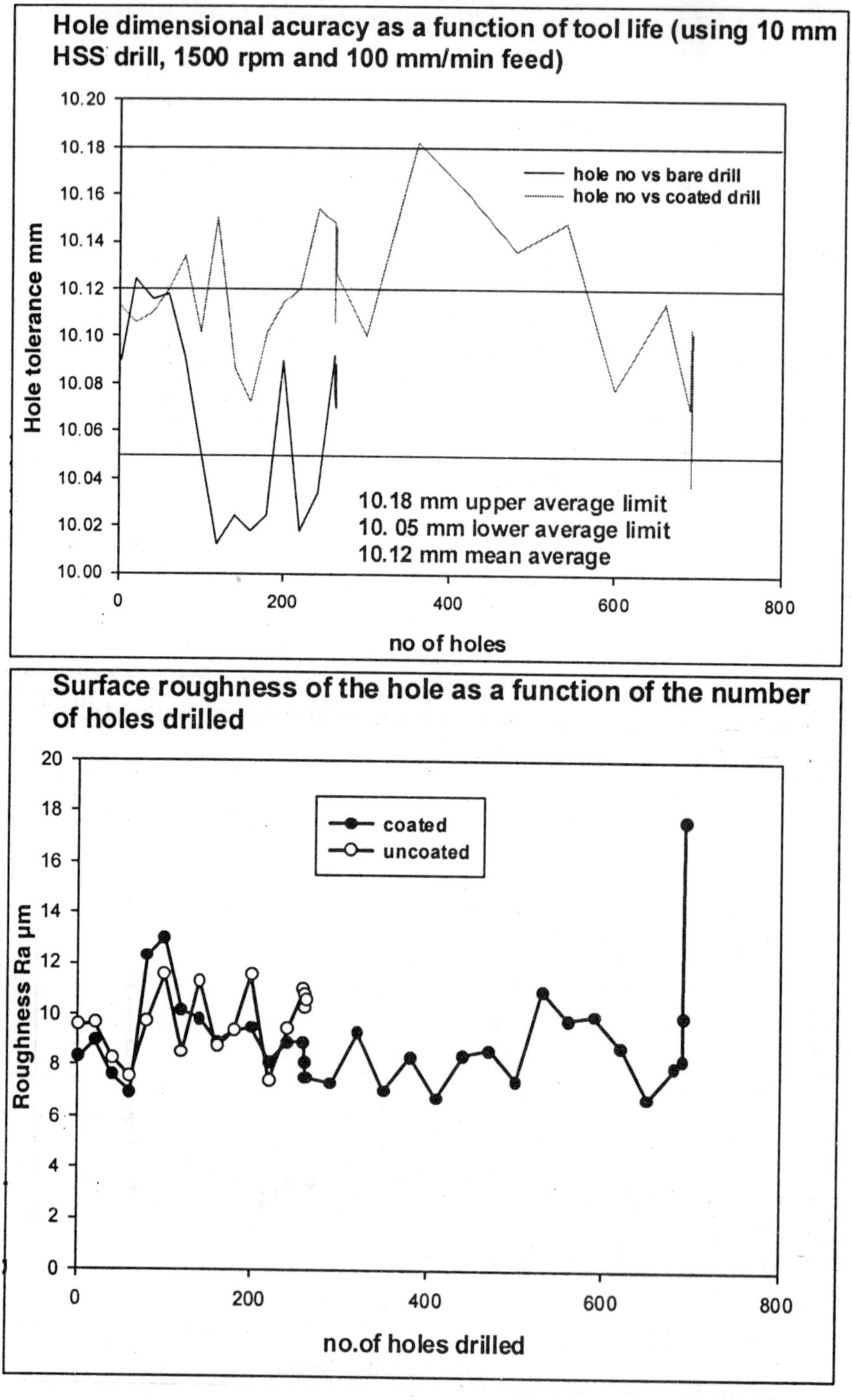

Figure 11.20: Hole Dimensions and Roughness as a Function of Tool Life

The interface of two surfaces forming the lap joint is the point of greatest resistance and hence is the point of greatest heat. In simple resistance welding a low voltage high amperage current flows from one adjoining plate to the other until the metal at the interface is heated to a high enough temperature to cause localized fusion (Figure 11.21). Thus under the applied pressure squeegees the molten metal from two parts to a homogenous mass called "nugget".

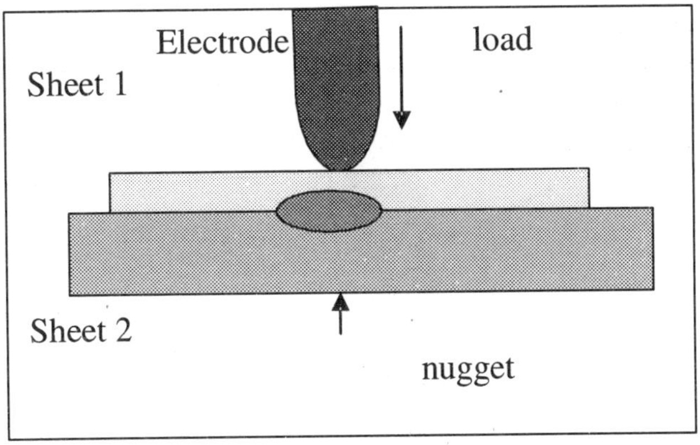

Figure 11. 21: Schematic Representation of Thin Sheet Welding

Welding variables: Current, time of current flow and the pressure applied

Both a.c and d.c are used to produce spot, seam and projection welding.

Control of up slope helps to avoid over heating and expulsion of molten metal at the beginning of the weld time as the interface resistance at that time is high. Down slope helps to control weld nugget solidification to avoid cracks in weldments.

For low carbon steels

I: $192 + Ke^{-t}$ A/mm^2

t: Sheet thickness, mm

k: A constant equal to 480 for mild steel

e: 2.718

Voltage requirement in the secondary is 1 to 25 V

Current requirement in the secondary is 1000 to 100 000 amps

Advantages of this Process (Micro Spot Welding)

It can be projected as a tool for refurbishment. Since in all other coatings either ESC or other thermal spray coatings, refurbished area is either poorly bonded (thermal spray) or of non uniform in thickness (ESC), this process comfortably can assume a place in the refurbishment area because of its metallurgical bond and the uniformity in thickness. Theoretically any amount of refurbishment can be done using thin wires, films and or powders. Carbon or alloyed steels sheets of very low thickness are employed as the consumable material and Tungsten rods ground to hemispherical end is used as non consumable electrode in such equipment.

While steel sheets are welded (0.1 to 0.3 mm), the deposited or refurbished area hardens due to the heat that is generated at the joint (nugget) and hence it will have an improved wear resistance in case of steels depending on the hardenability of the consumable as well as the substrate.

Powder Welding or Powder Over Lay

Operating principle and technique is same as that of the above except here that the powders, usually carbon and alloyed steel, are used as the consumable materials. The schematic representation of this technique is as presented in the Figure 11.22.

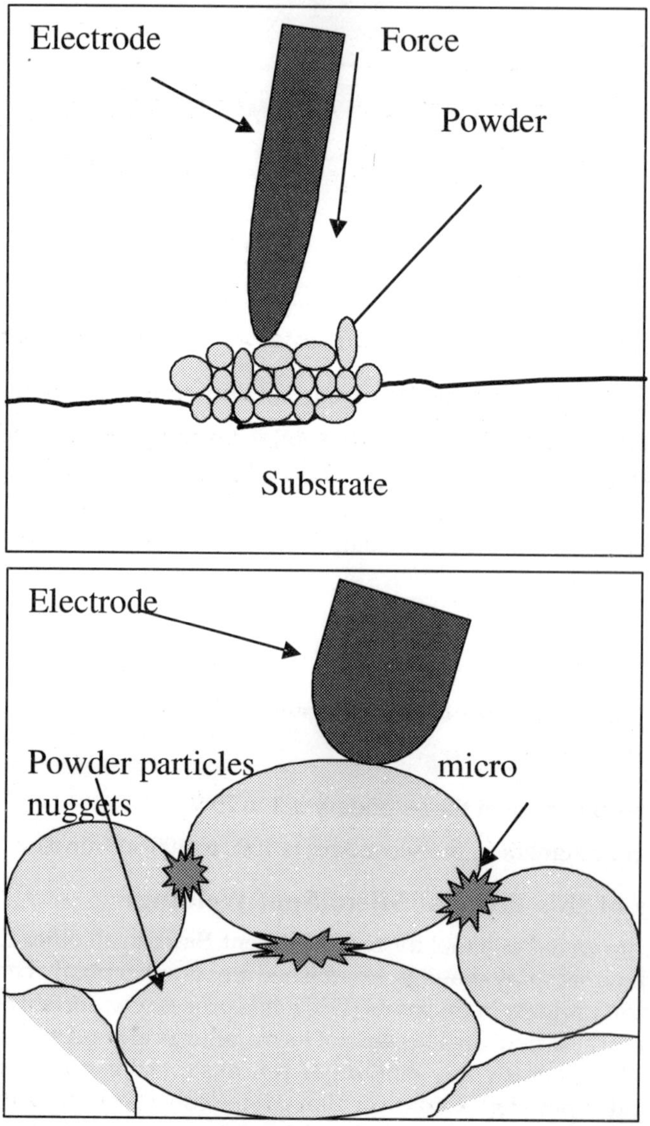

Figure 11.22: Schematic Representation of the Powder Coating

Advantages and Limitations

The ESC process has several inherent advantages associated with it. The more prominent among them are listed below:

- Coatings are metallurgically bonded
- Do not require expensive environment chambers
- Low heat input eliminates distortion or metallurgical changes in the substrate
- Special properties and amorphous layers are possible through rapid solidification
- Coatings typically exhibit higher load tolerance, lower wear rate, and lower corrosion rate than similar materials applied by other processes
- Little or no substrate preparation required
- Little or no surface finishing required
- Environmentally benign
- Reproducible process
- Operators can easily be trained
- Portable process and equipment
- Easily amenable for automation
- Applicable to I.D. surfaces (down to 3/8″ ID) and complex shapes
- Can apply most electrically conductive materials (metals and cermets)

The limitations of the ESC process can be listed as following:

- Coatings are associated with certain non uniformity in thickness as well as continuity
- Coating roughness increases with the increase in the thickness
- Both electrode and substrate material should be electrically conductive

Applications

Recognizing the varied above-mentioned advantages that the ESC technique offers, the process is being increasingly utilized in the engineering industry for a variety of applications. Based on the published literature, a majority of which has emanated from the erstwhile Soviet Union, the ESC process has clearly become a commonly used method to enhance the life of engineering components routinely subjected to wear. Some of the well-established applications include improving the wear performance of such components as valves, valve seats, turbine blades and pump parts. Other applications in nuclear, fossil and geothermal energy environments, high temperature turbine coatings and refurbishment for aerospace applications, industrial cutting tools, waste reclamation and water treatment plants, agricultural and textile equipment, modern sport equipment, petrochemical and pharmaceutical industries have also been reported. Apart from the above applications, the electro-spark deposited coatings are being widely explored for possible application to enhance the durability and performance for a variety of tools. Reported examples pertaining to Electro-spark coating of lathe tools, drills, milling cutters and hacksaw blades have confirmed ESC as a promising surface modification technique for improving tool life. Many exotic material combinations have been developed for engineering the surfaces of many critical components that include duplex and multi-layered coatings which allow special properties to be designed into the coating. Refractory metals such as niobium and tantalum are also being applied to steels as diffusion barriers and then are followed with coatings of corrosion resistant alloys.

The following are some of the typical applications where ESC is successfully implemented for batch production:

Nuclear

The need to develop irradiation damage and thermal fatigue resistance, wear and corrosion resistance with lower friction coefficient surface has eventually culminated in adoption of ESC coatings for some of the demanding nuclear applications. Over 120 commercially available materials and almost all of the surface engineering process combinations were evaluated and only ESC coated Cr_3C_2-NiCr could withstand the most severe acceptance criteria. This process is in production for the last 10 years without even a single failure. ESC applied Cr_3C_2-NiCr coating is metallurgically bonded and does not affect the 20 per cent cold work condition which is demanded for irradiation swelling resistance of 316 SS reactor core components.

Figure 11.23: Chromium Carbide Coated Core Components

Other applications are:

(a) Anti galling wear resistant chromium carbide coatings on 304 SS sodium valve guides against stellite 6 valve seats; (b) Steam generator tube supports; (c) Burnable neutron absorber coatings; (d) Electromagnetic pumps; (e) Transducers; (f) Component positioning hardware and control rods

Gas and Steam Turbine Applications

ESC is successfully implemented even in highly sensitive aerospace applications. The commercial applications include: 1. Deposition of hard facing alloys on "Z-notch" surfaces of turbine blade, 2. Application of corrosion resistant coatings to

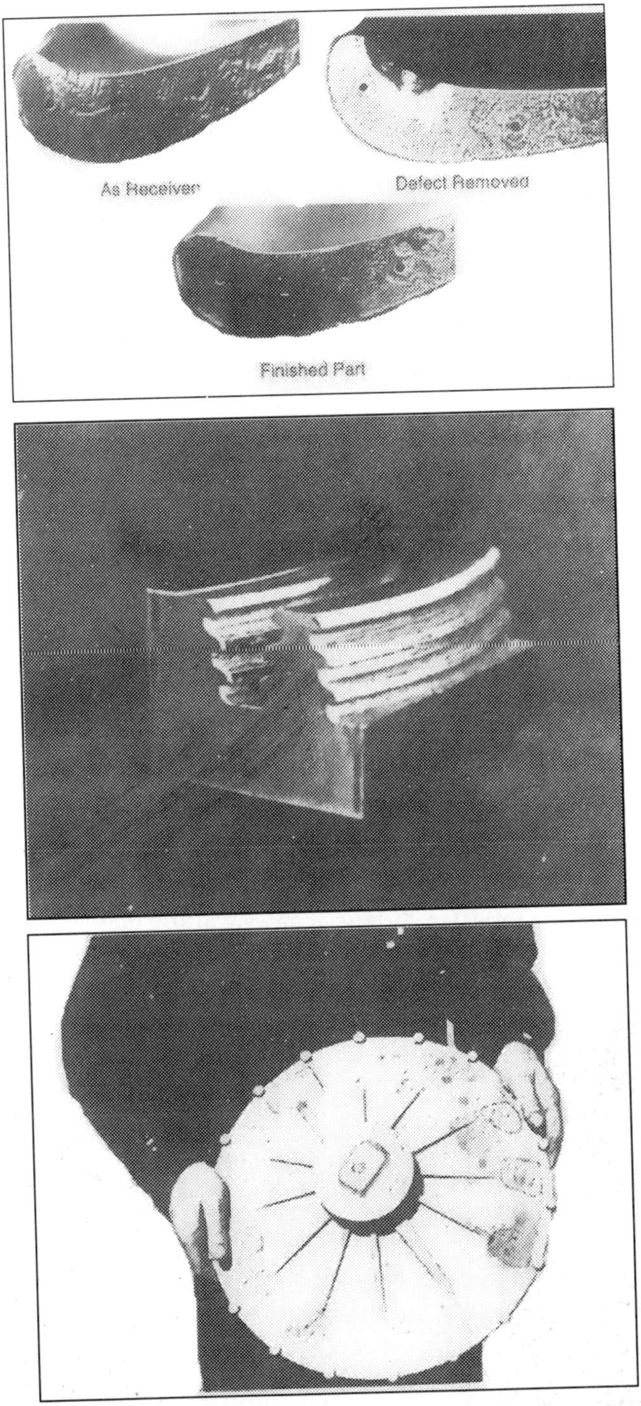

Figure 11.24: Repair of Single Crystal Turbine Blade, Turbine Blade to Prevent Stress Corrosion and Plenum Components to Repair Casting Defects

turbine blade tips where protective diffusion coatings were removed on finish machining operations, 3. Application of platinum to selected areas of turbine blade as a first step in forming platinum modified aluminide diffusion coatings, 4. Build up of nickel base super alloys to reclaim close tolerance part, 5. Coatings to resist particle erosion from ingestion of sand in military helicopter turbines and 6. Repair of thermal fatigue cracks in single crystal turbine blades.

ESC chromium carbide coatings on SS turbine blades could even with stand more aggressive and severe conditions of erosion by abrasive particles, hydrogen sulfide, carbon dioxide and ammonia in wet steam at temperatures upto 200ºC in geothermal steam turbines.

Automotive Industry

ESC applied titanium carbide coatings are used on titanium valve stems and guides of high performance IC engines of automobiles. They are also used to build up splines on critical shafts and to rebuild oil seals in close tolerance over-running clutches. Oil pumps and transmission mechanism are another areas of the ESC applications.

Die Casting Moulds

Die casting moulds are another area of successful ESC applications. Coatings are applied in the interior of the die for hot liquid metal erosion resistance and to rebuild the worn out parts by overlay coatings. ESC coatings ensure low friction for the liquid metal flow. Tungsten or titanium carbide based cermet coating protects the die with anti-wetting and extends the die life through prevention of heat crack, scuffing and improves in the release, liquid flow and gas escape.

Other areas that could be benefited by ESC are runners, over flows and vents. Soldering of the zinc or aluminum casting material to the die surface also could be eliminated by applying tungsten carbide at gates, parting lines, vents and die cavity etc. ESC applied WC-6Co coating has doubled the service life of the moulding sand crusher.

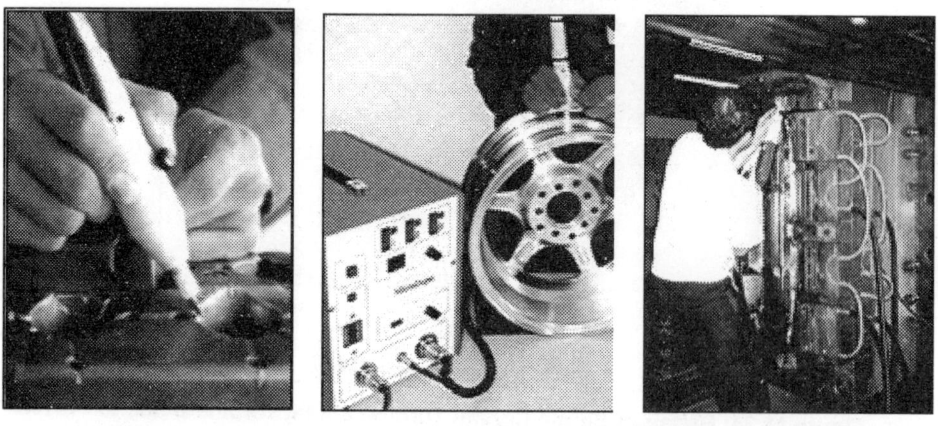

Figure 11.25: Localized Repair of Mould, Pattern and Components

Figure 11.26: Moulding Sand Crusher Coated with WC-Co

Figure 11.27: ESC Coated Die for Aluminum Wheel

Metal Working Tools

The very first commercial application of the ESC coatings were reported to be cutting tools like lathe bits, drills, milling cutters and end mills etc and metal working dies such as shears forging tools etc. The advantage of ESC coatings is that they can be applied on heat treated tools without the necessity to re-harden after coating. They are tough and their metallurgical bonding ensures no delamination in service. Tool steel dies used in the hot extrusion of titanium for turbine blade components were reported to be ESC coated with a mixture of refractory metal and molybdenum that resulted in the extension of the tool life by 3 times.

Medical Industry

The medical industry uses expensive carbide brazed insert tips for surgical tools and needle holders. These items are now routinely being coated with inexpensive chromium carbide which enhances the corrosion resistance. The applications include orthopedic drills, dental tools, hemostats and orthotic leg and ankle brace components.

Figure 11.28: Cutting Tools

Milling Cutter

Contd...

Figure 11.28–Contd...

End Mill

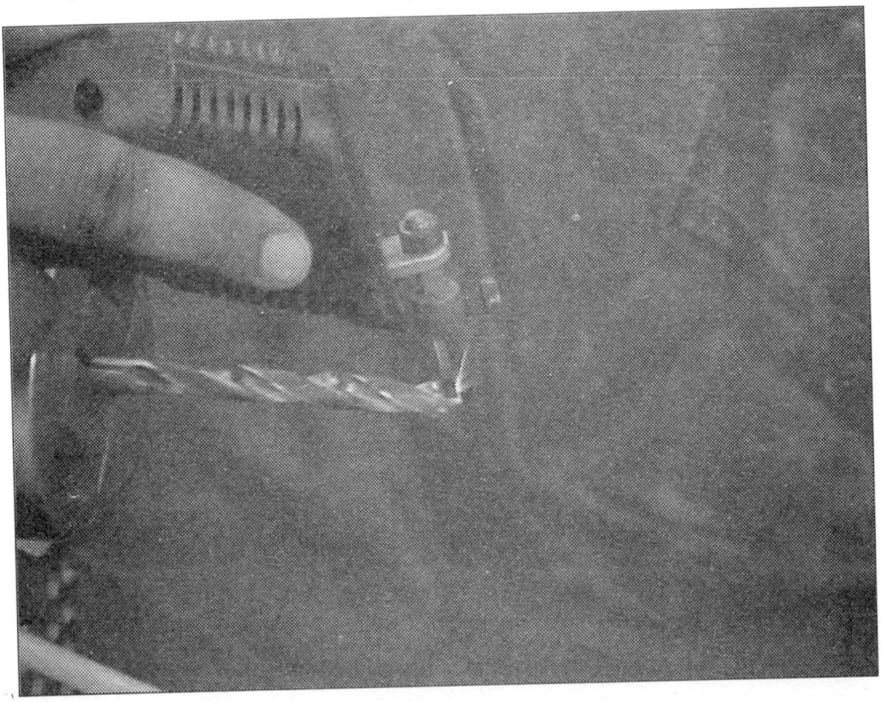

Twist Drill

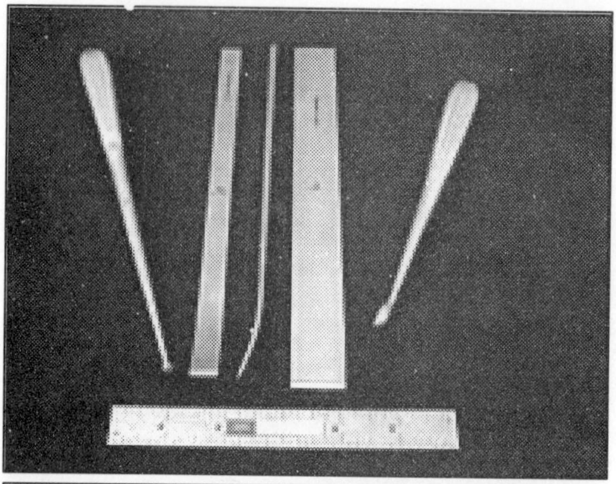

Figure 11.29: Medical Tools for ESC Coatings

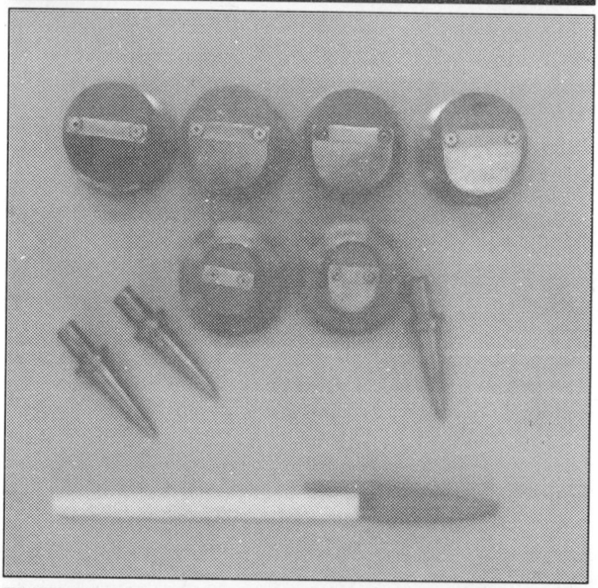

Figure 11.30: Ag Coated SS Nozzle

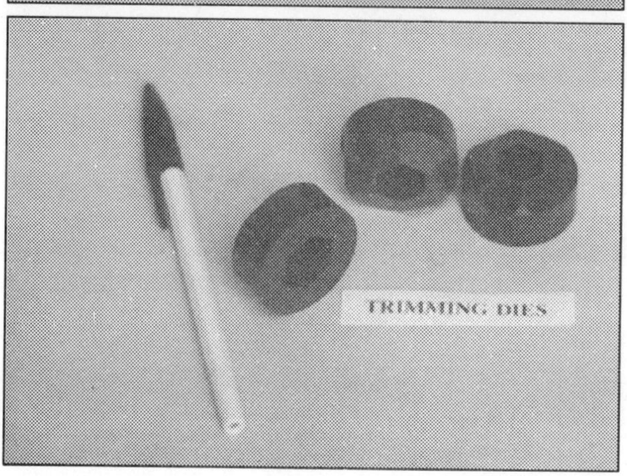

Figure 11.31: ESC Coated Die

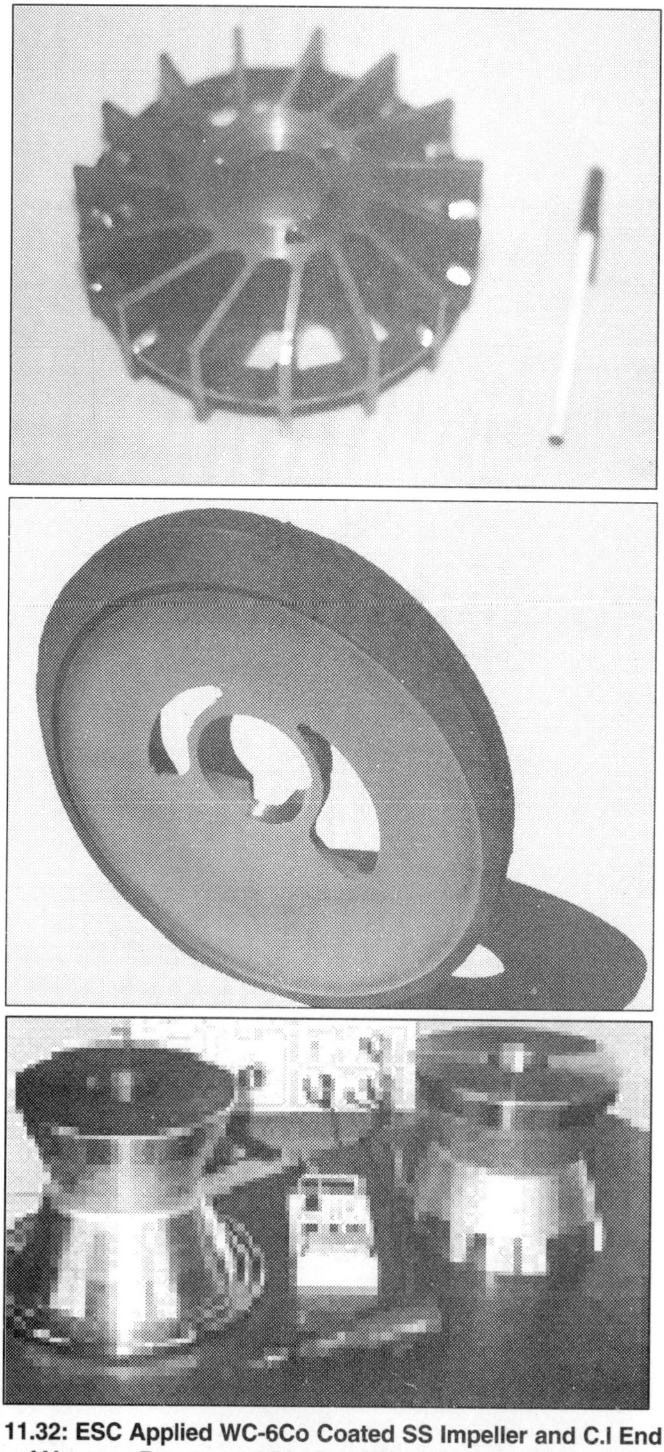

Figure 11.32: ESC Applied WC-6Co Coated SS Impeller and C.I End Cover Plates of Vacuum Pumps and Die Steel Pipe Bending Rollers which are Now Regularly Coated for Production

Figure 11.33: Al Die for Scuffing Resistance

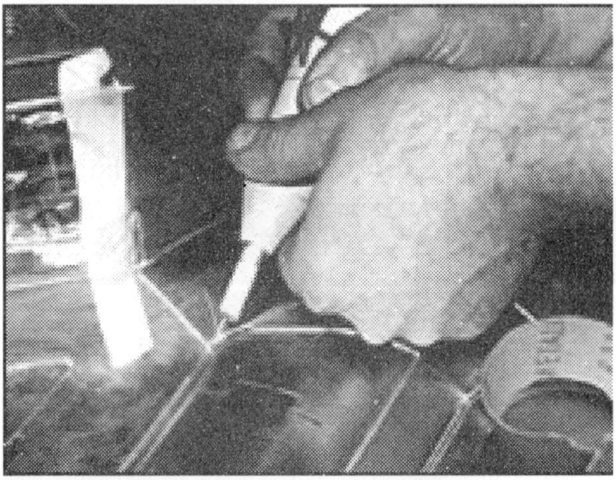

Figure 11.34: Localized Repair

Figure 11.35: C.I Valve for Erosion Resistance

Typical Cutting Tool Applications (Figures 11.36–11.39)

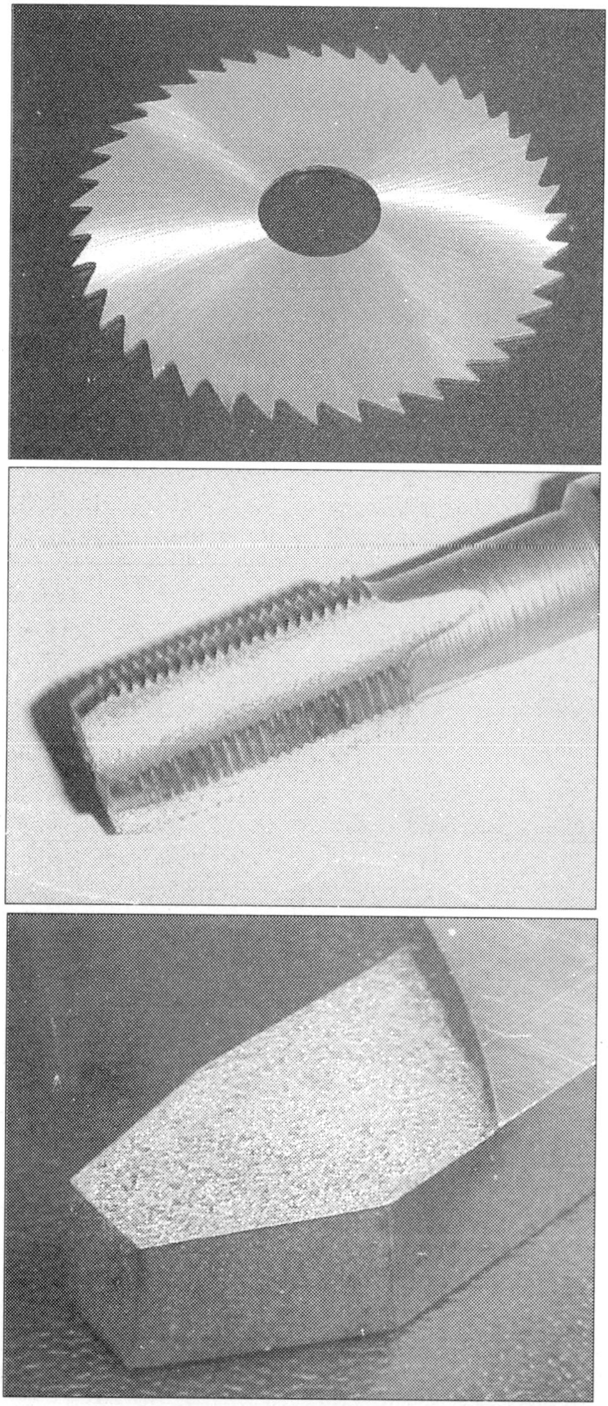

Figure 11.36: HSS Cutter, Tap and Lathe Tool are Some of the Proven Tooling Application for ESC

Figure 11.37: Cutting Tools

Figure 11.38: Disc Saw

Figure 11.39: De-barkers

Miscellaneous Applications (Figures 11.40–11.47)

Figure 11.40: Electrician Tools

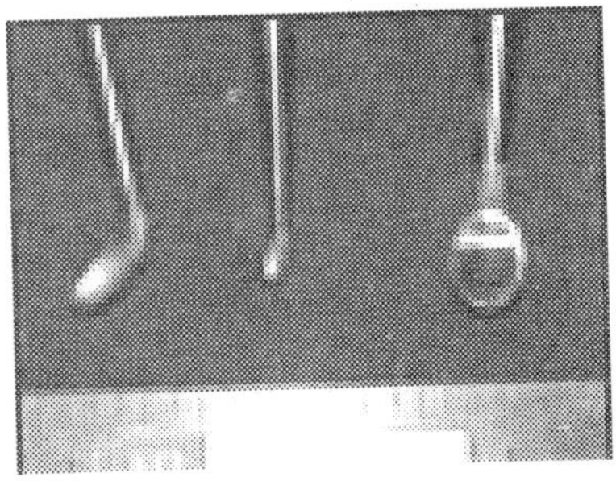

Figure 11.41: Medical Tools

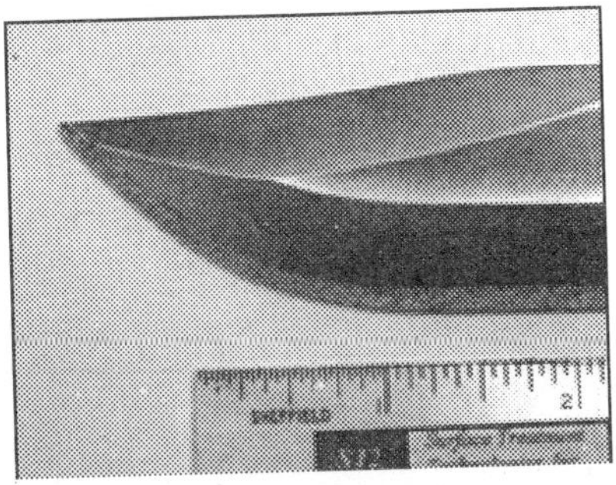

Figure 11.42: Knife-edge

Figure 11.43: ESC Coated Crusher Blow Bars for Increased Life

Figure 11.44: Rebuilding of the Fan and Floor Mill Motor Shafts

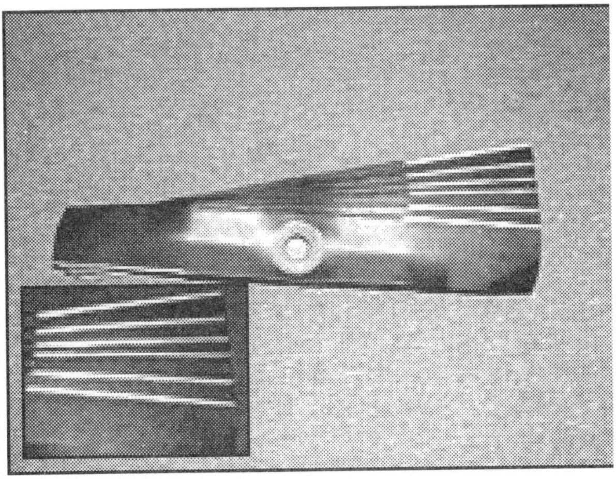

Figure 11.45: Power Mover Blades

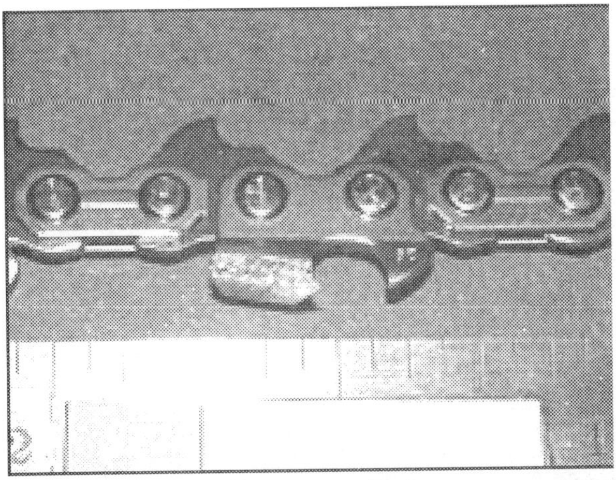

Figure 11.46: Chain

Figure 11.47: Sod Cutter

Military Applications (Figures 11.48–11.51)

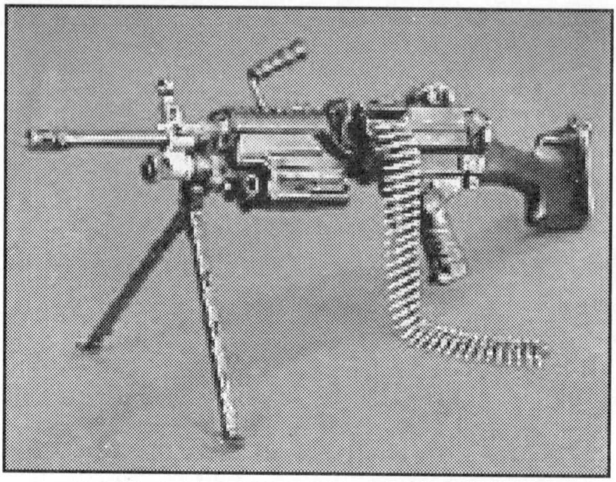

Figure 11.48: Squad Automatic Weapon

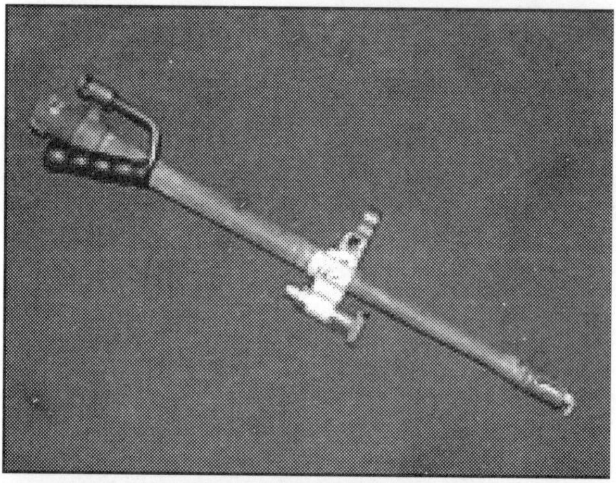

Figure 11.49: M246 Barrel

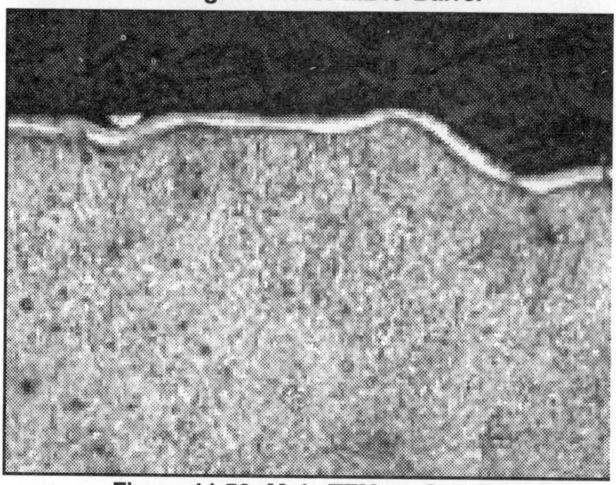

Figure 11.50: Moly TZM on Gun Barrel

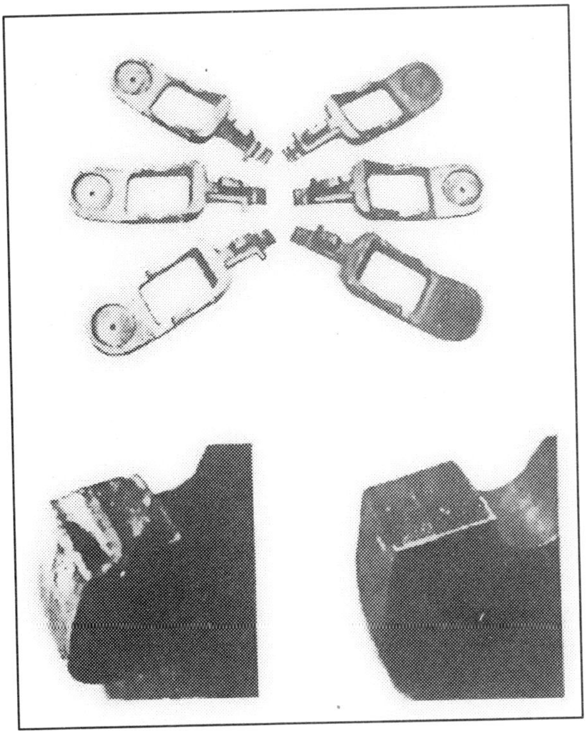

M249 Squad Automatic Weapon (SAW)

Coated barrel blanks before forging

ESA coating forms to barrel and remains after forging

coatings supplied and tested:

Haynes 25, Moly-TZM, Stellite 20, MoRe, Stellite 21, Tribolite, Ta-10W

Figure 11.51: Helicopter Missile Launcher Latch

Scope for Future R&D

Electro-spark coating technology is, although simple to operate, involves a complex physical process occurring at the electrodes. A complete understanding of the process can only be based on a sound knowledge of electro-physics, micro-metallurgy and solid state physics. Investigations have established a direct relation between these processes and properties of the layers formed and, consequently, service characteristics of the working surfaces of the parts being treated.

The future work can be aimed at achieving the following:

- Increasing the coating thickness, uniformity and properties while keeping the stress levels at negligible levels. This is required for expanding the use of ESC technology to applications that demand more severe abrasion conditions. Progress on this front could involve design of improved new circuitry.
- Formulating new electrode compositions that are economical and capable of significantly improving the coating performance. Introducing ultra dispersed diamond into the coating is one such possibility which is likely to considerably broaden the field of ESC applications.

References

1. Roger N.Johnson, "Electrospark Deposition: Principals and Applications", Society of Vacuum Coaters, 45th Annual Technical Conf. Proceedings, Lake Buena Vista, FL, p.87-92, April 13-18, 2002
2. E.A.Brown III, G.L.Sheldon and A.E.Bayoumi, "A Parametric Study on Improving Tool Life by Electrospark Deposition", Wear of materials vol2 pp.547-555, 1989
3. C.S.Kahlon, et. al. "Electric spark toughening of cutting tools and steel components" International journal of machine tool design and research, Vol.10, pp.95-121
4. Roger N. Johnson and G.L.Sheldon, "Advances in the electrospark deposition coating process" J.vac.sci.technology. A 4(6), Nov/Dec 1986
5. N.C.Welsh, " Spark-Hardening of Metals", Journal of Institute of Metals, vol.88, 1959-60 (103-111).
6. N.C.Welsh and P.E.Watts, "The wear resistance of spark-hardened surfaces", Wear, 5 (1962) 289-307.
7. J.Stiglich *et al.*, "Wear characteristics of a WC/6 wt. per cent Co coating deposited using the pulsed electrode surfacing(PEC) technique", Surface modification technologies XII pp.271-278, 1998
8. Arvind Agarwal and Narendra B. Dahotre, "Pulse Electrode Deposition Of Super Hard Boride Coatings On Ferrous Alloy", Surface coatings technology p242-250,1998
9. Shemegon V.I, Zhuk M.V, "Industrial experience of using Electric spark alloying to improve the wear resistance of cutting and die tools", Russian article 23, 96-99,1989
10. Shemegon V.I, "Hardening of cutting edge tools by the method of electrospark alloying", Stanki Instrum., (4), 19-20, 1986.
11. Kovel chenko M.S *et al.*, "Electrospark alloying and subsequent laser processing a high speed tool steel", Poroshkovaya Metallurgiya 5-7, 11-15, 1996
12. Shemegon V.I., "Effect of Electrospark Coating on the Cutting Properties of Twist Drills", Soviet Surface Engineering and Applied Electrochemistry (3), 112-115 1990
13. Rybalko A.V., Grichuk D.M., "Interdependence between the electric parameters of pulses and cathod mass increase under electric spark doping of metal surfaces" translated from Russian language, publication unspecified.
14. John A Vacval, "Die Life In Die Casting Extended by Tungsten Carbide Deposition(TCD) Process" presented at the third national casting exposition and congress, Cobo Hall, Detroit,1964
15. "Increased Die Life through T.C.D." Die and Stamping News July, 1963
16. Technical brochure, "Tungsten Carbide Depositing by Spark Machining" of M/S Union Carbide-Linde division.

17. Arvind Agarwal and Narendra B. Dahotre, "Synthesis of Boride Coating on Steel Using High Energy Density Processes: Comparative Study of Evolution of Microstructure", Materials characterization p31-44,1999

18. E.A.Le Vashov *et al.*, "About the Method of Thermoreactive Electrospark Surface Strengthening", Journal of Material synthesis and processing", vol. 7,p23-33,1999

19. D.Srinivasa Rao *et al.*, "Characterisation and application of Tungsten Carbide Cobalt (92-8) coatings deposited by electrospark coating technique", Proceedings of 50th annual technical meeting of IIM, New Delhi, November 1996.

20. N.Parkansky *et al.*, "Anode mass loss during pulsed air arc deposition", Surface and coatings technology 108-109 (1998) 253-256

21. N.Parkansky *et al.*, "Electrode erosion and coating properties in pulsed air arc deposition of Wc-based hard coatings", Surface and coatings technology 105 (1998) 130-134.

22. I.V. Gailinov and R.B. Luban, "Mass transfer trends during electrospark alloying", Surface and coatings technology 79 (1996) 9-18

23. G.L.Sheldon, "Galling resistant surfaces on stainless steel through electrospark alloying", J.of Tribolloy volume 117, 1995, 343-349.

24. E.Kudryashov *et al.*, "On the perspectives of application of new SHS materials to the technology of Electrospark alloying die steels", International journal of self-propagating high-temperature synthesis Volume 6, Number 4, 1997.

Chapter 12

High Temperature Coatings

A. S. Khanna

Professor, Corrosion Science and Engineering,
Indian Institute of Technology, Powai, Mumbai – 400 076, India
E-mail: khanna@iitb.ac.in

ABSTRACT

The paper briefly deals with the various coatings available for high temperature corrosion protection, the principle of the method, its advantages, limitations and applications. Examining various methods of coating substrates for their use in corrosion protection applications, it is pointed out that PVD methods are good but usually have a poor mechanical bond and the thickness is often very small. The method requires a vacuum, and coating of large or intricate parts is difficult. Diffusion coatings such as CVD, pack and slurry are very common and form a good metallurgical bond but because of high substrate temperature, the bulk properties are affected because of a change in the microstructure as a result of heating. Thermal spray coatings are perhaps being used extensively these days, but high porosity, non-uniform surface and mechanical bonding with the substrate, often limit their use in several aggressive environments. The recent development of LPPS and vacuum plasma has to some extent overcome these problems but the size of the specimen is limited by the vacuum chamber, and specimen movement has to be controlled by sophisticated maneuvering from the outside. Thus new methods which are more universal and do not require vacuum are very much required. One of the methods is laser surface alloying. One of its applications is for coating the outside of heat exchanger tubes to enhance their fireside corrosion resistance. Finally, it is highlighted that the Thermal Barrier Coatings (TBC) are insulative and oxidative resistant coatings that are finding increasingly broad utilization in the hot section of gas turbine engines. The prime benefit of thermal barrier coatings is the reduction of heat transferred into air cooled components. Other potential benefits include increased resistance to hot corrosion, erosion and oxidation

Keywords: *Alloys, Mechanical strength, High temperature corrosion, Surface modifications, Composite systems, Surface microstructure, and Corrosion protection.*

Introduction

With the increasing requirement of energy, it has become necessary to make the operating systems highly efficient. Take for example the case of gas turbines, in which there is a trend to continuously increase the gas inlet temperature to make it run efficiently. With an increase in the temperature, material specification becomes quite stringent. The requirement not only includes strong corrosion resistance at these temperatures, but also stable and excellent high temperature mechanical properties. Achievement of both these requirements simultaneously, sometimes cannot be met by alloy development alone. For example, in case of super alloy development, high mechanical strength is achieved by an alloy with high Al and relatively low Cr content. But this is at the cost of corrosion resistance, which requires high chromium content. It is, therefore, necessary to use an alternative approach to meet the stringent requirement of strength and corrosion resistance. The alternative is a composite system in which mechanical strength is achieved by alloy development and corrosion resistance by surface coating or surface modification. Experience has shown that this strategy has worked out very well in many industrial applications. In a gas turbine itself there are more than fifty different components which require efficient coating to resist the corrosive-erosive environment. Similarly, additional protection is required on materials when cheaper fuels with high impurity contents such as S, alkali salt or vanadium are used.

Surface modification in the wider sense includes all types of coatings and surface treatments that result in change in surface microstructure and composition. This has now become an independent discipline in itself and requires a whole volume to discuss every aspect of it, which is not the aim of this book. This chapter in fact briefly deals with the various coatings available for high temperature corrosion protection: the principle of the method, its advantages, limitations and applications.

Classification of Various Coatings

There are several ways in which coatings can be classified. The simplest is in terms of the mechanism of the coating process. According to this, the coatings are divided into two categories:

1. Overlay coatings
2. Diffusion coatings

In overlay coatings, the desired material is placed over the substrate by several techniques such as Physical Vapour Deposition (PVD), thermal spray methods–flame, arc or plasma. The coating in these cases has a mechanical bond with the substrate, without any diffusion into the substrate. In diffusion coatings, the coating material forms a chemical bond with the substrate. These coatings generally involve high temperatures. Chemical Vapour Deposition (CVD), pack cementation, hot dipping and some kind of surface modifications such as laser treatments are examples of diffusion coating.

Another classification is on the basis of the sophistication of the coating process:

- Conventional coatings
- Modified coatings
- Advanced coatings

Conventional coatings include the most common coating methods such as CVD, PVD, Pack Cementation, thermal spray methods, hot dipping and slurry methods.

Modified coatings are basically the conventional coatings modified for a particular application. Some examples are:

- Cr modified aluminide coating for enhancing the hot corrosion resistance. This is a modified pack cementation method, applied either by modifying the pack composition or pre-plating Cr before aluminizing.
- Yttrium modified aluminide coatings. Here the pack is modified to include some rare earth elements which impart high adherence to the coating.
- Platinum and palladium modified aluminide coatings. These are again used to impart high spallation resistance as well as higher hot corrosion resistance. These are applied by pack aluminization after pre-plating platinum or palladium.

Advanced coatings now include a large number of modern coating processes which mainly involve highly sophisticated equipment and provide excellent properties such as strong bonding, low porosity and are smooth and homogenous. Some examples are:

- Vacuum Plasma Coating (VPS) or Low Pressure Plasma Coating (LPPS)
- Laser Surface Cladding or Laser Surface Alloying
- Laser Plasma Coatings (LPC)
- Plasma Powder Welding (PPW)

Physical Vapour Deposition Methods

The simplest process is the thermal evaporation method. Here the metal to be coated is heated in a resistance coil or crucible and converted to vapour. In an evacuated chamber, these vapours then get deposited on the substrate. Simple evaporation is not a practical method for high temperature coatings as the coatings have very poor bonding and low thickness and the overall process is very slow. Therefore, simple evaporation is hardly ever used for high temperature coatings, however, several modifications of it such as: electron beam deposition, ion-sputtering and ion plating are the methods applied for coating specific components.

Ion Sputtering

When the gas pressure in a chamber is reduced to a fraction of a torr and a cold cathode discharge is excited between the metal electrodes in the chamber, the cathode electrode begins to emit atoms and atomic clusters. This process is called sputtering. Cathode material is ejected by the bombardment of positive ions and energetic particles in the electrical discharge and the sputtered substance is condensed on the chamber wall or the desired specimen to be coated. A wide range of substances can be sputtered to form coatings: metals, alloys, metal oxides, carbides and complex mixtures of inorganic substances such as cermets. Many types of systems have been developed for depositing different kinds of material with different sizes and shapes. Films of mixed substances can be prepared by sputtering simultaneously from several targets on a single receiver or multi layers prepared by sputtering the targets in sequence. Ceramic coatings such as oxide, carbide or nitride can be made by passing reactive gases such as oxygen, methane and nitrogen. The simplest form of a sputtering system is the cold cathode diode arrangement, which may then be modified to reduce the operating pressure of the cold cathode discharge by making the electrons follow a curved path in crossed electrical and magnetic fields (crossed field system). When a body is immersed in plasma and insulated from the ground, it quickly acquires a negative charge from the electrons which bombard its surface. Plasma can be excited by an R.F. (radio frequency) field giving an ion density comparable to that of a cold cathode discharge but at a tenth of the gas pressure (R.F. Sputtering). The R.F. sputtering has proved useful for sputtering a range of substances regardless of their individual electrical conductivity, and systems have been built containing several targets for depositing multilayer metals.

Ion Plating

When partial ionization of the melt vapour is used to increase the adherence of the film to the substrate, the thermal evaporation process is called ion plating. Metal atoms, evaporated from a heated source, have energies which are given by a Maxwellian distribution and are proportional to the absolute temperature of the source. Hence a good evaporation technique such as the electron beam, discussed above significantly increases the average energy of the metal atoms which results in improvement of adherence as compared to simple thermal evaporation and sputtering. The ion plating process can be considered to be divided into three stages: pre-plating, ion plating and evaporation. Contamination is removed during the pre-plating so that a clean surface is exposed for the subsequent deposition. The initial argon glow discharge will clean the surface and thus improve the adhesion. The quality of the coating depends very much on the substrate temperature and the current depending upon which the coating can be columnar, equiaxed or dense as shown in Figure 12.1. This variation is in correspondence to the variation in pressure in simple evaporation coating. Ion plating has several advantages. It is intrinsically clean as the film is deposited directly without using a carrier. An alloy film can also be deposited provided the vapour pressure of the two constituents is not very different. Deposition rates of microns per minutes are easily achieved. The main disadvantage of the process is that it works in vacuum and only small parts can be coated.

Diffusion Coatings

There are two processes in this technique.

- Chemical vapour deposition coating, also known as "out of contact gas phase deposition", in which the chemical reaction occurs in the vapour

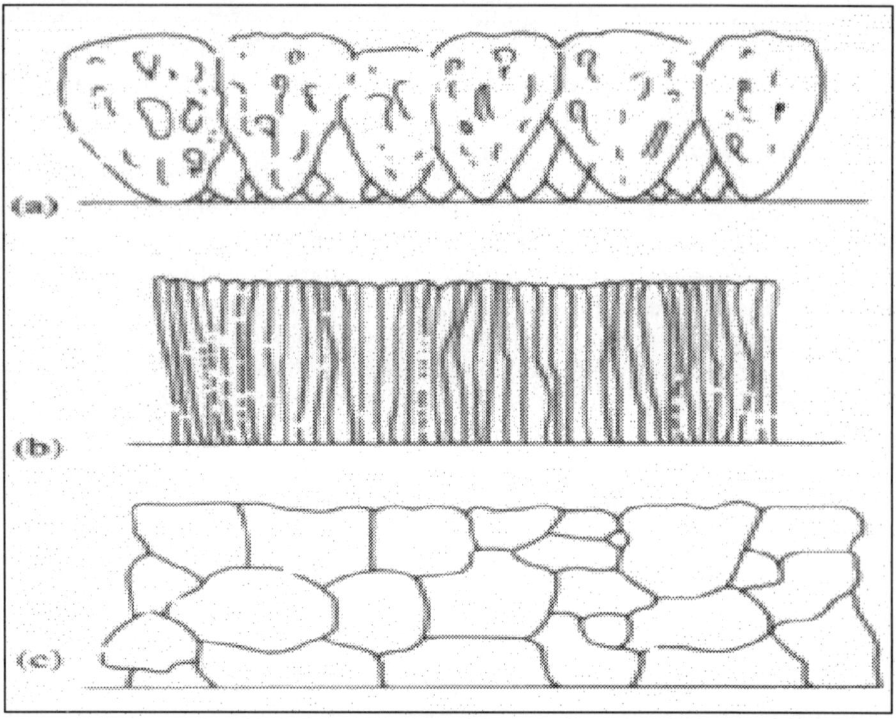

Figure 12.1: Effect of Temperature on Coating Structure

(a) columnar, at a substrate temperature less than 0.3 Tm; (b) fine columnar, at substrate temperature between 0.3 and 0.5 Tm; (c) equiaxed, with substrate at a temperature above 0.5 Tm.

phase followed by production of metal vapour which finds its way to the substrate which is usually maintained at high temperature.

- Solid state diffusion process: Here, the component to be deposited is treated in contact with the powder coating material in an inert atmosphere. Pack cementation and other slurry methods fall in this category. In this process, the vapour is produced by the chemical reaction in a dense pack containing the specimen and the desired metal produced after the decomposition reaction diffuses through the pack to the surface of the substrate.

Chemical Vapour Deposition (CVD) is a versatile process that can be used to deposit nearly any metal as well as non metals such as carbon or silicon. Compounds such as carbides, nitrides, oxides and inter metallic can also be deposited. The technology has become very important in: coatings on tools, bearings, and other wear resistant parts; optical and opto-electronic and corrosion resistance applications. The first step is the production of metal vapours. Several chemical reactions can be used: thermal decomposition, pyrolysis, reduction, oxidation, nitridation etc. Several reactions are listed in Table 12.1. The main reaction is carried out in a separate reactor. The vapours thus formed are transferred to the coating chamber where the sample is mounted and maintained at a high temperature. Aluminum coating by the halide route is very commonly carried out by this method and the reaction steps are shown below:

M + Halide = M-Chloride → decomposes to M vapour → deposit on the specimen

When the reaction is carried out by simple heating, it is usually termed as thermal CVD. The temperature of the reaction is usually high around 900-1100°C. Also the substrate needs to be maintained at high temperatures. One of the limitations of the CVD method is the high substrate temperature which in many cases changes the microstructure of the substrate, and another limitation is the size of specimens, wherein often smaller parts are used due to limitation of chamber size. There are some modifications in the basic technique such as use of plasma for heating. This limits the temperature of the process to between 300–700°C, instead of 900–1100°C in thermal CVD. A recent modification is Laser CVD where a laser source is used for heating.

Table 12.1: Different Compounds Formed Using the CVD Method

Materials	*CVD Method*	*Temp. (°C)*
Nitrides		
BN	$BCl_3 + NH_3$	1000–2000
HfN	$HfCl_x + N_2 + H_2$	950–1300
Si_3N_4	$SiH_4 + NH_3$	950–1050
	$SiCl_4 + NH_3$	1000–1500
TaN	$TaCl_5 + N_2 + H_2$	2100–2300
TiN	$TiCl_4 + N_2 + H_2$	650–1700
VN	$VCl_2 + N_2 + H_2$	2000–2500
ZrN	$ZrCl_4 + N_2 + H_2$	2000–2500
Oxides		
Al_2O_3	$AlCl_3 + CO_2 + H_2$	800–1300
SiO_2	$SiH_4 + O_2$	300–450
SnO_2	$SnCl_4 + H_2O$	
TiO_2	$TiCl_4 + O_2$ + Hydrocarbon	
Silicides		
V_3Si	$SiCl_4 + VCl_4 + H_2$	
MoSi	Mo (substrate) + $SiCl_2$	800–1100
Borides		
AlB_2	$AlCl_3$ + BCl	~1000
HfB_X	$HfCl_4 + BX_3$ (X=Br, Cl)	1900–2700
SiB_X	$SiCl_4 + BCl_3$	1000–1300
TiB_2	$TiCl_4 + BX_3$	100–1300
VB_2	$VCl_4 + BX_3$	1900–2300
ZrB_2	$ZrCl_4 + BBr_3$	1700–2500

Contd...

Table 12.1–Contd...

Materials	CVD Method	Temp. (°C)
Carbides		
B_4C	$BCl_3 + CO + CH_4$	1200–1800
Cr_7C_3	$CrCl_2 + H_2$	~1000
Cr_3C_2	$Cr(CO)_5 + H_2$	~1000
HfC	$HfCl_4 + H_2 + C_7H_8$	2100–2500

Pack Cementation

The pack cementation method is the most common technique, utilized for many high temperature resistance coatings. The process is very simple. The part or component to be coated is placed inside a pack consisting of a filler–usually a ceramic powder to prevent sintering of the mix at high temperatures *e.g.* alumina, the desired metal in powder form and an activator–a volatile halide, usually ammonium chloride or sodium fluoride. The pack is kept in the furnace and heated in argon. The aluminum powder reacts with ammonium chloride to form aluminum chloride which decomposes immediately and diffuses in the pack to the sample surface and gets coated. The different steps for an aluminide coating on a steel sample are shown below:

Decomposition $NH_4Cl(s) = NH_3(g) + HCl(g)$

Formation of volatile aluminum chloride

$HCl(g) + Al\,(pack) = AlCl_3(g) + H_2(g)$

Deposition on steel

$AlCl_3(g) + Fe\,(substrate) = AlFe\,(alloy)$

For a reasonable thickness, the heat treatment is carried out for several hours. The final thickness of the coating depends upon the temperature, time and the activity of the pack. A pack is called a low activity pack when the concentration of a depositing metal in the pack is low, a few percent. Under these conditions, a coating is formed with the outward diffusion of one or more substrate elements. In case it is a high activity pack, aluminum diffuses inward and a complex coating is formed which needs further heat treatment.

Table 12.2 lists different packs which can coat various elements with the corresponding temperatures and heat treatment durations. Figure 12.2 shows pack aluminization of IN 100 nickel base alloy.

The *slurry cementation process* is based upon the preparation of slurry where metal powders and the halides are suspended in an organic binder. The slurry is then applied to the part to be coated through dipping or spraying depending upon the size and geometry of the specimen. This is then subjected to heat treatment to allow the desired element to diffuse into the substrate.

Table 12.2: Different Pack Mixes and Processing Parameters for Various Pack Cementation Coatings

Coating Type	*Source Composition*	*Processing Parameters*
Pack aluminizing, inward diffusion in Ni_2Al_3 in nickel alloys	5–20 per cent Al (Al–10Si), 0.5-3 per cent NH_4Cl, balance Al_2O_3 powder	1 to 4h at 650 to 680°C in air, argon, H_2; heat treat 4 to 6h at 1095°C in argon.
Pack aluminizing, inward diffusion in NiAl in nickel alloys	44 per cent Al, 56 per cent Cr, NH4Cl balance Al_2O_3 powder	5 to 10h at 1040°C in vacuum (argon, H_2)
Pack aluminizing, outward diffusion in NiAl in nickel alloys	2–3 per cent Al, 20 per cent Cr, 0.25 per cent 25 per cent NH_4HF_2, balance Al_2O_3 powder	At 680°C in argon
Pack aluminizing of cobalt alloys	8 per cent Al, 22 per cent Cr, 1 per cent NH_4F, 4 to 20g, balance Al_2O_3 powder	At 980 in to 1150°C in argon
Gas-phase aluminizing, outward diffusion in NiAl in nickel alloys	10 per cent Co_2Al_5, 2.5 per cent NaCl, 2.5 per cent $AlCl_3$, balance Al_2O_3 powder	3h at 1095°C in argon
Gas-phase aluminizing, outward diffusion in NiAl in nickel alloys	30 per cent Al-70 per cent Cr alloy granules, NH_4F	4h at 1150°C in argon
Pack or gas phase chromizing of nickel alloys	15 per cent Cr, 4 per cent Ni, 1 per cent Al, 10.25 per cent NH_4Cl, balance Al_2O_3 Powder	3h at 1040°C in argon

Figire 12.2: Optical Micrograph Showing the Cross-Section of the Structure of Low Activity Park Aluminization of IN 100 Nickel Base Alloy

Hot Dip Coatings

The principle of hot dip coating is relatively simple. The substance to be coated is dipped into the hot molten metal. At first the liquid reacts with the metal forming inter metallic compound. This is followed by some diffusion. Usually hot dip coatings are not kept in liquid for a long time. Thus extensive diffusion as in the case of pack and slurry coating is not expected in this case.

Thermal Spray Techniques

The main difference between the various thermal spray techniques, namely flame spray, arc spray, detonation gun and plasma is the source of heat. Flame spray uses an acetylene oxygen flame, where a temperature of only 2500–3000°C can be achieved (Table 12.3). In an arc, a temperature of 3500°C can be achieved while in a plasma source a temperature of about 20,000°C can be achieved. A higher temperature improves the quality of coating *e.g.*, increased homogeneity. The second factor is the speed at which the particles are injected on the surface. Both in the flame and arc spray, compressed air is used to force the melted particles to the substrate surface. In a detonation gun, the shock created enhances the speed of melted particles to around 1000 m/sec which is higher than that even for plasma coated specimens. Hence the bond strength is highest and porosity is lowest. Different properties achieved by various coating methods are summarized in Table 12.3.

Table 12.3: Mechanical Properties of Different Thermal Spray Coatings

Parameter	*Type of Coating*					
	Tungsten-carbide				*Alumina*	
Nominal composition wt. per cent	W-7Co-4C	W-9Co-5C	W-11Co-4C	W-14Co-4C	Al_2O_3	Al_2O_3
Thermal spray process	Detonation gun	High-velocity combustion	Plasma	Detonation gun	Detonation gun	Plasma
Rupture modulus 10^3 psi	72	–	30	120	22	17
Elastic modulus, 10^6 psi	23	–	11	25	14	7.9
Hardness, kg/mm^2, HV3–	1300	1125	850	1075	>1000	>700
Bond Strength, 10^3 psi	>10,000	>10,000	>6,500	10,000	>10,000	>6,500

The schematic of the setup is shown in Figure 12.3. In the flame spray, a fine powder is carried in a gas stream and is passed through an intense combustion flame where it melts. In this method, the coating material can be fed as wire also. The gas stream which expands rapidly because of heating then sprays the molten droplets on to the substrate where they solidify. The fuel used is generally acetylene and hydrogen. The carrier gas may be oxygen, air or some inert gas. The quality of the coating depends upon the spray parameters: choice of the gas, geometry of the spray head, gas and powder flow rates, etc.

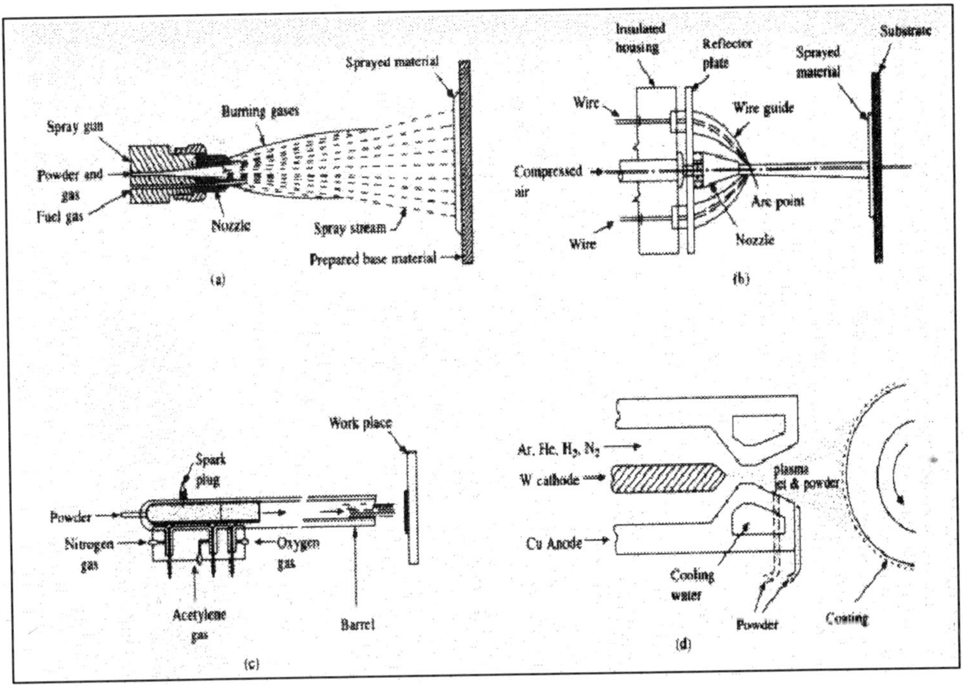

Figure 12.3: The Schematic Set-up Diagrams for Four Different Thermal Spray Techniques (a) Flame spray gun, (b) Electric-arc spray, (c) Detonation gun process and (d) Plasma spray process

The flame spraying process is a cold process as the temperature rise of the substrate is hardly 200°C. Since the coatings are not very dense and have poor mechanical bond, they are usually subjected to a fusing operation, *i.e.* heating the coated material up to 950-1100°C. A particular group of alloys, the so called self fusing alloys are generally coated by this method. These are Ni-Cr alloys with boron and silicon as fluxing agents. After fusing, the coating is more homogenous and dense, however, distortion of the substrate is a usual problem.

Arc coatings use the metal in wire form. This process does not use any external heat source. Heating and melting take place when an arc is struck between two electrically opposed charged wires. One of the wires is the metal to be coated. The molten metal on the wire tips is atomized and propelled onto the substrate by a stream of compressed air or other gas. The electric arc spray offers several advantages over the flame spray process: higher bond strength, higher deposition rates and lower substrate heating. It is a less expensive process as power requirements are generally low. No expensive gas such as argon or hydrogen is required. One generally requires a ductile, electrically conductive wire of about 1.5 mm diameter. Although, the electric arc process for oxides, nitrides and carbide coatings is currently not possible, recent developments of cored wires permit the deposition of some composite coatings containing carbides and oxides. Pseudo alloy coatings can be made by using dissimilar metal wires. Electric arc coatings are being used for many corrosion resistant coatings such as zinc base, aluminum and stainless steel coatings.

In the *detonation gun method*, a measured amount of powder is injected into the gun along with a controlled mixture of oxygen and acetylene. The mixture is ignited and a shock wave is generated, which lasts for microseconds and attains a velocity of 3000 m/s. The high velocity of the gas stream imparts energy to the powder which reaches the substrate with a velocity of 800 m/s. During the short duration of particle acceleration, heat is also generated which melts the particles. The coating process is repeated several times in a second depending upon the type of coating. The coating structure is built up from a series of such detonations. The particles impact on the surface flatness or splatter into thin overlapping platelets such that the diameter is many times greater than their thickness. The high kinetic energy of the particles (about 25 times that of the flame spray) is converted to additional heat. This additional heat keeps the particles in molten form and completely wets the substrate surface. The result is a continuous coating with very high bond strength.

A *plasma spray torch* is shown in Figure 12.3d. A gas, usually argon but occasionally including nitrogen, hydrogen and helium, is allowed to flow between a tungsten cathode and a water cooled copper anode. An electric arc is initiated between the two electrodes using a high frequency discharge and is then sustained using dc power. The arc ionizes the gas, creating a high pressure gas plasma. The resulting increase in gas temperature, which may exceed 30,000°C, in turn increases the volume of the gas and hence its pressure and velocity as it exits from the nozzle. Power levels in plasma spray torches are usually in the range of 30–80 kW, but they can be as high as 120 kW. Argon is usually chosen as the base gas as it is chemically inert and for its best ionization characteristics. The enthalpy of the gas is increased by the addition of diatomic gases such as hydrogen and nitrogen.

Powder is usually introduced in the gas stream either outside from the torch or in the diverging exit region of the nozzle. It is both heated and accelerated by the high temperature, high velocity plasma gas stream. The coating characteristics depend a lot on the torch design and operating parameters: Gas flow rates, power levels, powder feed rate, carrier gas flow, powder size and the distance of the torch from the substrate. Powder velocities usually achieved are from 300–500 m/s. The density is much higher than the flame spray coatings, typically 85–95 per cent. Coating thickness usually ranges between 50 to 500 mm and bond strengths are between 35–70 MPa.

One more very important factor on which coating quality depends is the environment surrounding the sample and that between the torch and the sample. The normal plasma spraying described above is called Atmospheric Plasma Spraying and here sometimes the oxidation of particles sometimes results in substantial reduction in the coating density, cohesive strength and bond strength. This can be avoided either by creating a shrouding of inert gas around a sample or modifying the coating either in vacuum or a low pressure inert atmosphere.

A simple inert atmosphere chamber spray system may include a jacketed, water cooled chamber, an air lock, a plasma system, work piece handling equipment, and a vacuum pumping system. Usually the chamber is pumped down to a pressure of

0.001–0.01 Pa then backfilled with high purity dry argon gas (with oxygen levels less than 30 ppm). This technique has several advantages over the conventional atmospheric plasma method. Because of the lower pressure, plasma gas stream temperature and velocities are extended to greater distance. The substrate can be pre-heated without oxidation, which allows better control of residual stress and better bond strength. Deposition velocity is increased because of increased particle dwell time in the longer heating zone of the plasma and the higher substrate temperature. The closed system also minimizes environmental problems such as dust and noise. One limitation of the closed chamber is that it restricts the size of the substrate and since the whole operation is to be carried out in the closed chamber, careful specimen stage movement is required.

In the transferred plasma arc process, it is possible to heat and melt the substrate surface (Figure 12.4). This can be achieved by a secondary arc current that is established through the plasma and substrate that controls surface melting and depth of penetration. It has several other advantages such as metallurgical bonding, high density, high deposition rate and larger thickness per pass. Coating thicknesses of 500–5000 mm can be achieved and width up to 32 mm can be made in a single pass at a powder feed rate of 9 kg/h. Further less power is required in this process than non-transferred plasma arc method.

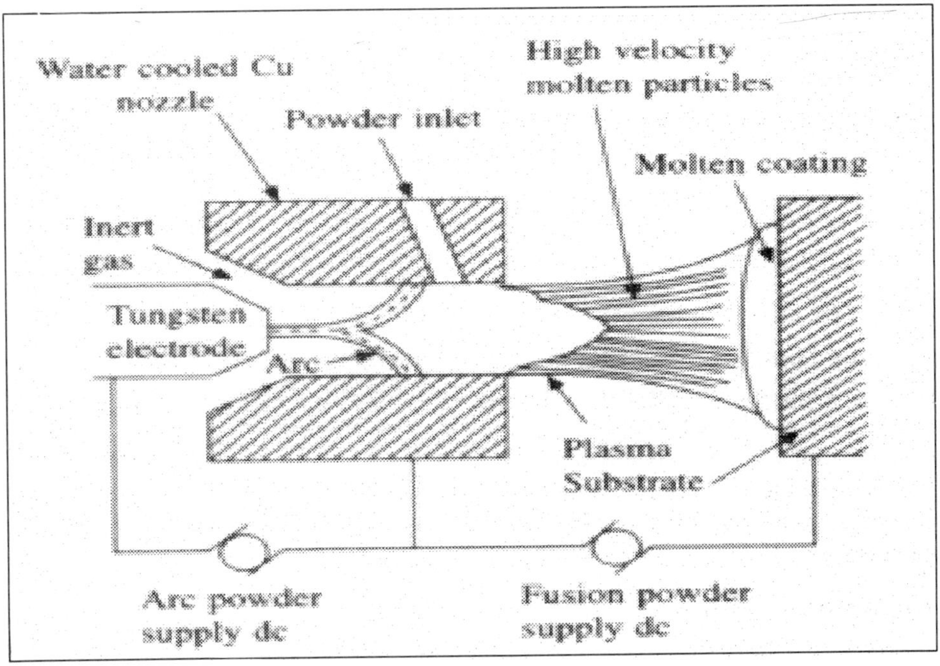

Figure 12.4: A Schematic Diagram of a Transferred Plasma-Arc Spraying Device

High Velocity Oxy Fuel Coating (HVOF)

A schematic of a HVOF device is as shown in Figure 12.5. Fuel, usually propane, propylene or hydrogen is mixed with oxygen and burned in a chamber. In some cases

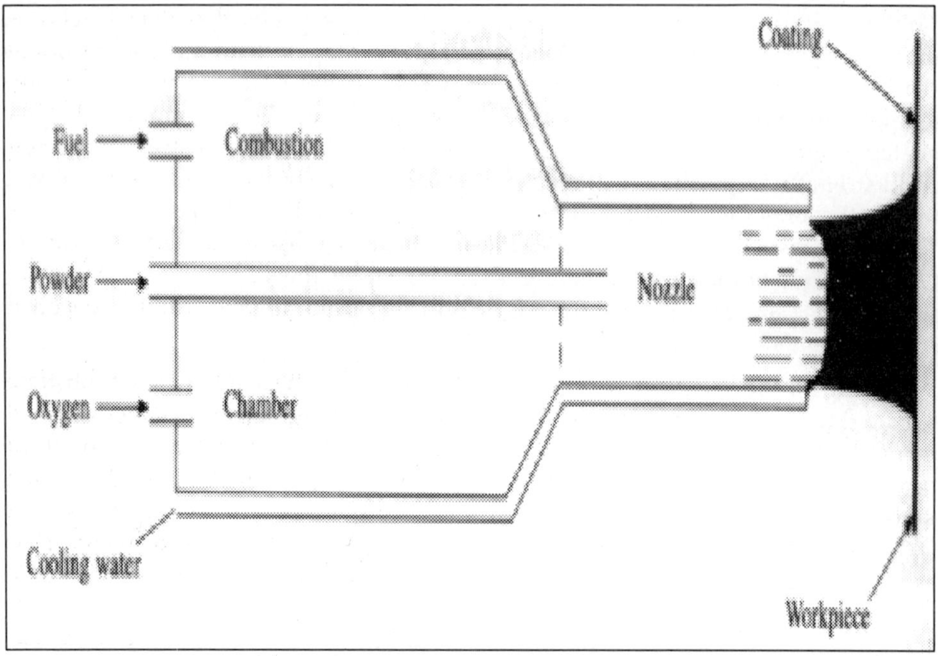

Figure 12.5: A Schematic Diagram of a High Velocity Oxy Fuel Device Process

liquid kerosene may be used as a fuel and air as an oxidizer. The products of combustion are allowed to expand through a nozzle where the gas velocities may become supersonic. Powder is usually introduced axially in the nozzle and is heated and accelerated. The powder is usually fully or partially melted and can attain velocities up to about 550 m/s. Because the powder is exposed to the products of combustion, they may be melted in either an oxidizing or reducing environment and significant oxidation of metallic and carbides is possible.

With appropriate equipment, operating parameters and choice of powder, coatings with high density, and with bond strengths frequently exceeding 69 MPa (10,000 psi), can be achieved. Coating thicknesses are usually in the range of 0.05–0.50 mm, but substantially thicker coatings can occasionally be used when necessary with some materials. HVOF coating can produce coatings of virtually any metallic or cermet material and for some HVOF processes most ceramics, *e.g.* a zirconia coating can be applied with acetylene as fuel.

Advanced Coatings

So far we have discussed various methods of coating substrates for their use in corrosion protection applications. Although, many of the above methods are in practice, they are not perfect and have several limitations. PVD methods are good but usually have a poor mechanical bond and the thickness is often very small. The method requires a vacuum, and coating of large or intricate parts is difficult. Diffusion coatings such as CVD, pack and slurry are very common and form a good metallurgical bond but because of high substrate temperature, the bulk properties are affected because of a change in the microstructure as a result of heating. Moreover the CVD method is used only for a limited number of coatings and the method becomes complicated if alloy coatings or ceramic coatings using a reactive pack are used. Thermal spray coatings are perhaps being used extensively these days because they are easy to apply, fast and can be made available at the site, however high porosity, non-uniform surface and mechanical bonding with the substrate, often limit their use in several aggressive environments. The recent development of LPPS and vacuum plasma has to some extent overcome these problems but the size of the specimen is limited by the vacuum chamber, and specimen movement has to be controlled by sophisticated maneuvering from the outside. Thus new methods which are more universal and do not require vacuum are very much required. One of the methods is laser surface alloying.

Laser Surface Alloying

Laser surface alloying uses a laser beam as the heat source. Special properties of the laser beam, such as its beam directionality, high intensity and high spatial resolution make it an excellent heat source. It heats a specific area very fast followed by faster cooling resulting in a novel microstructure. The desired material can either be added simultaneously along with laser irradiation or laser irradiation is carried out on the surface where the material has already been placed by some of the coatings discussed above. Several lasers are now available but CO_2 and Nd:Yag lasers are the two which are mainly used for laser surface processing. The schematic of the direct powder feed method as well as the two step coating method is shown in Figure 12.6.

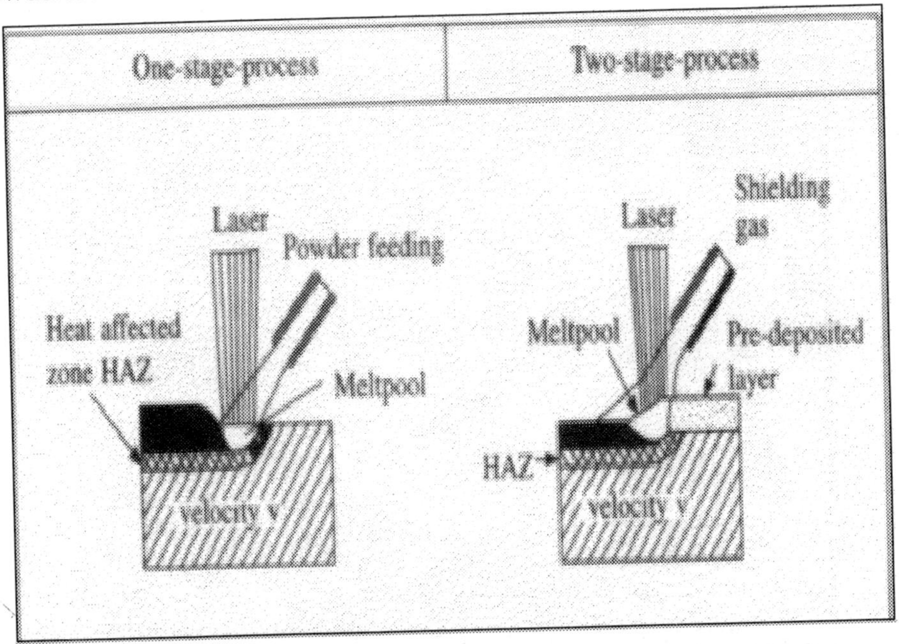

Figure 12.6: One and Two Stage Processes of Laser Surface Alloying

The very rapid heating and cooling cycle allows a great extension of alloying possibilities, as conventional limits such as solubility limits seem to be overcome. Typical structures are extremely fine dendritic which contain the added elements or compounds in the inter dendritic regions. The metallurgical stability of such structures is quite good and has been demonstrated by several applications. An example of modifying mild steel by chromium deposition is given here.

Figure 12.7 shows a cross-section of the Ni + Cr coating (75 mm each) by atmospheric plasma coating. As can be seen, the coating has a very poor mechanical bond and has several continuous cracks and voids. This was then laser treated with a 3 kW continuous CO_2 laser. The results after laser treatment are shown in Figure 12.8 for 4 different laser parameters. The surface alloy formed depends upon several factors: laser power, sweep speed or interaction time and the laser spot size. The most important is the interaction time which can be varied by varying the sweep speed of the laser beam by moving the specimen stage. In this figure the four different treatments

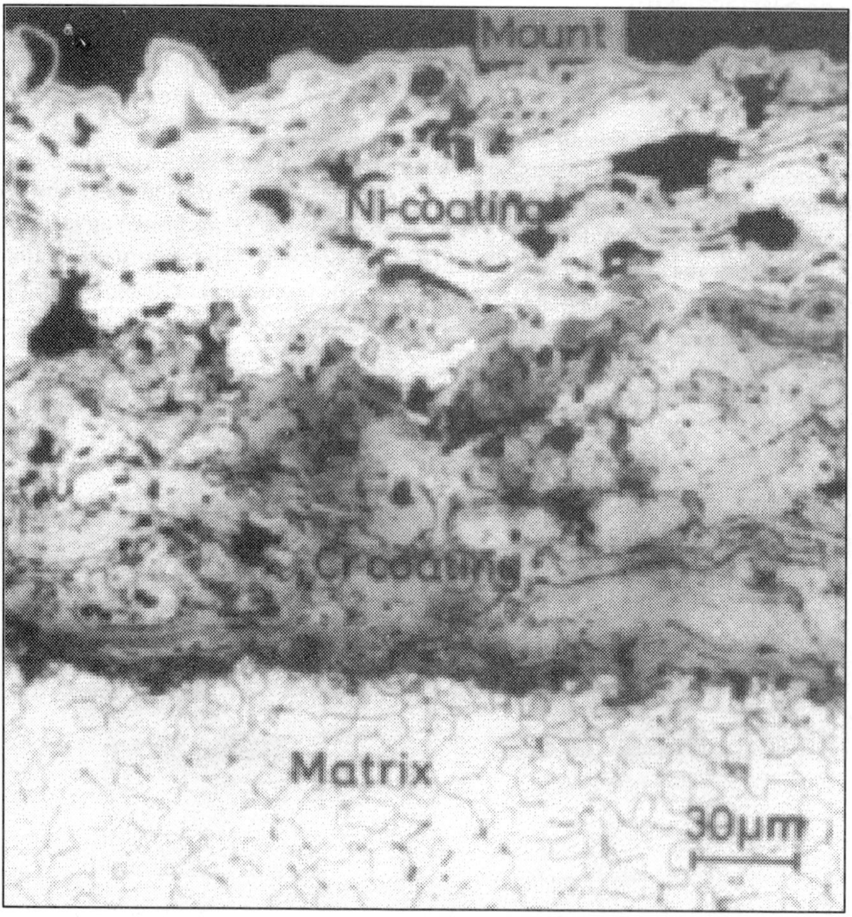

Figure 12.7: An Optical Micrograph Showing a Cross-Section of the Plasma Ni + Cr Coated Mild Steel

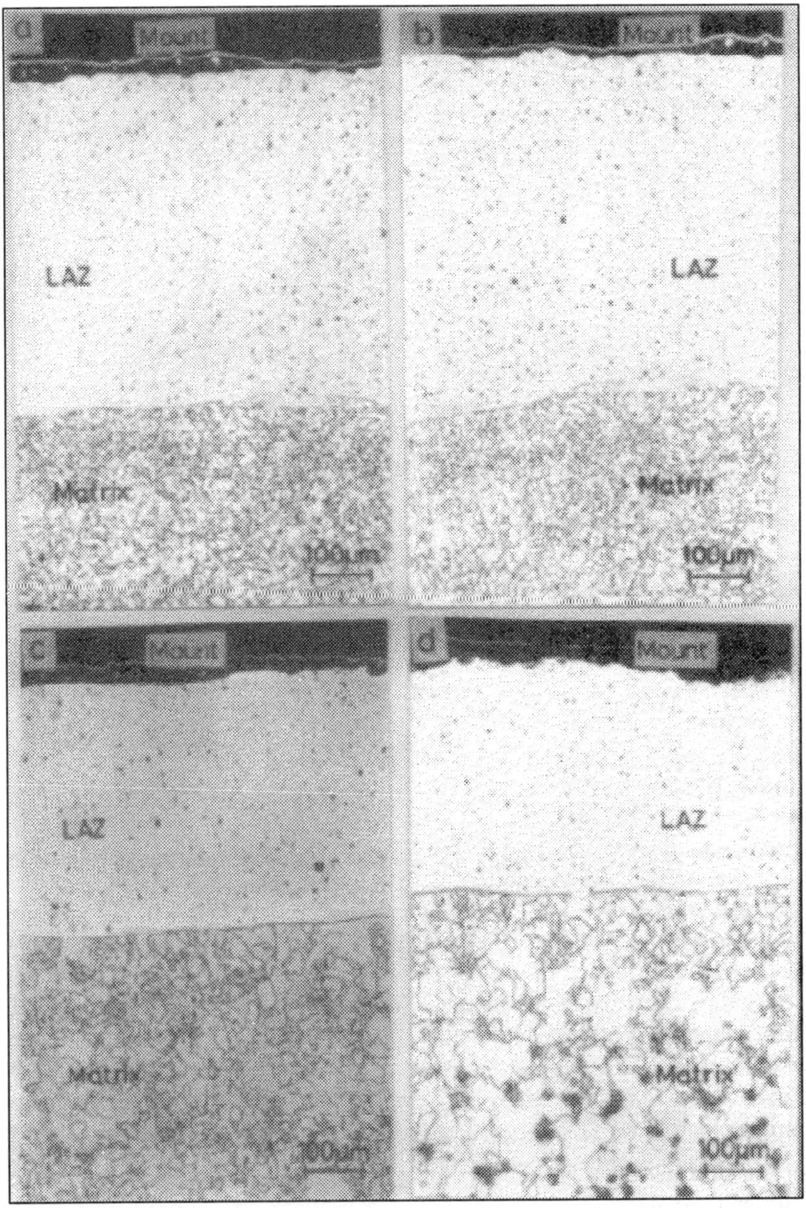

Figure 12.8: Optical Micrographs Showing the Cross-Sections of the Laser Ni + Cr Alloyed Mild Steel Specimens with Varying Laser Speeds, (a) 500 mm/min, (b) 575 mm/min, (c) 625 mm/min and (d) 750 mm/min

involve the same power (3 kW), same beam spot (0.5 × 6 mm) and sweep speeds of 725, 650, 500 and 250 mm/sec. The higher the sweep speed, the lower is the interaction time and hence the time for the dilution of deposited elements, and the laser melted zone or laser surface alloying thickness are both lower. The important thing to note is that the poor mechanical bond has been changed to a strong metallurgical bond and the coating appears as part of the main specimen. EPMA analysis showed perfect distribution of alloying elements throughout the LAZ. When tested for oxidation, the oxidation rate reduced remarkably over the plasma coated specimen (Figure 12.9).

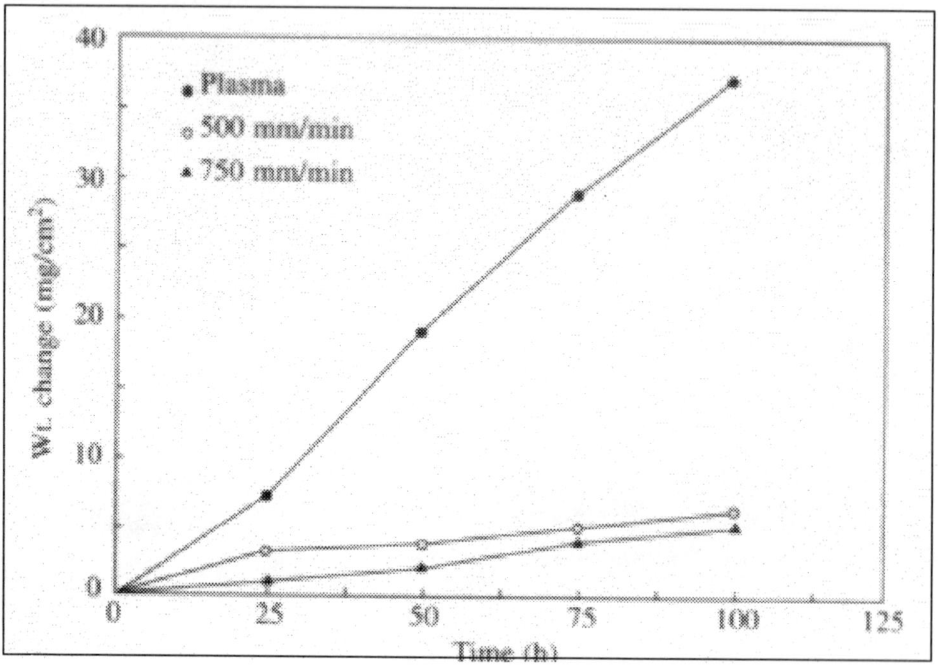

Figure 12.9: Linear Plots of Weight Gain Vs Time for the Oxidation of Ni + Cr Plasma Coated and Laser Treated Specimens

One of the applications of laser surface alloying is for coating the outside of heat exchanger tubes to enhance their fireside corrosion resistance. This can be achieved by cladding the outer surface of the 21/4Cr-1Mo steel with high Cr alloy such as Ni-25 Cr. This was successfully carried out and tested as shown in Figure 12.10. Cladding was carried out by 3 kW CO_2 laser. Testing was carried out in actual heat exchanger for one year. There was no change on the surface nor were there any fire cracks which were usually seen in the case of uncladded specimen (Figure 12.11).

Recent modifications in the laser have been made for cladding the inside of the tubes. Stress Corrosion Cracking (SCC) of a nuclear component was stopped by cladding it with Ni-25Cr-Mo alloy by a 5 kW Nd:Yag laser which was made to reach inside the tubes by fiber optics, which is not possible with a CO_2 laser. The design of the probe and the specimen as coated is shown in Figure 12.12.

Hybrid LPC Plasma Coating

LPC plasma spraying is used for the production of high quality, non-oxidizing, refractory alloy coatings. The main aim of such coatings is to simultaneously heat

Figure 12.10: Photographs Showing, (a) CO_2 laser cladding procedure and (b) unexposed and (c) exposed 2.25 Cr- 1Mo steel cladded with Ni-25 Cr specimens in actual heat exchanger conditions for one year

treat the plasma coating by laser heat and thus get a pore free dense coating. It is basically a direct feed method, in which instead of powder feeding, a jet of plasma adds the desired material, simultaneously processed by laser. The apparatus consists of a big stainless steel tank with vacuum arrangements, a port for the laser torch and laser beam. In addition a very sophisticated sample handling system is needed which can be moved in the X-Y direction or rotated at various angles. A photograph of the setup with a single plasma torch and 1 kW CO_2 laser is shown in Figure 12.13. It can work in vacuum (50-700 torr) and in atmospheric gases such as argon and nitrogen. The TiN coating carried out by LPPS using nitrogen gas and that carried out by LPS plasma spray are compared in Figure 12.14. Another application of LPC plasma spraying is to form inter metallic coatings, which are difficult to form by plasma spraying alone. A hard Mo,Ti,Ni coating on SS41 was formed with very good hardness.

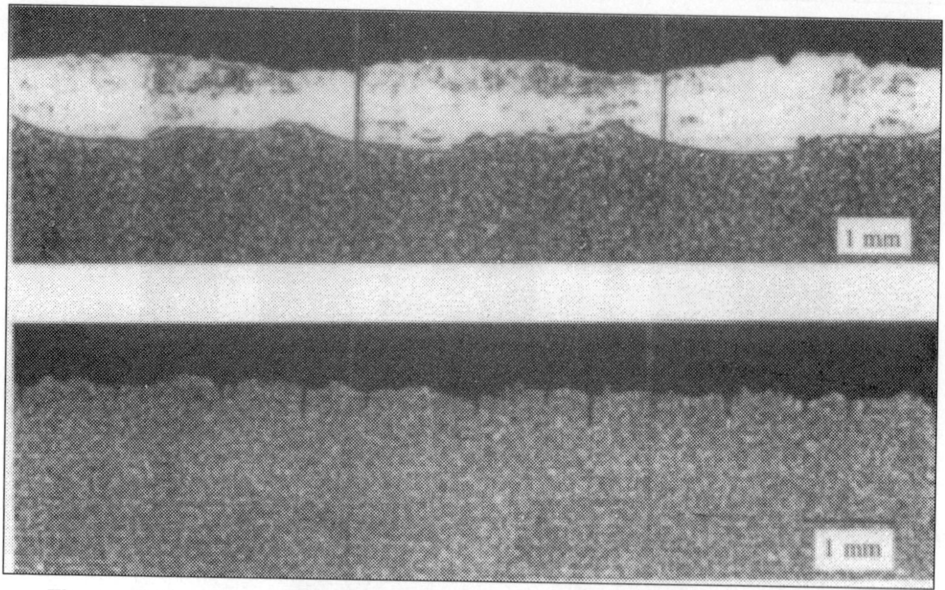

Figure 12.11: Optical Micrographs of the Cross-Sections of the Exposed CO_2 Laser Cladded 2.25 Cr-1Mo Steel with Ni-25 Cr Specimens in Actual Heat Exchanger Conditions for One Year

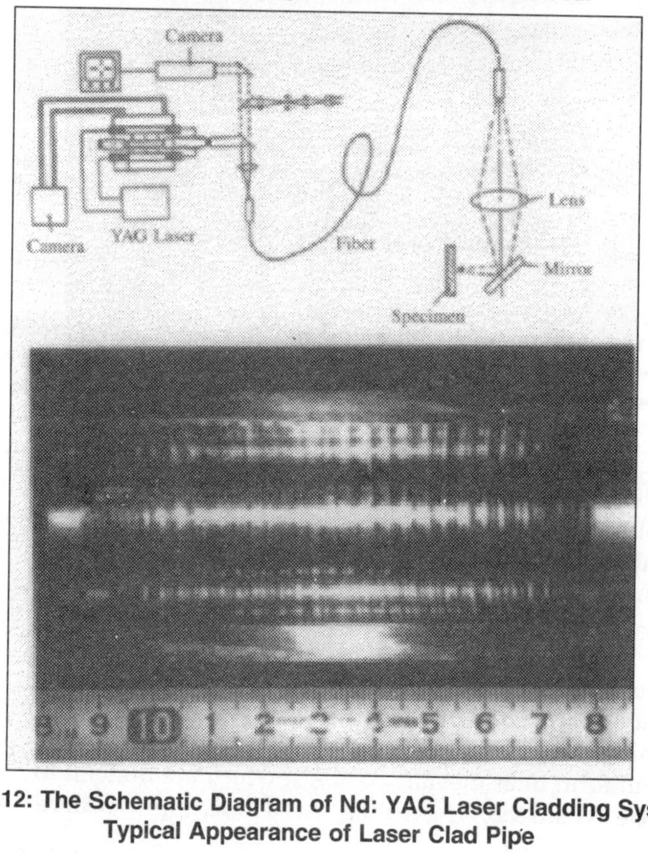

Figure 12.12: The Schematic Diagram of Nd: YAG Laser Cladding System and Typical Appearance of Laser Clad Pipe

Figure 12.13: Photograph Showing the Set-Up of the LPC Plasma Laser Thermal Spray Apparatus

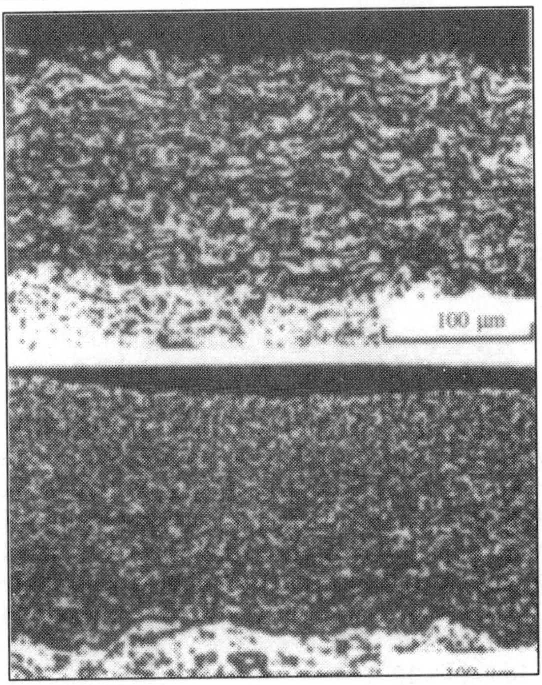

Figure 12.14: Optical Micrographs Showing,
(a) TiN (Ti) coating sprayed by N_2 plasma in a N_2 atmosphere and
(b) Laser irradiated TiN coating in a N_2 atmosphere

Plasma Powder Welding

The PPW system is receiving attention as a new surface modification technique because of its applicability to a wide range of materials and the possibility of obtaining high strength surface layers. It is a powder welding system utilizing the plasma arc. The welding material in powder form is introduced into a transferred plasma arc generated between the work (+ electrode) and a tungsten electrode (-ve electrode) and is deposited as a metal layer on the surface of the work. A schematic of the setup is shown in Figure 12.15.

The technique has now been set up for mass production for surface modification, coatings and claddings for corrosion protection, wear resistance and high temperature corrosion resistance. It is also possible to carry out the coatings inside the pipes.

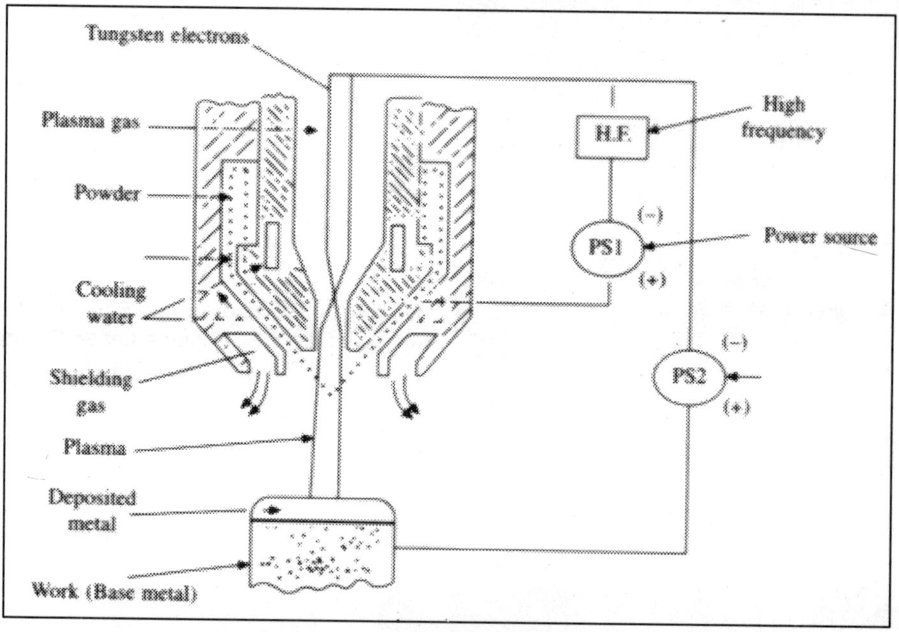

Figure 12.15: The Schematic Diagram of a Plasma Powder Welding Process Device

Ion Beam Assisted Deposition (IBAD)

These coatings refer to the process wherein evaporated atoms produced by physical vapour deposition (PVD) are simultaneously struck by an independently generated flux of ions. The extra energy imparted to the deposited atoms causes atomic displacements at the surface and in the bulk as well as enhanced migration of atoms along the surface. These resulting atomic motions are responsible for improved film properties, including better adhesion and cohesion of the film, modified residual stress, and higher density, when compared with a simple PVD deposited coating. Refractory coatings such as (Si_3N_4) can be synthesized at very low temperatures when reactive ion-beams or an evaporant is used. The two most common geometries used in IBAD processing are shown in Figure 12.16.

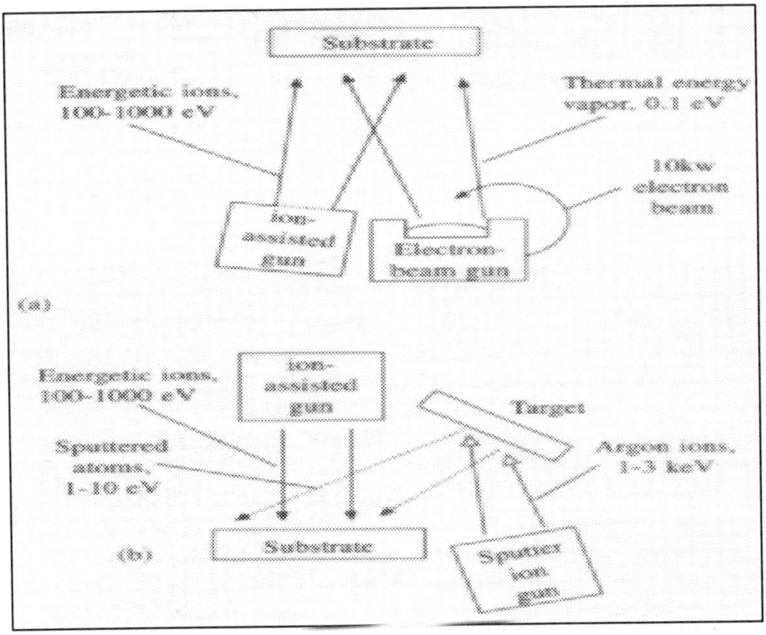

Figure 12.16: Two Common Ion-Beam Processing Techniques
(a) Ion-beam assisted deposition (IBAD) and (b) Dual-ion beam sputtering (DIBS)

Normally, a broad beam from an ion source such as a Kaufman type ion gun, impinges on a substrate simultaneously with the deposited atoms. The PVD source is usually an electron beam source, but it could also be a thermal source or a sputter ion source. The simplest geometry for the generation of uniform films and the treatment of complex geometries is a small angle (< 30°) between a vapour and ion source.

Typical ion beam energies are from 50–1000 eV for the Kaufman type ion source. It has been found that the adhesion of IBAD films is excellent between 500–3 keV but additional improvement in adhesion can be obtained with ion beam energies from 20–40 keV. Most of the laboratory equipment is designed for a sample having a maximum size of 30 mm in diameter. For many commercial applications, for large components, linear ion guns are made with lengths up to a 1000 mm and a width of 20 mm whereas circular-aperture ion guns have diameters ranging from 10–380 mm. These systems can handle work pieces with 1 m diameter in a continuous manner as shown in Figure 12.17. Dual ion beam assisted deposition systems which utilize ion beam sputter deposition are also available commercially.

Several examples of IBAD coatings, listed in Table 12.4 can be placed in three categories:

- Mode 1–in which inert heavy ions, such as argon are used to modify the coatings of an elemental film or a compound coating. Here the main precaution is to control the amount of argon incorporation in the film, which increases with arrival ratio and the beam energy.

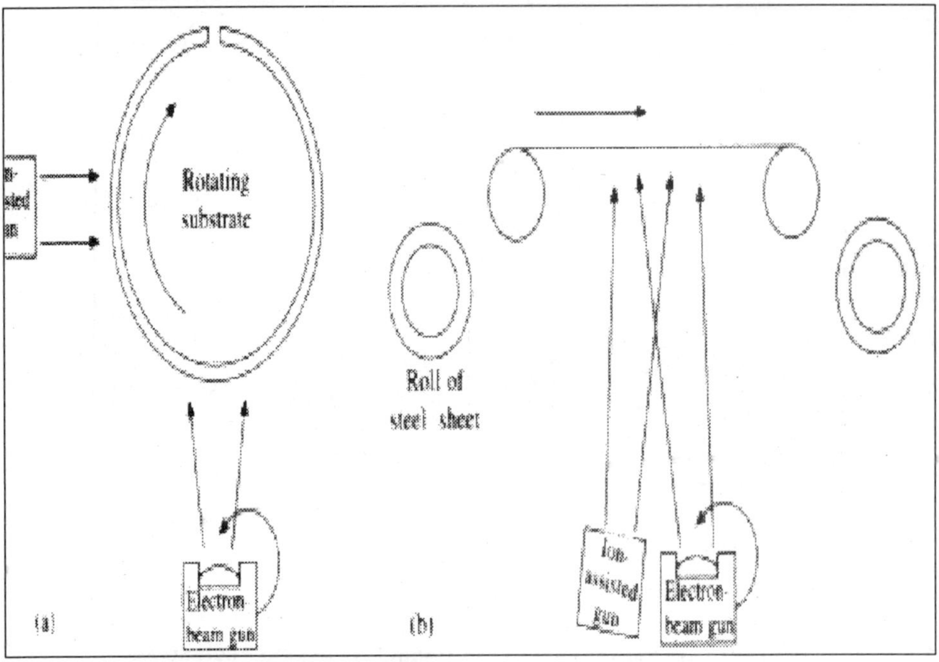

Figure 12.17: Two Methods Used for Large Area, High Volume Implementation of Ion Beam Assisted Deposition (a) for optical films and (b) for steel sheet

Table 12.4: Different Modes of Film Formation for Ion Beam Assisted Deposition (IBAD)

Vapour	*Ion or ion/gas*	*Film*
	Mode 1: IBAD	
Ge	Ar	Ge
Ag	Ar	Ag
CeO_2	O	CeO_2
Ta_2O_3	O	Ta_2O_3
	Mode 2: IBAD (Compound synthesis)	
Si	N	Si_3N_4
B	N	BN
Si	CH_4	SiC
Cu	O	Cu_2O
	Mode 3: Reactive IBAD	
Ti	N/N	TiN
Ti	Ar/N	TiN
Nb	N/N	NbN
Al	O/O	Al_2O_3

- Mode II–an ion assisted and compound synthesis group in which there is little reaction probability of the ambient gas with the evaporant. All the material that forms the compound in this deposition mode comes from the evaporation source as well as directly from the ion beam.
- Mode III–In this compound synthesis is possible only for very reactive materials. It can occur either with the gas associated with the ion beam (*e.g.* N_2 for TiN) or when inert ions such as argon are used in conjunction with a secondary (back ground) gas supply of nitrogen. In the latter case, the purpose of the ions is to activate and control surface chemical reactions. Typical deposition process variables and advantages and limitations of IBAD coatings are listed in Tables 12.5.

Table 12.5: Advantages and Limitations of IBAD

Advantages	*Limitations*
• Low deposition temperature • High Adhesion deposition • Control of stress level • Control of microstructure (nano crystalline, meta stable crystalline or amorphous) • Reproducible • Precise modulation of composition with depth • Highly versatile for metals, ceramics, semiconductors, dielectrics	• Moderately higher cost than physical vapour • Line-of-sight processing • Technology in commercial infancy

Thermal Barrier Coatings

Thermal Barrier Coatings (TBC) are insulative and oxidative resistant coatings that are finding increasingly broad utilization in the hot section of gas turbine engines. They are typically composed of a metallic bond coating layer, which inhibits oxidation and is an insulative ceramic coating layer. The prime benefit of thermal barrier coatings is the reduction of heat transferred into air cooled components (Figure 12.18). Other potential benefits include increased resistance to hot corrosion, erosion and oxidation. To achieve these benefits, the Thermal Barrier Coating systems must be reliable. Process-microstructure relationships are considered to be the key factors in the achievement of thermal barrier coating durability.

The role of coating in aircraft gas turbine applications is increasingly important. A higher engine operating temperature, new alloys and the components' life requirements all derive the need for more durable systems. Engine component designers are using newer coating systems to help provide the environmental protection needed for their more stringent demands. TBCs for turbine airfoils in high performance engines represent advanced materials technology that has both performance and durability benefits. Modern TBCs are composite coatings and may be constructed of two or three layers. The foremost benefit of TBCs is reduction of heat transferred into air cooled components. The zirconia ceramic layer of the TBC system

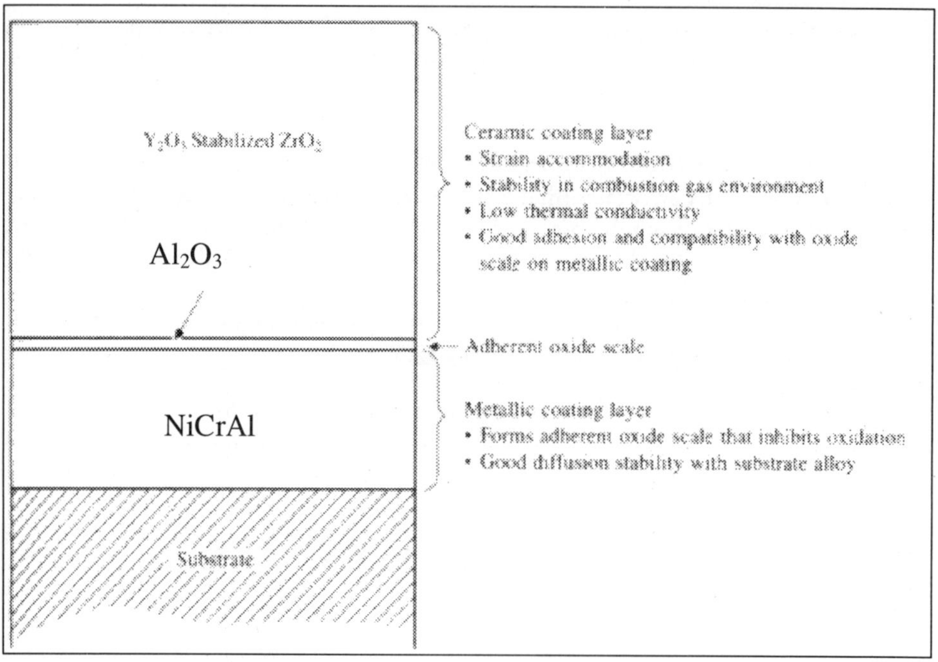

Figure 12.18: Constituents of TBC System

has a thermal conductivity that is about c Substrate ized Zirconia gnitude lower than that of a typical super alloy. Consequently on the basis of this large difference in thermal conductivity, a zirconia (ZrO_2) layer 50 to 250 µm thick can effectively insulate air cooled combustion and turbine components. The reduced heat transfer characteristic of TBCs can be used to increase the life of cooled components and engine performance. Because these insulative coatings extend life by reducing both metal, temperatures and the severity of transient thermal strains in the metal, the use of TBCs on combustor system components is prevalent. A ZrO_2 coating 125 mm thick can reduce blace colling requirements about 36 per cent as indicated in Figure 12.19. From the standpoint of the material the potential performance benefits of TBCs are comparable with those provided by substituting single-crystal alloys for directionally solidified alloys.

Plasma sprayed ceramic layers of zirconia stabilised with 8wt. per cent Y_2O_3 and of $ZrSiO_4$ over a metallic coating (CoCrA1Y) have exhibited significantly improved hot-corrosion resistance compared with conventional diffusion aluminide or CoCrA1Y coatings. Because of improved resistance (reduced solubility of sulphur oxides), thin zirconia and zircon surface layers have the potential to significantly improve component durability in aggressive gas turbine environments. Salt infiltration into the microstructure of a thick ZrO_2 layer increases coatings stress sufficiently to compromise strain tolerances; therefore densified surfaces, which inhibit molten salt penetration into the microstructure, are necessary when thicker TBCs are used in

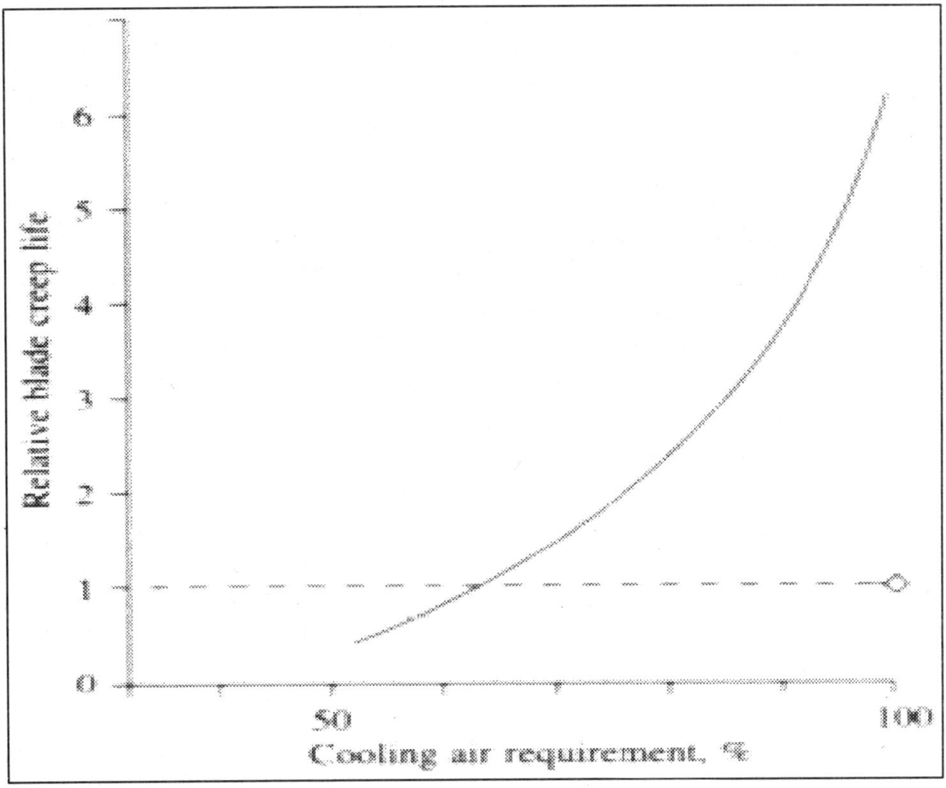

Figure 12.19: Improvement in Creep Life and Reduction in Cooling Air Requirements for a High Pressure Turbine Blade Resulting from the Use of a TBC

corrosive marine and industrial gas turbines. In aircraft gas turbines, the higher surface temperature of components coated with TBCs should also increase salt film evaporation and thus inhibit the retention of the corrosive salt film on the component surfaces. Significant improvements in the resistance of ceramic coatings to erosion by fine high velocity particles are anticipated for TBCs with densified surfaces. Thus ZrO_2 coatings with densified layers, which have a melting point around 2700°C, may be more erosion resistant than a super alloy with the incipient melting point of 1260°C (Figure 12.20).

Two distinct processes developed for application of TBCs to gas turbine components are: (1) plasma spraying; (2) electron beam evaporation-physical vapor deposition (EB-PVD).

Plasma-sprayed Coatings

The plasma spray coating process is a form of thermal spray that uses ionized gas plasma to melt and propel the powdered coating alloy towards the substrate. In this process a gas mixture (nitrogen, hydrogen, argon and helium) is ionized to the plasma state by passing it through a high frequency electrical current. Powder

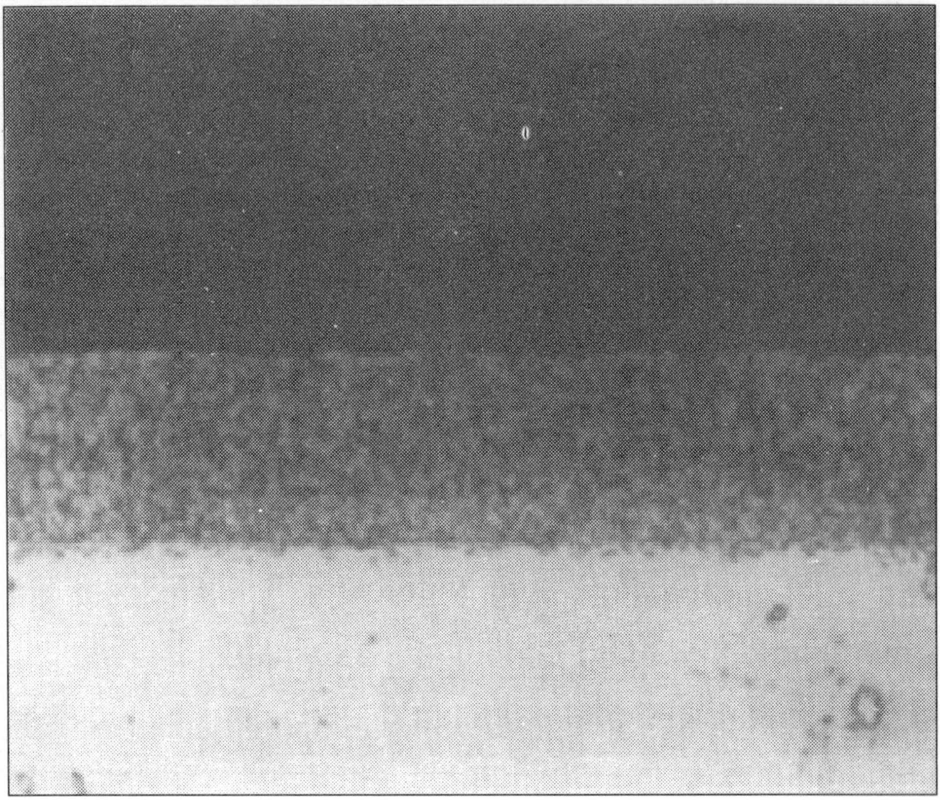

Figure 12.20: A Ceramic-metal Interface with Micro Roughness which Maximizes the Ceramic Layer Adhesion in Plasma-separated TBCs

particles are accelerated by the high velocity plasma (to a speed less than the gas velocity) and are then impacted to build up the coating. Ceramic layers of a plasma-sprayed TBC system are typically applied at metal temperature in the range of 100–300°C to avoid compressive thermal expansion mismatch stresses which could buckle the coating. Consequently, mechanical interlocking is the initial adhesion mechanism. This interface provides a tortuous path for a crack to follow, and cracks are forced to propagate within the ceramic layer. Although, ZrO_2 and some other ceramics such as $ZrSiO_4$ exhibit significant chemical resistance to molten salt attack, thick ceramic coatings will fail mechanically if the salt infiltrates the strain-accommodating porosity and micro cracks.

Electron Beam Physically Vapour-deposited Coating (EB-PVD)

This method is a modification of the high rate vapour-deposition process for metallic coating that has been successfully used to coat millions of turbine airfoils. The power to evaporate the ceramic coating material is provided by a high energy electron beam gun. This process feature is required since ZrO_2 becomes somewhat oxygen deficient as a result of partial dissociation during evaporation in a vacuum.

Vapour from this cloud condenses onto the turbine airfoil to form the coating. This type of coating depends on a chemical bond for adhesion of the ceramic and metallic layers. Higher deposition temperatures in the range 871–1093°C and/or post coating heat-treatments in the range 982–1093°C are typically achieved. Ceramic layer microstructure modification for strain accommodation has been most successful with an EBPVD applied TBC system. As with the plasma-sprayed coatings, small inter-columnar gaps in the microstructure permit the elastic modulus to be reduced to zero through the bulk of the coating. Strain within the coating is accommodated by free expansion (or contraction) of the columns into gaps, which result in negligible stress build-up in the coating.

Durability of Thermal Barrier Coatings

The durability of the TBC coatings depends on the strain tolerance of the ceramic layer, the toughness of the ceramic-metal interface and the oxidation resistance for the metallic bond coating layer. For TBC systems with adequate strain tolerance, oxidation becomes the life-limiting factor. Since the ZrO_2 layer is almost transparent to oxygen, oxidation resistance is provided by the metallic coating layer. Compositions of these coatings are tailored to form thin adherent aluminium oxide scales at the boundary between the metallic and ceramic coating layers. The alumina layer grows very slowly and inhibits additional oxidation.

One more factor contributing to the oxidation resistance for the coating is the presence of small amounts of elements such as yttrium, hafnium, zirconium and silicon which enhance adhesion of the alumina scale to the metallic layer. Another factor affecting the durability of plasma-sprayed TBCs is the quality of the metallic bond coating layer. In particular, spraying the bond coat in air results in significant oxidation of the aluminium and other reactive elements such as hafnium, yttrium and zirconium, which facilitate adhesion of the alumina scale at the ceramic-metal interface.

References

1. A.S.Khanna and Smita Kumari, Proce. ITSC Conference, May 15-18, 2006, Seattle, USA, organized by ASM International.

2. Smita Kumari and A.S.Khanna, Proc. Int. Surface Engg. Congress, ASM Int. Minnisota, August 2005.

3. M.G. Hocking, *Surface and Coating Technology*, Vol.62 (1993) pp 460–466.

4. Robert B. Heimann, *Plasma-Spray Coatings, Principles and Applications* (1996) p184.

5. A.K. Pattanaik, A.S. Khanna and K. Wissenbach, Oxidation and Corrosion Behaviour of Mild Steel Plasmac Coated with Cr-Ni followed by Laser Treatment, Proceedings of the 14th International Thermal Spray Conference, 22-26 May, 1995, Kobe, Japan, p. 993.

6. F.A.Smith, International Materials Reviews 35 (1990) 61.

7. A.S.Khanna *et al.*, Applied surface science, 156(2000) 47-64.

Chapter 13

Synthesis, Tribological Properties and Applications of Carbon Based Coatings

V.D. Vankar* and S.K. Arora

Department of Physics, Indian Institute of Technology Delhi, Hauz Khas, New Delhi – 110016, India

**E-mail: vdvankar@ohysics.iitd.ernet.in*

ABSTRACT

The authors here examine a new class of materials called carbon nanotubes (CNTs) which have been found as novel fibrous materials to be used for a variety of composites. Their nano-metric size and their structure without dangling bonds makes it a unique material which is chemically inert and offers low friction and dry lubrication. Developments in the synthesis of different types of carbon based coatings including CNTs and Fullerens have been discussed. Looking at the tribological properties of micro-crystalline diamond thin films and DLC coatings it is pointed out that the diamond films have much to offer for future tribological applications. Smooth diamond films described in this study exhibited low friction and wear against sliding Si_3N_4 counterfaces. Friction coefficients in open air were ~0.1, but fluctuated substantially in dry N_2 (*i.e.* varied between 0.06 and 0.55), especially during long duration tests. The applications of diamond, DLC coatings and CNTs have been discussed.It is concluded that the carbon family of materials at the nano-scale, with its wide diversity of forms, is a rapidly developing area from both the point of view of fundamental research as well as the current and prospective nanotechnological applications.

Keywords: *Carbon based coatings, Diamonds, Tribological properties, Surface engineering, Surface modification, DLC coatings, Carbon nanotubes.*

Introduction

In the recent past, a large number of carbon coatings such as those of diamond, diamond like carbon, carbon nitride etc. have emerged, which have tremendous technological importance[1-13]. The extreme hardness, chemical inertness, low friction of coefficient, high wear resistance, high thermal conductivity and chemical inertness are the outstanding combination of properties, which makes diamond as an ideal material for wear resistance applications. In many terrestrial applications ranging from electrical contacts to metallurgical dyes and abradable seals, graphite has been found to be very useful. The layered structure of graphite makes it possible for the basal planes to slide across each other and further adsorption of oxygen and water makes it a very suitable material where lubricating properties are of concern. The extreme and different properties of graphite and diamond have been found to be useful for many applications, where in composites of these two forms of carbon have been used. Recently considerable interest has been found in a new class of material called diamond like carbon (DLC), which has been produced by chemical vapour deposition (CVD) processes. The DLC basically consists of a mixture of sp^3 and sp^2 bonds, which are characteristics of diamond and graphite structures respectively. DLC may also contain different amount of hydrogen in it, which can be varied to change the material properties to large extents. Carbon can also bonded to nitrogen by similar CVD processes to produce carbon nitrides (C-N), which have been found to posses very different and useful properties. The recent developments in C-N coatings have shown the potential to produce harder than diamond material. A new class of materials called carbon nanotubes (CNTs) has also emerged in recent times. Carbon nanotubes can be thought of as a hexagonal network of carbon atoms that can be rolled up to make a seamless cylinder. Depending on the process the results are either multiwalled nanotubes (MWCNTs) or single walled nanotubes (SWCNTs). These structures may be open ended or closed by C_{60} type hemispherical caps. Typically the structure has length of the order of several micrometers and a diameter of 1-5 nm or up to 50 nm for single walled nanotubes or multiwalled nanotubes respectively. CNTs have found interesting tribological applications related to their nano-metric size, high strength, high elastic modulus, tremendous flexibility and unique conductivity. CNTs have been found as novel fibrous materials to be used for a variety of composites. Their nano-metric size and their structure without dangling bonds makes it a unique material which is chemically inert and offers low friction and dry solid lubrication.

Synthesis of Carbon based Coatings

Natural diamonds have been known for a long time and have remained as a material of great mistry until 1917 when it was invented as a form of carbon. Since then several efforts have been made to synthesize diamond. In 1955, GEC of USA described a process for synthesis of diamond at a pressure of ~ 55Kbar and temperature of ~1600K. Since then high pressure and high temperature route has been used very successfully to obtain reasonably large size diamond at large scale. The experimental beginning of low pressure phase synthesis of diamond can be traced way back to 1911 when von Bolton claimed to have grown diamond seed crystals from the decomposition of illuminating gas (acetylene) in the presence of mercury vapours. However, the first successful synthesis of diamond from the gas phase was reported by Eversole in 1958. In this method a carbon containing gas was passed at a temperature of about 1000C and a pressure of a few torr. New diamond was formed on the seed crystals until it was hampered by the accumulation of black carbon at about 1000C and 50 atmosphere pressure. For continuous diamond growth, this process of deposition followed by cleaning was repeated. A growth rate of about 1A/hour could be achieved in this way and increasing the growth was one of the major issues to be solved.

The beginning of new diamond thin film technology is often attributed to the work of Derjaguin and Spitsyn and their co-workers. The major initial advances were in the formation of diamond thin films by the reaction of hydrocarbon with hydrogen, where it was suggested that the hydrogen suppressed the formation of graphite, the stable crystalline form of carbon at normal pressure and temperature. Later Angus and his co-workers confirmed Eversole's results and found that diamond could be formed at low pressure where it is in metastable phase. However, the growth rates were too low for these processes to be used fore commercial applications.

Derjaguin and Fedoseev in 1975 were the first to report the growth of diamond on non-diamond substrates with reasonable high growth rates (few hundred Å/hr). This work attracted tremendous interest all over the world and several workers started growing diamond using thermal decomposition of hydrocarbon gases along with hydrogen. The important of atomic hydrogen as a selective etchant for removing graphite and its utility to dissolve continuously any graphite phase that forms during the deposition process, was soon realized. It became clear that diamond growth could be accelerated by introducing even higher concentrations of atomic hydrogen than the equilibrium value.

Intensive research in USSR (Spitsyn), USA (Angus, Messier, Badzian, Butler, Bunshah) Germany (Bachmann, Koidl), Japan (Matsumoto, Kamo, Sato, Setaka) and several others all over the world has been carried out in recent years. The first report for rapid growth of diamond at low pressure was published by Matsumoto (1982) where successful diamond was achieved using a hot filament to activate CH_4 and H_2 gas mixture. Growth rates as high as several micrometers per hour were achieved. Polycrystalline diamond (or CVD diamond) thin films are now produced by several chemical deposition processes on a variety of substrates and growth rates as high as 100μm/hr have been achieved. Free standing polycrystalline diamond discs of several centimeter in diameter have been obtained.

The discovery of an entirely new class of materials known as DLC and diamond like carbon nano composites (DLN) have emerged out of the above work. DLC is an excellent base coating to be alloyed with different elements. The amorphous nature of DLC opens the possibility to introduce certain amounts of additional elements, such as Si, F, N, O, W, V, Co, Mo, Ti and their combinations, into the film and still maintain the amorphous phase of the coating. By this technique, different film properties such as tribological properties, electrical conductivity and surface energy in contact with the surface can be continuously adapted to a desired value.

Since their discovery in 1991, Carbon Nanotubes (CNTs) have attracted tremendous interest. During the last decade, much attention has been paid towards synthesis and characterization of the nano-structured carbon materials such as carbon nanotubes, carbon nanoparticles, carbon nano-sheets, and carbon nano-horns etc. Several techniques including arc discharge, laser ablation, various types of CVD with or without plasma assistance have been used for the growth of carbon nanomaterials. Among them CVD technique is a simplè and low cost method and can be operated at low temperatures. Low temperature synthesis of carbon nanostructures is still a challenge to the researchers. The structure of CNTs confers them interesting properties and many potential applications. In addition to their application as field emitters and nanoscale electron devices, the CNTs have found several applications with respect to their mechanical properties. CNTs have recently been considered as reinforcing elements in ceramic matrix composites. The tensile strength and young modulus of CNT have been measured to be as high as 200Gpa and 1Tpa respectively, which are much high higher than that of whiskers. Tremendous improvements in the fracture toughness of CNT added alumina composites have been reported.

Recently it was predicted that covalently bonded carbon-nitrogen compound ß-C_3N_4 would have a silicon nitride structure and have bulk modulus greater than that of diamond. A number of attempts have been made to grow ß-C_3N_4 coatings using sputtering, laser ablation or ion assisted dynamic mixing processes. The films could be produced with approximately correct composition had predominantly amorphous structure. The small amount of sp^2 bonded amorphous carbon was always found in these coatings and affected the tribological properties of the films. Hardness values up to 67Gpa for such films have been reported. These films coated on Zr substrates showed effectively no wear and were found to be stable up to 670C. To produce crystalline ß-C_3N_4 coatings is still a challenge.

The tribological properties of another carbon material called Fullerenes (C_{60}) have also attracted lot of interest. The fullerenes are basically spherical shaped molecules containing distorted sp^2 bonded carbon atoms in pentagon and hexagon shapes. Within the sphere the carbon-carbon bonding is strong, but the intermolecular bonding is weak Van-der Wall's attraction, thus in sliding under load the molecules in a C_{60} layer are expected to slide over one another easily and the material could be a very good solid lubricant. Studies on the frictional properties of C_{60} coated metallic substrates have, however not met the predicted properties. The fullerenes have been synthesized by vacuum arc discharge using hydrocarbon gases and employing liquid nitrogen cooled substrates. Laser ablation of graphite has also been used to obtain C_{60} material.

Tribological Properties of Microcrystalline Diamond Thin Films

The advent of growing microcrystalline diamond thin films by CVD has been followed by evaluating microcrystalline diamond for a large number of applications(14-18). Each end-use tends to favour a particular crystallographic surface texture, purity and thickness matched to substrates that offer the most compatible properties (*e.g.* high adhesion, low interface stress or the desired Schottky barrier height at a metal-microcrystalline diamond interface). High quality microcrystalline diamond thin films exhibit most of the desired properties of natural diamonds. They are made of large columnar grains that are highly faceted and generally rough. They tend to grow continuously rougher as the thickness of the deposited films increases. The generally rough surface finish of these films precludes their immediate uses for most machining and wear applications. When used in most machining and wear applications, such rough films cause high friction and very high wear losses on mating surfaces. The surface morphology of conventionally grown, microcrystalline diamond films is poorly suited for the modification of bearing surfaces. Columnar growth with voids is ubiquitous, with grains as large as 10μm terminating on the surface. Nearly all of these films are too rough to use without polishing. Even if grown with the few crystallographic textures typical of microcrystalline diamond, there is always some off-axis tilt of the surface grains from their ideal alignment. The protruding grains are unavoidably miscut to a certain degree with their polished plane parallel with the substrate. The high purity and better finish of the microcrystalline diamond films are necessary but not sufficient for determining the fundamental tribological effects of any residual crystallographic disorder, especially where microcrystalline diamond is made to slide against materials other than itself (*e.g.* ceramics and metals) and/or where the tests are run in a lab-ambient environment only.

The rough microcrystalline diamond films can be polished by laser beams, fine diamond powders, or by rubbing against a hot iron plate. Over the years, great strides have been made in overcoming problems associated with surface roughness. Currently, methods are available to deposit smooth diamond films on a various substrates. A variety of effective methods are also available for polishing rough microcrystalline films. Friction coefficients of the polished films can be comparable with the friction coefficient of natural diamond. However, the polishing processes are tedious, take a very long time, and, in the case of complex geometries, they are very impractical. Despite high interest in using diamond films for diverse tribological applications, their widespread utilization in the industrial world has not yet met expectations. Until now, only certain cemented carbide (*i.e.* WC–Co) and ceramic (*i.e.* Si_3N_4) tools have been coated successfully with diamond and made available commercially. Although prototypes of other tribological parts (such as mechanical seals) have been prepared with diamond films, their large-scale utilization has not yet been realized.

On the other hand, nanocrystalline diamond films, which are directly grown as and/or composed of nano-sized grains from a substrate, are desirable in application fields of tribological and optical materials because of low friction, wear and scattering properties. These films exhibit a very flat and smooth surface. It has been reported that nanocrystalline diamond film could be grown using a high CH_4 concentration in CH_4–H_2 microwave plasmas or by applying a d.c. bias voltage to the substrate and from Ar- C_{60} (or CH_4) and/or Ar–H_2–CH_4 (or C_{60}) microwave plasmas. The growth mechanism of nanocrystalline diamond film has not yet been fully understood.

Diamond films have much to offer for future tribological applications. Unlike most other engineering materials, diamonds offers a combination of low friction and high wear resistance under a wide range of sliding-contact conditions. The roughness of microcrystalline diamond films limited their use in commercial applications. Smooth diamond films can be used in mechanical-seal applications, and for protection against wear and as abrasion resistant coatings in several demanding tribological applications. The results of tribological studies suggest that, depending on the tribological and environmental constraints, adsorption and desorption of gaseous species, surface reconstruction at high temperatures, and graphitization in inert test environments can occur at sliding contact interfaces of diamond films and control their friction and wear behavior. Smooth diamond films described in this study exhibited low friction and wear against sliding Si_3N_4 counterfaces. Friction coefficients in open air were ~0.1, but fluctuated substantially in dry N_2 (*i.e.* varied between 0.06 and 0.55), especially during long-duration tests. Combined results from Raman, electron diffraction, EELS, and TEM studies led us to conclude that the structure of wear debris particles generated at sliding interfaces in dry N_2 is that of a disordered graphite. Graphite is not a good lubricant in dry environments; hence is responsible for large fluctuations in the friction of nanocrystalline diamond films in dry N_2.

Tribological Properties of DLC Coatings

Under tribological conditions, usually the softer of the two materials will be worn. In the case of DLC, a different behavior is often observed. Wear products from the DLC coatings, which have a graphitic nature, are transferred to the partner surface forming a so-called transfer layer on the partner surface. The DLC then slides on this transfer layer that protects the softer partner surface from wear and the harder DLC coated surface wears off at an extremely low rate. Additionally, the graphitic wear products of DLC act also as a solid lubricant. The build up and especially the adhesion of this transfer layer depend critically on the chemical condition at the surface of the counterpart and the tribological and the environmental conditions.

The tribological behavior of DLC is strongly influenced by the film deposition conditions, the tribological conditions and especially the atmosphere during operation. Another possibility of improving the properties of DLC is the deposition of multilayer or nanocomposite coatings.

Applications of Diamond and DLC Coatings

In many tribological applications such as food processing, chemical pumps, biological applications, space technology, hard disks, etc. the use of liquid lubricants is not possible. In these cases self-lubricating coatings are used. Self-lubricating low friction coatings such as a-C:H/diamond deposited by plasma activated chemical vapor deposition and MoS_2 deposited by physical vapor deposition in recent years became available[19–21].

DLC is used in different industrial applications like protection of magnetic media, VCR head drums, diesel injection parts, sliding bearings, woodcutting tools, car valve rockers, gears, tappets of racing motorcycles, laser barcode scanner windows in supermarkets, fiber guidance systems in textile fabrication, motor cycle forks, razor blades and sensor protection etc.

Food Industry

Moving parts in pharmaceutical and food processing systems have to be operated under grease-free conditions. DLC on these parts acts as a dry lubricant, makes them scratch resistant (food as a natural product may always contain some traces of dirt including quartz particles), protects them against wear and is corrosion resistant against fruit acids etc. For example, in chemical pumps the sliding bearings are made of SiC because of its chemical inertness. In these pumps, lubricants cannot be used due to contamination or reactions with the pumped chemical. Therefore, the pumped chemical is used at the same time as lubricant for the bearings. Dry running due to temporary lack of the pumped media, results in seizure of the bearing due to overheating. It is therefore important to reduce the dry friction of SiC, which is in the range of 0.5–0.6, by applying a low friction coating such as DLC. A new application of very thin (~30nm) DLCs is as oxygen barrier layers on food wrapping foils. Such thin DLC protective layers compete favorably with protective layers made from SiO_x. However, with respect to plastic films, there is no coating thickness available below 80 nm.

Magnetic Storage Technology

As an application of high hardness value up to (6000HV) and low sliding coefficients (~0.02) of diamond like carbon coatings and a-C:H layers have been used as protective layers over magnetic storage medias like hard disc, rotating slitter knives used for the slitting of magnetic tapes.DLC layers as thin as 5-10 nm have increased protection against wear, better oxidation resistance, prevent shedding and therefore have a better reliability.

Razor Blades

Very hard DLC with tetrahedrally bonded amorphous carbon has been used very successfully on the razor blades. These coatings are deposited either by sputtering or by cathodic arc deposition. An additional layer of Teflon or material is generally sprayed on top of the DLC to reduce friction against the human skin.

Car and Engine Parts

In recent years, the fastest growing market for DLC application has been the automotive industry. In the automotive industry, different parts coated with DLC have already been investigated and demonstrate the excellent performance of DLC coated car parts, for example gears, wrist pins, valve lifters, fuel injector parts, piston rings and pump plungers of diesel injection systems. It is claimed that development of modern high pressure injection systems was only made possible through the use of diamond-like coatings. In tribological systems with full oil film lubrication (separation of the two counterparts by the hydrodynamic film), the DLC coatings do not show a beneficial influence. However, during boundary or mixed lubrication (*i.e.* when the two surfaces are temporarily in contact) due to stopping/starting, severe load conditions, or during temporary loss of lubricant, DLC and metal containing DLC coatings can significantly reduce wear. The maximum load carrying capacity of gears can be increased by 10–40 per cent due to the DLC coating. Additionally, the

DLC coating can reduce friction and therefore the gasoline consumption by up to 1 per cent.

Now a days, in the commercial motor vehicle industry, different parts are coated with DLC in large series. Several motorcycles are sold with fork tubes coated with DLC to lower friction and wear. The coating companies are routinely using different car engine parts, which have been coated with DLC (a-C:H, Tungsten containing a-C:H, ta-C and WC/C multilayer coatings). Today, worldwide, more than half of the diesel fuel pumps and fuel injectors have critical parts coated with DLC. Additionally, DLC (WC/C multilayer) coatings are already used in some commercial gear applications.

Textile Industry

Today, many components of textile machines in contact with yarn or textiles are protected with DLC against wear and corrosion, for example spinning rings, rapiers, needles, guides, grippers, yarn storage disks and weft tongs in weaving looms. Additionally, the chemically inert surface of DLC prevents the build up of deposits from the yarn, resulting in a better product quality and reduced tool cleaning.

Biological Applications

Due to its bio and haemo-compatible nature, there is a growing interest in the application of DLC on orthopedic (like hip joints, valves etc) and other implants. Today, two main fields of biological applications have emerged. One field of application is the coating of implants that are in direct contact with blood. Among these are heart valves, blood pumps and stents. Another field of application is the use of DLC coatings to reduce wear in load bearing joints. Whereas DLC coated heart valves and stents are successfully used, the results obtained with DLC coated femoral heads sliding against polyethylene cups were contradictive and revealed unexpected problems. However, joints where both sides are coated with DLC show promising results in laboratory tests.

Cutting Tools

DLC coated carbide ring sealing surfaces, knives, transport systems, valves, metering plungers, piston pumps, etc. are among some of the commercial applications. Machine components such as worm gears, lead screws, roller bearings, compressors and air bearings, are also running successfully in different applications. DLC on cutting tools has only a limited number of applications since DLC deteriorates when the temperatures exceed 350°C, a value which can easily be reached when working steel. However, there are a few commercial application of DLC coated end mills and drills used for working on fiber reinforced polymers, wood or aluminum.

Micro-electromechanical Systems (MEMS)

MEMS devices are currently fabricated primarily in silicon because of the available surface machining technology. However, because of the poor flexural strength and fracture toughness of Si, and the tendency of Si to adhere to hydrophilic surfaces, even these simple devices have limited dynamic range. Future MEMS applications that involve significant rolling or sliding contact will require the use of new materials with significantly improved mechanical and tribological properties, and the ability to perform well in harsh environments, Diamond is a super hard material of high mechanical strength, exceptional chemical inertness, and outstanding thermal stability. The brittle fracture strength is 23 times that of Si, and the projected wear life of diamond MEMS moving mechanical assemblies (MEMS MMAs) is 10,000 times greater than that of Si MMAs. However, as the hardest known material, diamond is notoriously difficult to fabricate. Conventional CVD thin film deposition methods offer an approach to the fabrication of ultra-small diamond structures, but the films have large grain size, high internal stress, poor inter-granular adhesion, and very rough surfaces and are consequently ill-suited for MEMS-MMA applications. Diamond like also being investigated for applications to MEMS devices. However, they involve mainly physical vapour deposition methods that are not suitable for good conformal deposition on high aspect ratio features, and generally they do not exhibit the outstanding mechanical properties of diamond. Future MEMS applications that involve significant rolling or sliding contact, MEMS MMAs will require the use of new materials with significantly improved mechanical and tribological properties and the ability to perform in harsh environments. Because the feature resolution in polycrystalline MEMS is limited by grain size, the use of MEMS is limited by grain size, the use of MEMS made by conventional CVD diamond methods is limited. In addition, the conventional CVD diamond films typically have large grain size ($\approx$ 1 mm), high internal stress, poor inter-granular adhesion, and rough surfaces (rms $\approx$ 1 μm).

Ultra Nano-crystalline diamond (UNCD) coatings possess morphological and mechanical properties that are ideally suited for MEMS applications. The roughness of the film is about 20 to 40 nm and the friction coefficient can be as low as 0.01. The surfaces are very smooth (rms~30 to 40 nm) and the hardness is as high as ~100 GPa. Three-dimensional MEMS structures fabricated from UNCD material, including cantilevers and multilevel devices, acting as precursors to micro-bearings and gears have been demonstrated.

The nano-crystalline diamond has been considered for a variety of applications, including MEMS and moving mechanical assembly devices, surface acoustic wave (SAW) devices, biosensors and electrochemical sensors, coatings for field emission arrays, photonic and RF switching, and neural prostheses.

Nanocrystalline diamond coatings on suitable substrates are promising materials for medical implants, cardiovascular surgery, and for the coating of certain components of artificial heart valves due to their extremely high chemical inertness, smoothness of the surface, and good adhesion of the coatings to the substrate. Nanocrystalline diamond coatings (reported grain size ~ nm) deposited by RF-PCVD of methane with nitrogen on mm-sized steel implants that had been inserted to tissue and bones for up to 52 weeks demonstrated excellent biocompatibility and biostability.

Using so-called 'flatland' (two dimensional) technology, diamond MEMS have already been produced in the forms of a seismic mass membrane accelerometer and a micro spot heater for a liquid ejector and an electro statically actuated micro switch.

Selective deposition micro patterning of diamond on Si and SiO_2 has been used to realize a movable micro gripper and a V-cantilever and tip for an AFM. Approaching the problem from the other direction, micromachining of diamond has been demonstrated using pulsed excimer laser irradiation. Reactive ion etching of diamond through an Al mask has been used, in conjunction with traditional lithographic techniques, to pattern diamond films. The technique has been used to produce beam-like and tuning fork-like resonators with sidewall angles of ~ 75°.

Miscellaneous Applications

Scratch resistant DLC coated glass plates are available for laser barcode scanner window applications. Scratch resistant DLC coated sun glasses are routinely used. DLC coated golf club heads cause the hit ball to have a minimal spin, because of their low friction surfaces.

Applications of Carbon Nanotubes (CNTs)

Application of single walled carbon nanotubes when they become available in large quantities, would be useful to develop high strength, low weight composites for a variety of structural applications[22–24]. A SWCNT or MWCNT grown directly on an AFM cantilever has been shown to be a robust, high-resolution tip for atomic scale imaging. Whereas conventional silicon probes either wear out quickly or even break, CNT tips wear only slowly. Imaging of metallic, semiconductor and dielectric surfaces, DNAs and proteins with atomic scale resolution has been demonstrated. The use of MWCNT tips for profilometry in IC manufacturing has also been demonstrated where the need exists to map the depth and shape of holes and trenches. These demonstrations have led to the active development of scale up techniques to produce hundreds of CNT tip cantilevers on a wafer compared to the current one at-a-time production.

Some of the medical applications of diamond are important from a completely different perspective. Within the concepts of molecular nanotechnology, carbon based materials could play an important role in building "nano-robots", structures that have been envisioned for applications in nano-medicine. Nano-medicine may be defined as the monitoring, repair construction, and control of human biological systems at the molecular level using engineered devices and nanostructures. Fullerene nanotube based rotors, shuffles, and gears have been computationally designed as well as those based on so called diamonoid materials.

The carbon family of materials at the nanoscale, with its wide diversity of forms, is a rapidly developing area from both the point of view of fundamental research as well as the current and perspective nanotechnological applications.

References

1. H. C. Barshilia, B. R. Mehta and V. D. Vankar, J. Mater. Res. 11 (1996) 1019.
2. H. C. Barshilia, B. R. Mehta and V. D. Vankar, J. Mater. Res. 11 (1996) 1119.
3. H. C. Barshilia and V. D. Vankar, J. Appl. Phys. 80 (1996) 3694.

4. H. C. Barshilia, S. Sah, B. R. Mehta, V. D. Vankar, D. K. Avasthi, Jaipal and G. K. Mehta, Thin Solid Films 258 (1995) 123.

5. D. K. Avasthi, D. Kabiraj, Jaipal, G. K. Mehta, H. C. Barshilia, S. Sah, B. R. Mehta and V. D. Vankar, Vacuum 46 (1995) 633.

6. H. C. Barshilia, B. R. Mehta and V. D. Vankar, Thin Solid Films, 302 (1997) 250.

7. Nita Dilawar, Somna Sah, B. R. Mehta, V. D. Vankar, D. K. Avasthi and G. K. Mehta, Vacuum 47 (1996) 1269.

8. V. D. Vankar and Nita Dilawar, Vacuum 47 (1996) 1275.

9. Nita Dilawar, V. D. Vankar, D. K. Avasthi, D. Kabiraj and G. K. Mehta, Thin Solid Films, 305 (1997) 88.

10. Nita Dilawar, Rahul Kapil, Braham Prakash, D. K. Avasthi, D. Kabiraj, G. K. Mehta and V. D. Vankar, Thin Solid Films 323 (1998) 163.

11. Rahul Kapil, B. R. Mehta and V. D. Vankar, Appl. Phys. Lett. 68 (1996) 2120.

12. Somna S. Mahajan, H. C. Barshilia, B. R. Mehta and V. D. Vankar, Thin Solid Films 302 (1997) 250.

13. V D Vankar, Surface Engineering Process Fundamentals and Applications, 1 (2003) 10-1.

14. M. N. Gardos, Surface and Coating Technology 113 (1999) 183.

15. A. Erdemir, G. R. Fenske, A. R. Krauss, D. M. Gruen, T. McCauley and R. T. Csencsits, Surface and Coating Technology, 120-121 (1999) 565.

16. O. Glozman, G Halperin, I Etsion, A. Berner, D. Shectman, G. H. Lee, A. Hoffman, Diamond And Related Materials, 8 (1999) 859.

17. H. K. Tonshoff, A. Mohfeld, C. Gey, J. Winkler, Surface and Coating Technology, 116-119 (1999) 440.

18. P. Hollman, O. Wanstrand, S. Hogmark, Diamond And Related Materials, 7 (1998) 1471.

19. Manfred Schlatter, Diamond And Related Materials, 11 (2002) 1781.

20. F. J. G. Silva, A. J. S. Fernandes, F. M. Costa, V. Teixeira, A. P. M. Baptista, E. Pereira, Wear, 255 (2003) 846.

21. Ian S. Forbes, John I. B. Wilson, Thin Solid Films, 420-421 (2002) 508.

22. W. X. Chen, J. P. Tu, L. Y. Wang, H. Y. Gan, Z. D. Xu, X. B. Zhang, Carbon, 41 (2003) 215.

23. M. R. Falvo, R. M. Taylor, A. Heiser, V. Chi, F.P. Brooks Jr, S. Washburn, R Superfine, Nature, 397 (1999) 236.

24. Aleskey N. KolMogorov and Vincent H. Crespi, Physical Review Letters, 85 (2000) 4727

Chapter 14

Diode Laser Treatment and HVOF Coating to Combat Cavitation and Silt Erosion in Power Industry

B.K. Pant, Vivek Arya[1], S.P. Gupta[2] and B.S. Mann[1]*

[1]*BHEL R&D, Hyderabad*

[2]*BHEL Haridwar, India*

**E-mail: bkpant@bhelrnd.co.in*

ABSTRACT

Examines the problems of cavitation and jet erosion which mostly occur in (i) propeller, hubs and rudders in case of ship; (ii) pumps of all types; (iii) valves, regulators, gates; (iv) diesel engine cylinder lines; (v) bearings of all types; (vi) torpedoes and missiles; (vii) sudden enlargements, pipe bends etc; and (viii) chemical process plants where reactions are decelerated. This study highlights some of the results of our findings and BHEL's future approach to overcome these problems

Keywords: *Cavitation, Jet erosion, Surface engineering, Surface coatings, Power industry, Hydroelectric power stations, High Velocity Oxy Fuel coatings, Diode laser.*

Introduction

Silt and cavitation erosion in hydroelectric power stations and machinery is a worldwide problem. Silt erosion is more aggressive in India, especially in the Himalayan region. India has six big river systems with a tremendous hydro potential of over 84,000 MW as per recent estimates (at 60 per cent load factor) and has an extensive programme for installation and commissioning of hydro projects. The erosion is mainly due to excessive silt content having particle size in excess of 90μm (ASTM 170 mesh). During monsoon it becomes impossible to control the silt content passing through the turbine. It increases from 250 to 10,000 ppm and size exceeds 1000μm. The concentration of quartz in Indian silt is also very high (~90 per cent). This is responsible for very quick erosion. The components, which are exposed to erosion in existing power stations are guide vanes, labyrinth seals, runner blades, lower rings and top covers and guide vane bushings. These are made of corrosion-resistant martensitic (13Cr-4Ni and 12Cr steels), austenitic (18Cr- 8Ni steel) and manganese steels (1.5 Mn steel). All these steels are considerably less resistant to silt erosion, and require suitable protective coatings. If unattended, the erosion of these critical components can lead to loss of turbine efficiency as high as 5-10 per cent resulting in a revenue loss of the order of 500 to 600 crores per year for all the power stations affected due to silt and is likely to be much more for coming new projects. These hydro turbines are designed to overcome the problem of cavitation and jet erosion. The cavitation problem occurs due to operating the hydro machine under off-load conditions, which is very common. Figures 14.1–14.4 shows the eroded hydro components.

This problem of cavitation and jet erosion also occurs in Boiler Feed Water Pumps which is of great importance to BHEL. These pumps, which operate on a continuous basis, are employed in Thermal Power Plants and hence need greater attention to ensure 100 per cent availability. These pumps are designed and built with "wear rings" to control the flow of recirculation from the discharge side of the impeller to its suction side. These non-contact type seals which also provide for rotor stability are generally mounted in pairs, one on the rotating impeller (impeller wear ring) and the other in the pump case (case wear ring).

These wear rings are usually made for various grades of steels. During operation, since friction caused by steel running on steel is so high, large wear ring clearances are required so that the pumps rotor and stator do not touch each other. Also, because of steel's particular tendency to gall/scuff, even greater wear part running clearances are required. However, these large clearances lead to internal fluid recirculation with consequences such as loss of efficiency, recirculation cavitation wear and high vibration levels.

The cavitation and jet erosion problems also occurs in (i) propeller, hubs and rudders in case of ship; (ii) pumps of all types; (iii) valves, regulators, gates; (iv) diesel engine cylinder lines; (v) bearings of all types; (vi) torpedoes and missiles; (vii) sudden enlargements, pipe bends etc; and (viii) chemical process plants where reactions are decelerated.

This paper highlights some of the results of our findings and BHEL's future approach to overcome these problems.

BHEL in Hydro

Various Indian as well as international agencies were trying to over come the problem of silt erosion of hydropower stations but could not succeed. In 1989, BHEL adopted plasma coating on BairaSuil guide vanes. However, the technique involved vacuum heat treatment and hence could not be commercialized.

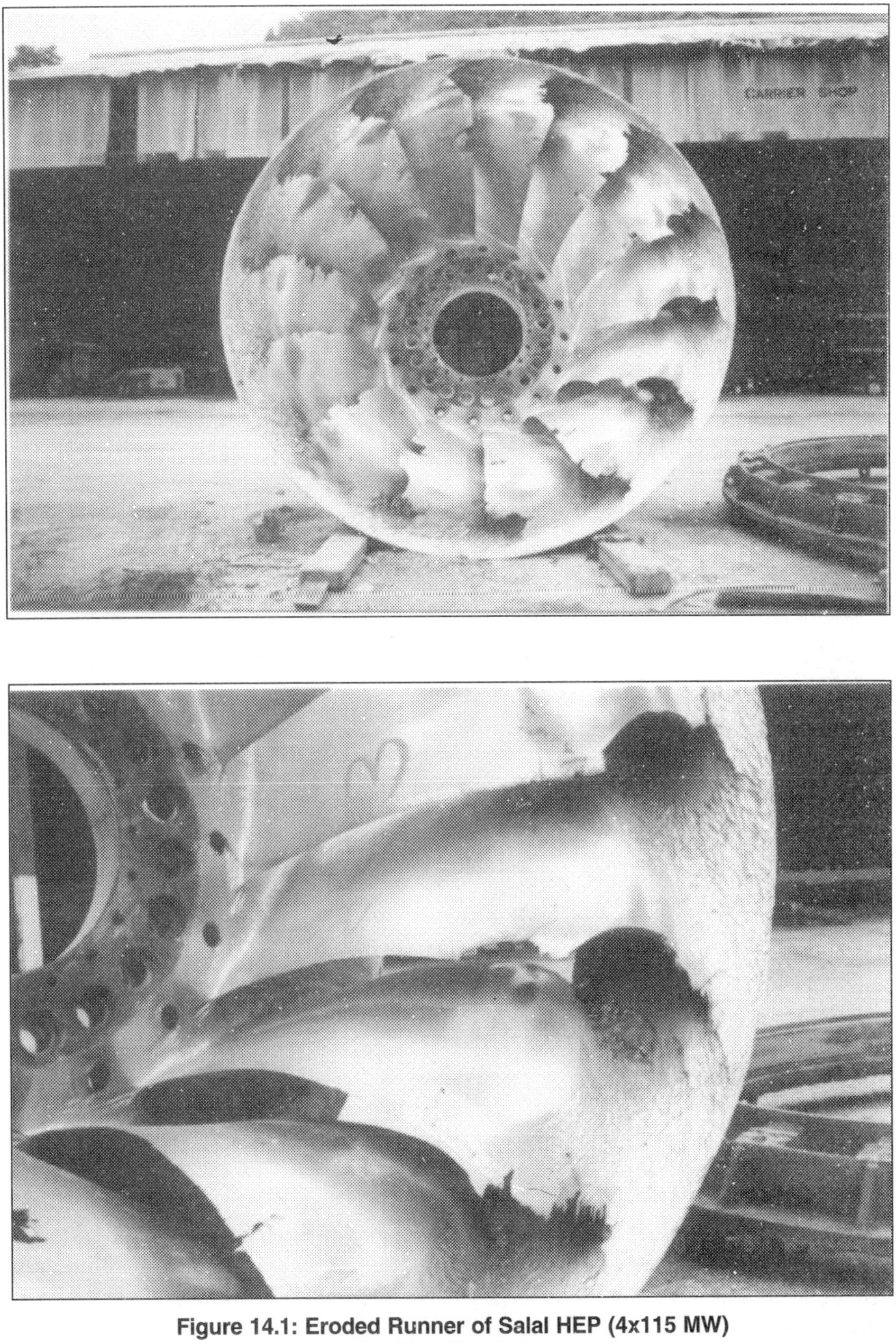

Figure 14.1: Eroded Runner of Salal HEP (4x115 MW) after Two Monsoons

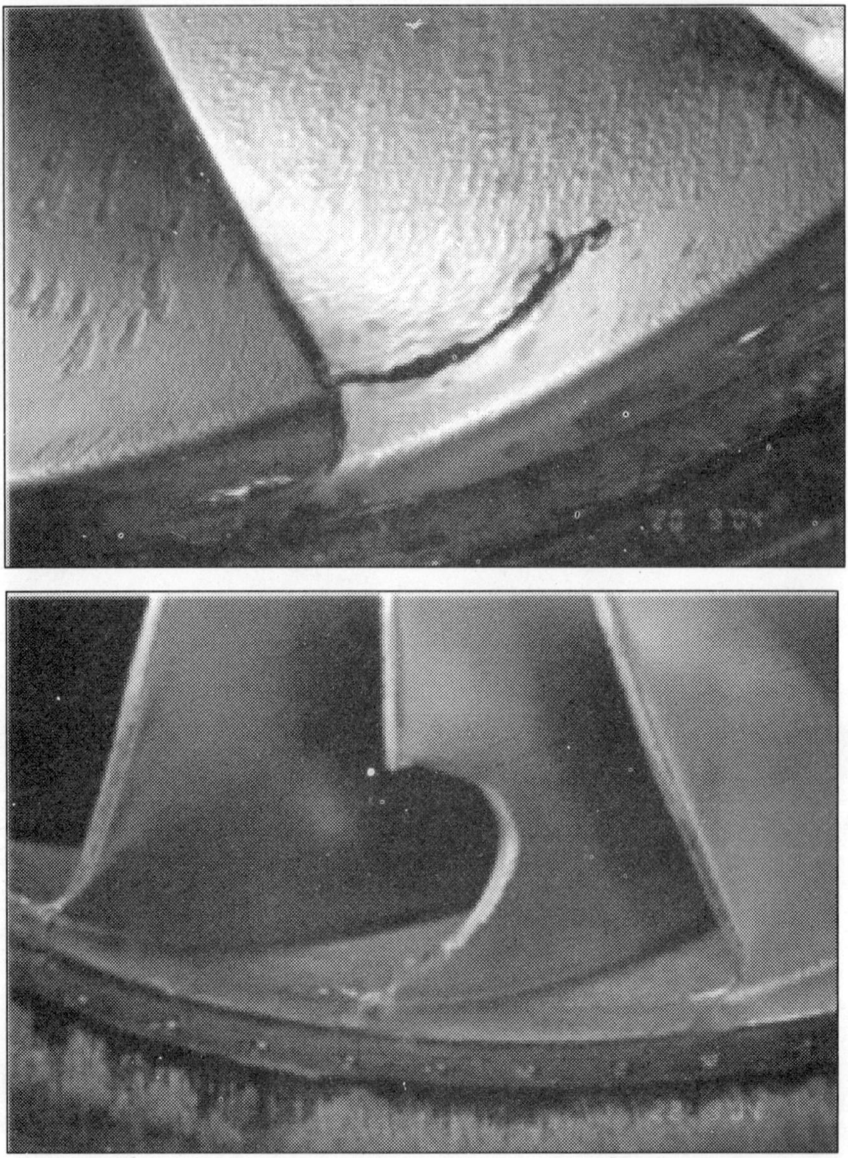

Figure 14.2: Eroded Runner of Nathpa Jhakri HEP (6 × 255 MW) after One Monsoon

Finally coatings by High Velocity Oxy Fuel (HVOF) technique were found to be suitable for tackling the problem of silt erosion. These coatings have been optimized at BHEL R&D and were proven on the top cover and lower ring (weighing 16 ton and 12 ton respectively) of BairaSuil project. The coatings were done at BHEL Haridwar utilizing their facilities. The coating process requires automatic movement of the component and the HVOF gun. The components were mounted on a rotator and the

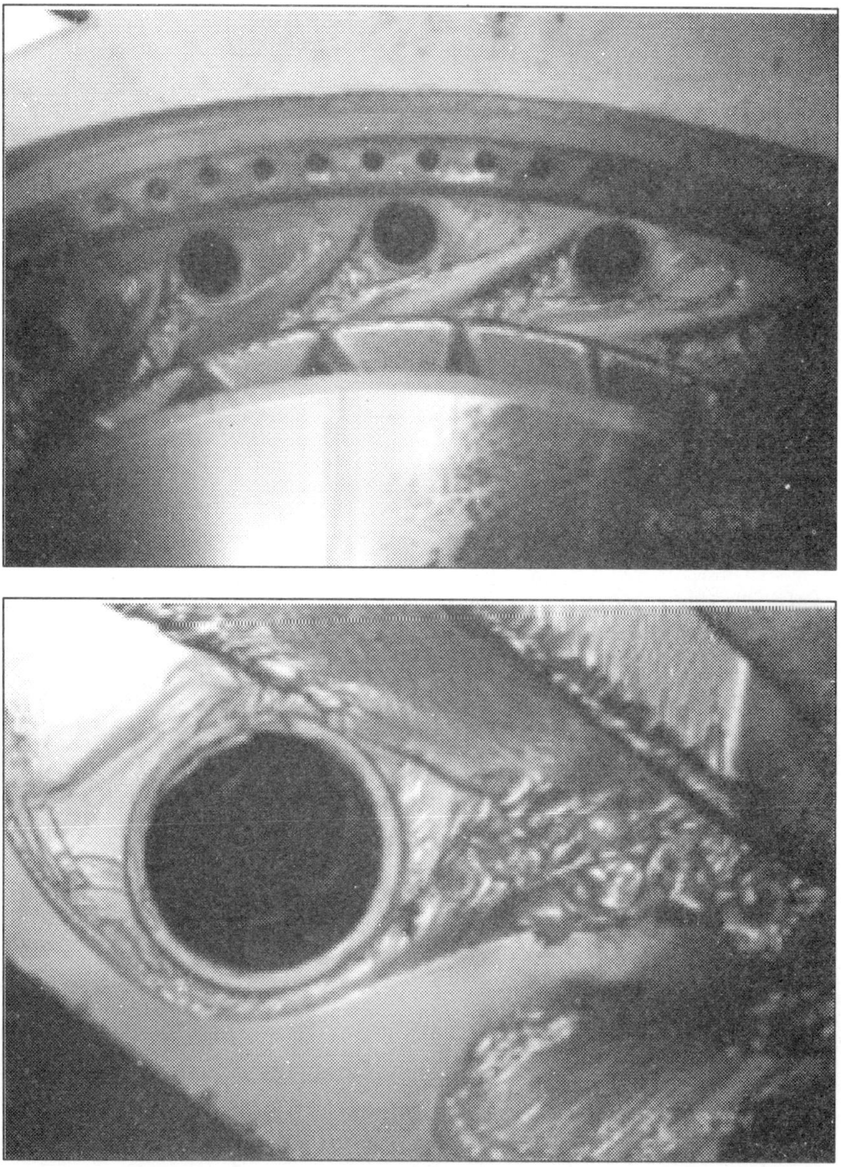

Figure 14.3: Eroded Cheek Plates of Nathpa Jhakri HEP (6 × 255MW) after One Monsoon

HVOF gun on the boom components, which were rotated at controlled speed to get a uniform coating thickness upto 300 micron. Other components such as guide vanes were also coated. With this coating, service life is enhanced minimum two to three seasons as compared to just one season without coating. This technology has resulted in a substantial benefit to the customer typically "power output" in the form of increased efficiency and reduction in shutdown and maintenance time.

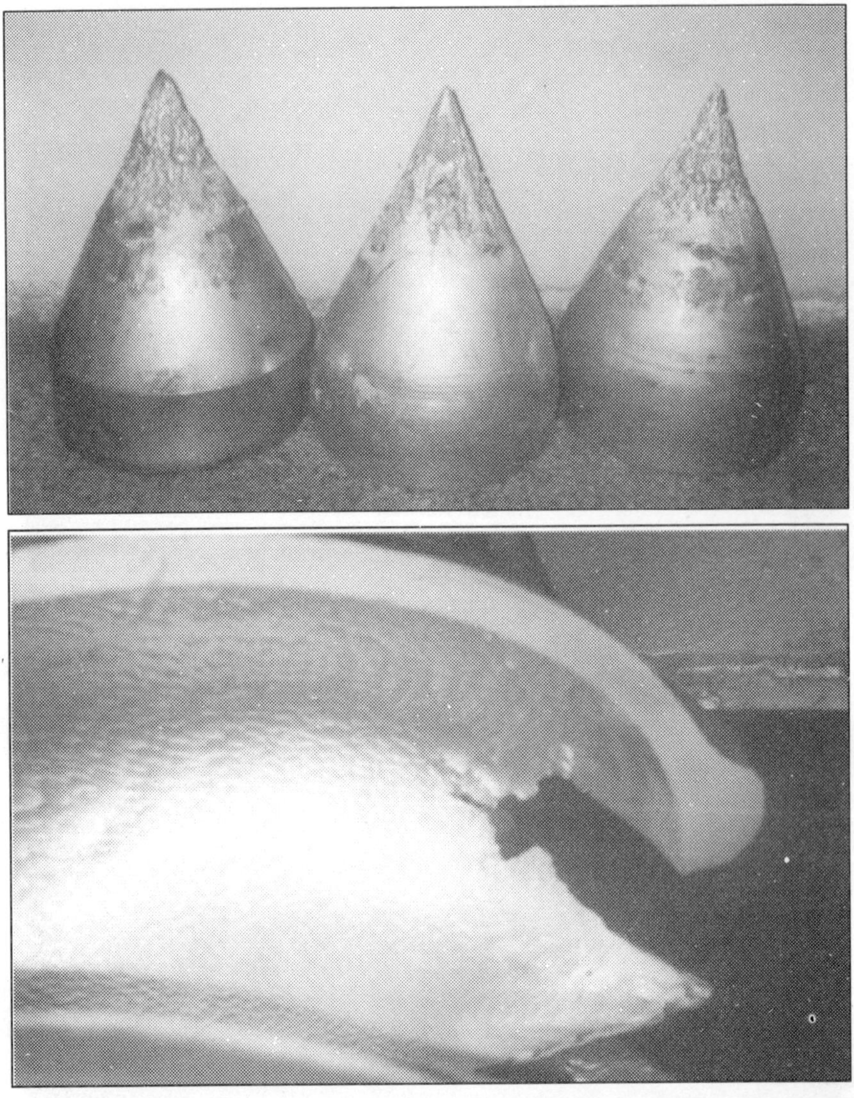

Figure 14.4: Eroded Needles and Runner Blade of Shanan HEP (4 × 15 + 1 × 50 MW) after One Monsoon

Based on this feed back, UJVLN, Dehradun placed an order worth of Rs. 10.06 Crores on BHEL for coating underwater components of all its four units of Maneribhali Hydro project (4 × 74 MW). At present HVOF coating is being carried out at BHEL, Hardwar.

Figure 14.5–14.7 show that HVOF coating being done at BHEL, Haridwar for BairaSuil and Maneribhali-II hydro projects. BHEL is also carrying out HVOF coatings for NHPC's Parbati Hydro power station (4 × 200 MW) located in H.P., for NTPC project at Kol Dam (4 × 200 MW) and many more hydro power stations. There are

Figure 14.5: Maneribhali HEP (4 × 76 MW) Lower Ring and Labyrinth with Top Cover while HVOF Coating

Figure 14.6: Ceramic Lined HVOF Coated Runner MB-II at BHEL, Haridwar (Ready for dispatch)

Figure 14.7: HVOF Coating in Guide Vanes of Baira Suil HEP in Progress

likely to be another 30 hydro power stations requiring HVOF coating. The details of these hydro power stations are given in Table 14.1.

GE Hydro, USA has supplied plasma nitrided underwater hydro components such as guide vanes, top cover, lower ring, labyrinth and runner of Nathpa Jhakri Hydro Power station (6 × 250 MW). It was unfortunate that plasma nitrided above components could not last more than one season. Nathpa Jhakri customer is contacting BHEL to apply HVOF coating for all the above components so that this problem can be controlled and substantial national saving can be achieved. Based on the HVOF proven technology, CEA has made it compulsory to adopt HVOF coating for all the hydro power stations located in Himalayan region and there is a plan to allocate 14 Crores to BHEL, Haridwar to set up a national facility for further research on hard coating for turbine underwater components.

Table 14.1: List of Hydropower Stations to be Commissioned by 2007 and Likely Affected Due to Silt Erosion

Sl.No.	*Hydropower Station*	*State*	*Power*	*Remarks*
	The Great Indus System			
1.	URI 2nd phase	J&K	280 MW	On Jhelum river
2.	Bursar	J&K	1020 MW	On Jhelum river
3.	Pakul Dul	J&K	1000 MW	On Jhelum river
4.	Baglihar	J&K	450 MW	On Chenab river.
5.	Dulhasti	J&K	390 MW	On Chenab river.
6.	Chamera 2nd phase	HP	300 MW	On Ravi river
7.	Chamera 3rd phase	HP	231 MW	On Ravi river
8.	Parbati 1st phase	HP	800 MW	On Parbati river
9.	Parbati 2nd phase	HP	470 MW	On Parbati river
10.	Baspa stage II	HP	300 MW	On Satluj river
11.	Karcham Wangtoo	HP	1000 MW	On Satluj river
12.	Malana	HP	86 MW	On Satluj river
13.	Alain Duhangan project	HP	192 MW	On Satluj river
	The Ganga Basin System			
14.	Tehri Hydro Phase I	UA	1000 MW	On Bhagirathi river
15.	Tehri Hydro Phase II	UA	1000 MW	Pumped storage
16.	Koteshwar dam	UA	400 MW	On Bhagirathi river
17.	Lohari Nagpala	UA	520 MW	On Bhagirathi river
18.	Tapovan Vishnugad	UA	360 MW	On Alakananda river
19.	Vishnu Prayag	UA	400 MW	On Alakananda river
20.	Dhauliganga stage1	UA	280 MW	On Alakananda river
	Brahmputra River System			
21.	Subansiri upper	AP	2000 MW	On Subansiri river
22.	Subansiri Middle	AP	1600 MW	On Kamla river
23.	Subansiri lower	AP	2000 MW	On Subansiri river
24.	Siang middle	AP	1700 MW	On Siang river
25.	Siang lower	AP	700 MW	On Siang river
26.	Kameng dam	AP	600 MW	On Bichom and Tenga river
27.	Tipaimukh dam	Manipur	1500 MW	On Tuivai and Barak river
	Eastern Region			
28	Teesta stage 5	Sikkim	510 MW	On river Teesta
29	Teesta stage 4	Sikkim	168 MW	On river Teesta
30	Teesta stage 3	Sikkim	132 MW	On river Teesta
	Total Power		21389 MW	

Cavitation and Jet Erosion Problem in Hydro Turbines and Boiler Feed Pumps

Cavitation

If the pressure of a fluid drops below its vapour pressure due to an increase in its velocity or otherwise, vapour bubbles tend to form. These bubbles are at low pressure and when they come in contact with regions of high pressure, *i.e.* on the runner surface or draft tube, etc. they collapse and implode. Over a period of time, small pits are developed on the surface of the component exposed to repetitive collapse of the bubbles thereby causing erosion of the base material. This phenomenon is called cavitation.

In hydro turbines there are basically two types of erosion. The first is due to the silt comprising hard quartz particles of size ranging from 10 micron to 200 micron striking on the exposed surfaces of the hydro components. This problem is particularly acute during rainy season when excessive land slides on to the riverbed cause the number of silt particles to increase in excess of 5000 ppm. This kind of erosion on the components is called *silt erosion*. To counter this type of erosion, hard coatings of a tungsten carbide based alloy are deposited on the exposed surfaces by HVOF process to enhance their life. In addition to the problem of silt erosion which occurs predominantly on the pressure side of these components, the suction side faces the problem of *cavitation erosion* due to the sudden drop in water pressure as it flows across the guide vane or runner blade resulting in the formation of air bubbles and the subsequent collapse and implosion causing erosion. This is proposed to be tackled by using laser hardening of the affected part.

Laser Treatment: An Emerging Technology

Lasers have been in industrial use from the late sixties. Since then, many advances have been made in the quality and power of available laser beams. The commonly used laser systems till date have been the CO_2 and NdYAG lasers. However, both these lasers suffer from many shortcomings. The CO_2 laser is extremely bulky and requires complex optics, which make them very maintenance intensive. Also, since their characteristic wavelength is 10.6 microns, their absorption co-efficient are very poor for most ferrous-based alloys. Special arrangements have to be made to enhance their absorption making the entire process very cumbersome. The NdYAG lasers are more compact than CO_2 ones and also have a better absorption on the substrate (wavelength 1.064-micron) but are very expensive. With the development of new diode lasers, many of the above drawbacks associated with the hitherto available laser systems have been eliminated. Diode lasers are extremely compact and operate at a wavelength typically 0.8 and 0.94 micron. This greatly improves the absorption characteristics of the laser and hence the efficiency. Also, diode lasers require very low maintenance and are very rugged. A comparison of the three laser systems is given in Table 14.2.

Table 14.2: Comparison Chart for Different Laser Technologies

Sl.No.	Property	Diode Laser	CO_2 Laser	Nd:YAG Flash Pumped	Nd:YAG Diode Pumped
1.	Wavelength (micron)	0.8	10.6	1.064	1.064
2.	Absorption per cent steel	40	12	35	35
3.	Operating cost (dollar) 100 per cent power and 4000 work hr/yr	1.5	10	30	6
4.	System efficiency (operation at 100 per cent incl. chiller)	25	6	1	6
5.	Maximum kW available commercially	6	50	4	4
6.	Plant space required for 4 kW laser (sq. ft.)	15	1000	NA	NA

Unlike the conventional systems, diode lasers do not require warm up time to stabilize. Power can also be turned on and off instantaneously. The instantaneous power control of the diode lasers realizes additional energy savings. Also, there is no key-hole and no plasma formation in diode lasers as is the case in CO_2 and Nd:YAG lasers. Keyhole leads to turbulence in the melt and quality issues.

Diode lasers are already being exploited commercially for a variety of applications such as laser welding, laser hardening, laser cutting, laser cladding, brazing Laser sintering, Cladding, Plastic welding, Soldering Paint Stripping and Heat treating. A typical diode laser facility is shown in the Figure 12.8.

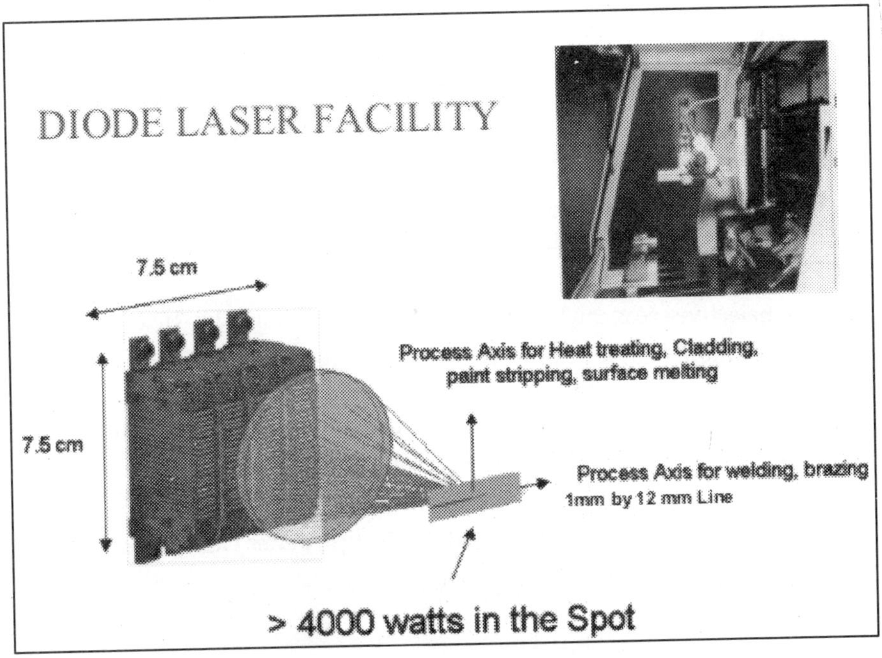

Figure 14.8: Diode Laser Facility

Laser Hardening

Laser hardening is a relatively new technique, which has the potential to overcome the problem of cavitation erosion in hydroturbines. The principle of laser hardening (shown in Figure 14.9) is similar to flame hardening except that it has several advantages over the latter. In laser hardening, as the laser beam scans the sample surface, it causes the base material to heat up rapidly to the austenitic zone. Once the beam moves forward, this heated zone gets self-quenched and results in the formation of a case of the hard martensitic phase. The case depths vary with the laser beam power and scan rate. The size of the laser on the sample can be adjusted using special optics to convert a spot beam into a slit. Laser hardening allows good hardenability of local well-defined areas of various ferrous alloys. The hardened case has a Vicker's hardness of over 700 HV.

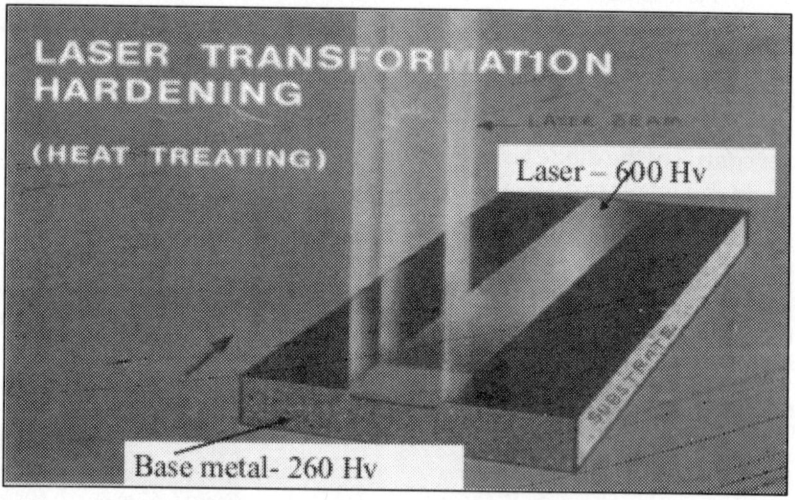

Figure 14.9: Principle of Laser Hardening

The laser is most appropriate for heat treating (hardening) small areas on sensitive high value components. Specific advantages compared to the conventional processes are

- Selective areas can be hardened without affecting the surrounding material.
- Minimal heat input results in limited distortion and reduces the need for additional machining.
- Treatment depth can be accurately controlled and is highly reproducible. Case depths upto 2 mm are possible using laser.
- No external quenching is required.
- The process can be automated resulting in better yield and greater accuracy.
- Environmentally friendly process.

BHEL Approach

BHEL has done extensive work in the area of protecting components against silt erosion. Some of the findings in this regard are given in Figures 14.10 and 14.11. Highly accelerated jet and silt erosion test rigs have been used for the study as shown in Figures 14.12 and 14.13. The investigation shows that while HVOF coating is

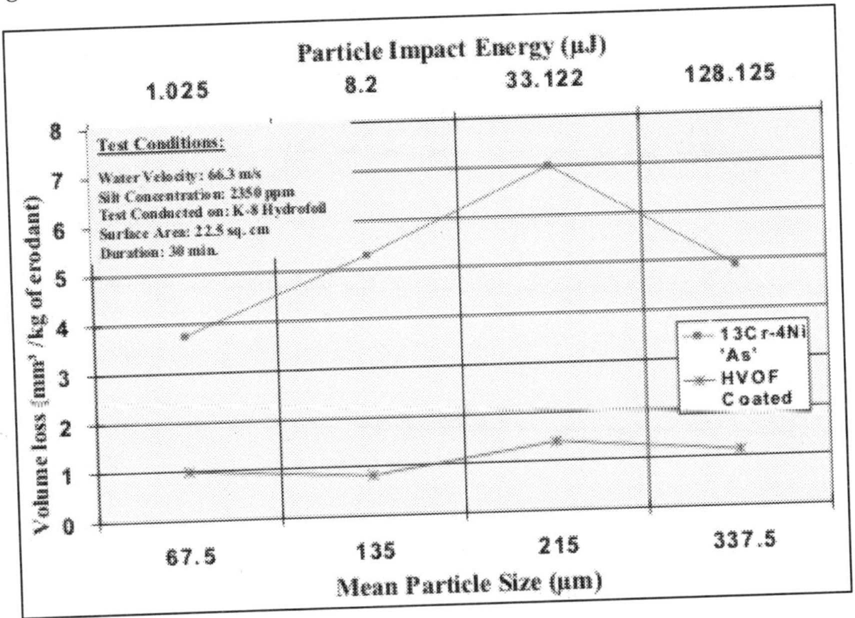

Figure 14.10: Simulated Silt Erosion Test Results of HVOF Coating

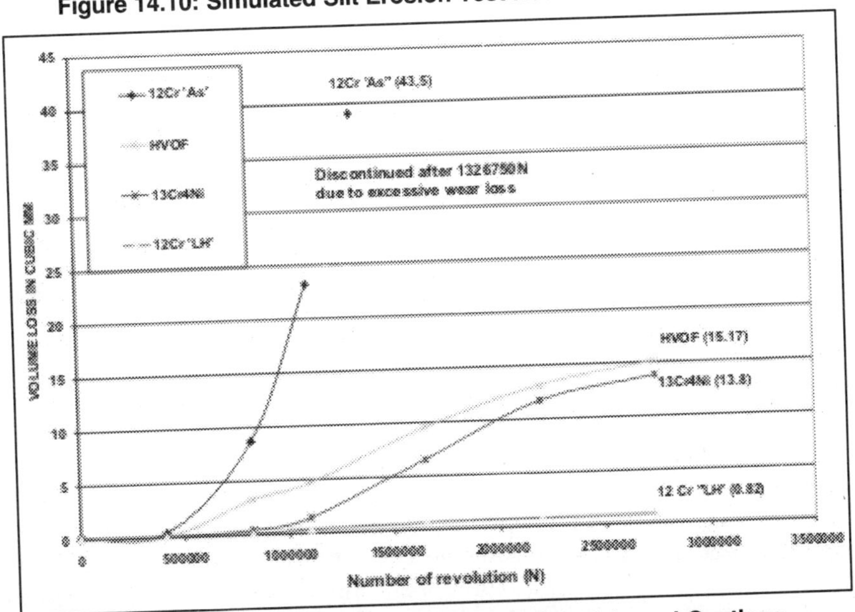

Figure 14.11: Volume Loss of Different Materials and Coatings at Impact Energy of 0.0633351J

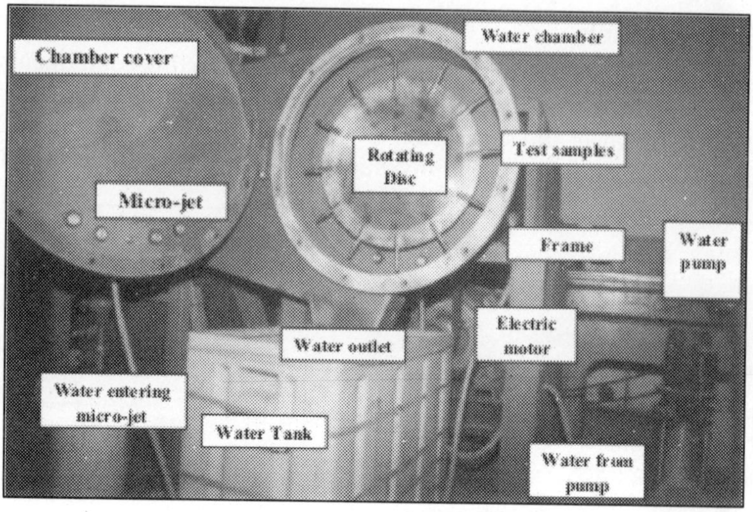

Figure 14.12: Water Jet Impingement Erosion Test Facility

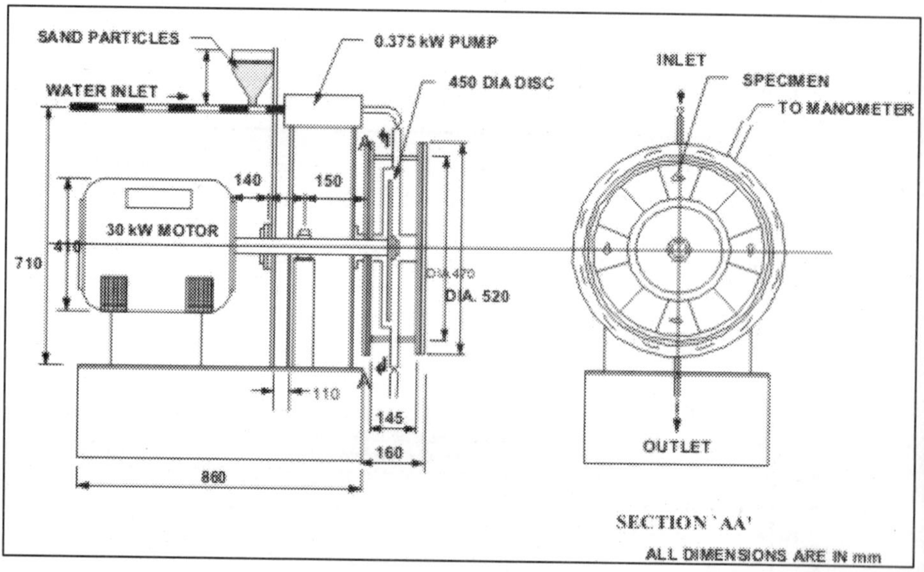

Figure 14.13: Cross-Sectional View of the Silt Erosion Test Facility

excellent for silt erosion, it is not suitable for jet and cavitation erosion. Laser hardening has given highly encouraging results as shown Figure 14.11. Since there are chances that both the phenomena occur simultaneously on hydro components, HVOF coating alone can not provide an appropriate solution. Situation as above, combination of laser hardening and HVOF coating would be used to provide an effective and long lasting solution. While HVOF would be useful against silt erosion, laser hardening is the appropriate solution for the components prone to cavitation and jet erosion. As hydro potential in India is growing significantly, the combination treatment of HVOF and laser would result in substantial savings for the utility by enhancing the life of components.

Another area of great importance to BHEL is the incorporation of coatings in components exposed to wear in Boiler Feed Water Pumps (Figure 14.14). These pumps, which operate on a continuous basis, are employed in Thermal Power Plants and hence need greater attention to ensure 100 per cent availability. Boiler feed pumps are

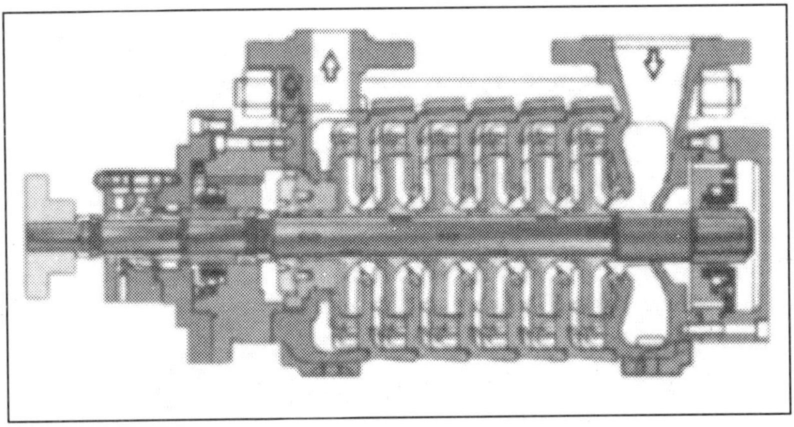

Figure 14.14: Boiler Feed Pump

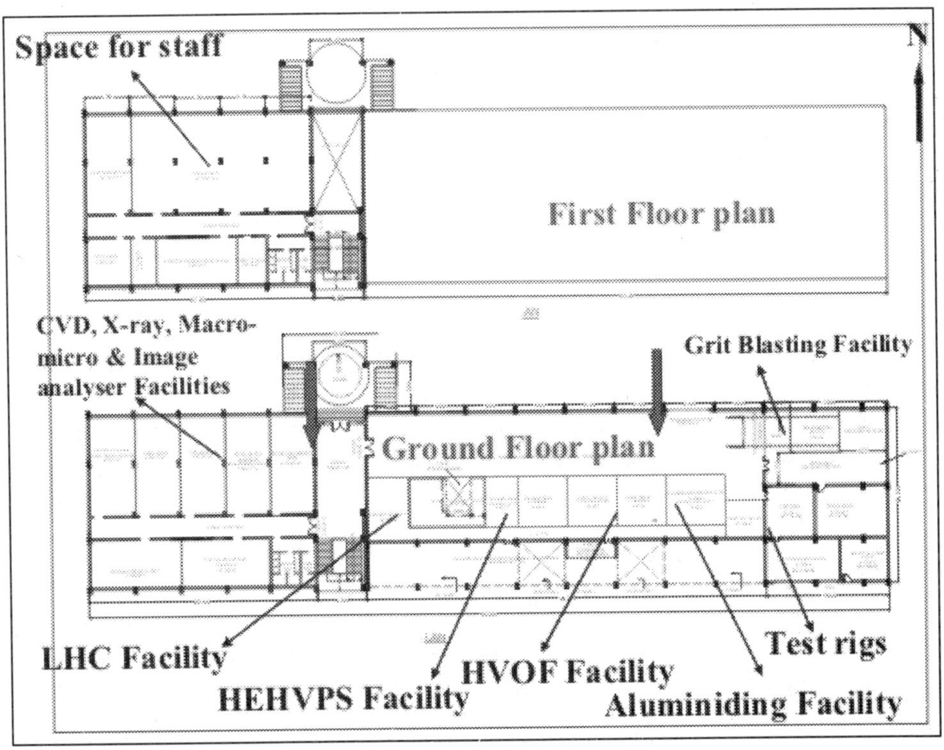

Figure 14.15: Layout of Centre of Excellence for Surface Engineering

generally eroded due to high speed of water jet impingement in addition to cavitation erosion. Laser treatment can effectively overcome this problem by enhancing surface hardness upto 700 Hv. BHEL (R&D) has done extensive study to overcome the problem of water jet erosion and cavitation erosion. BHEL is establishing a Diode laser facility worth 300 Lakhs as part of Centre of Excellence in Surface Engineering. This centre will be able to provide a comprehensive solution to such problems. In addition to hydro, the problem of cavitation erosion occurring in pumps and impeller will also be overcome. The centre of Excellence on Surface Engineering worth 840 Lakhs will have all the advanced facilities of material evaluation such as high velocity jet erosion, cavitation erosion and highly accelerated jet and silt erosion facilities and also study of the material damage mechanism using image analyzer and micro-macro hardness system. The centre will become fully operational by mid-2006. Figure 14.15 shows the schematic layout of Centre of Excellence for Surface Engineering.

Acknowledgements

Authors are thankful to the management of BHEL Corporate R&D, Hyderabad for giving permission for delivering guest lecture at 'International Training Course and Business Opportunities Workshop on Surface Engineering' at ARCI Hyderabad.

References

1. I.M. Sahai, May 2003, Indian Intention, Large Hydro India, International Water Power and Dam Construction, pp. 16-20.
2. Erosion by Cavitation or Impingement, ASTM STP 408, 1967, 3-158.
3. ASTM Standard for liquid Impingement Erosion Testing, ASTM-G73-82.
4. Erosion and cavitation, 1961, ASTM, STP pp. 307.
5. B.S. Mann, Vivek Arya, 2003, HVOF coating and surface treatment for enhancing droplet erosion resistance for steam turbine blades, Wear 254 pp. 652-667.
6. Janez Grum, Roman Sturm, 1998, Influence of laser surface melt hardening conditions on residual stresses in thin plates, Surface and coating technology 100-101 pp. 455-458.
7. F. J. Heymann, 1979, Conclusions from the ASTM Inter laboratory test programme with liquid impact erosion facilities, Proceedings of the 5th International Conference on Erosion by Liquid and Solid Impact, Cambridge, England.
8. H. G. Feller and Y. Kharrazi, 1984, Cavitation erosion of metal and alloys, Wear 93, pp.240-260.
9. N. Frees, 1983, Cavitation erosion of titanium carbide coatings on cemented carbide and other substrates, Wear, 88 pp. 57-74.
10. B. S. Mann and Vivek Arya 2001, Abrasive and Erosive Wear Characteristics of Plasma Nitriding and HVOF Coatings: Their Application in Hydro Turbines, Wear 249 pp.354-360.

Chapter 15

An Introduction to Characterization of Coatings

Ravi C. Gundakaram

Centre for Mechanical and Microstructural Characterization,
International Advanced Research Centre for Powder Metallurgy and New Materials
Balapur P.O., Hyderabad – 500 005, India

ABSTRACT

The aim of this article is to provide a brief introduction to some of the characterization techniques with special emphasis on coatings. The techniques have been classified into different sections, namely microstructural, structural, chemical, mechanical and surface. The focus is on those techniques that are suitable for the length scale from a fraction of a micrometer to several hundred micrometer. The facilities available at ARCI are also discussed in some detail.

Keywords: Surface engineering, Coatings, Materials characterization, Microscopy.

Introduction

Surface Engineering is the modification of surfaces for a variety of reasons such as to enhance the corrosion resistance, decrease wear or to provide electrical or thermal insulation. One of the important ways of modifying the surface is by providing a coating.

In terms of length scales, coatings come in between thin films and bulk materials. The former are often several Ångstroms thick whereas bulk materials can range from a fraction of a millimetre to any thickness. The thickness of coatings is thus a micrometer or less to probably a few hundred micrometer, the thickness being dictated among other factors by the application on hand.

Materials characterization is an important field of study. To ensure the performance of a coating or a device, its inherent characteristics and response to external stimuli need to be well understood. Thus, the necessity of standard characterization techniques and properly calibrated facilities can never be overstated. In the case of a coating, the material is everything and failure of a coating in a particular application can often directly be traced to the failure of the material to perform under the required operating conditions.

The term 'characterization' has a broad purview. The aim of this article is to provide a brief introduction to some of the characterization techniques with special emphasis on coatings. There are several levels at which a material can be characterized and Table 15.1 provides some of these, along with a sampling of the experimental techniques to achieve the objective. The abbreviations used in this Table are explained as they occur in the text.

Table 15.1: Different Ways of Characterizing a Material/Coating

Sl.No.	*Type*	*Representative Experimental Techniques*
1.	Microstructural	Optical and Electron Microscopies, OIM
2.	Structural	XRD, EXAFS
3.	Chemical	EDS, EPMA, EXAFS, SEXAFS, NEXAFS, XRF
4.	Mechanical	Microhardness, Nanoindentation
5.	Surface	SPM

The choice of the experimental technique depends on the length scale. As mentioned above, the range of thickness of typical coatings is from a fraction of a micrometer to several hundred micrometers. Thus, the focus of this article on those techniques that are suitable for this length scale. The facilities available at ARCI are discussed in some detail. Table 15.2 lists the techniques that are discussed in this article, along with the main information that can be obtained from the technique. It is to be borne in mind, though, that some of the techniques can also be used for the study of bulk materials and some appear under more than one category as is to be expected. The grouping of Table 15.2 follows that in Table 15.1. We now proceed to discuss each of these techniques in detail.

Techniques for Microsctructural Characterization

Optical Microscopy

The optical microscope is more often the first instrument of choice for visual observation and characterization of coatings. The instruments are relatively inexpensive and take up little laboratory space. Magnifications ranging from less than 10x to over 1000x are possible with optical microscopes. Variations such as the stereomicroscope provide additional capabilities. Any solid sample can be viewed and sample preparation, if any, is minimal. For material analysis, it is more useful to use polarized light instead of white light. The reason for this is that white light consists of several colours (*i.e.*, wavelengths), in addition to the light vibrating in all directions, which provides composite information making it difficult to analyse the sample features.

Table 15.2: Characterization Techniques Discussed in this Chapter, along with the Main Information that can be obtained from each Technique

Category	*Techniques*	*Experimental Information*
Microstructural	Optical Microscopy	Surface imaging, microstructure
	Scanning Electron Microscopy	Surface imaging, microstructure
	Orientation Imaging Microscopy	Surface imaging, grain and grain boundary information
Structural	XRD	Structure, phase, defects
	EXAFS, SEXAFS, NEXAFS	Local structure
Chemical	EDS	Elemental information, imaging
	EPMA	Elemental information, imaging
	EXAFS	Elemental information, local structure
	SEXAFS	Elemental information, local structure (surface)
	NEXAFS	Elemental information, local structure, valence state
	XRF	Elemental information
Mechanical	Microhardness testing	Hardness
	Nanoindentation	Hardness and elastic modulus of surfaces (10 μm or less), surface imaging
Sufrace	SPM	Surface roughness, topography, defects, imaging

By nature, coatings consist of pores and voids that occur in the coating process, which cannot be avoided completely. In addition, oxides may be present, as also unmelted particles. Contamination of the coating as an artefact of preparation is also possible. All these features can be studied by optical microscopy. The pore size can sometimes be of the order of several micrometer or more. In such cases, the optical microscope makes it quite convenient to observe the features. Figure 15.1(a) shows the cross sectional image of a WC-Co coating. Some of the features mentioned above can be seen in this image.

The capability of an optical microscope can be greatly enhanced by the attachment of an Image Analyser. Quantitative measurements of the microstructural features mentioned above can then be made. Basically, an image analyzer converts the optical image into an electronic one (*i.e.*, 'digitizing' it) by converting the optical image into a two-dimensional array of numbers representing the position and brightness of the different features of the coating. By using the image analysis software, the dimensions of the different features can then be measured accurately. The digital image also serves as a record of the observation, obliterating the need for a film camera and the associated film processing. Indeed, the images in Figures 15.1(a) and (b) have been obtained with an image analyzer attachment. Figure 15.1(b) shows a digital image of a spherical indentation showing the measurement of the indent dimensions.

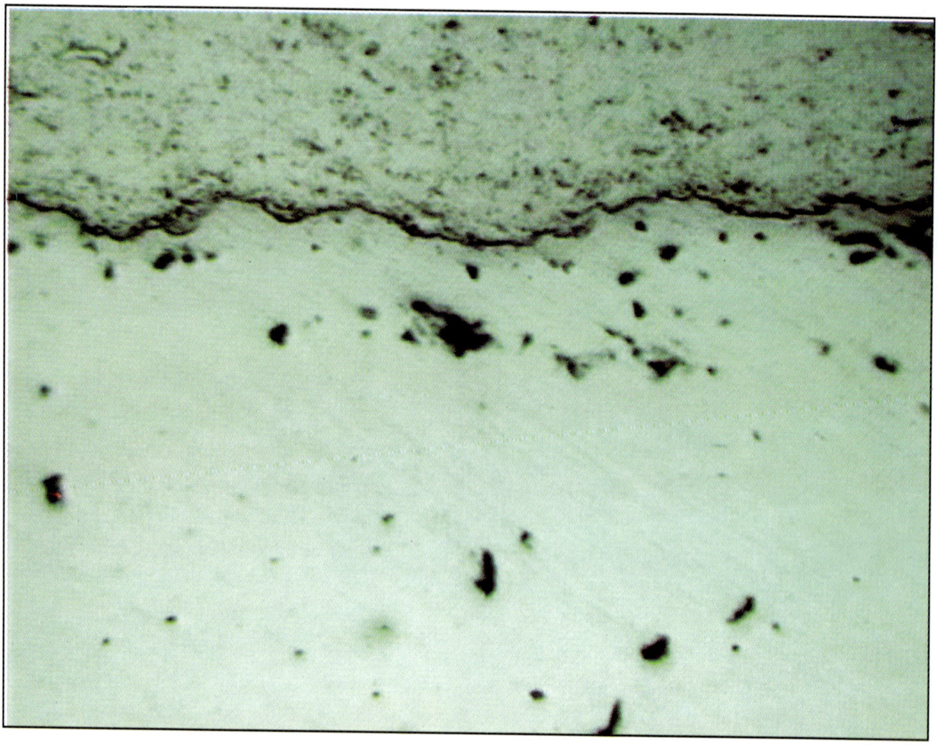

Figure 15.1(a): Optical Micrograph of a WC-Co Coating in Cross Section

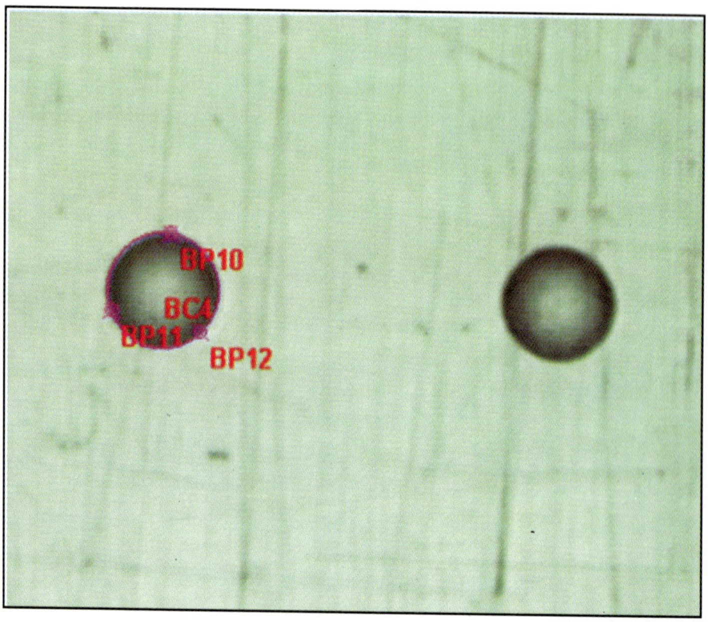

Figure 15.1(b): Optical Micrograph of a Spherical Indentation
Using the image analyser, the dimensions of the indent can be measured.

Scanning Electron Microscopy (SEM)

The scanning electron microscope is often the instrument of choice next to the optical microscope for the study of the microstructural features of a coating. The SEM is arguably much more expensive and needs maintenance. Furthermore, some restrictions apply to the sample, and some sample preparation may be needed. However, the magnification that can be attained using an SEM is way beyond that of the optical microscope (1400x), and magnifications in excess of 100,000 are easily attainable in an SEM. Furthermore, by the use of attachments such as EDS and EBSD (to be discussed later), elemental and phase information can be obtained, which is not possible with an optical microscope.

The SEM consists of the following main parts:

1. A source of electrons
2. A set of apertures and condenser lenses
3. Stigmator and deflection coils
4. A final lens and aperture that focuses the electron beam on to the sample
5. Detectors for secondary and backscattered electrons, as also x-rays emitted by the sample
6. Vacuum systems to maintain the sample chamber in high vacuum

Figure 15.2 shows the SEM facility at ARCI and Figure 15.3 shows the schematic of the working of a scanning electron microscope.

When an accelerated electron beam impinges on the sample surface, different interactions occur, giving rise to secondary electrons, backscattered electrons and elemental x-rays. The mechanism of each of these interactions is different and a brief description of each is given below:

When the energetic primary electron interacts with an atom in the sample, it is scattered either elastically (*i.e.*, there is little loss of energy), or inelastically. In an inelastic collision, the incident electron transfers its energy to an electron in the atom, leading to the ejection of the electron from the atom. If the energy of the ejected electron is less than 50 eV, it is referred to as a secondary electron. In case the electron is scattered elastically by the nucleus of the atom, there is little energy loss and such an electron leaves the surface with energy greater than 50 eV, which is termed as back-scattered. Elements with higher atomic numbers are more likely to cause backscattered electrons.

In the third type of interaction, the primary electron knocks off a core electron from the atom of the material. The excited atom returns to its ground state by emitting an x-ray photon, the energy (or wavelength) of which is characteristic of the element that produced it. Thus, by analysing the energy with an energy dispersive detector or the wavelength using a wavelength dispersive detector, elemental analysis can be performed in the SEM. This is the principle of working of Energy Dispersive Spectroscopy (EDS), which was referred to in Table 15.2.

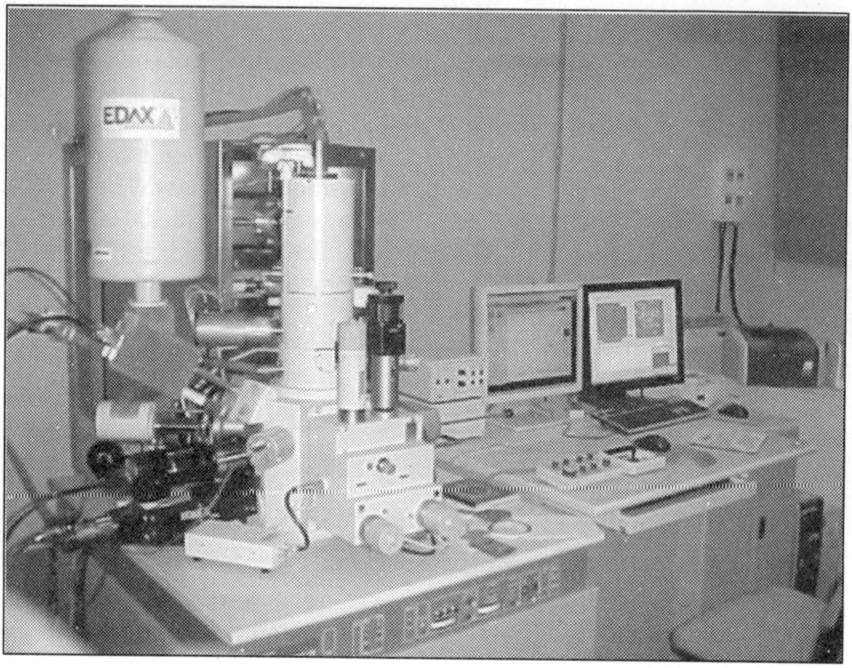

Figure 15.2: The SEM Facility at ARCI Showing also the EDS and EBSD Units

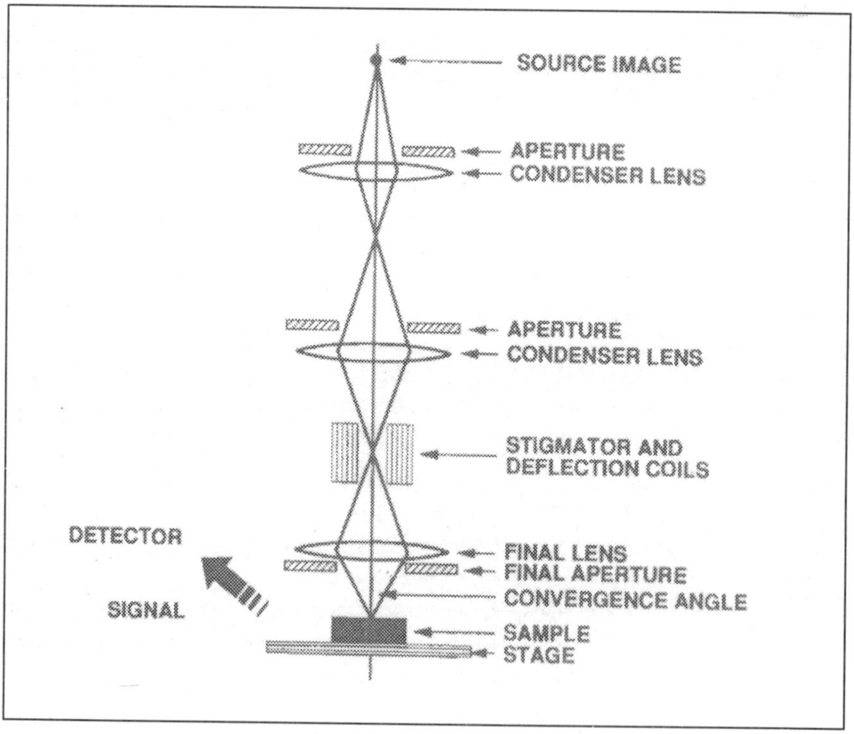

Figure 15.3: Schematic of the Working of the SEM

Secondary electrons are useful for the detection of surface morphology due to the sensitivity of the secondary electron yield to the local variations of the sample surface. Backscattered electron yield on the other hand is proportional to the concentration of the heavier elements in the sample surface and the backscattered electrons are hence very useful to determine the atomic number contrast.

During the operating of the SEM, the source of electrons is focussed on to the sample as a fine probe that is rastered over the sample surface. This rastering of the beam gives rise to the scanning function in the SEM. The output electrons are collected using suitable detectors, the output of which is used to modulate the brightness of a cathode ray tube whose x- and y-inputs are driven in synchronism with the x-y voltages of the scanning electron beam. In this way, every point that the source beam strikes on the sample has a corresponding point on the CRT screen. If the magnitude of the saw tooth voltage applied to the x- and y- deflection amplifiers in the SEM is reduced by some factor while the CRT saw tooth voltage is kept fixed at the level necessary to produce a full screen display, the magnification will be increased by that factor.

One of the limitations of a standard SEM is that the surface should be electrically conducting. If not, it is necessary to provide a thin conductive coat on the surface. Gold and carbon are often the materials of choice for this purpose. This limitation can be overcome in SEMs (the so-called environmental microscopes) that can work at higher pressures (*i.e.*, at lower vacuum levels).

Figure 15.4 shows the coating-substrate interface of an Al_2O_3 coat on an Al substrate. An additional feature of the EDS unit available at ARCI is that it can generate elemental maps showing the distribution of each element on the sample surface. The spatial variation of the various elements across the coating is shown in Figure 15.5. In addition to the detection of elements over an area, a spot elemental analysis can also be carried out. Figures 15.6(a) and (b) show the SEM images of a WC-12 per cent Co coating where the EDS analysis has been performed on the WC particle and Co-rich areas respectively. The corresponding EDS spectra show strong W peaks [Figure 15.6(c)] and Co peaks [Figure 15.6(d)]. It should be kept in mind that the interaction between the electron beam and the material occurs in three dimensions so that all elements within the interaction volume are detected. This is especially important when working with thin coatings.

Orientation Imaging Microscopy (OIM)

OIM is a technique using which a wealth of information can be extracted about grains and grain boundaries. When carried out within an SEM, the technique is referred to as Electron Back Scatter Diffraction. As mentioned earlier, the SEM can be used to study the microstructure. What is missing, however, is the information concerning the orientation of each grain, and the statistics concerning grain boundaries. The study of orientations is important because most materials show anisotropy in their properties. Texture analysis is thus the characterization of the crystallographic orientation in a microstructure and EBSD is a powerful technique to study texture. Large areas of the sample (as compared to X-ray Diffraction and

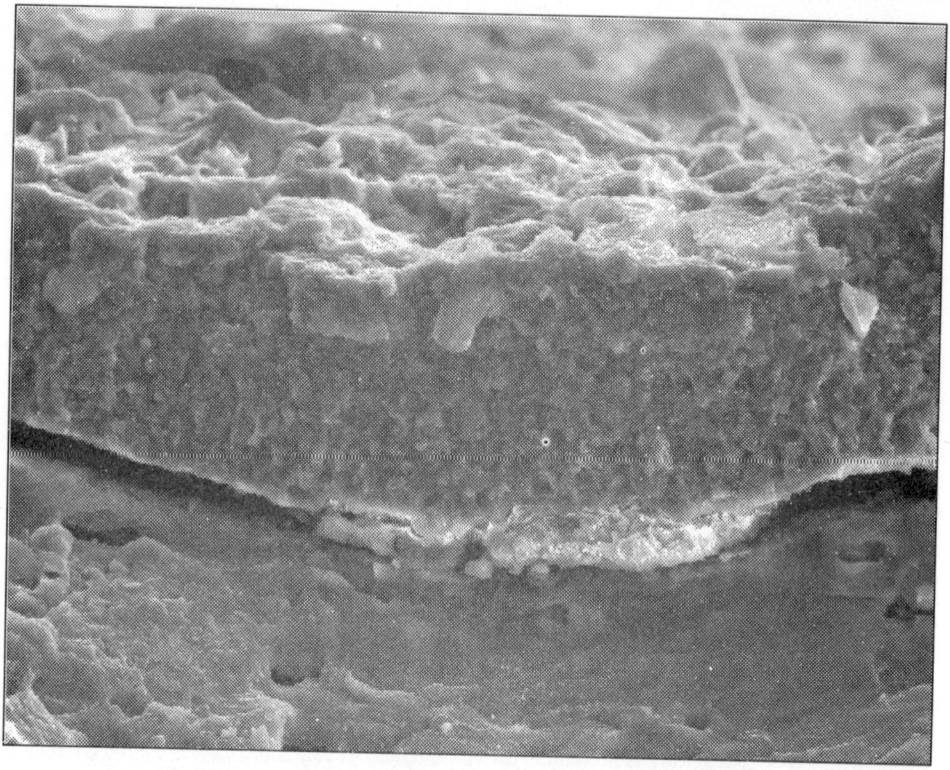

Figure 15.4: SEM Picture Showing the Interface Between an Al_2O_3 Coat (Top) with the Al Substrate (Bottom)

Transmission Electron Microscopy) can be studied in a relatively short duration of time and furthermore, statistics can be obtained easily. While an SEM *per se* is not necessary for EBSD, it is nevertheless very convenient to have an EBSD attachment to an SEM since the electrical and electronic components for the generation and raster scanning of the electron beams as well as the vacuum systems are in place.

Basically, the EBSD unit consists of a phosphor imaging screen and a low-light television or CCD camera, in addition to the normal working components of the SEM. The sample is tilted to an angle of 70º and exposed to the electron beam. EBSD patterns occur due to the interaction between the electron beam and the sample, imaged on the phosphor screen and captured using the camera. Similar to the peaks generated in an x-ray diffraction measurement, the bands in the pattern represent reflecting planes in the diffracting crystal volume. Thus, by studying the geometric arrangement of the bands, the orientations of the diffracting crystal lattices can be determined. Knowing the crystal structure, the orientation can be determined automatically by indexing the experimentally observed patterns to obtain the corresponding orientations.

In an OIM scan, the electron beam is moved across the area of interest of the sample in a regular (usually hexagonal) grid. At each point, the pattern is captured

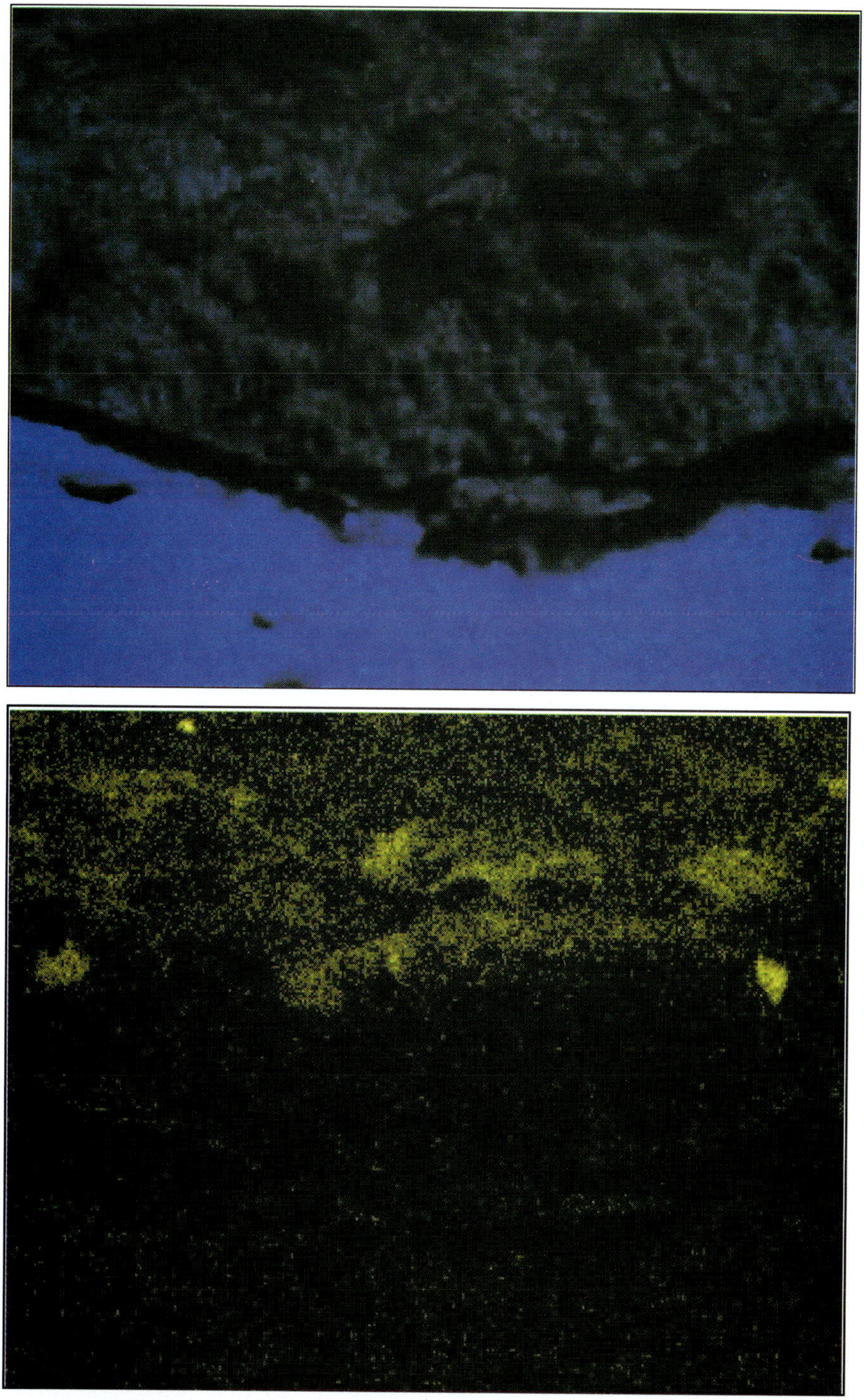

Figure 15.5: EDS Elemental Maps of Al (Top) and Si, the Corrosion Product (Bottom) for the Area Shown in Figure 15.4. The brightness of the colour indicates the relative concentration of the element.

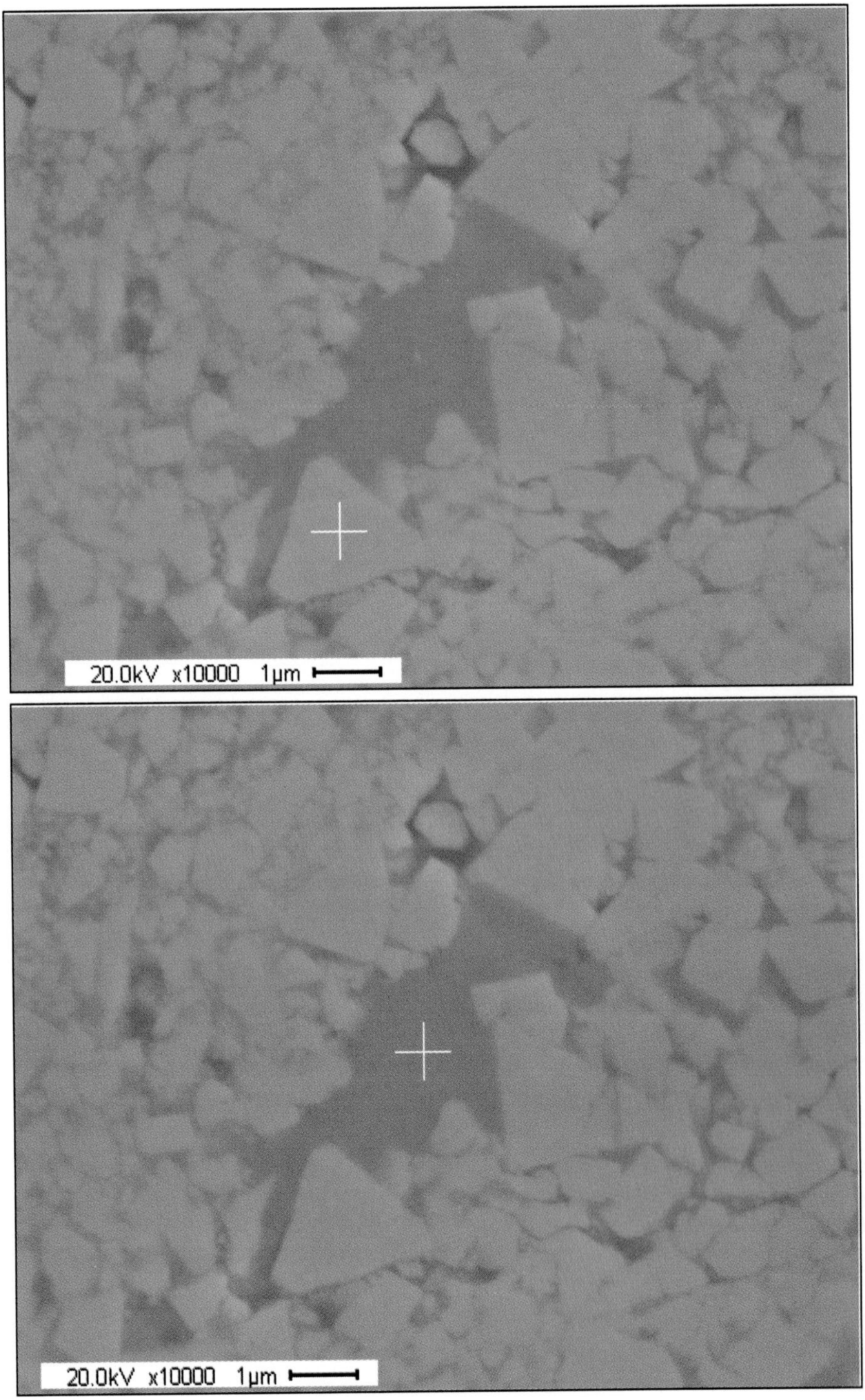

Figure 15.6(a) and (b) [Top and Bottom images, respectively], Showing the SEM Pictures of WC-12 per cent Co Coating. The cross wires indicate the spots where EDS data were collected.

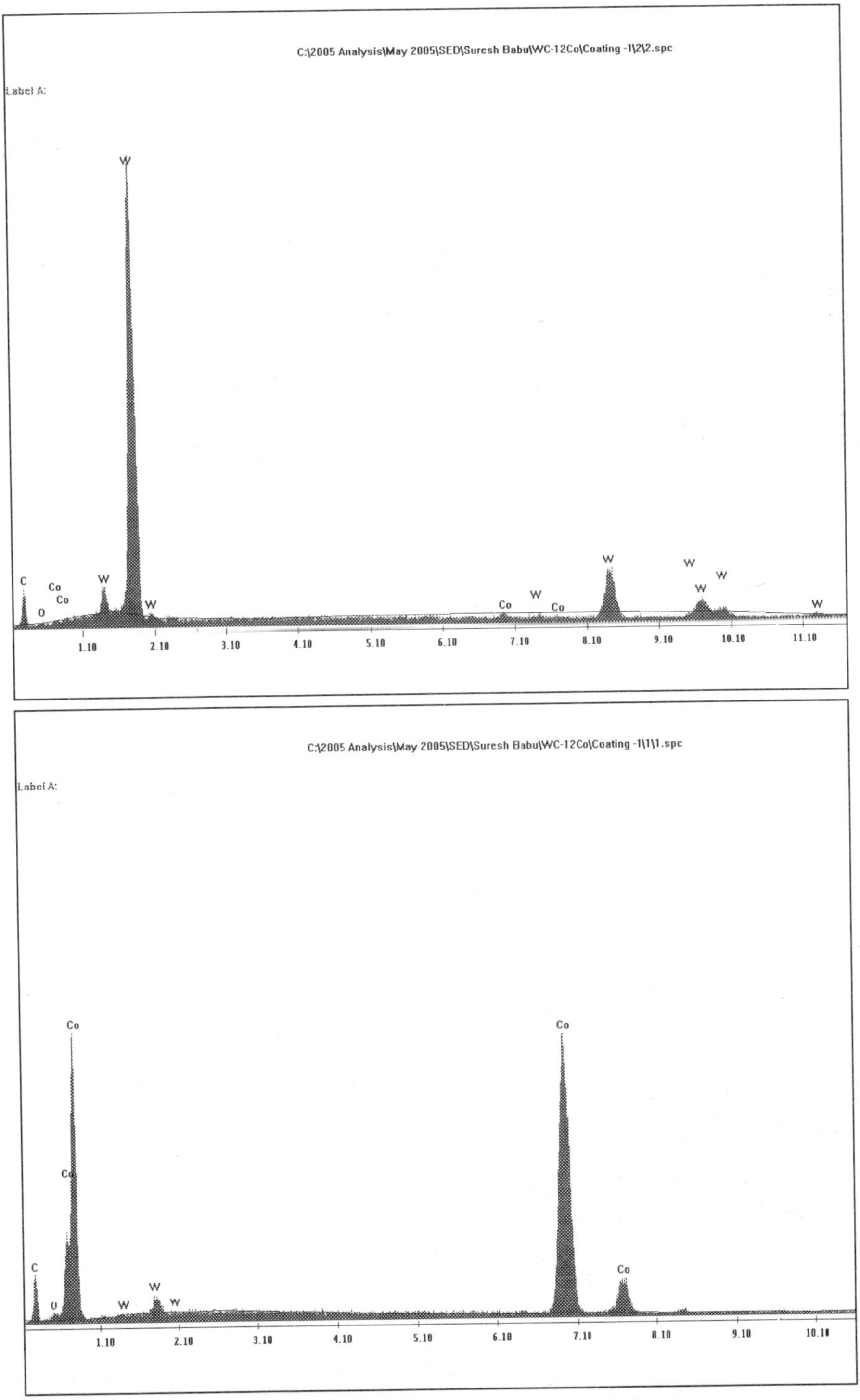

Figure 15.6(c) and (d) [Top and Bottom images, respectively], Showing the EDS Data at the Spots Indicated in Figures 15.6(a) and (b) Respectively

and automatically indexed, and the orientation at that point is determined. When the beam moves from one grain to the next, the change in orientation is detected and in this way, the entire area of interest on the sample surface is mapped to determine the extent of each grain.

Figure 15.7 shows the schematic of the EBSD measurement and Figures 15.8(a) and (b) show the experimentally observed pattern and the indexing, respectively. It is to be kept in mind that although crystallographic inputs are essential to index the patterns, EBSD is basically a technique to study microstructure.

Sample preparation is crucial to obtain EBSD patterns. Since the information depth of the technique is of the order of tens of nanometres, the sample surface should be clean and free from surface strains. Initially, the sample is sectioned and mounted in a bakelite or resin mount using a mounting press. The surface is then polished using polishing papers of increasing grit. In the next stage, the sample is polished on rotating wheels over which cloths are stretched and charged with a fine abrasive of progressively decreasing particle size with an appropriate liquid. The final polishing is carried out using colloidal silica or alumina with particle size of 0.05 µm or less. Colloidal alumina has a pH of about 4 and provides aggressive polishing and etching while the pH of colloidal silica is around 8 and gives a gentle etch which is suitable to enhance the EBSD pattern contrast. As mentioned above, the sample is inclined at a steep angle of 70º, due to which too much etching of the sample surface causes the pattern intensity to go down.

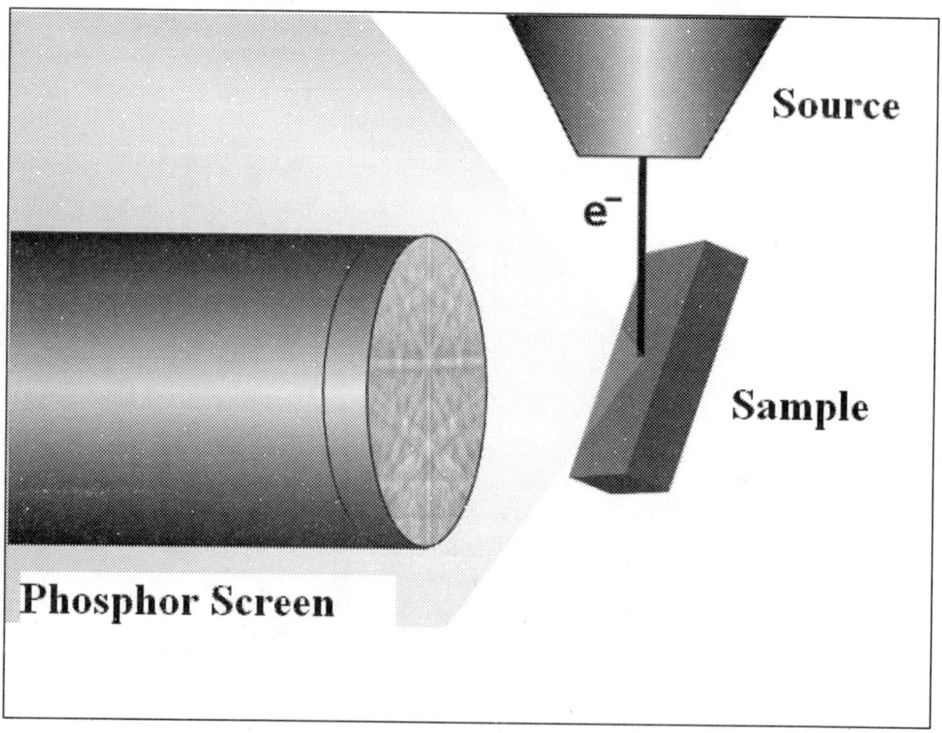

Figure 15.7: Schematic of the EBSD Measurement

Figure 15.8(a): EBSD Pattern as Recorded on the Phosphor Screen

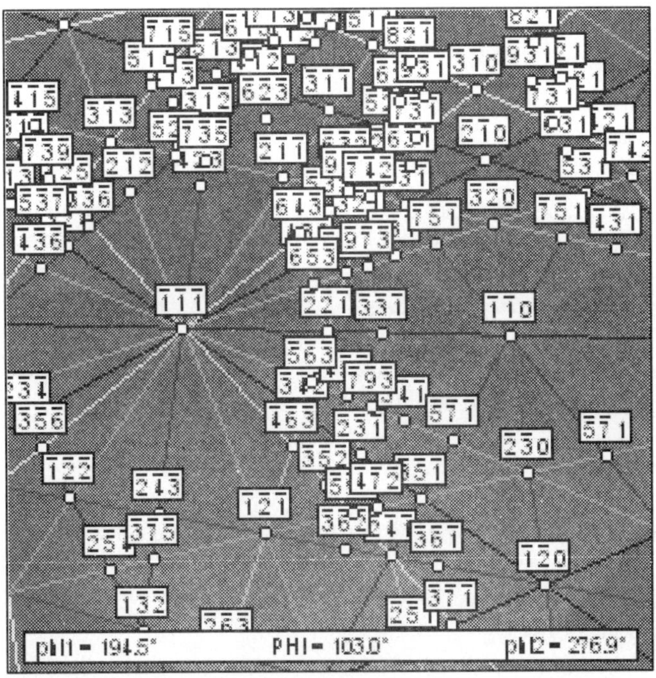

Figure 15.8(b): Indexing of the EBSD Pattern Shown in Figure 15.8(a)

We have carried out EBSD measurements on bulk ceramics such as $BaTiO_3$ and ZnO, and on several coating systems such as WC-12 per cent Co prepared by the DSC technique and Al_2O_3 coatings by MAO. Figure 15.9(a) shows the crystal orientation map (COM) of a ZnO varistor sample showing the data from ZnO while Figure 15.9(b) shows the COM of the pyrochlore phase. The colour-coded inverse pole figure for each phase is shown below the respective COM. The different colours show the crystallographic orientations of the grains. It is seen from these figures that EBSD can be used to study multiphasic samples.

Information concerning the grain size can also be extracted from EBSD data. Figure 15.10 shows the grain size distribution in the ZnO phase.

Using the SEM facility at ARCI chemical information from the EDS unit and microstructural information from EBSD can be collected simultaneously. This is very advantageous for studying small amounts of phases that may be present in the sample. With the sample in position for the EBSD measurement (*i.e.*, tilted to 70°), an EDS spectrum can be collected. After identifying the different elements present, a search can be performed using the ICDD database to identify the possible phases with the constituent elements. The EBSD patterns can then be indexed using the crystallographic information and the right phases can be detected. A word of caution, though, that this procedure may not be trivial, and usually involves a lot of instrument time. The data acquisition time increases with increase in the number of phases. EDS information can be collected simultaneously along with the EBSD data, and can be used to further the analysis.

Techniques for Structural Characterization

X-ray diffraction (XRD)

Of the techniques to determine the crystal structure of a material, x-ray diffraction is by far the most common. XRD units have come to be the most popular structural characterization tools in laboratories around the world.

X-rays are generated when a stream of electrons is accelerated (typically to 30 keV) and are made to strike a metal target such as Cu. The incident electrons are sufficiently energetic to ionise some of the Cu 1s electrons. An electron in an outer orbital (2p or 3p) drops down to take the place of the vacant 1s level and the energy released in the transition appears as x-rays with wavelength of the order of 0.1 nm. When the x-rays so generated are incident on the sample under study, diffraction takes place by the crystal planes according to the Bragg law

$$2\delta \sin\theta = n\lambda$$

where,

δ is the spacing between neighbouring crystal planes, θ is the angle of incidence of x-rays on the sample and λ is the wavelength of the radiation. The diffracted pattern thus obtained is used to identify the crystalline phase(s) in the sample and the structural properties.

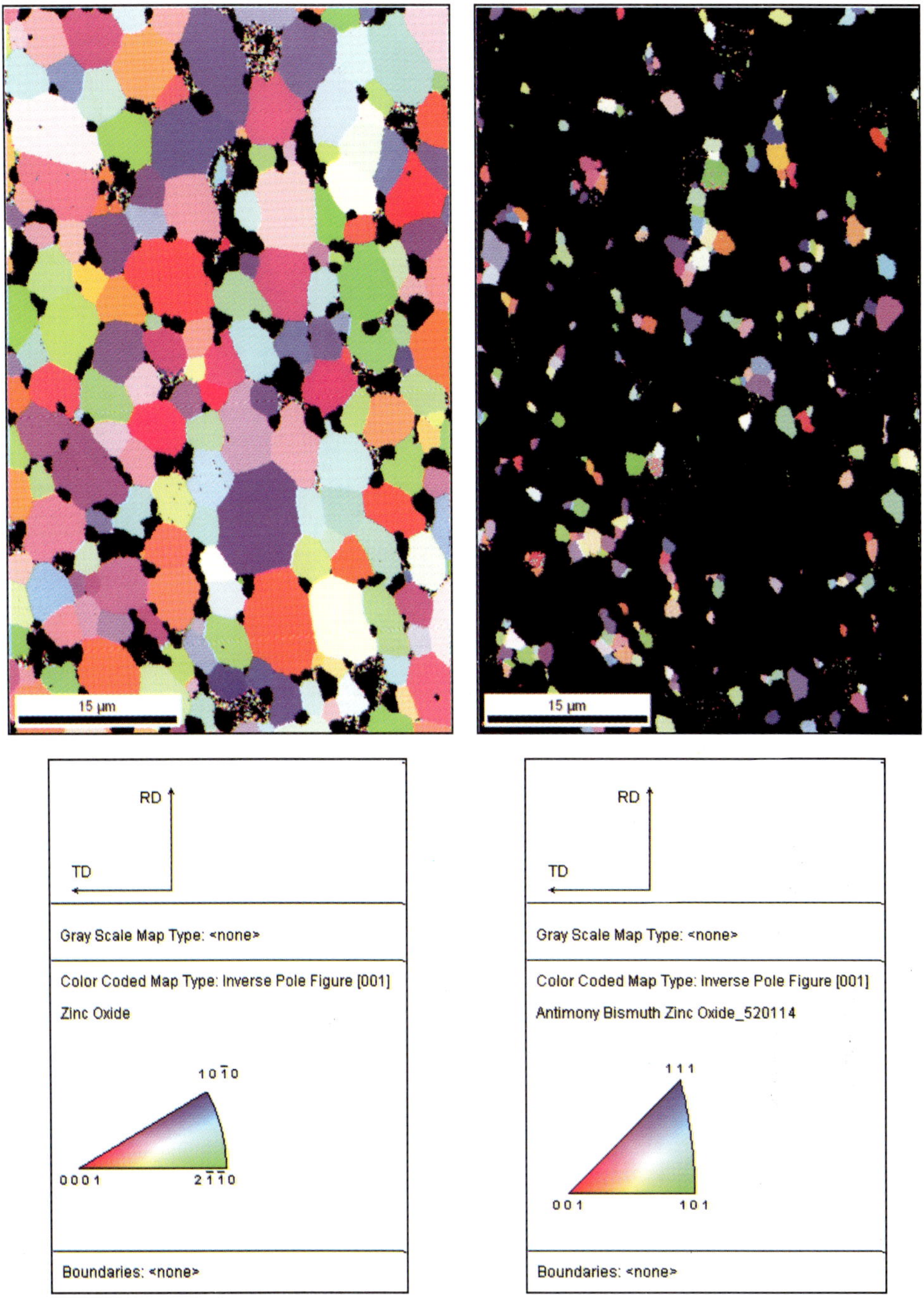

Zinc oxide **Pyrochlore**

Figures 15.9(a): Left, and (b): Right, Showing the Crystal Orientation Maps for a ZnO Sample Showing the Data from ZnO and Pyrochlore Regions Respectively. The colour-coded inverse pole figure for each phase is shown below the respective maps.

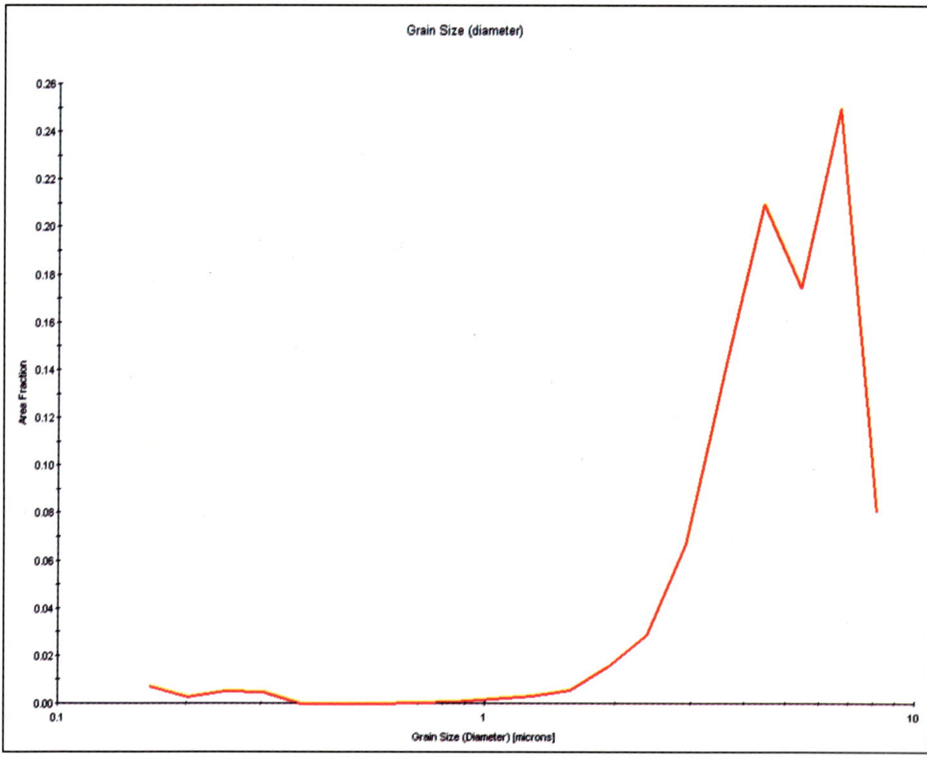

Figure 15.10: Grain Size Distribution of the ZnO Grains Shown in Figure 15.9(b). The x-axis represents the grain size (diameter, micron) and the y-axis shows the area fraction.

The XRD facility at ARCI is shown in Figure 15.11. In an experiment, the sample is rotated by an angle θ and the detector by 2θ. When there is constructive interference between adjacent x-ray beams, a diffracted beam occurs in that direction, as shown in Figure 15.12. It is also common to keep the sample fixed and move both the source and the detector by θ. Modern units facilitate the very rapid collection of XRD peaks in a few seconds by using a curved detector that can simultaneously collect diffracted data at various angles. An XRD pattern recorded on a WC-12 per cent Co coating is shown in Figure 15.13.

Extended X-ray Absorption Fine Structure (EXAFS)

When a material absorbs x-rays, both inner shell and outer shell electrons can be excited. At an electron binding energy characteristic of an element, the x-rays are absorbed and a steeply rising absorption edge is observed. When the energy of the incident beam is greater than the binding energy, the core electron is excited to the continuum. The resulting photoelectron wave propagates from the x-ray absorbing atom and is scattered by neighbouring atoms. The EXAFS spectrum arises from the constructive and destructive interference between the outgoing and incoming

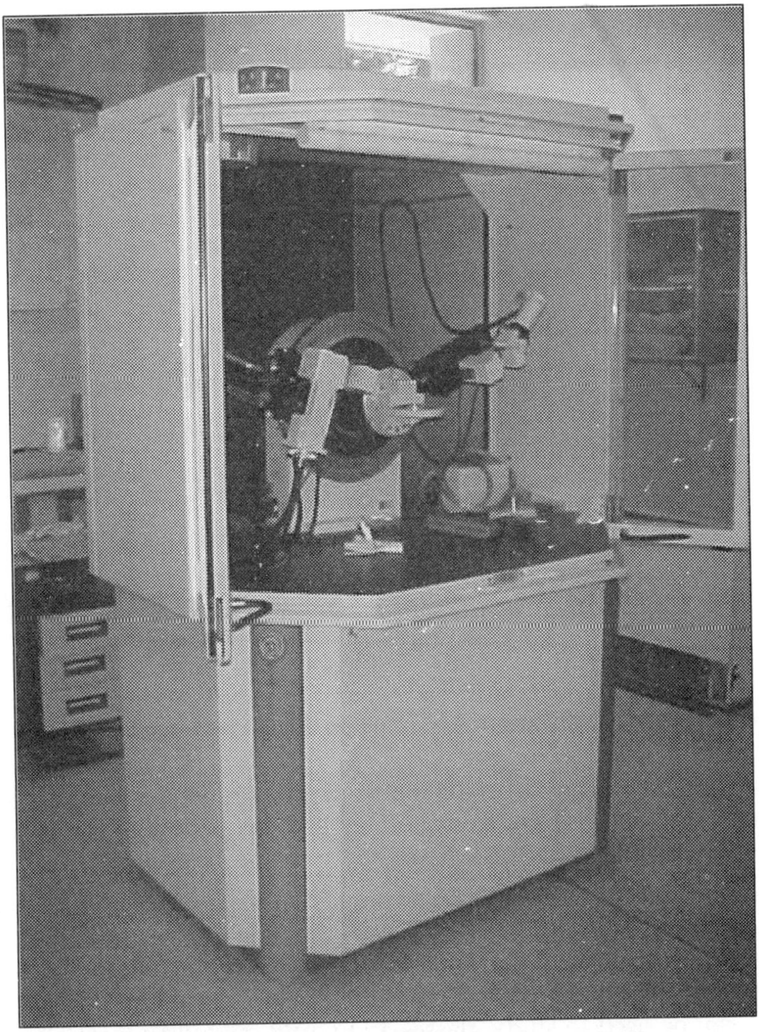

Figure 15.11: The X-Ray Diffractometer at ARCI

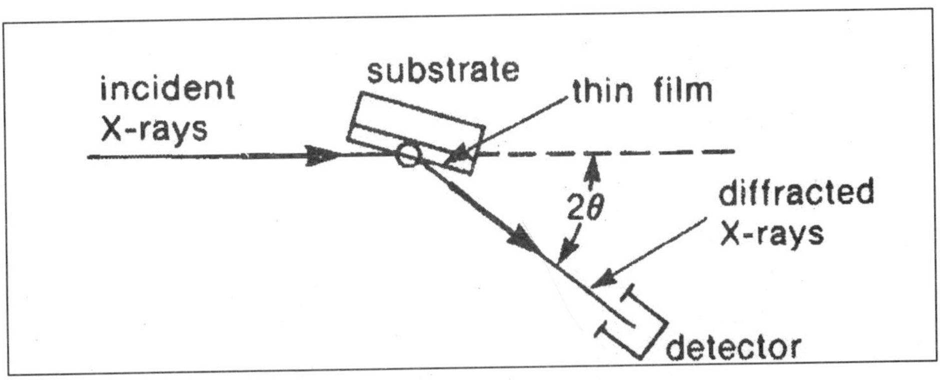

Figure 15.12: Schematic of the XRD Measurement

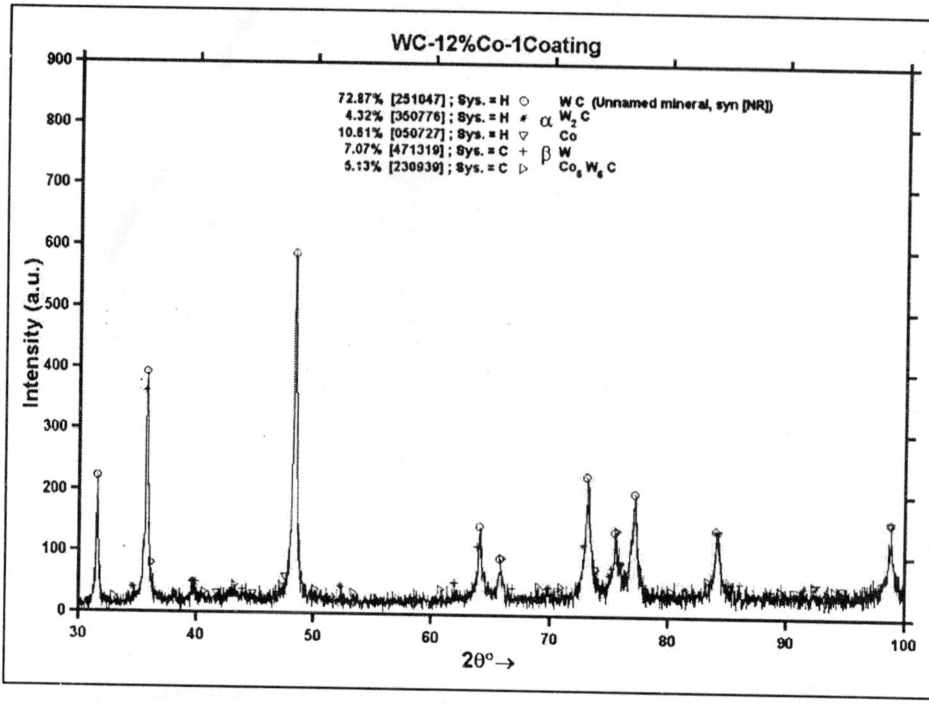

Figure 15.13: XRD Pattern Recorded on a WC-12 per cent Co Coating

photoelectron waves. The part of the spectrum prior to and up to the edge as shown in Figure 15.14 is called NEXAFS or XANES. EXAFS as modified to study surfaces is known as the Surface-EXAFS or SEXAFS.

The information that can be obtained from EXAFS includes interatomic distances, numbers of neighbouring atoms (*i.e.*, the coordination number) and the identity of atoms in the vicinity of the x-ray absorbing atoms. An important difference between XRD and EXAFS is that the former is applicable only to crystalline material whereas the latter can be used to study glasses and partly crystalline materials. Another difference is that XRD provides structural information over larger length scales whereas EXAFS provides local structural information. The two techniques thus complement each other to provide information about long and short range ordering.

EXAFS data are normally recorded on a thin foil of the material in the transmission mode but for coatings, detection by X-ray fluorescence (XRF, to be discussed later), or by electron or ion yields is more appropriate. The analysis of EXAFS data is, however, not trivial and specialized software is needed.

Techniques for Chemical Characterization

Of the several techniques listed in Table 15.2 to obtain chemical information, EDS and EXAFS have already been discussed. Details concerning EPMA and XRF are presented below.

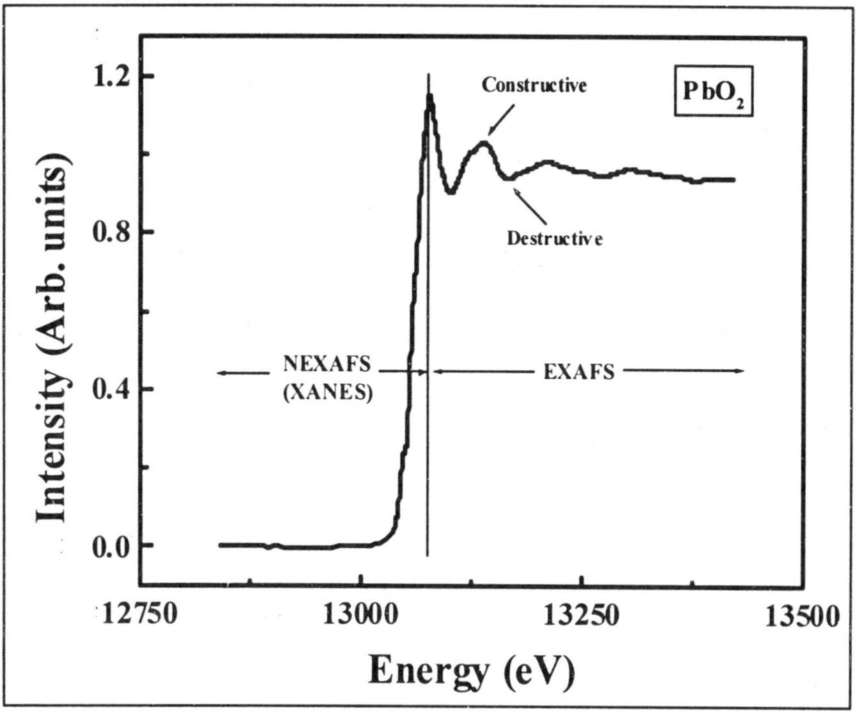

Figure 15.14. X-Ray Absorption Spectrum of PbO_2, Illustrating the Extent of the Techniques EXAFS and NEXAFS

Electron X-ray Microprobe Analysis (EPMA)

EPMA is a technique for quantitative elemental analysis that is based on the generation of characteristic x-rays when the sample is exposed to a focused beam of electrons.

In the discussion on the working of a SEM, some of the interactions that can occur between the source electron beam and the specimen were presented. When the incoming beam transfers sufficient energy to an atom in the sample, a bound atomic electron is ejected, leading to inner shell ionisation. To fill the atomic vacancy, a transition occurs involving one or more outer shell electrons. The energy difference between the shell levels gives rise to an x-ray photon. Since the atomic shells have precisely defined energies, the energy of the emitted x-ray photon is of the same energy and can be used as an accurate detector of the element involved in the transition.

The detection of the x-ray photon is performed using energy dispersive or wavelength dispersive modes and an EPMA unit is equipped with detectors of both types. The principal advantage of EPMA is its ability to perform quantitative analyses and detect the concentration of the constituent element in the sample in comparison with that of a standard.

X-ray fluorescence (XRF)

In the case of EPMA, a core electron is ejected by the interaction of the source electron beam with the sample. In XRF, x-rays are used as the source, and when the x-ray photon has energy greater than the binding energy of the inner shell electron, one of the electrons from another shell fills up the vacancy, giving rise to an x-ray photon. The emitted photon–or fluorescent radiation–has the characteristic energy of the difference between the inner and outer shell electrons, and is an indicator of the element that underwent the transition. The principal difference between electrons and x-rays is that the penetration depth of the latter is much higher than that of the electron beam due to its uncharged nature.

The technique of XRF is governed by the same Bragg's law as that of XRD, *viz.*

$$2\delta \sin \theta = n\lambda$$

For detection, the characteristic x-rays go through a single crystal detector such as LiF (200) or (220), the 'δ' spacing of which is fixed. Thus, in the case of XRD, λ is kept fixed whereas it varies in XRF based on the element being detected, while the 'δ' spacing varies in XRD whereas it is fixed in XRF. Depending on the wavelength of the characteristic x-ray photons, diffraction occurs at different values of θ. The detector is rotated as in the case of XRD and the characteristic x-rays are detected. By calibrating the intensity of the diffracted signal against a standard, quantitative analyses can be performed. Figure 15.15 shows the XRF spectrum of an Au coating on a Ni/Fe substrate.

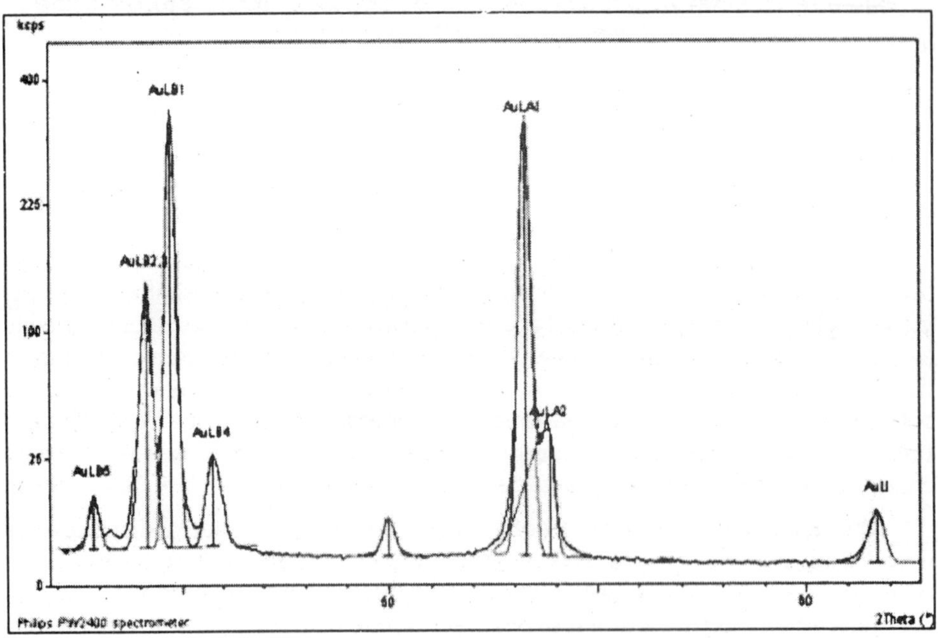

Figure 15.15: XRF Spectrum of a Gold Coating on Ni/Fe Substrate

Techniques for Mechanical Characterization

Coatings are quite often used in applications where there is significant wear and tear and characterization of their mechanical properties is hence very important. Hardness measurements are easy to perform and are routinely used as the first level of mechanical characterization.

Microhardness Testing

Microhardness measurements are used to determine load variations in mechanical properties of coatings. They are also used for profile measurements in the case of diffusion and functionally graded coatings. The measurement consists of indenting the material by applying a known load on an indenter and then imaging the resultant indentation. One of the more common indenters for the purpose is the Vicker's indenter, which has the tetrahedral pyramidal shape with an angle of 136° between opposite faces. The load is typically applied for 10-15 seconds and after the removal of the load, a square impression is formed. The length of each diagonal is measured with the aid of an optical microscope and the hardness is measured using the formula

$$HV = 1.854\ P/L^2$$

where,

HV is the (Vickers) hardness of the material, P is the applied load and L is the average length of the two diagonals. In general, hardness is a function of the applied load and it is customary to mention the load that was used for the measurement. For instance, HV_{500} indicates that a load of 500g was used. Figure 15.16 shows the Vickers impression on a Ni-20 per cent Cr coating. The material used for the indenter is diamond, since during the indentation process there should be no deformation in the indenter itself.

Nanoindentation

In the Vickers method, typical loads are in the range 1-2000g. However, for coatings that are thin (of the order of a few microns), smaller loads needs to be applied so that the indent remains in the material and does not enter the substrate. Indeed, it is important to avoid substrate effects in measurements on coatings. For very small loads, the impression is very small and it may not be easy to image the same. This situation is overcome by the use of the Instrumented Indentation Technique. Basically, a load is applied and the displacement of the indenter into the sample is continuously monitored by a capacitance gauge. The hardness and modulus values are calculated from the resultant load vs. displacement curves. This method thus eliminates the need to image the indent and facilitates the measurement of the properties in the submicron scale.

Figure 15.17 shows the schematic of the working of a Nanoindenter and Figure 15.18 shows the load versus indenter displacement data for an indentation experiment. Here, P_{max} is the peak indentation load, h_{max} is the indenter displacement at peak load, h_f is the final depth of contact impression after unloading and S, the initial unloading stiffness.

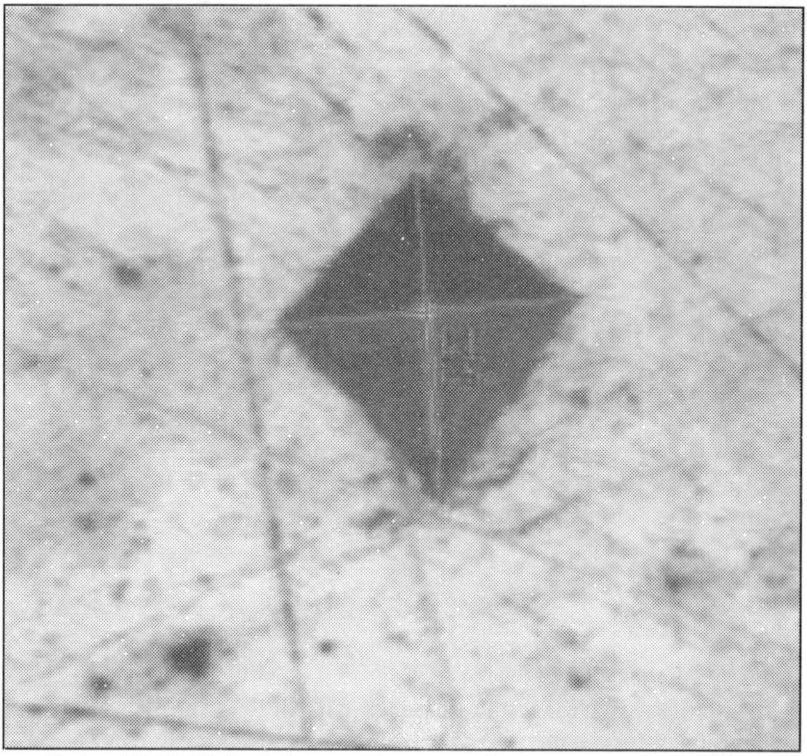

Figure 15.16: Vicker's Impression on a Ni-20 per cent Cr Coating. The diagonals are indicated.

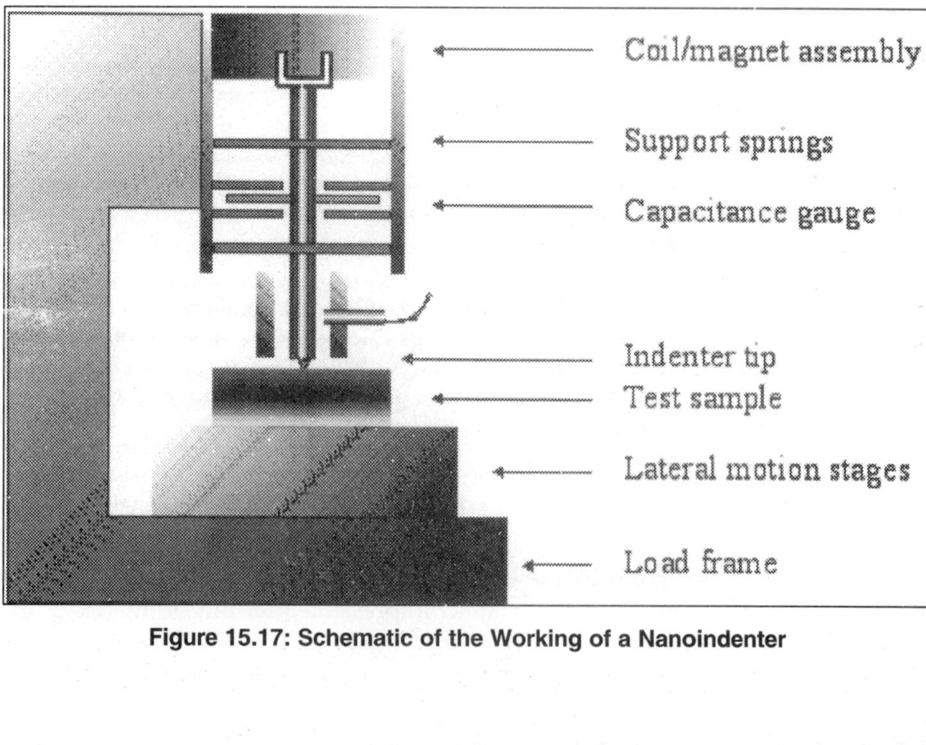

Figure 15.17: Schematic of the Working of a Nanoindenter

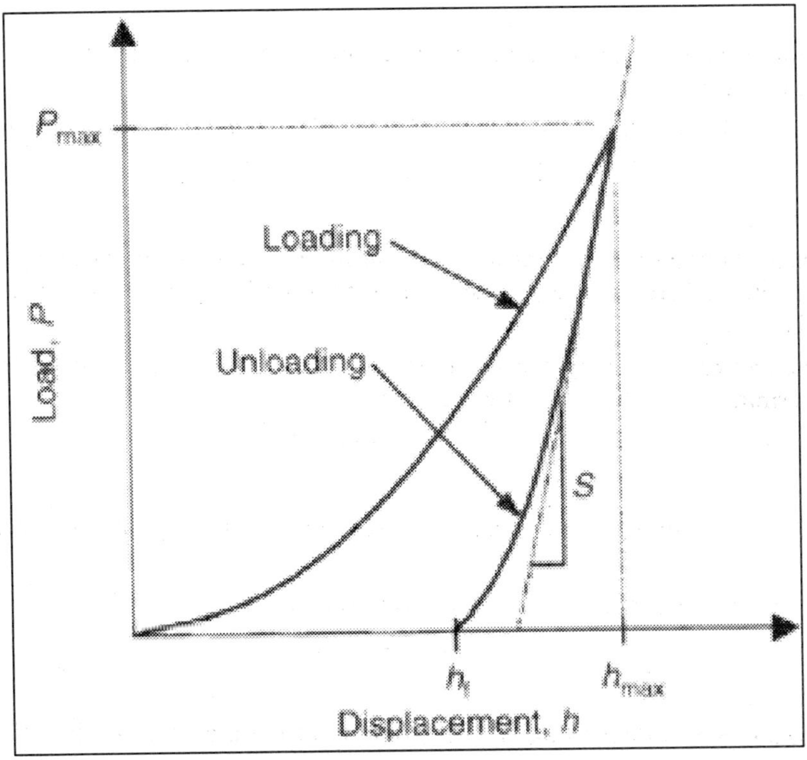

Figure 15.18: Typical Load-Displacement Curve

Both elastic and plastic deformation cause the formation of a hardness impression conforming to the shape of the indenter as the indenter is driven into the material to some contact depth. As the indenter is withdrawn, the elastic portion of the displacement is recovered. This recovery allows the determination of the elastic properties of the material.

For performing the indentations, a Berkovich tip is used. This is a three sided pyramidal tip made of diamond. The total included angle on this tip is 142.3 degrees, with a half-angle of 65.35 degrees. This tip is used primarily for bulk materials and thin films greater than 100 nm thick. In practical indentation tests, a Berkovich tip with a radius of about 100 nm is used.

From the load-displacement curve hardness can be obtained at the peak load as

$$H = P/A$$

where,

P is the applied load and A is the projected contact area.

The elastic modulus E is determined from the reduced modulus E_r, given by

$$E_r = (\sqrt{\pi}.\, S)/(2\beta.\, \sqrt{A})$$

where,

β is a constant that depends only on the geometry of the indenter, and is 1.034 for the Berkovich tip.

The elastic modulus of the test material, E, is determined using the expression

$$E_r^{-1} = [(1-\upsilon^2)/E] + [(1-\upsilon_i^2)/E_i]$$

where,

υ is the Poisson's ratio of the test material and E_i and υ_i are the elastic modulus and Poisson's ratio respectively of the indenter material. For diamond, E_i is 1141 GPa and υ_i is 0.07.

From the contact theory developed by Oliver and Pharr, the unloading stiffness S is determined at $h = h_{max}$ using the formula

$$S = B.m.(h-h_f)^{m-1}/h$$

where,

B and m are empirically determined constants and h_f and h_{max} have been defined earlier.

The contact depth is estimated using the relation

$$h_c = h - \varepsilon.P/S$$

where,

ε is a constant depending on the geometry of the indenter and is 0.75 for a Berkovich indenter.

The contact depth is the depth over which the test material makes contact with the indenter. In general, it is different from the total depth.

Finally, the projected contact area is calculated by evaluating an empirically determined area function at the contact depth h_c.

Figure 15.19 shows the load vs. displacement data for a thin stainless steel film approximately 100 μm thick. As can be seen from this figure, the hardness has been calculated only at the maximum penetration depth. To obtain the hardness as a function of depth of the specimen, the Continuous Stiffness Measurement (CSM) is used. In the CSM option, a small oscillation is superposed on the primary loading signal and the resultant response of the system is analyzed. CSM requires the ability to accurately model the dynamic response of the system so as to allow the isolation of the material response. Figure 15.20 shows the schematic of the CSM and Figure 15.21 shows the hardness as a function of depth for a SS film. The CSM option is especially useful to study the properties of coatings on substrates where a change in hardness is expected as a function of displacement into the surface. It is also useful in the case of polymers where both the storage and loss modulus are important.

Techniques for Surface Characterization

Surface characterization is basically to study the surface roughness of the material. The technique is known as Surface Probe Microscopy (SPM), comprising, among

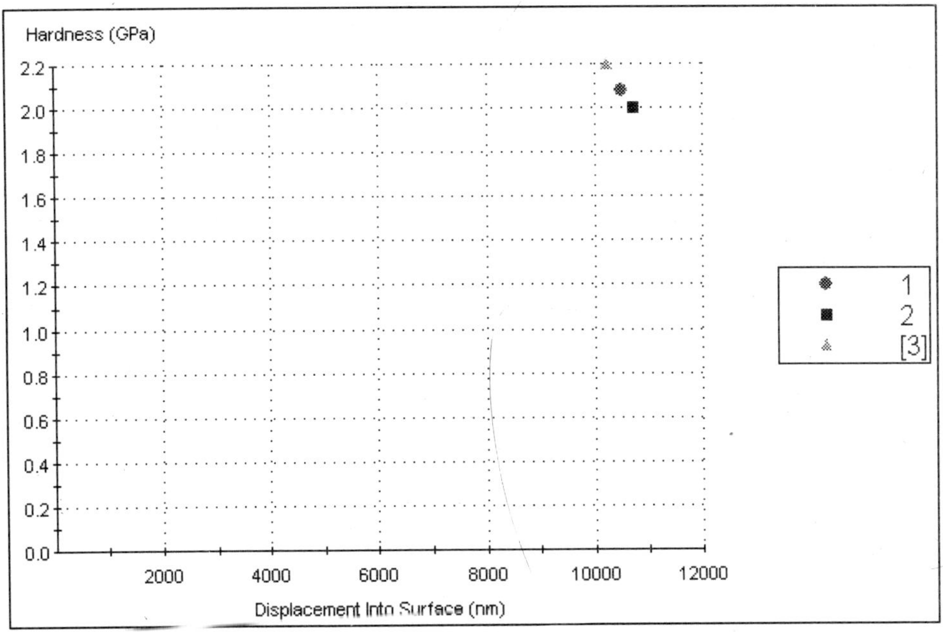

Figure 15.19: Load Vs. Displacement Data for a Thin Stainless Steel Strip

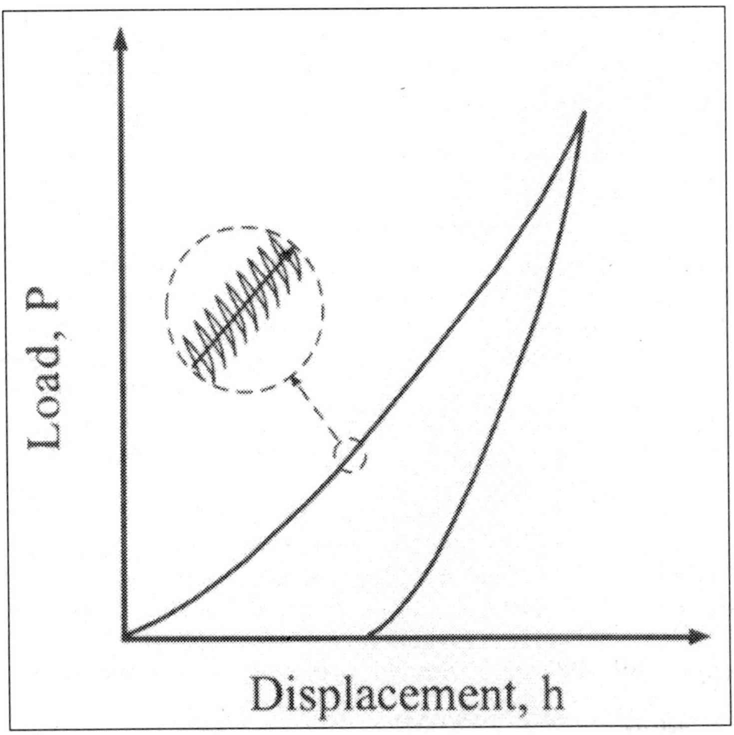

Figure 15.20: Schematic of the Working of the CSM

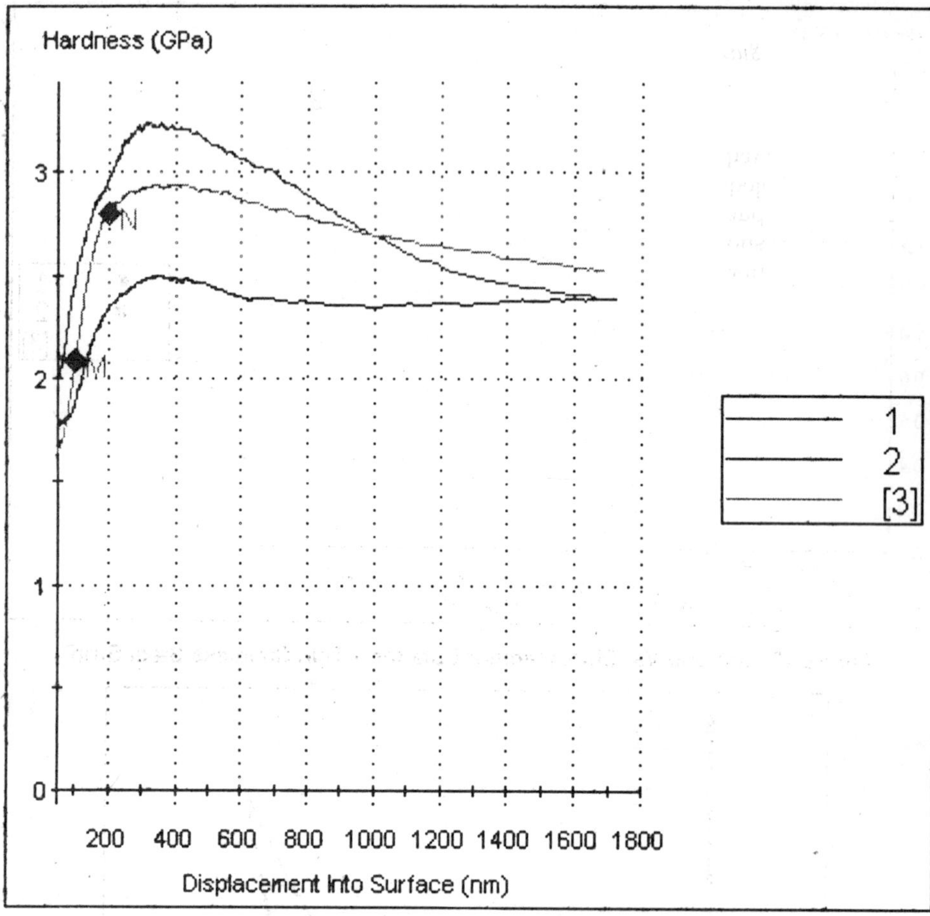

Figure 15.21: Variation of Hardness as a Function of Depth for the SS Film. The data were collected using the CSM option.

others, of scanning profilometry, atomic force microscopy and scanning tunneling microscopy. Surface roughness is the two-dimensional measurement of the variation in height of a coating, whereas topography is the three-dimensional measurement of the height variation. We provide a brief discussion of the techniques here.

Scanning Profilometry

In a scanning profilometer, a sharp diamond stylus with a tip radius of 2 μm or larger is kept in contact with the sample surface by the application of a small load. The surface is scanned in 2D and the vertical position of the stylus is recorded digitally. Figure 15.22 shows the surface roughness of coatings prepared by various techniques and Figure 15.23 shows the surface profile of a coating obtained using the profilometer.

Atomic Force Microscopy (AFM)

The AFM works on the same principle as the scanning profilometer, but with a greater resolution that comes about from the much sharper tip that is used (tip radius

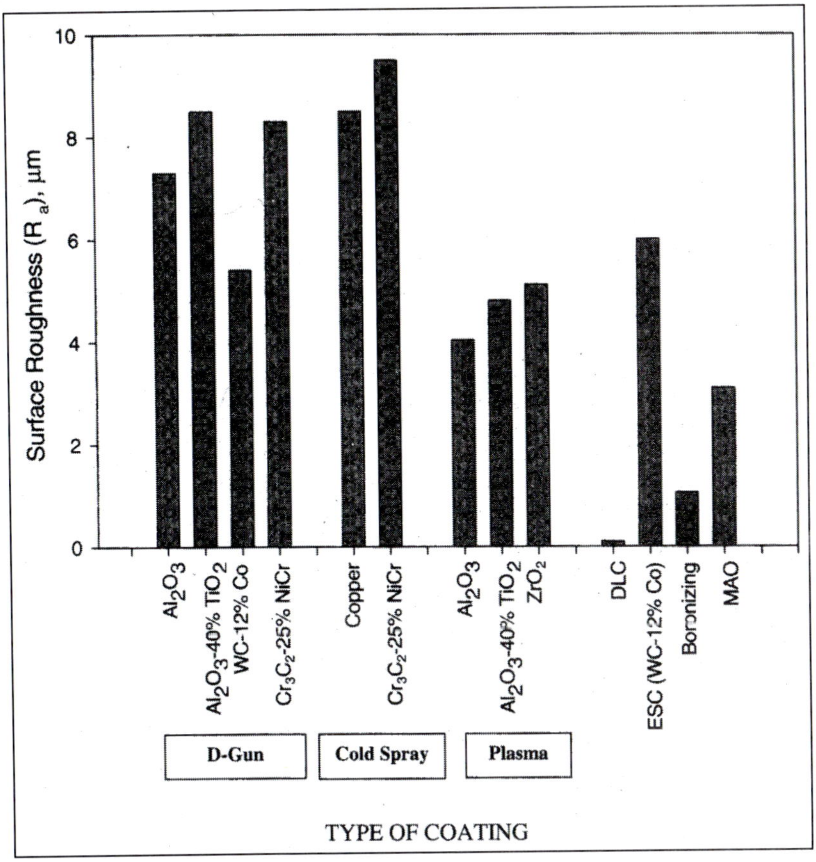

Figure 15.22: Surface Roughness of Different Coating Systems

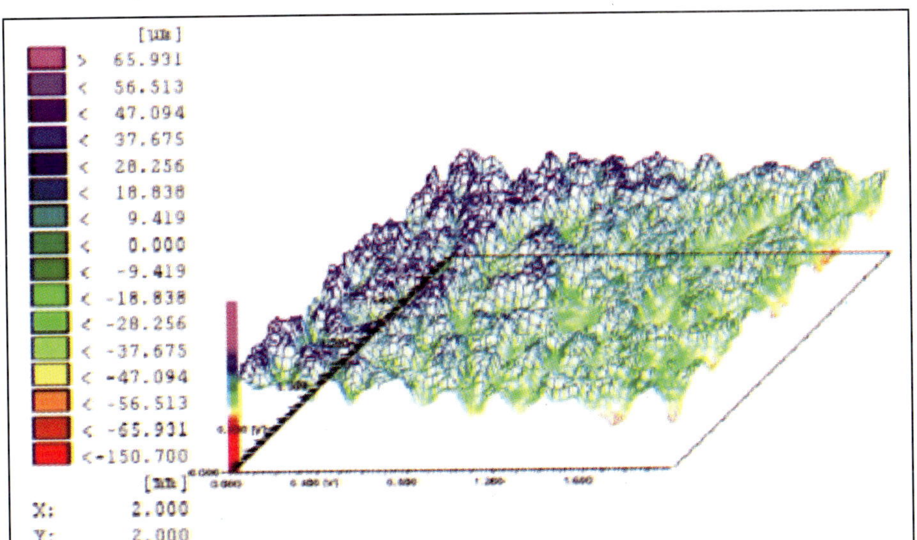

Figure 15.23: 3D Rendering of the Surface Roughness of a Coating

of 50 nm or less). The tip is dragged across the sample surface with a small force, ensuring that the tip remains in constant contact with the surface. The change in the vertical position directly measures the topography of the surface. Figure 15.24 shows the schematic of the working of an AFM.

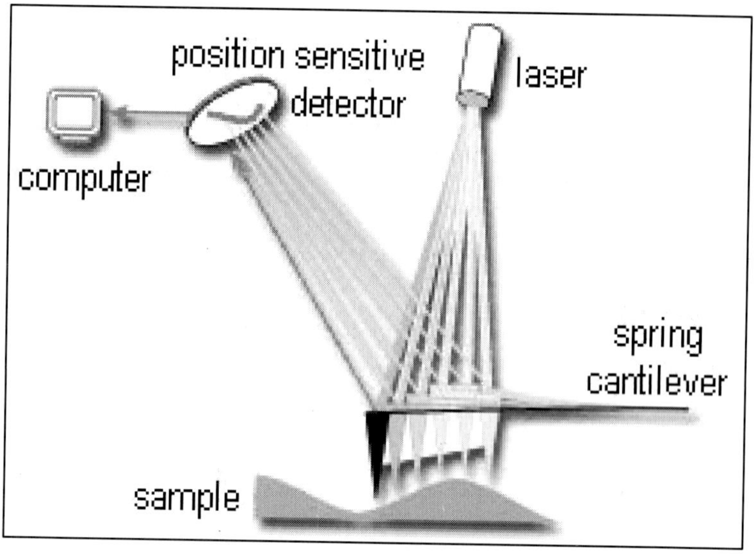

Figure 15.24: Schematic of the Working of an AFM

The tip mentioned above is actually at the end of a flexible cantilever. The top position of the cantilever is coated with gold, making the surface highly reflective. A laser beam is shone on to the top surface of the cantilever and the reflection is detected by a two-segment photodetector. The difference between the signals of the two photodiodes indicates the angular deflection of the cantilever. Since the distance between the cantilever and the detector is a few orders of magnitude greater than the length of the cantilever, the optical lever greatly magnifies motions of the tip and thus provides sub-nanometer resolution. In addition, a feedback loop helps not only in measuring the forces on the sample by the tip, but also regulates it, allowing images to be acquired at very low force levels. Thus, the AFM is the instrument of choice for truly atomic resolution in surface imaging.

Scanning Tunneling Microscopy (STM)

The STM works on the quantum mechanical principle of tunnelling, wherein a small electric current flows across two metals that are brought quite close to but not exactly touching each other. The current increases with the decreasing distance between the metals and is maximum on contact. The sample is one of the metals mentioned above, while the microscope tip is the other. The sample is quite flat and spread out as compared to the tip. The sharp tip is likely to likely to have only one atom at the end and all the tunnelling current passes through the single atom, providing atomic resolution. However, the disadvantage is that the sample surface needs to be electrically conducting, which is not the case with most coatings. Thus, the AFM seems a better choice for the surface imaging of coatings.

The Nanoindenter as a Surface Profilometer

As mentioned in the discussion on the Nanoindenter, the displacement into the surface is measured using a capacitance gauge (Figure 15.17). By applying a small force of the order of a few micro Newton on the tip, thus keeping it in contact with the sample surface and scanning the surface in 2D, the Nanoindenter can also be used as a surface profilometer. However, the method of measuring deflections of the tip in an AFM is much more sensitive than the capacitive method used in the Nanoindenter and hence the resolution in this case is of the order of only a few nanometer. This is however not a serious problem while working with coatings, since in most cases the surface roughness approaches a micrometer or more. Thus, we have been able to use the Nanoindenter tip for surface profiling. Figures 15.25–15.27 show some of the scans obtained.

As can be seen from these figures, the variation in height is indicated by the variation in the colour. This method, however, is not quantitative and it is desirable to have suitable software for quantitative analysis of the indent. Measurement of the depth of the indent from the 3D plot would make it possible to compare the depth obtained from the load vs. displacement curves, although the latter are expected to be much more reliable. However, the 3D scans are certainly of value to determine the blunting of the tip due to regular use. As can be seen from Figure 15.22, considerable surface roughness is present in coatings. Due to this, it is only to be expected that the indenter tip undergoes non-uniform wear. Tip calibration procedures are regularly

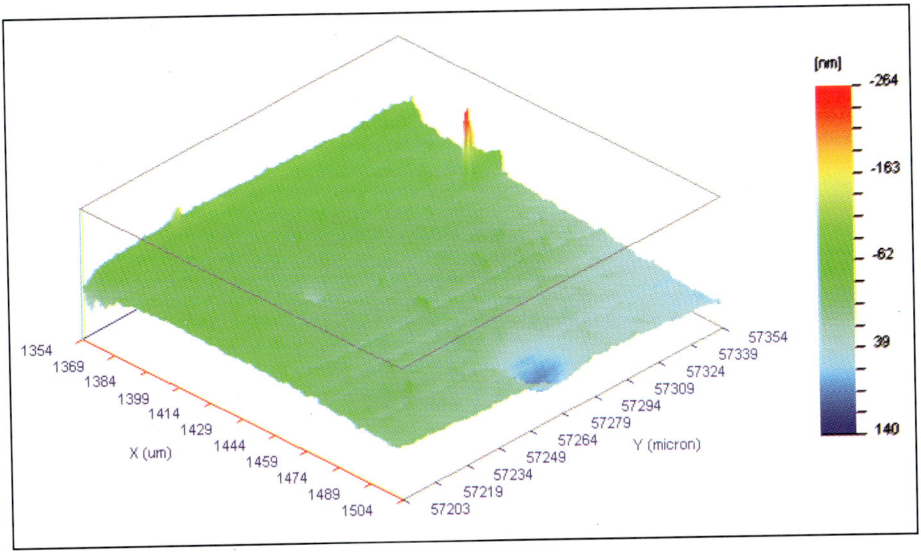

Figure 15.25: Profile Obtained by the Nanoindenter on a High Speed Steel Substrate. The dip in the front was obtained using a spherical indenter under a load of 500g. The variation in height is indicated by the color legend on the right.

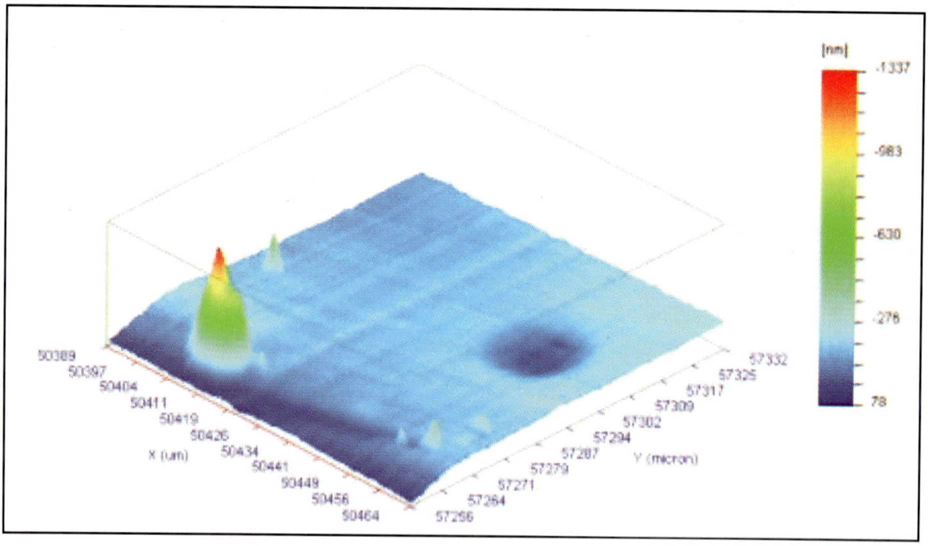

Figure 15.26: Profile Obtained by the Nanoindenter on a DLC Coating on an Aluminum Substrate. The dip is due to spherical indentation. Fine ridges are seen on the surface.

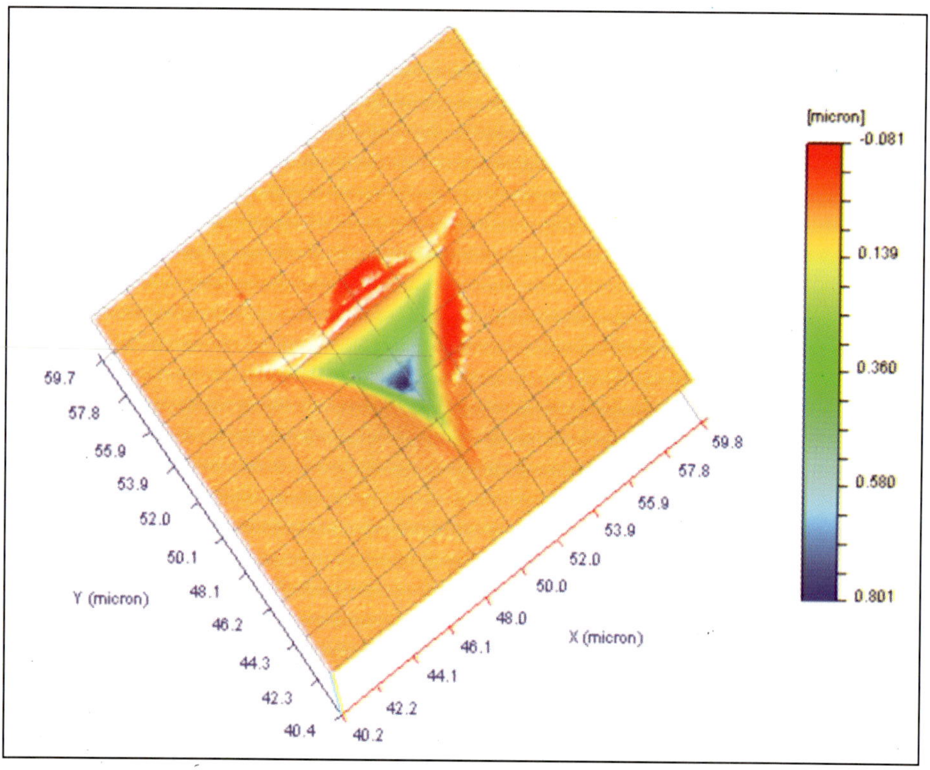

Figure 15.27: Profile of a Berkovich Indent. The perspective is as looked into the indent.

needed to estimate the changes in the tip geometry and the necessary corrections are applied to the hardness and modulus measurements.

Recently, special software has been developed for the analysis of Nanoindentations using an AFM. The load vs. displacement data and the 3D profile data are used as inputs by the software to calculate the elastic properties of the sample. As the indenter is driven into the material, both elastic and plastic deformations occur but when the material is withdrawn, there is elastic recovery in the material only the plastic deformation remains. The software has the ability to use the data to reconstruct the fully loaded indentation, and this can be used to study the response of different materials.

Conclusions

Materials characterization is arguably a vast field of study with continuous developments occurring at a steady pace. This article attempts to introduce some of the techniques that can be used for characterization of coatings. The techniques have been classified into different sections, namely microstructural, structural, chemical, mechanical and surface characterization. Due to the vast nature of the subject and the limited scope of this article, several other techniques could not be included. Especially missing are the methods of measuring the interface toughness and bond strength of coatings. Details about these, and indeed about all the techniques presented in this article, are available in the literature. It is nevertheless hoped that this article serves as a preliminary introduction, and as a pointer to the possible techniques that can be used to address the characterization issue on hand.

Acknowledgements

The author thanks his colleagues Mr. P. Suresh Babu for data on the WC-12 per cent CO coatings, Mr. A. Venkatesh for the SEM and EDS images, Mr. K. R. C. Somaraju for the surface profile of a coating and Dr. Radha K. for the XRF spectrum. The help of Ms. R. Meenakshi and Ms. A. Parvati in the preparation of the manuscript is also sincerely acknowledged.

Chapter 16

Evaluation of Tribological Coatings

B. Venkataraman
Defence Metallurgical Research Laboratory,
Kanchanbagh P.O., Hyderabad – 500 058, India

ABSTRACT

Tribological modes are classified based on the manner with which the contacting surfaces interact. The present article highlights various tribological modes that are normally encountered in engineering applications and the material characteristics that influence the wear behaviour for each of the tribological mode. Various characterizations including evaluation of tribological behaviour are also described.

Keywords: Tribological modes, Tribological coatings, Surface engineering, Surface coatings, Tribological behaviour, Surface wear.

Introduction

Tribology is the field of science and technology dealing with contacting surfaces in relative motion, which means that it deals with phenomena related to wear, and friction. Wear is defined as the removal of material from solid surfaces as a result of one contacting surface moving against another. Wear can be defined as the progressive loss of material from the operative surface of the body, occurring as a result of relative motion of the surface with respect to another body. The components that are subjected to frictional contact are prone to wear in some way or other. There is a growing need to control friction and wear for extended life of the components and to increase efficiency of the system. This is to some extent achieved by design changes, by lubrication or by selecting improved bulk materials. There is another approach to control friction and wear- that is by employing suitable surface modifying technique.

This approach is increasingly becoming popular and this has led to a growth of new discipline called surface engineering.

A single surface engineering technique cannot be utilized for all tribological problems because the tribological behaviour of the surface is governed by numerous factors and it is not unique to any particular material or coating and largely controlled by the tribo system. The tribo system is primarily characterized by the manner with which the interaction takes place between the mating surfaces For example, the interaction can be sliding or impacting against each other. The mechanism of wear is classified based on type of interaction. For standard terminology related to wear Annual Book of Standards (1987).

Wear Classifications

Sliding Wear

The loss of material associated with sliding of two bodies is termed as sliding wear. If the sliding takes place in presence of lubrication, then it is termed as lubricated sliding, otherwise it is termed as unlubricated sliding.

The mechanism of wear under sliding conditions is strongly influenced by operating variables such as normal stress, sliding speed and environmental condition. For example at very low speed and load, adhesion effects dominate. With moderate increase in load or speed, the bulk material deformation below the worn surface also contributes to the wear process. It is often observed that the mechanical property and microstructure of the worn surface is distinctly different from either of contacting material, as the wear surface is a result of mechanical mixing of contact asperities. Environmental effects temperature, humidity also contributes in the modification of structure and property of the wear surface. The layer (called mechanically mixed layer) formed on the surface is found to control the wear behaviour rather than bulk mechanical properties of contacting materials.

A wear-mechanism map, or a wear map proposed by Lim and Ashby (1987), Figure 16.1 shows the regimes of different wear mechanisms depending on two parameters along the coordinate axes. This is of course a simplified and approximate way to present wear mechanism regimes,

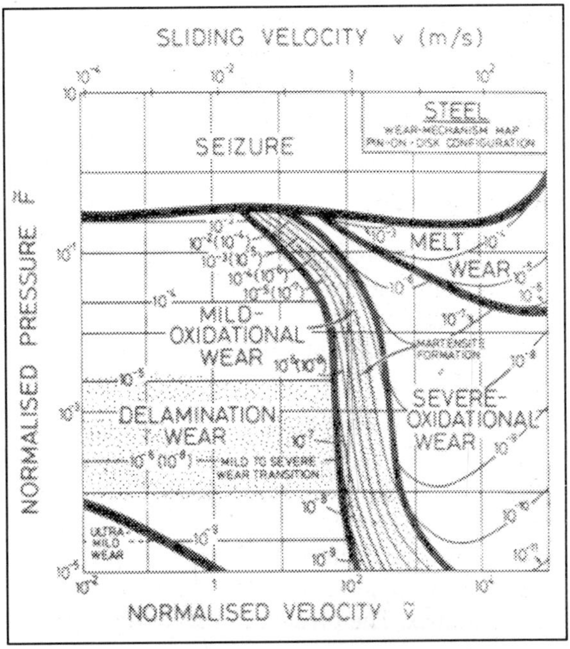

Figure 16.1: Wear Map for a Sliding Steel Pair

because only two parameters are varying and all the others are kept constant. If, however, the two parameters are those, which have a dominating influence on wear, the wear map gives a good indication of how the wear mechanisms change and where the more uncontrolled transition regimes between two mechanisms can be expected to appear.

Lim and Ashby (1987) have shown how wear maps can be developed for steel using a pin-on-disc configuration. The map, which is shown in Figure 16.1, is produced on the basis of a detailed analysis of a large number of published wear test results and on a theoretical analysis of the wear mechanisms. It shows how the dominating wear mechanism changes with the speed and load given as normalized velocity and normalized pressure, respectively.

Fretting Wear

The term Fretting denotes a small oscillatory movement between two solid surfaces in contact. The direction of the motion is usually, but not necessarily, tangential to the surface. When the amplitude of the motion lies in a range typically from 1 to 100 μm, surface degradation occurs which is called Fretting Damage or Fretting Wear.

Whereas sliding wear usually results from deliberate movement of the surfaces, fretting often arises between surfaces that are intended to be fixed in relation to each other, but which nevertheless experience a small oscillatory relative movement. These small displacements often originate from vibration. Typical examples of locations where fretting may occur are in hubs and discs press-fitted to rotating shafts, in riveted or bolted joints, between the strands of wire ropes, and between the rolling elements and their tracks in stationary ball and roller races. It may also occur between items packed inadequately for transport, where vibration can lead to fretting damage at the points of contact between the items. Fretting wear can lead to loosening of joints, resulting in increased vibration and a consequently accelerated rate of further wear. Adhesion between contacting asperities play a major role in fretting wear (Waterhouse, 1987)

Abrasive Wear

Abrasive wear arises from the penetration of one surface by a harder body or surface. Damage involves a cutting or ploughing action. It may involve particle moving over a surface (two body abrasion) or hard particle moving between the two moving surfaces (three body abrasion) as shown in Figure 16.2. The wear arising from impact of hard particles is called erosion which is described in the following section.

It is possible to think that wear rate should be inversely related to hardness. However, practical results on abrasive wear tend not to conform this. In material of simple microstructure, there may be a simple relation between hardness and wear rate, as has been shown for example for commercially pure metals (Figure 16.3). However, with materials of more complex microstructure, this not so. In steels, the relation of wear to hardness is affected by the carbon content and by the microstructure of the matrix. The presence of secondary phases in the structure is also important.

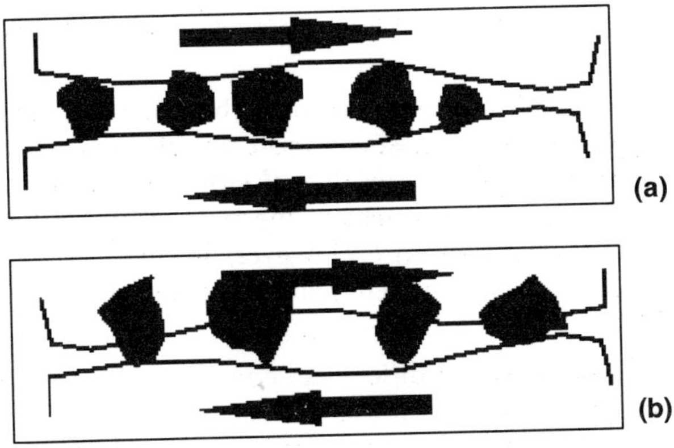

Figure 16.2: Illustration of the Differences Between (a) Two-body Abrasion, (b) Three-body Abrasion

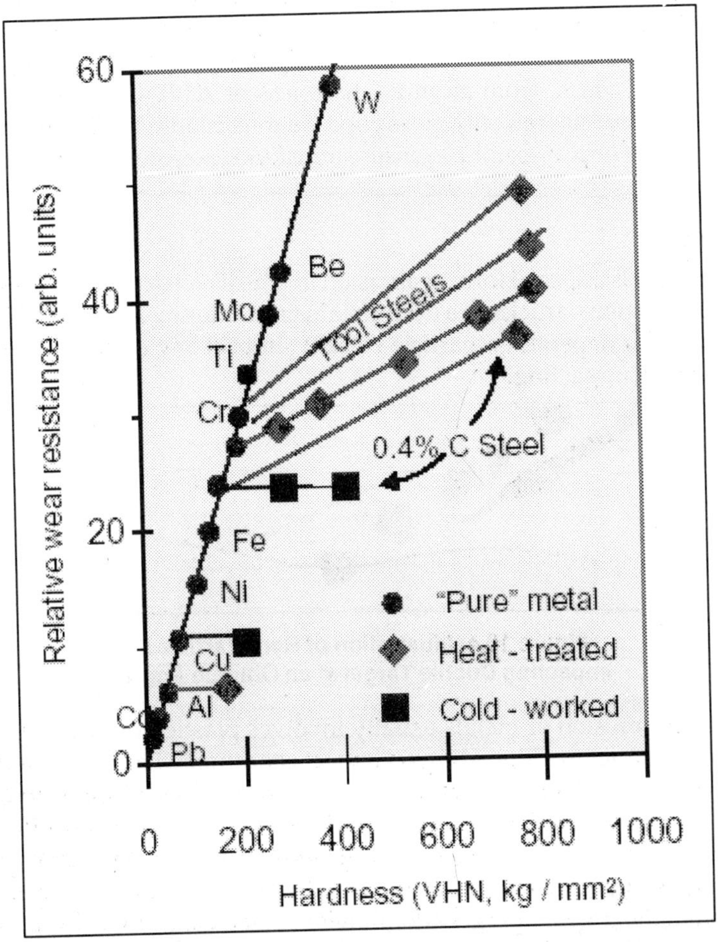

Figure 16.3: Relative Wear Resistance of Pure Metals, Steel and Cold Worked Metals

Carbides, borides and nitrides are widely used with success to provide resistance to abrasion.

With many materials, it is possible to reach a specified hardness by different means. Thus, improvement in resistance of steel to abrasion can be much greater if the matrix is hardened by alloying, than if either quenching and tempering or precipitation hardening is used to achieve the same result. Carbides in steel tend to improve the abrasive wear resistance. However, precipitation hardening is not very effective as precipitate sizes are much smaller as compared to abrasive particle size (Zum Ghar, 1979).

In considering hardness, the difference between the abrading body and the other surface is important. Wear of a surface tends to be progressively reduced if the ratio of its hardness to that of the abrading body increases over the range 0.5-1.3.

Most metallic surfaces work-harden, and a few transformation-harden, during wear. Discussions of the role of hardness have generally centreed on the hardness measured from unworn samples, whereas the relevant hardness is that after wear. These values can be quite different, the most significant example being austenitic manganese steel, which, from an initial hardness of 200HV, can be hardened to 600HV. Indeed, this material only gives good wear resistance to abrasive conditions if there is sufficient impact loading to ensure transformation of austenite to martensite during operation.

Erosive Wear

Erosive wear is a special form of abrasion in which contact stress arises from the kinetic energy of solid particles in a fluid stream encountering a surface (Figure 16.4). The extent of wear depends on particle velocity, impact angle, size and mass of the abrasive particles impacting.

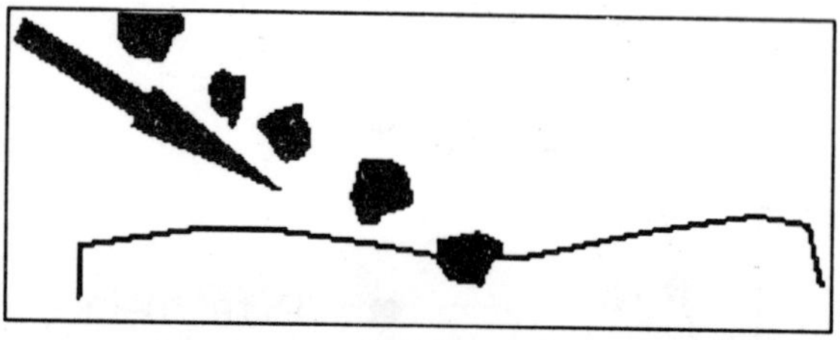

Figure 16.4: Illustration of Hard Particles Impacting Ductile Target at an Oblique Angle

The surface erosion of metals usually involves plastic flow while brittle may erode by flow or fracture depending on impact conditions. The deformation under erosion conditions involves very large strain rates. Therefore, strengthening mechanisms that are effective under low strain rate conditions are not effective. The sequence of events leading to wear particle generation when single abrasive particle strikes a ductile target material is as follows. Upon impact crater is formed and a volume around the crater undergoes deformation that is constrained by un-deformed material surrounding the deformed volume. With further penetration, at some critical strain, the deformation is localized and lip is formed around the crater. With subsequent impact the lip is removed from the surface. In ductile materials, the critical strain for the localization of deformation depends predominantly on the factors such as strain hardening ability of the material, melting point, specific heat capacity and density. The erosion resistance of ductile materials tends to increase with increased values of these factors (Sundararajan and Shewmon, 1983).

The erosion behaviour exhibits stronger angular dependence as shown in Figure 16.5. The ductile materials exhibit maximum erosion at an oblique angle as the shear deformation predominant. On the other hand, the hard brittle materials exhibit maximum erosion at normal angle, as the fracture intensity on the sample is maximum at normal angle. Since coated systems normally involve hard ceramic coating on ductile target, they may provide erosion resistance at oblique impact.At normal impact the coating may erode more than the substrate material (Murthy, Rao and Venkataraman, 2001).Therefore, care should be taken while selecting coating for erosion applications. If the application involves normal angle erosion, application hard coating may improve the erosion life. However, if applications involve erosion at oblique angle, hard coating is beneficial. The ratio of hardness of particle to target is also important. It should be ensured that the hardness of coating is at least 1.3 times greater than the hardness of the particle.

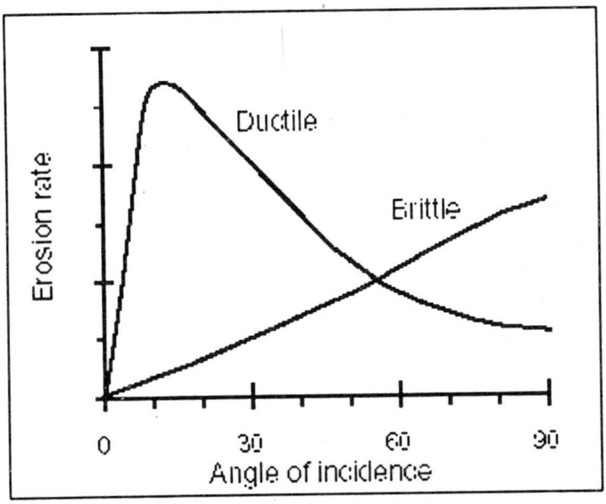

Figure 16.5: Dependence of Rate of Erosion on Angle of Attack of Impinging Particles

Evaluation of Coating Characteristics

Coating Thickness Measurement

The thickness of a thin layer on a surface can be measured by a large variety of different methods as shown in Figure 16.6. The bottom three of these give a composite reading, which relates to both thickness and density.

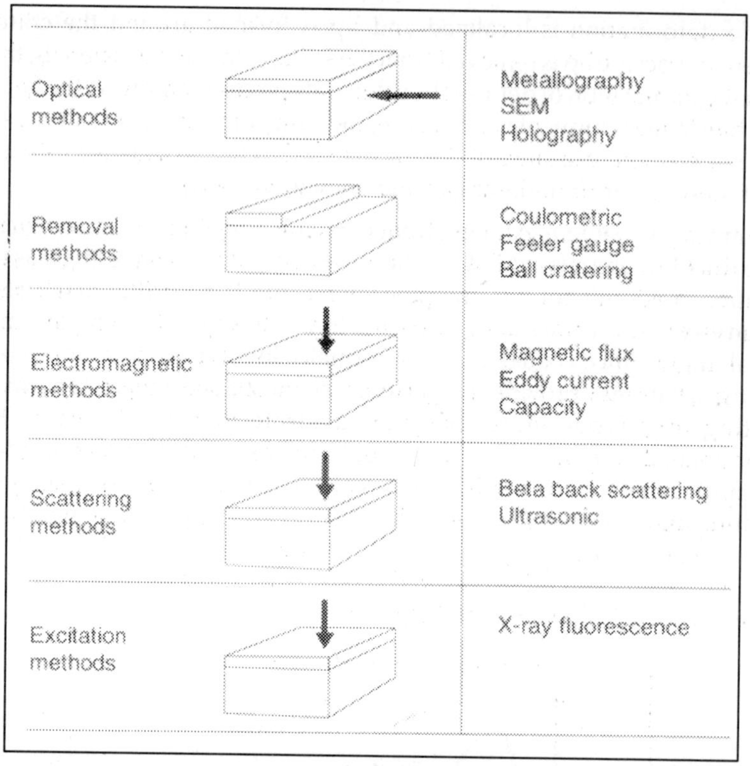

Figure 16.6: Different Methods of Measuring the Coating Thickness

Even if some of the thickness measurement methods can achieve great accuracies, they all have both strengths and weaknesses. X-ray fluorescence, for example, requires careful calibration, but is particularly useful in a coating facility where nominally identical coatings are being produced. Multilayered coatings can present problems due to differences in defining the substrate surface. Direct observation of sections by optical or scanning electron microscopy can avoid this, but this suffers from the drawback that the component must be cut to produce the section. One of the most accurate methods is to use a step-height measurement facility on a stylus profilometer– which of course necessitates leaving or producing an uncoated region, over which the step can be measured.

For thickness measurements the ball crater method also is being used (Hutchings, 1997). This method is simple, cheap and easy to use technique. The principle of the ball crater measurement is shown in Figure 16.7. The ball diameter used is normally 25.4 mm. A kerosene diamond suspension was used as the abrasive medium. The following expression is used to assess the thickness:

$$Thickness = \frac{x * y}{ball\ diameter}$$

Measurements of x and y are made using profile projector.

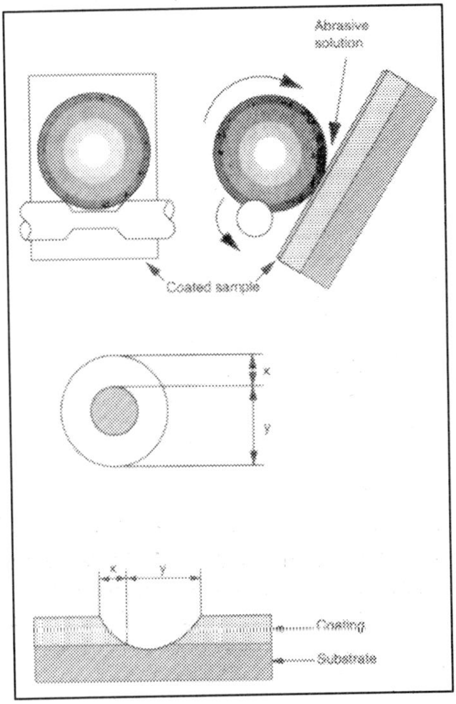

Figure 16.7: Schematic View of the Ball Crater Apparatus and the Crater

Adhesion

For thin coatings, the adhesion strength is normally evaluated by the scratch test (Conczy and Randall, 2005). Equipment similar to the one shown in Figure 16.8 used for scratch test consists of an indenter on which an acoustic emission detector is mounted (Henquist and Hogmark, 1997). The sample is kept on a movable stage, which is connected with a friction force transducer. A known normal load is applied on the indenter and friction coefficient is continuously monitored with increase in normal force. The friction coefficient is measured as a function of increase in normal force. The friction behaviour so obtained is schematically represented in Figure 16.9. In the plot lower and upper critical loads are indicated corresponding to different degrees of coating failure. Normally, the lower critical load is taken to indicate cohesive failure within the coating and the upper critical load signifies adhesive failure between the coating and the substrate. The coating adhesion strength is evaluated by the following equation

$$L_C = \frac{A}{v_C.\mu}\sqrt{\frac{2.E_C.W_a}{h}}$$

where,

L_c: Critical load

A: Cross-sectional area of the track

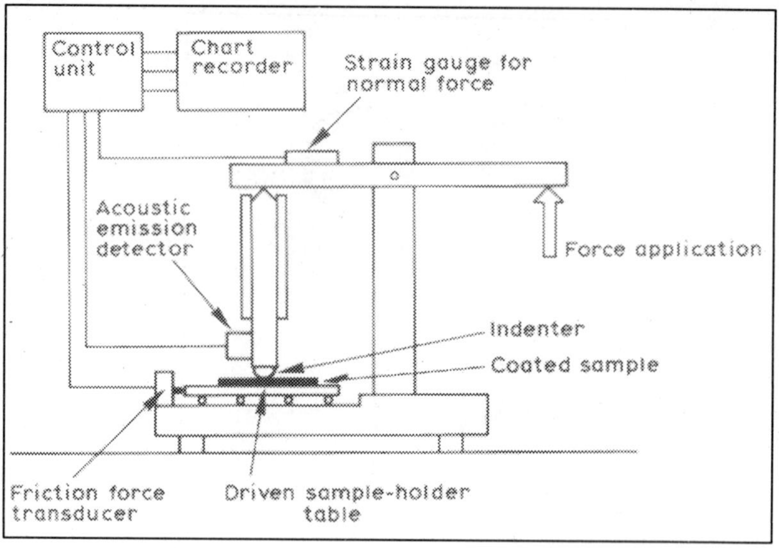

Figure 16.8: Schematic of Scratch Tester Used for Adhesion Strength of Thin Coatings

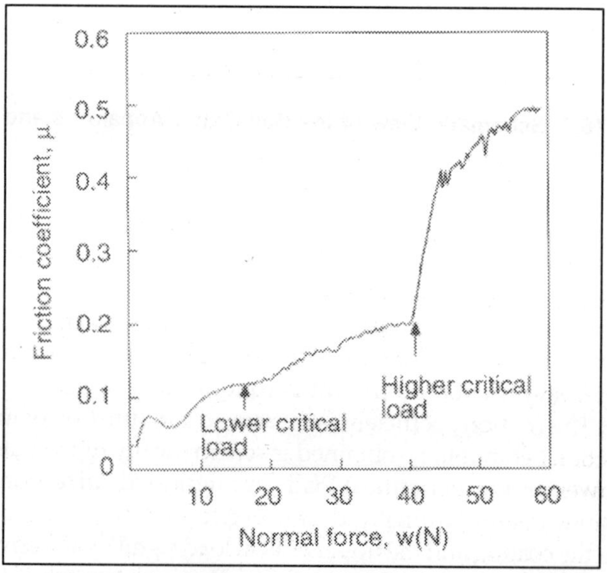

Figure 16.9: The Coefficient of Friction as a Function of Normal Force Measured as a Diamond Stylus is Sliding on the Coated Surface with a Linearly Increasing Normal Load

ν_c: Poisson's ratio for the coating

μ: Coefficient of friction

E_c: Elastic modulus of the coating

W_a: Work of adhesion and

h: Coating thickness

A method called tensile bond strength test is widely used for thicker coatings especially thermally sprayed coatings. The procedure is as per ASTM C-633. The coating for which the adhesion strength (also called bond strength) is to be evaluated is coated on a flat end of a cylindrical specimen. The coating is glued on to another cylindrical sample with suitable adhesive material using a jig and cured. The specimen assembly so prepared (Figure 16.10) is mounted on a tensile testing machine and pulled. If the glue is stronger than the coating bond strength, then the coating bond strength value can be obtained. If the glue fails and the coating remains intent on the surface, it is not possible to obtain the bond strength values. Therefore, it is important to use stronger glues. Heat curable epoxy adhesives are often used. However, their strength levels depend on various factors such as shelf life of the glue, surface conditions of the sample (roughness, flatness, etc) curing conditions, etc. This method is quite successful for the coatings deposited by air plasma spray, flame spray where the bond strengths are below 30 MPa. However, for stronger coatings such as the carbide coatings deposited by high velocity processes namely, detonation gun and high velocity oxyfuel (HVOF). For these coatings, the bond strengths are in the excess of 70 MPa.

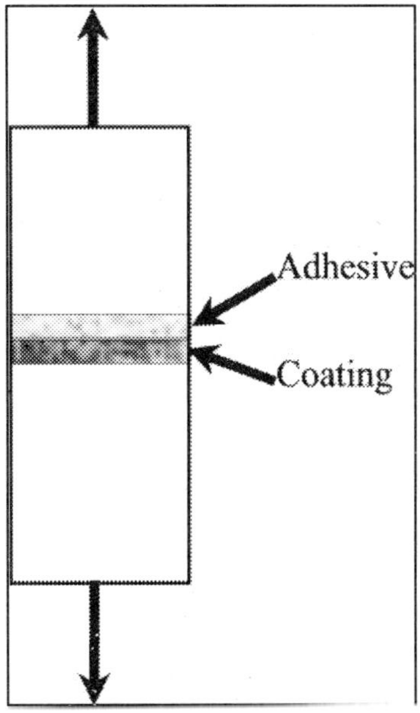

Figure 16.10: Schematic of Tensile Bond Strength Test Specimen

Evaluation of Hardness

Nano-Indentation Technique

A typical system layout is shown in Figure 16.11. Usually, Berkorich-type diamond indenter is used which has a triangular pyramid geometry permitting accurate tip profiling. With the knowledge of the indenter geometry it is possible to obtain a value for hardness from the depth of plastic deformation taking away the elastic contribution from the total depth as shown in a typical load-depth curve for an elastic-plastic material (Figure 16.12). By applying oscillatory load with extremely small time constant, it is possible to evaluate the hardness and modulus continuously during the loading cycle itself. For more details on nano-indentation Fischer-Cripps (2005) may be referred.

Tribological Evaluation

Sliding Tests

The variety of tribological test methods used only for sliding contacts is illustrated by Table 16.1 where some details of 6 test methods are given. Of these, the pin-on-disc test is very suitable as a material screening test and also for friction and wear studies. It is by far the most widely used. The other test methods may be very useful in cases

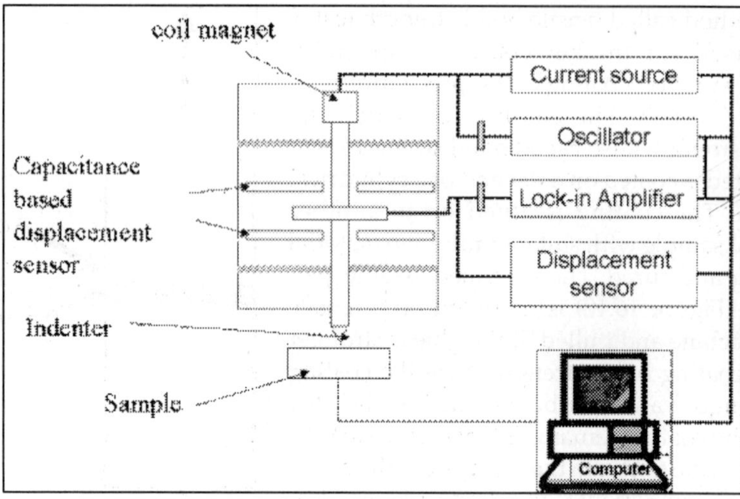

Figure 16.11: Schematic of Nano-indentation Equipment

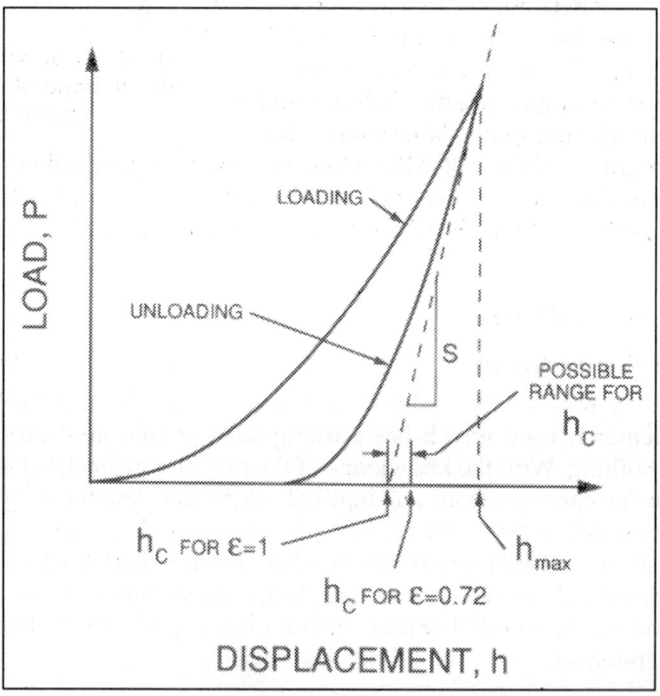

Figure 16.12: A Typical Load Displacement Curve for Elastic/Plastic Material (h_{max} is the total depth of penetration and h_c is the depth of plastic deformation)

where they better simulate the actual contact conditions and the results thus become easier to transfer to a specific application.

Engineers frequently tend to regard the friction coefficient, like hardness, as a kind of intrinsic material property. This is, however, not the case, since both are the result of quite complex interactive behaviour and are subject to statistical variability. A measured friction coefficient is the result of both physical and chemical interaction in a discrete system, which is quite difficult to repeat from one laboratory to another because of the large number of influencing parameters. In the case of pin-on-disc testing it has been observed that even the vibration characteristics of the test machine can have a marked influence on friction and wear behavior. As with the friction coefficient, wear rates also showed considerable variability. However, observations of the worn surfaces and transfer layers indicated that wear mechanisms were similar under identical conditions.

Table 16.1: Wear and Friction Test Methods for Tribological Evaluation of Coated Surfaces (Typical parameters are given but many variations of these are in use)

	Pin-on-Flat (ASTM)	*Pin-on-Disc*	*Block-on-Ring (ASTM)*
Test configuration	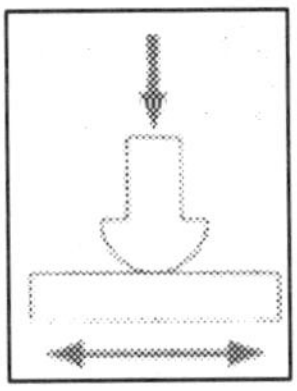	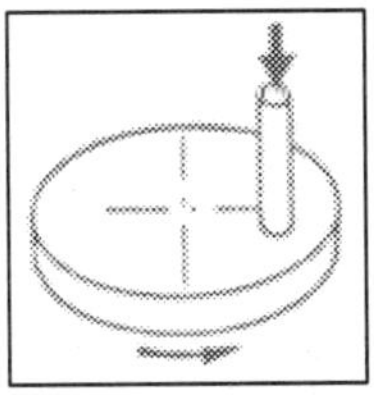	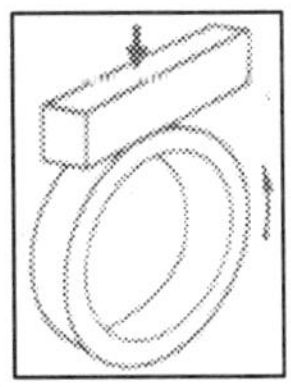
Test purpose	Determination of wear rate and coefficient of friction at low loads and speeds and with small specimens	Determination of sliding wear rate and co-efficient of friction.	Determination of adhesive wear rate of materials.
Specimens	A pin on ball rubbing against a plate.	A pin or a ball rubbing on a rotating plate.	A steel ring is rotated against a stationary block.
Test conditions (Typical)	Reciprocating sliding frequency 5 Hz, stroke length 10 mm, Load 25N, Temperature 18-23°C, Humidity 40-60 per cent RH specimen can be dry or fully immersed in lubricant.	Sliding speed 0.1 m/s, Load 10N, Sliding distance 1 km, Temperature 23°C, Humidity 50 per cent RH.	Ring speed (max) 700 rev/min., Load (max) 600 N, Duration 5400-24000 cycles. Specimen can be dry or immersed in oil or other fluid.
Measurements	Weight loss of disc, Height loss of pin, Wear track, Profilometry, Ball wear area measurement, friction force	Height loss of pin, Wear track, profilo-metry, ball wear area measurement, friction force	Test block weight loss. Scar width, Ring weight loss, Friction force, Lubricant film load carrying capacity.
Wear types/ comments	Mild adhesive wear suitable for tribological coating evaluation	Mild and severe adhesive wear. Suitable for tribological coating evaluation.	Mild and severe adhesive wear. Lubricated wear.

Contd...

Table 16.1–Contd...

	Block-on-Pin	*Twin Disc(AMSLER)*	*Fretting*
Test configuration	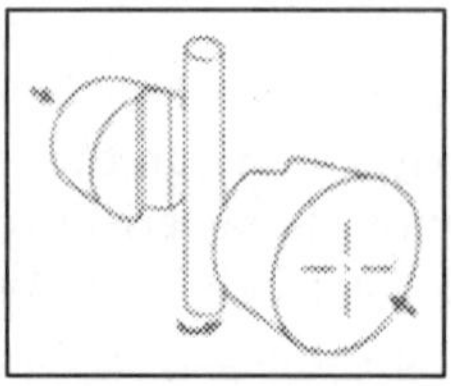	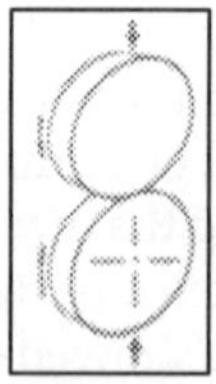	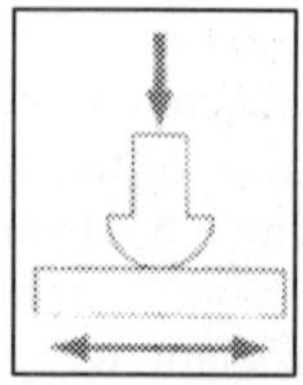
Test purpose	Determination of wear rate, load carrying capacity and friction coefficient for sliding contacts	Determination of sliding or rolling wear rate of test materials, treatments or coatings.	Determination of wear rate, load carrying capacity and friction co-efficient for sliding contacts
Specimens	A 6.35 mm diameter cylindrical journal is rotated within two loaded stationary V-blocks to give a 4-line contact.	Normally two discs 40 mm diameter and 10 mm thick. Alter-natively the discs may be 30 and 50 mm in diameter.	Ball or pin against a flat plate
Test conditions	Journal speed 290 rev/min., Sliding velocity 0.1 m/s, Load (constant or increasing) 89-20000 N, Hertz stress 242-3450 MN/m^2, Specimens can be immersed in oil or other fluid.	Lower disc speed 0-400 rev/min., Upper disc speed 440 rev/min., Load 200-2000N, Dry or lubricated.	Sliding amplitude 0.02-0.4 mm, Sliding frequency 5-20 Hz, Load 2-50 N, Temperature 22°C, Specimen can be dry or fully immersed in lubricant.
Measurements	Journal driving torque, Progressing wear depth, Wear life, seizure failure load, final specimen weights	Individual disc weight loss or profile, Total disc weight loss, wear life.	Volume loss by wear scar measurements, Friction force.

Abrasion Testing

The best-known tests in this category are the Taber test and the rubber wheel test (Figure 16.13). The former relies on the abrasiveness of the impregnated rubber wheels, whilst the latter utilizes a loose third-body particulate abrasive, either dry or in water slurry. A typical medium is quartz sand, and the specification code for the dry sand test is ASTM 65 whilst the wet test is known as the SAE wet sand rubber wheel test.

A good surface finish on a coating is important in such contact conditions; where asperity removal can lead to accelerated wear by revealing the substrate. The wet slurry test has also been useful, for example, in evaluating the wear resistance of carburized steel surfaces. The tests are quick, taking only a few minutes to produce measurable wear, and assessment is easily carried out by profilometry of the worn regions or by weight loss.

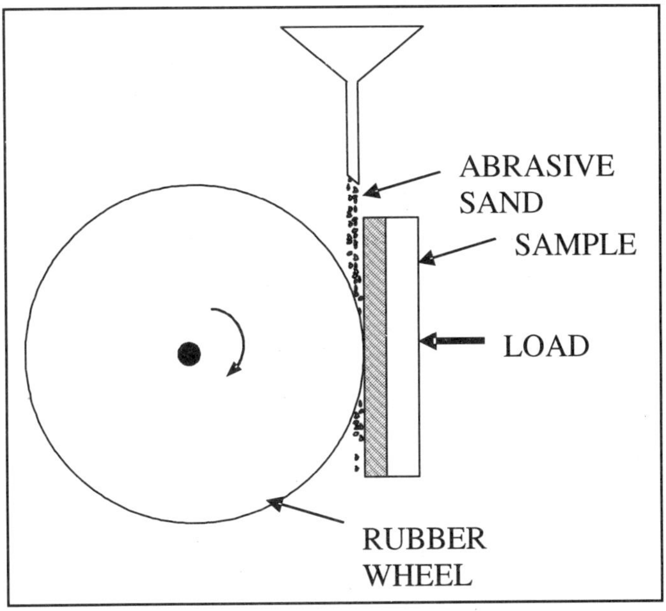

Figure 16.13: Schematic Illustration of Rubber Wheel Abrasive Wear Test Method (ASTM G-65)

Solid Particle Erosion Tests

For purely erosive wear a very effective method is to use a particle air-blast test. This has the benefit of providing a controlled erodent supply including speed and direction. The method can be used as a single particle test, whereby the damage due to one impact can be determined.

Erosion Test

The schematic diagram of solid particle erosion test rig is shown in Figure 16.14. The erodent particles are allowed to fall freely under gravity from a powder feeder arrangement consisting of a powder feeder and a conveyor belt as shown in the Figure 16.14. The particles are accelerated by the compressed air fed into the mixing chamber where they entrain the oil and moisture free dry compressed air coming from the air compressor. The particles are further accelerated as they move with the air stream through a tungsten carbide converging nozzle and then finally impinge the sample kept on a sample holder. Varying the pressure of the compressed air can vary the velocity of the particle.

The test involves the following procedure. Prior to the test, the feed rate of the erodent particles is determined. The erodent is allowed to fall freely onto a conveyor belt rotating at a constant speed. The erodent is fed into a small container by the conveyor belt for a predetermined time (say 5 min) and it is weighed to determine the erodent particle feed rate. The erodent particle velocity is determined by the well-known double disc method. After obtaining the desired feed rate and the particle velocity, the test sample (of dimensions 30 mm x 30 mm x 3 mm) cleaned with acetone

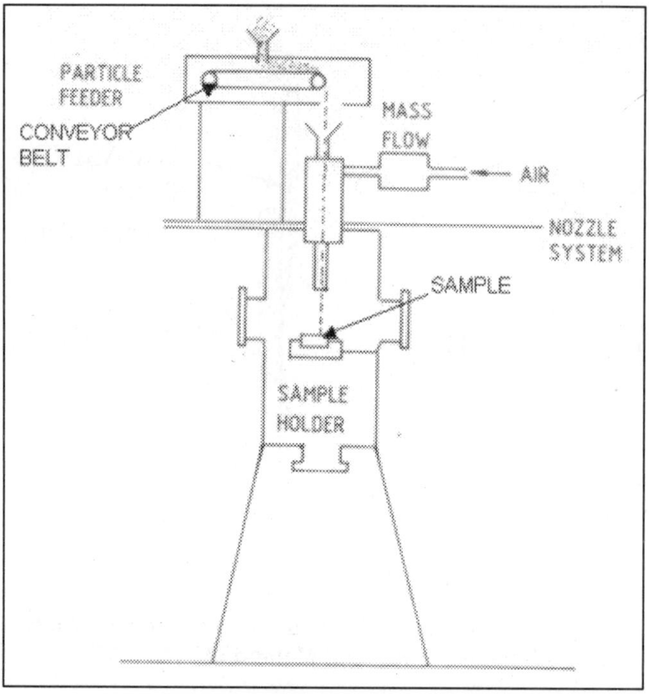

Figure 16.14: Schematic Diagram of Solid Particle Erosion Test Rig

in an ultrasonic cleaner for 30–45 secs, dried and weighed in an electronic balance with an accuracy of 0.01 mg is then placed on a sample holder in the rig at a certain distance (say, 10 mm) below a stainless steel tube through which the accelerated erodent particles exit. The pressure of the compressed air is set at a predetermined value, which in turn determines the impingement velocity of the particles. The air pressure and its flow rate was controlled by a regulating valve and monitored using a digital pressure and mass flow system.

The test samples are subjected to erosion by particulate impact for a fixed period of time (5 min.), ultrasonically cleaned with acetone, dried and weighed in the electronic balance and again placed in the sample holder. The weight loss of the test specimen during each exposure is determined till it attains a steady-state weight loss. The dimensionless erosion rate (E) is calculated as the ratio of the steady-state weight loss of the test specimen to the weight of the erodent particles causing the loss. The erosion rate obtained is referred to as the *'Steady State Erosion Rate'*. Changing the sample holder with a different configuration can also vary the angle of impact of erodent particles on the test specimens relative to the plane of the coated surface.

The parameters that influence the erosion rate are as follows: Erodent material, Particle size range, Particle feed rate, Particle velocity, Impact angle, Stand off distance and Environment.

Measurement of Particle Velocity

The velocity of the erodent particles is determined by a rotary double-disc method (Ives and Ruff, 1978) shown schematically in Figure 16.15. The double-disc arrangement consists of two metallic discs firmly secured on a common vertical shaft of a D.C. motor whose speed can be controlled by a variable D.C. power supply. A known distance, L, separates the two discs. The top disc has a narrow radial slit and the bottom disc is well polished. To measure the particle velocity, the nozzle exit of the test rig is kept just above the slit of the top disc and two erosion exposures are made. First, the discs are kept in the stationary condition and the particles exiting from the nozzle are allowed to impinge on the top disc where the single slit allows some of the particles to pass through and impinge on the bottom disc to form a wear scar. The centre of the wear scar is marked. Next, the motor is switched on and the discs are allowed to rotate at a known speed, N, by adjusting the dc power output. The erodent particles which fall through the rotating slit cause wear scar on the bottom disc at a certain distance 'S' from the first scar.

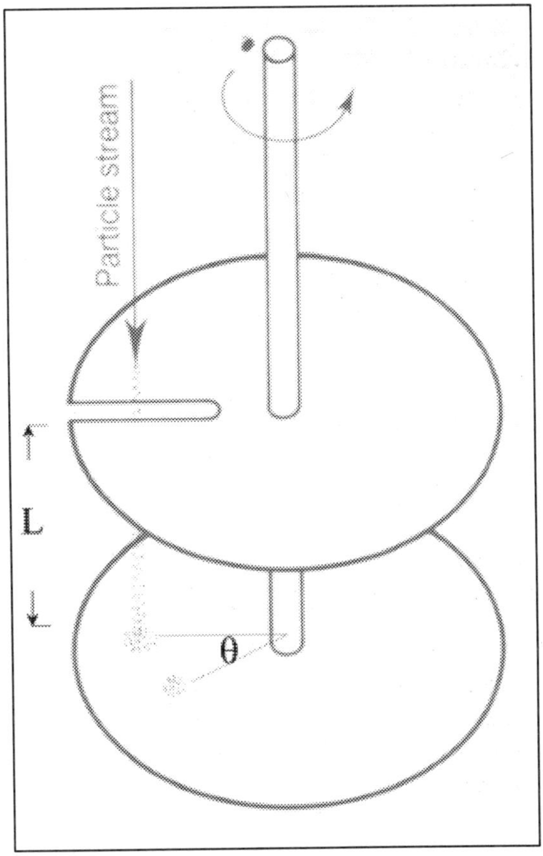

Figure 16.15: Schematic Illustration of the Double Disc Method for Measuring Particle Velocity in Erosion Experiments

If V_p is the velocity of particles, the time of flight to cross the discs separated by a distance L is L/V_p. The number of revolutions made during this time will be $(L/V_p)N$ or $(L/V_p)\,2\pi N$ radians which is equal to S/R radians, the angular displacement between the two wear scars, where R is the radius of the wear scar from the axis of rotation. Thus, the velocity of the particle is given as,

$$V_p = \frac{2\pi NLR}{S} \; or \; \frac{360\,NL}{\theta}$$

where,

θ is the angular displacement in degrees.

Another common tribo-contact in industry is the metal cutting tool to work piece interface. This is more difficult to simulate, because of the high contact forces and heat generated. Nevertheless, methods have been developed to evaluate coatings for these conditions.

Macro Mechanism Friction in Coated Systems

The macro mechanical tribological mechanisms describe the friction and wear phenomena by considering the stress and strain distributions in the whole contact (Holmberg and Mathews, 1994). In contacts with one or two contact surfaces following four main parameters which control the tribological contact process.

1. Coating to substrate hardness relationship
2. The thickness of the coating
3. The surface roughness, and
4. The size and hardness of the debris in contact

Thickness of Coating

Let us consider a soft film on a hard substrate with smooth surfaces and no debris present. When the film is thin enough, the ploughing of the film is small. The friction is then determined by the shear strength of the film and contact area, which is related to the deformation properties of the substrate. There are two important criteria in achieving low friction. The film must possess property of low shear strength and it must be applied on surface of high hardness and high elastic modulus. Therefore, there is an optimum film thickness when low friction is achieved. Below the optimum value, the friction increases with decrease in film thickness as the film may break down due to mating asperities. Above the optimum thickness, the friction may increase with increase in thickness due to ploughing. This has been schematically shown in Figure 16.16.

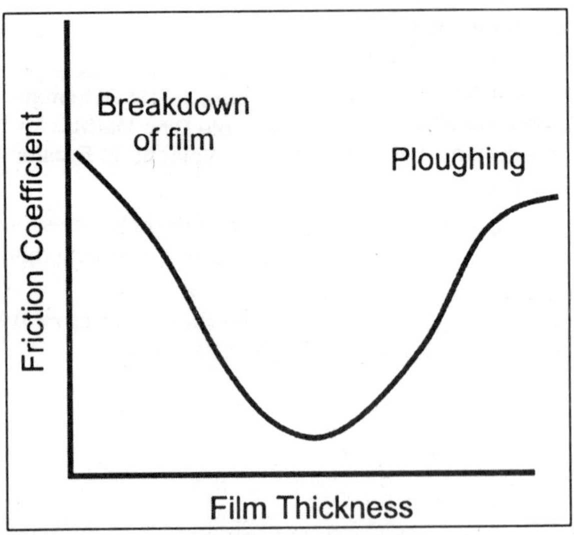

Figure 16.16: Schematic Representation of the Effect of Coating Thickness on Friction Coefficient for a Soft Film on a Hard Surface

Hardness of the Coating

One of the most important parameters influencing the tribological behaviour of a coated surface is the hardness of the coating and its relationship to surface hardness. It is common to consider hard and soft coatings separately. When a ball is sliding on a plate, the friction is ideally the product of shear strength and the contact area. A harder plate material will result in decreased contact area but increased shear strength and the effect on friction is only minor. A reduction in friction can be achieved by adding a thin soft coating on the plate, which reduces both contact area and the interfacial shear strength.

A hard coating on a softer substrate can reduce wear by preventing ploughing on both a macro and micro scale. Hard coatings are thus particularly useful in abrasive environment. Low friction can be achieved with hard coating if a low shear strength microfilm and the load are well supported by the hard coating. The films such as MoS_2 and graphite can provide low shear strength and hence they are used as solid lubricants.

Debris at the Interface

Debris is often present in sliding contacts. They can either originate from the surrounding environment or be generated by different wear mechanisms in the sliding contact. Their influence on friction may depend on the particle diameter, coating thickness, surface roughness relationship and the particle, coating and substrate hardness relationship. For example, if the hard particle entrapped is partly anchored to the soft film and partly protruding, it may abrade the counter face material, which will result in increase in friction coefficient. The presence of hard particles between hard surfaces may in some cases even reduce the coefficient of friction. If the particles are fairly round in shape, hard enough to carry the load and at least one of the surfaces is smooth, the particles may act as rollers and reduce the friction coefficient.

Summary Conclusions

1. A single coating cannot be used for all the tribological applications.
2. Selection of coating for a given tribological coatings depends on the operating conditions. For example in erosion, for normal impact, ductile coatings perform better than brittle hard coatings.
3. In abrasive wear, intrinsic hardness of the surface is beneficial. Hardness achieved by cold working or precipitation hardening does not have influence on wear resistance.
4. In sliding wear, the hardness on the surface rather than the bulk hardness controls the wear behaviour. The thickness and hardness of the coating should be selected based on the stress state, operating conditions and mating surface of the coating.
5. The present article also gives various tribological evaluation methods that are applicable for coatings.

References

1. Annual Book of Standards, 1987, Vol 03.02, ASTM, p. 243-250.
2. Conczy, S.T, Randall, N., 2005, International Journal of Applied Ceramic Technology, (5), p. 422-428.
3. Fischer-Cripps, A.C., 2005, Nano Indenation, Mechanical Engg.Series, Springer
4. Henquist, P., Hogmark, S. 1997, Tribology International, 30(7), p. 507.
5. Holmberg, K, Mathews, A., 1994, Coatings Tribology, Elsevier, Amsterdam, Netherlands.
6. Hutchings, I.M., 1997, New Directions in Tribology, Mechanical Engineering Publications, St Edmunds (Ed.Hutchings, I.M.) pp. 267-280.
7. Ives, L.K. and Ruff, A.W., 1978. Wear, Vol 46, p. 149-162.
8. Murthy, J.K.N., Rao, D.S., Venkataraman, B. 2001, Wear, 249, p. 592-600.
9. Sundararajan, G., Shewmon, P.G., 1983, Wear, p. 237.
10. Waterhouse, R.B., 1987, Wear, Vol. 45, p. 355-364.
11. Zum Ghar, C.H., 1979, Metal Progress, 116(4), 46-52.

Country Reports

Chapter 17

Development in Surface Engineering in Egypt

Randa Abdel-Karim
Associate Professor, Department of Metallurgy,
Faculty of Engineering, Cairo University, Egypt
E-mail: randa38@hotmail.com

ABSTRACT

This study presents state and prospects of development of surface engineering, both at research as well as at industrial levels in Egypt. Describes various methods of coatings being used for different purposes. Points out that there are at present 14 public universities, 6 private universities. In addition, there are 3 R&D institutes working in the field of Materials Science and Engineering. In the recent history, Egypt had its first Engineering School in the Middle East in 1816. The School was first housed in the Greater Citadel. The Faculty became an independent Entity in 1905 and settled where it is located now at Cairo University campus in Giza. Significant success has been achieved in the field of wear resisting and corrosion resisting coatings over a wide spectrum. Scientists of Egypt have made considerable achievements in the research relating to thin films, nano coatings, methods for their production and processing.

Keywords: *Surface engineering, Surface modification, Alloys, Plasma coatings, Thin films, Nano-coatings, Egypt.*

Introduction

Surface engineering techniques generally consists of surface treatments, where the composition or the mechanical property of the existing surface is altered, or a different material is deposited, as thin films or coatings to create new surface. Coatings offer a simple method of changing or improving the chemical, mechanical and even thermal surface properties of materials and components [Grunling, 1985]. The world market in surface coatings covers industrial, architectural, automotive, vehicle refinish, and heavy goods vehicles, aerospace, marine and offshore coatings. The market volume growth in 2002 was 26.72 million tones with a value of nearly US $ 470 billion [Technical report, 1997]. This study presents state and prospects of development of surface engineering, both at research as well as at industrial levels in Egypt.

Wear Resistant Coatings

Thermal spray technology and coatings are receiving increased attention as solutions to surface engineering problems involving wear, high temperature and aqueous corrosion and thermal degradation [Patt and Whitny Aircraft, 1974]. Plasma coatings are one group of metallic and/or ceramic coatings, which are applied to the surface of parts. The process is primarily designed to protect materials from wear, hot corrosion, and abrasion [El-Khatib, 2005]. The process was being applied for specific applications almost hundred years ago, although it was only in the late 1950s that it became viable and attractive metal processing route [ABECO, 1985]. Plasma coating is used widely in the machine component field for worn or inaccurately designed machine parts. More recent applications have been in the field of high temperature-resistant coatings for jet aircrafts, rockets and missiles.

Plasma coating process uses a high intensity electric arc, which heats an inert (or relatively inert) gas or gaseous mixture to the ionized state. This produces expansion of the gas resulting in considerable velocity. Powder metal is injected into the hot flame and deposited on the desired surface [Abdel-Karim, 1988]

Plasma spraying has certain advantages. The bulk plasma temperature is well above that obtained by combustion or resistance heating. Also the operator has control over the atmosphere to be either inert or reactive. Another advantage is the high output rates [ABECO, 1985]. Plasma equipment is readily available at reasonable cost [Patt and Whitny Aircraft, 1974]. In the plasma process, it is possible to spray metals, ceramics, plastics and combination of these. Spraying process is considered to be cold process which doesn't affect the substrate properties or cause distortion [El-Khatib, 2005].

Table 17.1: Requirements for Protective Coatings

Surface Properties	*Interface Properties*
Chemical stability with environment	Good adhesion-
Mechanical stability with environmental constituents	Good adaptation of mechanical and physical properties.
Bulk coating properties	Low inter-diffusion rate.
	Technical and economic properties.
Adequate thermal stability and properties	Reparability.
Low thermal conductivity	
High thermal shock resistance	Compatibility of heat treatments.
Sufficient toughness and impact strength	Sufficient life time.

The Arab British Engines Company (ABECo) is one of the biggest Egyptian companies working in the field of surface coatings. It is located at Helwan, about 60 kms south Cairo. The following applied types of plasma coatings are discussed:

Tungsten Carbides (W2C) is comparatively resistant to acids. It is however dissolved by hot NHO_3 and by HNO_3–HF mixture 1:4 at room temperature. It reacts at 200°C with chlorine with the formation of WCL6 and graphite. In a stream of oxygen it is oxidized to WO3 at 500°C. The Moh's hardness is 9. The electric conductivity is 7 per cent of that of pure Tungsten.

WC is a gray metallic powder. It is resistant to acids. It is not attacked at room temperature by HNO_3–HF mixture 1:4 and starts to react at temp 600 to 800°C. Against chlorine it is stable up to 400°C. Upon heating in air or oxygen, WC is slowly oxidized to W03. The Moh's hardness is + 9. The electrical conductivity amounts to 40 per cent of that of pure tungsten. WC becomes super conductive at about 2.5 K*. It is the most important constituent in applications requiring high wear resistance and high temperature applications.

Tungsten carbide deposits are generally brittle, but withstand low velocity impact (less than 50 ft/sec). These deposits have low resistance to oxidation above 800°C. Corrosion resistance is similar to carbon steel. Tungsten carbide surfaces resist low stress scratching and high stress grinding abrasion.

When plasma spraying carbides, pure carbide is seldom used, but almost alloy in company with a binder metal. This binder prevents any decomposition that might occur under the effect of high temp present in the process. For WC, the binder is cobalt. Co has specific gravity of 8.9 and melting point (m.p) 1490°C. The hardness level is below 200 kp/mm^2. Its crystal structure is hexagonal close packed.

WC + Co serves as anti-abrasion protection at room temperature. The presence of Co matrix prevents wearing by plucking out. Polishing work on components results in a severe abrasion of the equipment. Here, armoring with WC + Co results in an improvement of service life by 1000 per cent. Interesting uses occur in gas turbine instruction on the turbine blades that are provided with WC + Co overlay. The bond strength of WC-Co is 10,000 psi and the maximum thickness capable is 0.02 inch [Abdel-Karim, 1988].

Pure Molybdenum (Mo) is a refractory metal, having a melting point at 2620°C. It is non magnetic and has good thermal conductivity. Its resistance to shock is fairly good but not as good as tungsten. The density of this material is 9.06 g/cm^3. Molybdenum has B.C.C. Structure.

With sufficient power, Molybdenum may be sprayed to very dense, thick layers of several mm by plasma method. Mo has good antifriction characteristics, which can be transferred to metals with high coefficient of friction such as titanium by over spraying. It has excellent wear resistance. The bond strength is 3000 psi. Knops hardness is approximately 786.

Mo can also be used as a solder rejecting film, but the tendency to oxidation must be considered. Mo begins to oxidize at about 400°C, and at about 700°C the oxide vaporizes as rapidly as it forms. That means, molybdenum is suitable only for low temp applications. Because of its good spraying qualities, complicated forms can also be made of Molybdenum by spraying method. These spraying qualities are also the reason for the choice of Mo for machining repairs or re-conditionings of steel components. Mo shows good resistance to hydrofluoric, hydrochloric and sulfuric acid. High modulus of elasticity means it is much stiffer and deflects much less than steel under load [Abdel-Karim, 1988].

Ni-base alloys represent one class of the so-called supper alloys. In addition to aircraft, marine, industrial and vehicular gas turbines, supper alloys are now used in spacecrafts, nuclear reactors, steam power plants and other high temperature applications. Ni-base alloys coatings are of great importance for hard facings.

Supper alloys demonstrate outstanding impact resistance, and good resistance to high cycle and low cycle mechanical fatigue as well as thermal fatigue. The close packed F.C.C lattice of these austenitic alloys has great capability to maintain good tensile, good rupture, and good creep properties to homologous temperature much higher for equivalent B.C.C systems, because of several factors. They include excellent modules of the F.C.C lattice for secondary phase, and the ability to control precipitation of inter metallic compounds and solid solution hardening both for strengthening. The melting point of Ni is 1455ºC, with density being 8.90 g/cm^3. One example of Ni-base supper alloys is Ni-5 per cent Al alloy. Alloy Ni-5 per cent Al has good adhesion, wear and oxidation resistance properties. The maximum working temperature is 815°C. Good oxidation resistance is achieved by formation of a tight continuous surface scale $A1_2O_3$ that acts as a diffusion barrier and doesn't spill off during thermal cycling [Abdel-Karim, 1988].

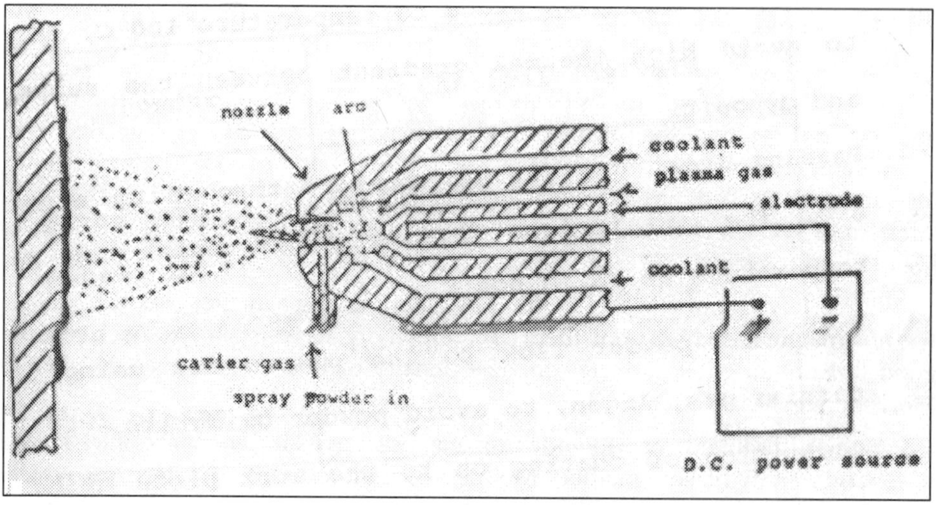

Figure 17.1: Plasma Spraying Gun

Morks *et al.* [Morks, 2005] studied the effect of plasma power on microstructure and hardness properties of sprayed cermets coating. The results showed that WC phase appeared at low power (3.9 Kw) and W2C phase appeared at plasma power 5-9Kw. The x-ray intensity peak of WC decreases and that of W2C phase appeared at plasma power increases. Maximum coating hardness increased with increasing plasma power.

Ion implantation technique is used for modification of engineering materials surfaces. In general this technique can increase the hardness and wear resistance of cutting. Physical vapour deposition (PVD) technology has been used to deposit metallic layer of TiAl. The results showed that the hardness of TiAl implanted by Boron increased by 300 per cent [Waheed, 2004].

Another type of coating for wear resistant applications was alumina ceramic containing 5 per cent porosity alloyed by adding HfO_2 or mixtures of HfO_2 with 5 or 10 mol per cent SiO_2, CeO_2 or cordierite glass using infrared laser radiation. Thickness of the alloyed surfaces ranged between 500 and 800 μm depending on the composition of the alloyed powders and laser parameters used. The resulting microstructures and worn surfaces were analyzed by electron microscopy. As a result of the alloying, the tendency of crack formation and grain growth was substantially reduced in the Al_2O_3 ceramic during laser treatment. The composite layers, showing no open porosity, contained 20–40 vol. per cent second phases, which were homogeneously distributed at the grain boundaries of the alumina matrix. Due to alloying, average grain size, hardness and Young's modulus of the composite layers were reduced, but resistance to cracking under the mechanical action of a sliding diamond (scratch test) abrasive wearing were substantially enhanced compared with the unmodified dense alumina ceramics [Zawrah, 2005].

Thermal Barrier Coatings

The ceramic barrier coatings (TBC's) are recognized to hold considerable promise for variety of applications involving high heat flux and/or moderately high temperature environments. These coating systems generally consist of plasma sprayed oxidation/corrosion resistant metallic bond coat MCrAlY (M=Ni and/or Co) and either a single composition ceramic overcoat, or a graded structure ceramic over coat to minimize the thermal expansion mismatch between ceramic and metallic substrate. A.F. Waheed *et al.* [Waheed, 1996] studied the behavior of duplex and graded TBC's in which partially stabilized zirconia was used as a ceramic. The thermal shock test using burner rig was employed to determine coating failure. Failure is considered at the appearance of the first bright spot during heating period.

Corrosion Resistant Coatings

The corrosion resistant coatings are one of the important methods for corrosion control. Since its inception in 1982, and as part of EPRI progtam for petroleum development, Egyptian Petroleum Research Institute (EPRI)–Metallic Surface Protection Centre has continued to strengthen and enhance capabilities to meet the challenges of the technological advancements in the fields of coating systems. Backed with eleven years of experience, Surfaces Protection Centre (SPC) now covers:

1. Inspection of three layer PE and PP coating systems and their field joints, inspection of single and dual layer FBE system, developments insulation of petroleum pipe-lines with foam systems.

2. Testing and evaluation of coating materials, internal lining and evaluation and protection of petroleum projects from corrosion.
3. Documentation of coating procedure and materials.
4. Corrosion protection plans.

Carbon steel alloy petroleum pipelines were protected by organo-silicon compound (tetra-n-propoxysilane (TnPOS), Zn powder, P_2O_5 with different ratios) and were mixed with polyisoprene waste (PI) and used as coating material. The coated specimens were subjected to heat and corrosive media individually and their effects were examined. Carbon steel specimens were coated by 5-30 parts of TnPOS, 0.1-0.6 parts of Zn powder, 3-18 parts of P_2O_5 and 65 parts of PI. The curing time of the coating films at 160°C was decreased by increasing the percentage of each of TnPOS and Zn powder and increased by increasing the P2O5 (approximately 50-60 sec). The dry film thickness (DFT) was 50 + or-10 mum. The hardness, adhesion, thermal stability and degree of coherence of the formed films were increased by increasing the TnPOS, Zn powder and P2O5. The coating films acquired high thermal stability and corrosion resistance against NaCl and acid solutions. The results revealed the presence of an effective layer at high percentage of TnPOS, Zn powder and P2O5 [Abou-El-Enain, 1999]

In another study [El-Mesikawi, 2000], different types of epoxy resins based paints have been selected for coating steel pipe test samples. The surfaces of these steel samples were pre-treated before application of different coatings by blast cleaning. The surfaces profile of the treated samples was checked by using the tested tape. The corrosion resistance of the coated samples, to different corroding media, has been determined via two main methodologies (weight loss method and potentio-dynamic and potentio-static electrochemical methods). The results showed that it was possible to rely on the accelerated electrochemical methods in order to predict the durability of the coating material against different corroding media. The study also revealed that some types of epoxy resins were superior in corrosion protection and durability against saline water and acidic corrosive media.

Aluminized coating is an important class of protective coatings for high temperature alloys and many nickel base supper alloys against severe surface degradation. A group at Suez Canal University produced the coating by heating Monel 400 970Ni-30 Cu) alloy at 760 0C within a chemical powder pack containing aluminum powder, aluminum oxide and halide salt. Aluminizing of Monel alloy lead to improvement of wear,corrosion and oxidation resistance [Abbas, 1997].

Another group at the same university [El Sabahy, 1997] studied the electrode position of Copper-Nickel multi-layers on mild steel. It was concluded that the corrosion behaviour of some diffusion treated electrodeposited layers proved to be similar to that of Monel–400 alloy.

Another group at El Menia University studied the effect of acidic and basic media (solutions of different pH values) on the corrosion resistance of both uncoated and coated steel using chromium diffusion coating. Moreover, the influence of solution bath temperature and coating temperature as process parameters on the corrosion rate were investigated. It was observed that the pH values, solution bath temperature and coating temperature have had significant effects on the corrosion resistance. The corrosion resistance has been much higher in the basic solutions (pH = 9 and 10) than that in the acidic solutions (pH = 1.5 and 3). The chromized specimens at 1200°C have had a higher corrosion resistance compared with that chromized at 1100°C. The uncoated specimens have had the lowest corrosion resistance. The weight loss varies according to a polynomial function with the time of corrosion and linearly with the solution bath temperature [Aly Ebrahim, 1999].

Thin Films and Nano Coatings

Over the past decade nanostructure materials have been the subjects of enormous interest, with the potential for wide ranging industrial, biomedical, and electronic applications. These can include metals, ceramics, polymeric materials or nano-composites.

Nano-crystalline materials are polycrystalline with grain sizes in the 1-100 nm range that are found to exhibit improved hardness, increased ductility and toughness, superior magnetic and optoelectronic properties. Nano-crystals are aggregates of atoms from a few hundred to tens of thousands that combine into a crystalline form of matter known as cluster.

Nano particles are less than 100 nm in size. At this level the percentage of exposed area in proportion to its total volume changes dramatically when compared to bulk materials. In particles of around 3-5 nm in diameter, about a third of the atoms are actually surface atoms; this means they behave a lot more like gas particles rather than solids. This gives them new properties, they float where bigger particles would sink, and dissolve more easily than the same material in bulk form. The extra exposed surface area also changes depending on how the element and particles interact with each other. Production techniques include electro-deposition and Gas Condensation Based Powder Production (Sputtering)

The Central Metallurgical Research and Development Institute (CMRDI) is a national institution serving foundries and metallurgical industries in Egypt. It is divided into four main divisions, including 18 departments as follows; minerals technology department, Metals technology department, advanced materials department, manufacturing technology department, technical services unit. Two departments are concerned with surface engineering namely: "Corrosion Control and Surface Protection; and Nano Materials and Technology.

The research activities include:

- Nano-coatings WC-Co
- Preparation and properties of $TiSi_2$ thin films from $TiCl_4/H_2$ by Plasma enhanced Chemical Vapor Deposition. This covers studying the effects of $TiCl_4$ flow rate and RF power, chemical reactions governing the $TiSi_2$ thin film growth. Effect of in-situ H_2 plasma cleaning.
- Comparison and application of electroless deposited and electroplated composite coatings.

- Preparation and characterization of Electroless Nickel Alloy–Teflon, Nickel-Phosphorus, Ni–MoSi2 composite and Nickel-Tungsten Phosphorus Alloy Coating.
- Performance of Zinc- Nickel Alloy Coatings as Anodes for Impressed Current Cathodic Protection of Steel
- Structure and Properties of Electrodeposited Ni- PTFE Coatings.
- Recycling of Hexavalent Chromium Bath in Metal Finishing Industry",
- Fabrication and Properties of Al- MMC Reinforced with Metallic Coated Fine Ceramic Particles.
- Preparation and investigation of Plasma Sprayed Cast Iron

Concerning the research activities at the Laser Institute–Cairo University, new project has been proposed to apply new technology in surface coating. It is proposed that new nano-materials such as SiO_2, Al_2O_3, Ba silicates will be synthesized. These materials can be used for restoration and conservation of Egyptian cultural Heritage and art-work made by stones.

The National Research Centre (NRC) was established in 1956 in Dokky- Giza. It consists of 120 Departments and 13 divisions. Some of the research activities in the field of surface engineering are as follows:

Thermal stability of nc-TiN/a-BN/a-TiB_2 Nano-composite coatings prepared by plasma chemical vapor deposition was investigated at the department of ceramics and advanced materials. It was shown that the thermal stability of nc-TiN/a-BN/a-TiB_2 nano-composite coatings is fairly high but lower than that of the nc-TiN/a-Si_3N_4 and nc-TiN/a-Si_3N_4/a-$TiSi_2$ coatings. The loss of boron from the coating together with the diffusion of chromium and other elements from the steel substrate into the coating during annealing are limiting the thermal stability of these coatings [Karvánkováa, 2005].

The study of selenium has attracted attention due to its weight in photometric applications and photovoltaic solar cells. Ca doped thin films were prepared by thermal evaporation technique by I. El Zawawi [El Zawawi, 2004] at the department of solid-state physics.

Nano-composite thin films of ZnSe/polymer nano-composites were prepared successfully at the department of solid-state physics by inert gas condensation method [El Zawawi, 2005]. Nano-composite systems could offer an enhancement in many electronic and photoelectron applications due to their improved optical and optoelectronic applications and solar cells.

The semi conductor CuGaxIn 1-x-Se2 is currently of interest as a thin film window material in conjunction with CuInSe used for photovoltaic solar cells. The structure and properties of thin polycrystalline films of such thin film prepared by vacuum evaporation were studied by B. A. Mansour *et al.* [Mansour, 2003].

A joint research study between Cairo University and Egyptian Atomic Authority was carried on preparation of cermet anode used for planar solid-oxide fuel cell with the required properties such as good mechanical strength, controlled amount of porosity, high electrical conductivity, beside being suitable for electrolyte layer application. The cermet anode was prepared by a wet powder mix process followed by cold pressing and sintering. The optimum process parameters such as compacting pressure, sintering temperature and porosity were determined [Ismail, 2005].

Mubarac City for scientific research and technology applications (Alexandria-Egypt) consists of 12 research institutes and technology centres. The focus sectors are:

- Biotechnology.
- Information technology
- Advanced Engineering.
- Nano technology (solar cells). Nanotechnology aims to improve the efficiency as well as cost aspects of solar cells. Some research was conducted on synthesizing anodes for Nano-crystalline solar Cells.

Thin films shape-memory alloys (SMAs) have been recognized as promising and high performance materials in the field of microelectronic mechanical systems owing to their strong recovery forces. Vacuum Plasma Spraying (VPS) and High Velocity Oxy Fuel (HVOF) are considered as preferred techniques for the production of dense TiNi coatings, due to their high deposition rate and high level of atomization. In this investigation, chemical composition, microstructure and phase transformation behavior of coated TiNi films on substrates of titanium and 316L stainless steel were studied. The surface and cross-section morphology of the coating were examined using chemical analysis and Scanning Electron Microscopy (SEM). The results from the Differential Scanning Calorimeter (DSC) showed martensitic transformation upon heating and cooling. X-ray diffraction analysis (XRD) revealed also the crystalline structure changing with temperature and heat treatment. A good Shape-Memory Effect (SME) was obtained through optimizing the conditions of spraying and heat treatment [Hamed, 2004].

Conclusions

Significant success has been achieved in the field of wear resisting and corrosion resisting coatings over a wide spectrum. Scientists of Egypt have made considerable achievements in the research relating to thin films, nano coatings, methods for their production and processing.

References

1. Abbas M.I, Ahmed M.S. and Ibrahim M.A., 1997, "Aluminizing of Monel Alloy by Pack Cementation", The Fifth International Conference on Pet., Mining and Metallurgical Engineering, 24-26 Feb., Suez Canal University, Suez, Egypt,.

2. ABECO, 1985,"Plasma Spraying, Annual meeting", Cairo-Egypt, October.

   Abdel-Karim R., 1988, "Wear Resistance of Plasma Coatings on Mild Steel", M.Sc thesis, Faculty of Engineering, Cairo University.

3. Abou-El-Enain Usama. Shaker M, Abou-El-Enain N.O., 1999, "A study of Petroleum Coating System trials", Mansoura Science Bulletin: Chemistry, vol. 26, No 6, pp 97-120.

4. Aly Ebrahim Hamed, Murad M., Abdel-Hamid I, 1999," Corrosion Protection of Steel Using Chromium Diffusion Coating "Engineering Research Journal. Vol. 64, Aug., pp 48-63.

5. El-Khatib A. F. and Rassoul S.M., 2005, " Thermal Spraying Processes", The 9th International Mining, Petroleum and Metallurgical Engineering Conference, 21-24 Feb.

6. El-Mesikawi, Bayoumi M.T., El-Zughbi M.A., 2000, "Investigation on Corrosion Protection of Internal Coated Crude Oil Pipelines", Engineering Research Bulletin Monoufia University, vol. 23, No 2, pp 67-88.

7. El Sabahy F., 1997, " Electro deposition, diffusion Treatment and Electrochemical Behavior of Copper-Nickel Multilayers on Low Carbon Steel Substrate", The Fifth International Conference on Pet., Mining and Metallurgical Engineering, 24-26 Feb., Suez Canal University, Suez, Egypt.

8. El Zawawi I.K., 2004, "Photo transformation Influence on Optical and Photoconductive Properties of Ca-Doped Se Thin Films", Phys. Low-Dim. Struct., vol. 7/8, pp 87-102.

9. El Zawawi I.K., 2005," Investigation and Optical Studies of Zn-Se Nanoparticle", Egypt. J. App. Sci., vol. 20(5B), pp584-592

10. Ismail K., 2005, "Ni-Ytria Stabilized Zirconia Anode for solid Oxide Fuel Cells", M.Sc thesis, Faculty of Engineering, Cairo University.

11. Grunling H.W., Scheider K., 1985, "Plasma Sprayed Coatings ", Thin solid films, 83, pp1-15.

12. Hamed O. A., 2004," Characterization of TiNi Shape Memory Alloy Thin Films formed by VPS and HVOF coatings", 8th Cairo University International Conference on Mechanical Design and Production., Jan 4-6, Giza (EGY).

13. Aly Ebrahim Hamed, Murad M., Abdel-Hamid I, 1999," Corrosion Protection of Steel Using Chromium Diffusion Coating "Engineering Research Journal. Vol. 64, Aug., pp 48-63.

14. Karvánkováa P., Vepřek-Heijmana M. G. J., Zawrah M. F. and Vepřeka S., 2005, " Thermal Stability of nc-TiN/a-BN/a-TiB2 Nanocomposite Coatings Deposited by Plasma Chemical Vapor Deposition (In Press).

15. Manosour B.A., Abdel-Hady S.A., El Zawawi I. K., 2003, "Structure and Optical Properties of Thin Polycrystalline Films", Fizika, vol. A12, No 2, pp75-88.

16. Morks M. F., Gao Y., Yingqing F. Y., 2005," Microstructure and Hardness Properties of Cermet Coating Sprayed by Low Power Plasma", International Surface Engineering Congress and Exhibition, August 1-3, St. Pull, Minnesota, USA.

17. The Surface Coatings Industry Worldwide 1997-2002, a market/technology report, Materials Technology Publications, Hertz, UK,
18. Patt and Whitny Aircraft, 1974, "Plasma Coatings, Standard Practice Manual", 70-46-4, 1-5 Nov.
19. Waheed A.F. and Soliman H. M., 1996," Characterization of Failure Mechanisms of Duplex and Graded Thermal Barrier Coatings Exposed to Thermal Shock Test", J. Mat. Sci. Tech., vol. 12, pp. 35-40.
20. Waheed A.F., El Awady G., Abdel-Samad A., 2004," Ion Implantation of PVD Deposited TiAl Layer by N,C and B Ions", 8th International Conference on Production Engineering Design and Control, December, Alexandria, Egypt.
21. Zawrah M., Schneiderb J. and Zum Gahrb K. H., 2005,"Microstructure and mechanical characteristics of laser-alloyed alumina ceramics", to be published.

Chapter 18

Status of Surface Treatment Industry and Evaluation of Characteristic Properties of Surface Coating Techniques

***S.S.K. Muthurathne*[1] *and Sarath Jayatileke*[2]**

[1]*Scientist, Building Materials Division,*
National Building Research Organisation, Colombo – 05, Sri Lanka
E-mail: sunethranbro@yahoo.com
[2]*Head Materials Laboratory, ITI, #63, Bauddhaloka Mawatha,*
Colombo – 07, Sri Lanka

ABSTRACT

Surface coating techniques are cost effective methods of protecting metallic and non-metallic products from degradation due to different mode of functional and environmental impacts. In Sri Lanka, application of surface coating techniques are becoming more popular as a result of depletion of natural products such as timber. With the shortage of naturally durable timbers, the trends are changing to the utilization of metallic, plastic, ceramic and innovative clay based products. The surface treatment methods currently practiced in Sri Lanka are Hot-Dip Galvanizing, Polyester Powder Coating, Anodizing, Wet Painting and Electro-Plating. Coating thickness is an important variable that plays a role in product quality, process control and cost control. Understanding the equipment that is available for coating thickness measurement, and how to use it, is highly related to every coating operation. Commonly used measuring techniques include non-destructive methods such as magnetic, eddy current, ultrasonic and micrometer measurement and also destructive methods such as cross sectioning or gravimetric measurement. Methods are also available for powder and liquid coatings to measure the coating thickness before it is cured. Eddy current and gravimetric methods are commonly used in Sri Lanka to determine the coating thickness. The correct gauge to use depends on the thickness range of the coating, the shape and type of substrate, the cost of the gauge and how critical it is to get an accurate measurement. The basic measure of a coating thickness gauge's performance is the accuracy with which the gauge takes readings.

Keywords: *Surface treatment, Surface coatings, Coating thickness, Corrosion resistance, Surface quality, Gauge performance, Sri Lanka.*

Introduction

Sri Lanka is an Island with an area of 65610 sq km. having 19.5 million population. Although, it is an agricultural country, manufacturing industry has grown significantly over the last few years, but no major smelting operation for the production of metallic materials such as iron, steel etc. is undertaken, except for a few small scale scrap melting industries. Although few ore deposits like magnetite etc, were found during the past few years, no industry was started to extract metals. Therefore, almost total requirement of metal is imported. During the last 15 years many small scale and medium scale industries were set up to manufacture metallic products using imported metal billets and sheets. With the development of metal forming industry, surface treatment and finishing became more important. Until recently, Sri Lanka used mainly primitive methods for surface preparation and surface coating. Surface treatment is needed to prevent corrosion, improve the surface quality, prevent contacts from other materials, improve scratch resistance, improve electrical conductivity etc. In Sri Lanka the most commonly applied surface treatment processes are electroplating and galvanizing. Polishing (mainly jewellery, braze and ornamental industry), Anodizing (chemical), powder coating, hardening and spraying methods also used in minor scale.

Surface coating techniques are cost effective methods of protecting metallic and non-metallic products from degradation due to different mode of functional and environmental impacts. The science of surface protection is now well established and the diffusion of the technology continues to be expanded to satisfy the changing customer requirement in respect of the applying environment, cost, quality and appearance. However, the technological improvements have inevitably increased the capital cost involved which is of little comfort to a business in an industrially developing country where access to investment capital is extremely difficult.

Present Status

In Sri Lanka, application of surface coating techniques are becoming more popular as a result of depletion of natural products such as timber. Traditionally, timber products like door and window frames, green houses, frontage panels, ceiling boards, partition walls were used for the various construction applications. With the shortage of naturally durable timbers, the trends are changing to the utilization of metallic, plastic, ceramic and innovative clay based products that have been saturated into the present market. The advancement of surface treatment technology has been the result of social and economic factors such as the limited availability of skilled labour and the liberal policies of the government towards the setting up of private factories. ISO Certified joint venture factories and several small and medium scale industrialists are presently engaged in the industry. The surface treatment methods currently practiced in Sri Lanka are Hot-Dip Galvanizing, Polyester Powder Coating, Anodizing, Wet Painting and Electro-Plating.

About fifteen steel processing industries, which carry out metal finishing processing processes, are in operating in Sri Lanka. All of them are in Western Province. There are also hundreds of small plants, machine industries, electro chemical industries and repair shops, where metal finishing operations are carried out.

Hardly any of these industries or shops have taken measures to reduce the discharge of wastes. Therefore, this industrial sector contributes heavily to the contamination of water sources, soil and ground water. Therefore it is necessary to introduce waste minimization programs, which eventually turns waste in to money.

Methods Used in Sri Lanka

Plating

Pre-treatment of the Metal Surface

Metal plating processes are preceded by pre-treatment processes, such as cleaning and stripping to remove grease, rust and dirt from the metal surface.

Cleaning

Cleaning is mainly carried out to remove dirt, oil and grease, which remained on the metal surface after machining, polishing and preservation stages. Grease is removed with an organic solvent such as benzene, gasoline, carbon tetrachloride or with alkalis available on the market as metal cleaners. These metal cleaners are usually mixtures of sodium carbonate, caustic soda, tri-sodium phosphate, sodium silicate, sodium cyanide and borax. After use the cleaning solutions are discharged as waste water.

Stripping

Stripping is used for removal of rust and scale from the metal surface. It is carried out by pickling with strong solutions of sulphuric acid or hydrochloric acid (in the case of iron articles). Recently developed techniques using electrolytic methods for stripping are also used. Fine sand particles from sand blasting can be removed from the surface with hydrofluoric acid. Spent acid solutions and rinsing water are discharged as waste water.

Electroplating

The principle of the electroplating process involves the use of an electrolytic cell for the deposition of dissolved metal ions on the surface to be coated. The metal surface is linked to the cathodes of the electrolytic cell.

Large quantities of metal and plastic parts are electroplated to produce a metal coating that imparts corrosion or wear resistance, improves appearance (through colour or lustre), or increases the overall dimensions. Virtually all commercial metals can be plated, including aluminium, copper, brass, steel and zinc-based die castings.

Plastics can be electroplated, provided that they are first coated with an electrically conductive material.

The most common plantings are zinc, chromium, nickel, copper, tin, gold, platinum, and silver. The electro-galvanized zinc plantings are thinner than the hot-dip coatings and can be produced without subjecting the base metal to the elevated temperatures of molten zinc. Nickel plating provides good corrosion resistance but is rather expensive and does not retain its lustrous appearance. Therefore, for lustrous appearance, a chromium plate is usually specified. However, chromium is seldom used alone. An initial layer of copper produces a levelling effect and makes it possible to reduce the thickness of the nickel layer to less than 0.0006 inch. The final layer of chromium then provides the attractive appearance. Gold, silver and platinum plantings are used in both the jewellery and electronics industries, where the thin layers impart the desired properties while conserving the precious metals.

The plating baths are acidic in nature and generally contain sulphuric, hydrochloric or nitric acid. Alkaline baths containing sulphide, carbonate, cyanide and chromium, nickel, zinc and silver. Normally these plating baths are reused and not discharged.

The electroplating of large articles such as motor car body parts (front, rare, buffer etc.) hand rail etc. were depleted with the introduction of PVC parts for motor cars and stainless steel products. Only very few companies are carrying out chrome and nickel plating. Both hard chromating and decorative chromating are being undertaken by these industries. Since new technique and upgraded methods are not being used, these industrialists struggle to survive. Most of these small scale industrialists use the traditional techniques given by older generation to next generation. They hardly study or invest in new techniques.

Electroplating is also used in jewellery and ornamental articles. For jewellery, gold, silver and rhodium plating are used. There are about thirty gold platters and about sixty silver platters in the country. Only very few gold platters are engaged in export oriented jewellery industry. Others are catering to local market. Plating techniques used by few exporters are comparatively better than others. Some of them use bright nickel plating prior to gold plating for better finish. Majority of the exporter still use the traditional polishing methods, which is not competitive with the new techniques used in developed countries. Therefore, it is very important to introduce new methods and equipments to Sri Lankan jewellers to be in international market.

Silver platters are also catering to jewellery industry. Most of these palters are confined to the southern province (Galle and Matara) where more tourist are attracted. In the central province (Ukuwela, Nattarampota etc.) silver plating is done for handicraft items.

It should be noted that most of the jewellery and ornamental items plating industries are catering for export market and tourist. Therefore, it is a must to educate them to use new technology for surface finishing, if they want to be competitive and survive in this industry.

Hot-Dip Galvanizing

Large quantities of metal products are given corrosion-resistant coatings by direct immersion into a bath of molten metal. The most common coating materials are zinc, tin, aluminium and Terne Metal (an alloy of lead and tin)

Hot-dip galvanizing is the most widely used method of imparting corrosion resistance to steel (The zinc acts as a sacrificial anode, protecting the underlying iron). After the products or sheets have been cleaned to remove oil, grease, scale and rust, they are fluxed by dipping into a solution of zinc ammonium chloride and dried. Next, the article is completely immersed in a bath of molten zinc. The zinc and iron react metallurgical to produce a coating that consists of a series of zinc-iron compounds and a surface layer of nearly pure zinc

Following galvanizing methods are commonly available in Sri Lanka (Figures 18.1–18.7).

Figure 18.1: Inspection Prior to Galvanize and Loading

Figure 18.2: Pickling

Figure 18.3: Fluxing

Figure 18.4: Drying

Figure 18.5: Hot-Dip Galvanizing

Figure 18.6: Quenching

Figure 18.7: Final Inspection

Benefits of Galvanizing

- Hot-Dip galvanizing is lower in cost than many other commonly specified protective coatings for steel

- It is more cost effective because it lasts longer and needs less maintenance
- A hot-dip galvanizing has a unique metallurgical structure
- Every part of a hot-dip galvanized article is completely protected
- Galvanizing enables a faster erection time

Powder Coating

Powder coatings are a cost effective solution for coating a diverse range of products. Electro-statically applied and oven cured, the process produces a tough durable finish with minimal impact on the environment. Since their introduction, powder coatings have continued to maintain a position as the fastest growing sector in industrial finishing.

Benefits of Polyester Powder Coating over Anodizing

- Wide colour range
- Consistency of colour match
- More economical
- Easier to repair
- Less reliant on quality of metal

Figure 18.8: Polyester Powder Coated Products Available in Sri Lanka

- Fabrications and welded assemblies can be coated
- For materials other than aluminium
- Different gloss levels are available

Others

Other than the metallic protecting coating, PVC coated metallic materials, such as chain links etc, are also produced. This gives protection as well as attractive colours to the product.

Polishing is extensively used in jewelry industry and also Cupper, braze, bronze and other alloy.

Evaluation of Coating Thickness

Coating thickness is an important variable that plays a role in product quality, process control, and cost control. Measurement of film thickness can be done with many different instruments. Understanding the equipment, that is available for film thickness measurement and how to use it, is useful to every coating operation.

The issues that determine what method is best for a given coating measurement include the type of coating, the substrate material, thickness range of the coating, size and shape of the part, and cost of equipment. Commonly used measuring techniques for cured organic films include non-destructive dry film methods such as magnetic, eddy current, ultrasonic or micrometer measurement and also destructive dry film methods such as cross-sectioning or gravimetric(mass) measurement. Methods are also available for powder and liquid coatings to measure the film before it is cured.

Magnetic film gauges are used to non-destructively measure the thickness of a non-magnetic coating on ferrous substrates. Most coatings on steel and iron are measured this way. Magnetic gauges use one of two principles of operation: magnetic pull-off or magnetic/electromagnetic induction

Following tests methods are used in Sri Lanka to check the coating thickness of the final products.

Eddy Current

Eddy current techniques are used to non-destructively measure the thickness of non-conductive coatings on nonferrous metal substrates. A coil of fine wire, conducting a high-frequency alternating current(above 1MHz) is used to set up an alternating magnetic field at the surface of the instrument's probe. When the probe is brought near a conductive surface, the alternating magnetic field will set up eddy currents on the surface. The substrate characteristics and the distance of the probe from the substrate (the coating thickness) affect the magnitude of the eddy currents. The eddy currents create their own opposing electromagnetic field that can be sensed by the exciting coil or by a second, adjacent coil.

Eddy current coating thickness gauges (*e.g.* PosiTector 6000 N Series) look and operate like electronic magnetic gauges. They are used to measure coating thickness over all nonferrous metals. Like magnetic electronic gauges, they commonly use a constant pressure probe and display results on an LCD. They can also have options to store measurement results or perform instant analysis of readings and output to a printer or computer for further.

Gravimetric Method

By measuring the mass and area of coating, thickness can be determined. The simplest method is to weigh the part before and after coating. Once the mass and area have been determined, the thickness is calculated using the following equation:

$$T = \frac{m \times 10}{A \times d}$$

where,

T is the thickness in micrometers, m is the mass of the coating in milligrams, A is the area tested in square centimetres, and d is the density in grams per cubic centimetre.

It is difficult to relate the mass of the coating to thickness when the substrate is rough or the coating uneven. Laboratories are best equipped to handle this time-consuming and often destructive method.

In this gravimetric method, the triple spot test or single spot test as per ASTM, BS or ISO is being carried out depending on the requirement and the results are reported as grams per square cm or as thickness of the coating.

Conclusions

At present surface treatment industry in Sri Lanka is not up to the required level compared to the regional level. Although few educational programs were conducted for these industrialists, no significant improvement was observed. Since the inferior quality of galvanized products produced by small scale industrialist, Ceylon Electricity Board, who is the major user of GI products set up their own plants for galvanizing. There are few other companies, where they produce good equality GI products. But only our company operate to produce Zn/Al coated sheets, using modern technology. Sri Lanka is a leading gem (Blue Sapphire, Star Sapphire etc.) producing country in the world. Therefore there is an opportunity to capture better market, if the jewellers were given training, technology and equipment. This is a must for Sri Lankan jewellery products to be competitive and survive in the international market.

In addition to metallic surfaces, it is important to treat surfaces of other materials such as ceramic, wood, plastic etc., to develop the surface treatment industry in Sri Lanka.

Finally it should be noted that new technology and knowledge should be given to Sri Lankan industrialists to improve the equality of their products.

Acknowledgement

We gratefully acknowledge M/S Lanka Transformers Ltd., M/S Alumex Marketing Service (Pvt.) Ltd., M/S Central environmental Authority, and Mr. A.W.W. Wijekoon for their support and information given to prepare this document.

Chapter 19

Preparation of Coating Films of Pb $(Mg_{1/3},Nb_{2/3})O_3$-$PbTiO_3$-$PbZrO_3$ by Sol-Gel Method and their Dielectric Properties Measurement

Silvester Tursiloadi

Research Centre for Chemistry, Indonesian Institute of Sciences, Kawasan Puspiptek Serpong – 15314, Indonesia
E-mail: tursiloadi@yahoo.com

ABSTRACT

The $Pb(Mg_{1/3},Nb_{2/3})O_3$-$PbTiO_3$-$PbZrO_3$ thin films were prepared by hydrolysis of Pb(iso-$OC_3H_7)_2$, Zr(n-$OC_4H_9)_4$, Ti(iso-$OC_3H_7)_4$, $Mg(CH_3COO)_2 4H_2O$ and $Nb(OC_2H_5)_5$ in propanol, and dipcoating. The film compositions were near the morphotropic phase boundary (MPB), xPMN-0.47PT-0.53PZ, (x = 0–0.35). The concentration of PMN-PT-PZ in coating solution was 0.7 M, and the pH of the solution was 4.5. The crystallization behaviors of the PMN-PT-PZ thin films showed that the formation of perovskite phase at low temperature becomes difficult with increasing the content of PMN. The amounts of pyrochlore and perovskite phase in PMN-PT-PZ films depended on the heating temperatures and PMN contents. The higher PMN addition retarded the formation of perovskite. When x = 0 and 0.125, films of single-phase perovskite was obtained after calcinations at 600 °C. The Curie temperature and dielectric constant were approximately 360 °C and 1615, 335 °C and 1615, 315 °C and 1664, 300 °C and 179, and 280 °C and 951 for the films coating 0, 12.5, 21, 30 and 35 mol per cent PMN respectively. The dependences of dielectric constant on temperature showed the broad maxima dielectric constant for the PMN containing films and the decrease in Curie temperature with increasing PMN content.

***Keywords**: Sol-gel, Coating film, PMN-PZT, Dielectric properties, Indonesia.*

Introduction

Sol-Gel processing involves hydrolysis and condensation of organometalic compounds with polymerization proceeding as the solution turns into gel form. In recent years, significant interests in the sol-gel processing are growing because their products are expected to be highly pure, homogeneous, and stoichiometric. In addition, it is a low temperature processing, low cost and possible to give various forms, bulk, film, powder, etc. [1–3]. For specific applications, *e.g.* electronic ceramics, low temperature processing for crystalline oxide films is required [4]. The development of thin films and coatings were the most interesting application of sol-gel technology [5]. The sol-gel processing is a relatively inexpensive method and it is compatible with a variety of substrate materials [6]. Recent years are marked by growing interest in sol-gel processes films for possible applications in microelectronics circuit, including FRAMs (ferroelectric random access memories) and DRAMs (dynamic random access memories) elements, and capacitor [7].

Lead magnesium niobate, $Pb(Mg_{1/3},Nb_{2/3})O_3$,(PMN) is a well-known ABO_3 type perovskite relaxor in which Mg^{2+} and Nb^{5+} are randomly distributed in the B site sub lattice [2]. It is well known that the perovskite relaxor ferroelectric PMN exhibits unusually high dielectric properties making it an attractive material for various dielectric and electrostrictive application. However, it is very difficult to prepare perovskite PMN thin films free from pyrochlore, [2, 3] and it has a low Curie temperature (–15°C) [4]. Little attention has been focused on its piezoelectric properties because of its low Curie temperature. In contrast, both $PbTiO_3$ (PT) and $PbZrO_3$ (PZ) have a perovskite structure similarly to PMN and high Curie temperatures, 490°C [4] and 230°C [8], respectively. The combination of the PMN and PT-PZ should confirm the following assumption: (a) Accelerate the formation of perovskite phase PMN, because perovskite phase PT-PZ is formed at lower temperature than perovskite phase PMN. (b) Elevated Curie temperature and increase in spontaneous polarization for PMN, because PT-PZ has a large spontaneous polarization and a high Curie temperature. (c) Elevated dielectric constant for PT-PZ, because perovskite phase PMN has a large dielectric constant.

The purpose of this study was to fabricate films with a ternary system $Pb(Mg_{1/3},Nb_{2/3})O_3$–$PbTiO_3$–$PbZrO_3$ composition near morphotropic phase boundary (MPB) at low temperature by sol-gel method. The crystallization behaviors and dielectric properties were investigated.

Experimental Procedure

Film Preparation

The film near the morphotropic phase boundary of the system, xPMN-0.47PT-0.53PZ, (x → 0–0.35), were prepared by sol-gel method. The starting solutions consisted of $Pb(CH_3COO)_2$..$3H_2O$, $Mg(CH_3COO)_2$.$4H_2O$, $Nb(OC_2H_5)_5$, $Ti(iso\text{-}OC_3H_7)_4$ and $Zr(n\text{-}OC_4H_9)_4$. The excess amount of 10 mole per cent Pb, more than the stoichiometric composition was used to prepare the solution, in order to reduce the crystallization temperatures for dense perovskite phase [9]. The lead acetate trihydrate was dissolved in acetic acid and then heated at 105°C for 2h to remove water. The dehydrated solution was cooled to below 80°C before the required quantity of zirconium tetra-n-butoxide was added. After adding zirconium tetra-n-butoxide, the solution was stirred at room temperature for 1h, and titanium (IV) tetra-i-propoxide in normal propanol was then added. After stirring at room temperature for 1h, niobium penta ethoxide in normal propanol was also added and stirring was continued at room temperature for 1h. Water, magnesium acetate, normal propanol and ethylene glycol were added to the solution and stirred at room temperature for 2h. Ethylene glycol was used as an additive in order to prevent cracking of the films during drying and to improve the surface smoothness of the films. The concentration of solution was kept 0.7 M, pH around 4.5, and 4 times of the theoretically required amount of water was added for the hydrolysis. By dip coating with a withdrawal speed of 10 cm/min, PMN-PT-PZ films were prepared on SiO_2 and Pt/Ti/SiO_2/Si substrates. In order to prepare thick films, dip coating and heating processes were repeated. The films were characterized by X-ray diffraction (RAD-C system, Rigaku Co.)

Dielectric Properties Measurement

Dielectric effects of PMN-PT-PZ on Pt/Ti/SiO_2/Si substrates have been characterized. The PMN-PT-PZ coating was prepared by dip coating with a withdrawal speed of 10 cm/min on the Pt/Ti/SiO_2/Si substrates. This procedure was repeated 9 times to obtained thick films. Calcinations of the sample were done at 750 °C for 2 hours using an electric furnace. The metalisation was made by sputtering with gold (Au) on upper surface of the sample using a mask system. The mask was a metal sheet punched a circle, 1 mm in diameter. The 2-terminal arrangement was employed. The structure of device is shown in Figure 19.1. The dielectric constant measurements were made at frequency 1 kHz, and temperatures in the range 25 to 400°C using an LCR meter (Ando Electric Co., type AG-430B).

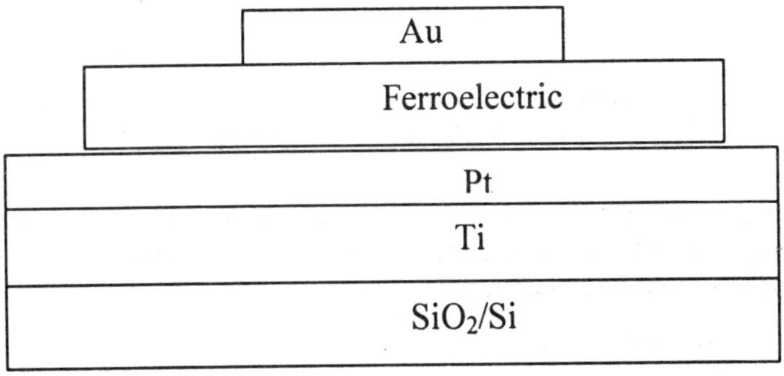

Figure 19.1: Devices Structure MOS

Results and Discussion

Crystallization Behaviors

All of the PMN-PT-PZ films on SiO_2 substrates are amorphous after heating at 450°C for 30 min (Figure 19.2). However, the PMN-PT-PZ films on Pt/Ti/SiO_2/Si substrate have pyrochlore structure (Figure 19.3). The amorphous films will be transformed into perovskite phase through pyrochlore phase, it transforms upon

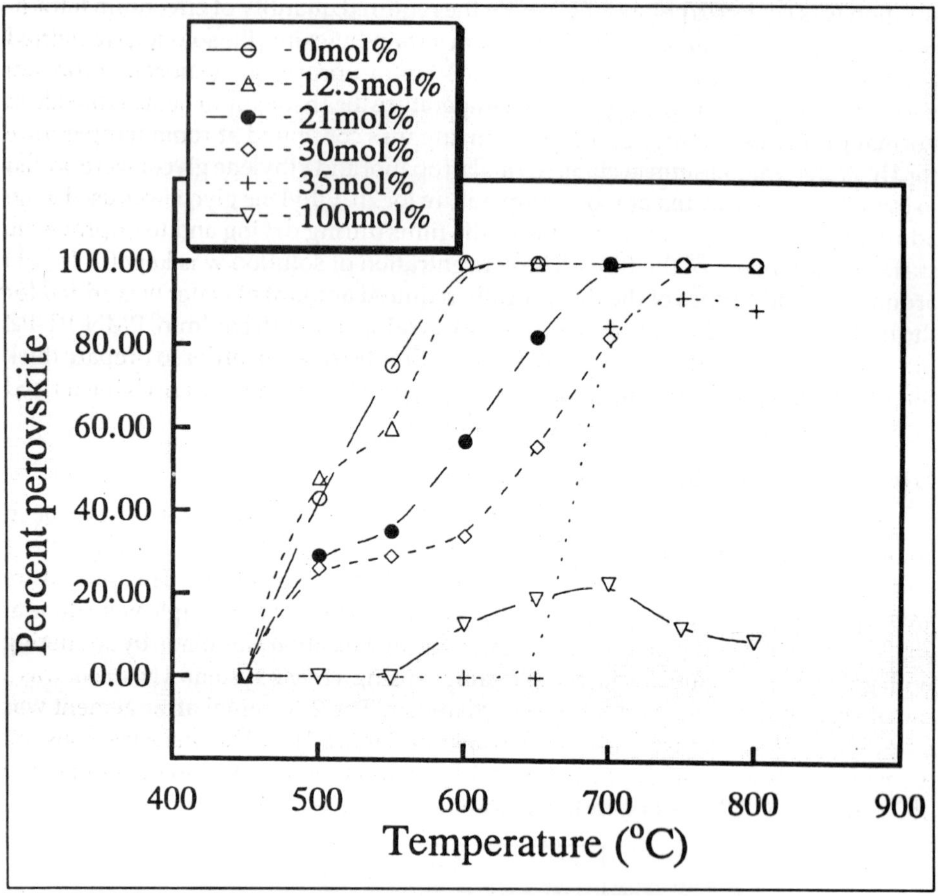

Figure 19.2: Percentage of Perovskite Phase of Coated Films on SiO_2 Vs Temperature Heating

increasing heating temperature. The transformation of the pyrochlore to the perovskite have been monitored by X-ray diffraction. The relative amount (mol per cent) of perovskite phase could be calculated from [4]

$$\text{"Percent Perovskite"} = I_{perov}/(I_{perov} + I_{pyro}) \times 100 \qquad (1)$$

where,

I_{perov} and I_{pyro} are the intensities of the major X-ray peaks, (110) at d = 3.056Å and (222) at d = 2.877 Å of the perovskite and pyrochlore phases, respectively. Coated films on SiO_2 substrates, after heating at 500°C for 30 min, the pyrochlore and perovskite phases began to appear for PMN content under 35 mol per cent. However only the pyrochlore phase appeared for PMN content at 35 mol per cent (Figure 19.2). This pyrochlore phase was stable until heating at 650°C. For the films coated on Pt/Ti/SiO_2/Si (Figure 19.3), the perovskite phase begins to appear after heating at 550°C and 450°C for 0 mol per cent and 12.5 mol per cent PMN content, respectively. However

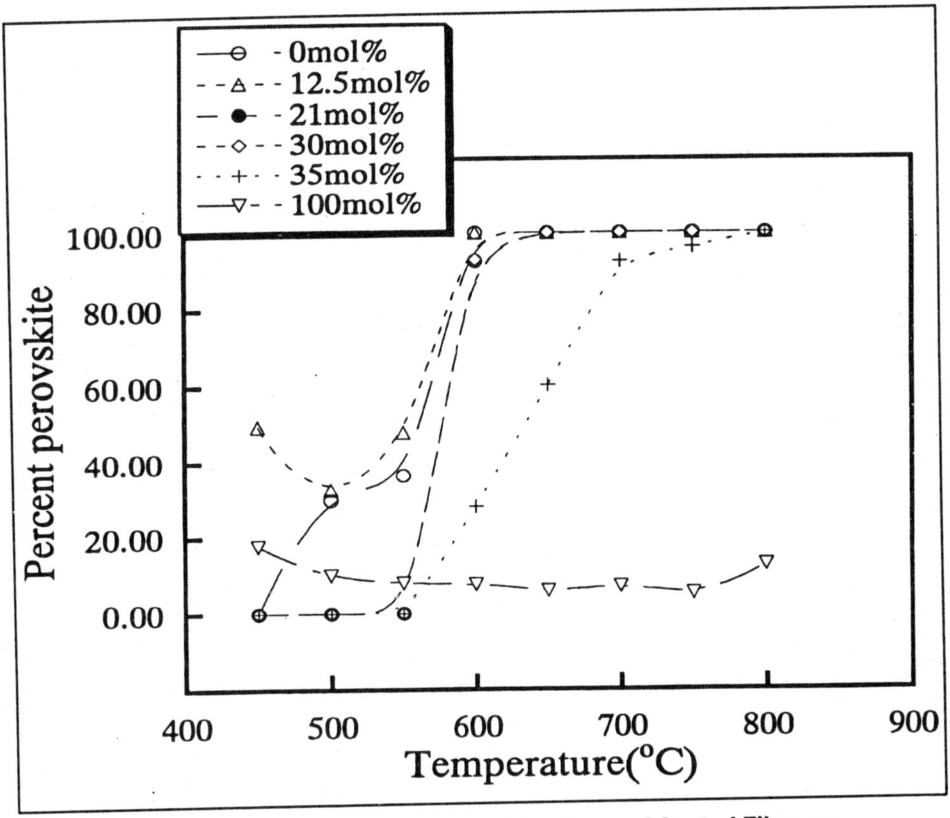

Figure 19.3: Percentage of Perovskite Phase of Coated Films on Pt/Ti/SiO_2/Si Vs Temperature Heating

until heated at 550°C, only pyrochlore phase appears for PMN content higher than 12.5 mol per cent. In Figure 19.2, single-phase perovskite was found for the coated films containing 0 and 12.5 mol per cent after calcining at 600 °C, 21 mol per cent after calcining at 700 °C, and 30 mol per cent after calcining at 750 °C. Single-phase perovskite of coated films will never be found when the content of PMN was 35 mol per cent.

The crystallization behaviors of the PMN-PT-PZ films showed that the formation of perovskite phase at low temperature becomes difficult with increasing the content of PMN. This was due to the temperature, which needed for the formation of perovskite phase free from pyrochlore phase of PMN was higher than that PT-PZ. As has been commonly known that the structure of a PMN-PT-PZ system goes towards pseudo cubic structure with increasing PMN content and at 35 mol per cent of PMN, the PMN-PT-PZ system belong to pseudo cubic domain. This fact implies that a single perovskite phase is difficult to be found at the 35 mol per cent of PMN. In Figure 19.3, single phase perovskite was found after heating at 600°C for 0 mol per cent and 12.5 mol per cent PMN content, at 650°C for 21 mol per cent and 30 mol per cent PMN content, and at 800°C for 35 mol per cent PMN content, respectively. Figures 19.2 and

19.3 show the temperature formation of pyrochlore and perovskite phase films coated on Pt/Ti/SiO_2/Si lower than that films coated on SiO_2. The crystallization behaviors of PMN-PT-PZ showed if the content of PMN increase, the formation of perovskite phase at low temperature became difficult. It is due to the temperature needed for formation of single perovskite phase of PMN, which is higher than that of PT-PZ. The temperature of formation of perovskite phase is 600°C for PT-PZ and above 700°C for PMN [10].

Single perovskite phase coated films on Pt/Ti/SiO_2/Si with content of PMN upper 12.5 mol per cent, can be occurred on temperature lower than that of films coated on SiO_2. This is due to the results from the crystallographic and d-spacing matching of PMN-PT-PZ (111) at d = 2.348 Å to Pt (111) at d = 2.259 Å, so the (111) plane of platinum may thus accelerate the crystallization of perovskite phase PMN-PT-PZ. On the other hand, this does not happen on SiO_2 substrate, due to the SiO_2 in amorphous phase.

Dielectric Properties

Characterization of dielectric properties, which is caused by PMN-PT-PZ ferroelectrics materials in the semi conductor device, has been observed. The relationships between dielectric constant and temperature measurement of PMN-PT-PZ films after heat treatment at 750°C for 2 hours are shown in Figure 19.4. The PMN content more than 12.5 mol per cent, resulted broad maxima of dielectric constant and decrease in Curie temperature (Tc) which occurred along with the increase of PMN content. The Curie temperature and dielectric constant were approximately 360 °C and 1615, 335°C and 1615, 315°C and 1664, 300°C and 179, and 280°C and 951 for the films coating 0, 12.5, 21, 30 and 35 mol per cent PMN respectively. The Curie temperature of PT-PZ around 360°C close to the Curie temperature of the PT-PZ in sub-solidus phase diagram at 0.47 of PT and 0.53 of PZ [8]. The value of dielectric constant increased with increasing of PMN content up to 30 mol per cent. However, for 35 mol per cent content of PMN has smallest dielectric constant.

It indicated that beside dielectric constant of PMN higher than PT-PZ, also the presence of pyrochlore phase, which is centrosymmetric, does not show ferroelectric properties.

Conclusions

Ferroelectrics ceramic films in the xPb ($Mg_{1/3}$,$Nb_{2/3}$)O_3-y$PbTiO_3$-z$PbZrO_3$ (x = 0–0.35, y = 0.47 and z = 0.53) system having the single perovskite phase can be prepared on SiO_2 if the mole fraction of PMN is $x < 0.3$. Single-phase perovskite was found for the coated films on both SiO_2 and Pt/Ti/SiO_2/Si substrates, containing up to 12.5 mol per cent PMN and heated at 600°C or higher temperatures. Single-phase perovskite of coated films on SiO_2 will never be found when the content of PMN was 35 mol per cent. The amount of pyrochlore and perovskite phase depended on the heating temperature, PMN content and kind of substrates. The formation temperature of perovskite phase coated films upper 12.5 mol per cent PMN on Pt/Ti/SiO_2/Si is lower than that on SiO_2 substrate.

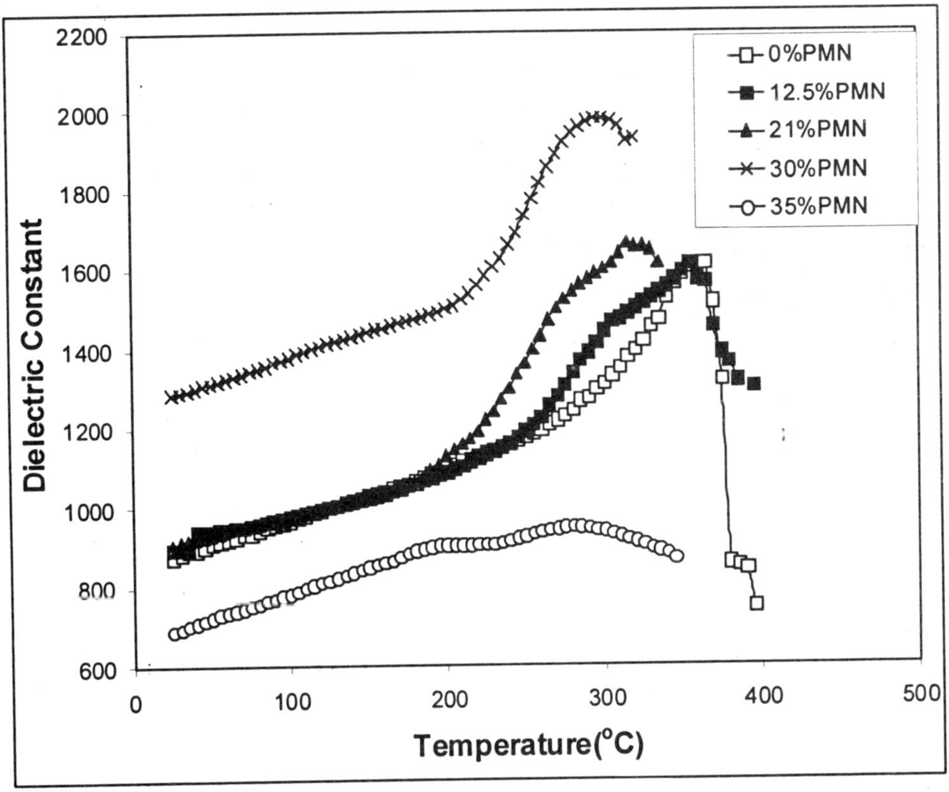

Figure 19.4: Dielectric Constant Vs Temperature Measure for PMN-PT-PZ Films with PMN Content after Heat Treatment at 750°C

The dependence of dielectric constant on temperature showed the broad maxima dielectric constant for the PMN containing films and the decrease in Curie temperature with increasing PMN content.

References

1. Azaroff, L.V., Introduction to Solids (McGraw-Hill, New York, 1960)
2. Jona, F., and Shirane, G., Ferroelectric Crystals (Macmillan, New York, 1962)
3. N. de Mathan, E. Husson, P. Gaucher and A. Morell, Mat. Res. Bull. 25, 427, (1990).
4. Chandler, C.D., Roger, C., and Hampden-Smith M.J., Chem. Rev. 93, 1205 (1993).
5. K. Okuwada, M. Imai and K. Kakuno, Jpn. J. Applied Physics, 28 (7) 1271, (1989).
6. S.K. Dey, K.D. Bubb, and D.A. Payne, "Thin-Film Ferroelectrics of PZT by Sol-Gel Processing," pp. 80 in IEEE Transactions on Ultrasonics, Ferroelectrics and Frequency Control., vol.35, No 1, January 1988.
7. Lorrâine Falter Francis and D.A. Payne, Mat. Res. Soc. Symp. Proc., vol. 200 (Material Reseach Society), 173 (1990).

8. J. Bernard, R. William, JR. Cook and J. Hands, Piezoelectric Ceramics 124 (1971)

9. S.D. Bernstein, Y. Kisle, J. M. Wahl, S.E. Bernacki, and S.R. Collins, "Effects of Stoichiometry on PZT Thin Film Capacitor Properties", pp. 373 in Ferroelectric Thin Films II, Proceedings of the Materials Research Society Symposium (Boston, MA, December, 1991). Edited by A. I. Kingon, E.R. Myers, and B. Tuttle. Materials Reasearch Society, Westerville, OH, 1992.

10. Lorraine Falter Francis and D.A. Payne, J.Am. Ceram. Soc. 74(12) 300, (1991).

Abbreviations

- PMN: $Pb(Mg_{1/3},Nb_{2/3})O_3$
- PT: $PbTiO_3$.
- PZ: $PbZrO_3$.

Chapter 20

Extending the Service Life of Oil Pumps by Detonation Coating and Status of Surface Engineering

Khalil Azimeh

Professor, University of Damascus,
Faculty of Mechanical and Electrical Engineering, Syria
P.O. Box 31262 Damascus
E-mail: khazimeh@aloola.sy, k.azimeh@mail.sy

ABSTRACT

Long life of the equipments for oil extraction has big importance in the national economy. This equipment breaks down due to wear and tear. The corrosion corrodes the active surfaces of the most downloaded equipment. Protection of the surface of new parts and its restoration after a specific period of use with the corrosion resistant coatings is of economic significance and is a new technology. The method of detonation precipitation has an important place among different methods of thermal spraying which allow to get high density gas coating and high strength of tenacity with basic metal. In this study, we coated steel with Al_2O_3, Al_2O_3+3 per cent TiO_2 and Al_2O_3+3 per cent Cr_2O_3, and tested this materials for friction contact and cavitation conditions and studied the operations on the surface during the tests, and compared this material with another material in accordance with test results. The coated equipment for oil extraction (parts of plunger oil pump and the pump's parts) were tested in normal work conditions which showed that the coating increases the service life of these parts up to 5-8 times than the parts without coating.

Keywords: *Surface engineering, Surface modification, Oil extraction pumps, Detonation coating, Syria.*

Introduction

Improve and increase reliability of oil extract equipment consider one of the most economic and technology issues in field of building machines which enables to economize fundamental materials, efforts and energy. Oil extract equipment usually break-down due to the wear of active surfaces and to mechanical loads on it and the effective chemical gas milieu or the liquid one and high pressure, temperature and velocity.

In conclusion,the reasons of break-down of surfaces are: friction, abrasion, cavitations, chemical and elect-chemical corrosion which can be classified according to basic phenomenon, which determine its effective as a mechanical corrosion (which happens due to plucking out molecules or injuring or breakdown surface layers) and the mechanical–molecular corrosion (molecular coupling is happened on separated places of coupling surfaces and form a contact) as well as a mechanical rust (the products of friction surface react with effective milieu and exclusion products from area). For example of exposed equipment to corrosion we mention operating elements in oil's extract pump (clutter, rings, sealers valves.etc.). Protect the operating surface of elements with resisting corrosion coating and restoration this elements after wear is considered the optimal solution.

Thermal spray coating is the active method of precipitation methods, and it is used widely, and this allows saving phys-mechanical and element shape thus improve the thermal spray coating precipitation and use it in preparing and restoring machines element will be interested for all industrial developing countries. Moreover detonation coating method is considered the most important method because it produces a coating with high quality for that it has been chosen for this research.

Detonation Coating Equipment

Precipitate the Coating by Detonation

Detonation coating precipitation has been performed as follows: Detonation machine with closed side barrel is filled with combustible gas mixture (*i.e.*. oxygen-acetylene) then we put some of powder required to be precipitate and put the piece required to be coated in front of the barrel. The gas mixture begin to combust due to electrical discharge combust products run out of the barrel rapidly carrying the powder and impact the piece to form the coating. The molecules speed reach to 2000-3000 m/sec and heat to temperature 3000-5000C°. The cohesion of molecules powder with piece due to formation different kinds of contact between the contacted surfaces.

This technology has the following characteristic:

1. Ability to get coating with good cohesion (160 Mpa) and high density (porosity 0.5 per cent)
2. Ability to precipitation strength coating on different types of metals
3. Precipitation equipments are simple

Disadvantages of this method

1. High noise level (120-140 dec)

2. Disability of precipitate the coating on un-rigid surfaces
3. Surface acceptable for coat should have hardness more than 60 HRC

Automatic equipment of precipitation have the following technical features:

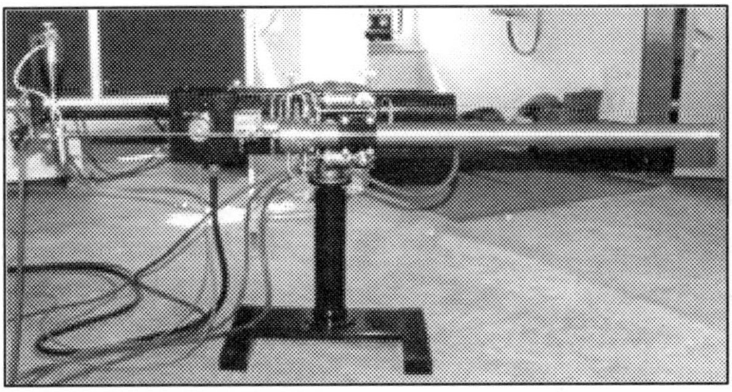

- Maximum temperature 3500ºC
- Operative gases are oxygen " azote " acetylene" argon" and methane
- Pressure of operative gases is 0.05 to 0.14 Mpa
- Consumption of operative gases is 1.0 to 4.5 m^3/h
- Caliber of barrel is 16–25 mm
- Speed of shooting is 1-5 cycle/sec
- Coating thickness for one shot is 8-15 µm
- The productivity is 1-5 kg/h
- The capacity is 350 V A
- Power supply 220A

The Materials

For precipitate the coating we use standard powder of Al_2O_3 have grains size 20-36 µm and mechanical mixture of Al_2O_3 +3 per cent Cr_2O_3 and Al_2O_3 + 3 per cent TiO_2.

Coating Resistance of Abrasion

Surface failures as due to abrasive molecules consider the most important kind of wear. The resistance of compacted materials against abrasion determine not only by hardness but by crystal structure too. Sometimes crystal structure consider less important than hardness for determine the wear resistance against abrasion. The materials hardness specifies the resistance of the materials against abrasion just if the hardness of the abrasing material more than the material being abrased. In accordance with above the oxides, which have high hardness consider the most resisting material against abrasive wear. Detonation coating has been performed from aluminium oxide in non-fixed abrasing conditions. Selenium oxide has been used as abrasing material fall in the coating surface which goes around rubber cylinder and this oxide caused abrasion to the sample.

The loss in the sample weight is determined by the difference between the weight before and after test on sensitive electron balance the following table shows some of test result :

Table 20.1: Wear Resistance of Some Materials and Coating

Sl.No.	Type of Tested Material	Composition of Coating or Material Hμ,Kg/mm²	Micro-hardness	Relativity Resistance of Wear
1.	Quenched steel	-	600	1
2.	Alomeseletsat	$Fe_3(Al,Si),Fe_2Al_5$	680-770	1.93
3.	Detonation coating	90% γ-AL_2O_3-+10% α-Al_2O_3	1150-1200	2.32
4.	Detonation coating	80% γ-Al_2O_3-+17% α-Al_2O_3+ 3% TiO_2	1300-1400	3.65
5.	Detonation coating	2% γ-Al_2O_3-+95% α-Al_2O_3+2% TiO_2+ 0.5% FeO+0.5% SiO_2	2100-2300	4.12

According to the table, the most resistance against wear by abrasion is detonation coating,which contains, in addition to Al_2O_3 Titanium, iron and selenium oxides the micro hardness of this coating four times more than the micro hardness of quenched steel, and two times more than the Alomeseletsat layer which we get from saturation the surface layer with Aluminium and selenium (diffusion, metallizing and non-metallizing). We conclude in this case that coating hardness has the importance decisive the analysis of abrased coating surface shows that the wear is occurred due to micro scratch \ micro cut \ in some special conditions dependant grains have been removed from coating structure.

As we know that in accordance with the relation between the hardness of abrasing material (H_A) and metal (H_M) makes a difference between three kinds of wear by abrasion:

1. Mode of soft wear when $H_A<H_M$
2. Mode of middle wear when $H_A=H_M$
3. Mode of strong wear when $H_A>H_M$

then in order to reduce the abrasion which occurs the wear, the material hardness should be 1,3 times more than hardness of abrasing material : H_M=1.3 H_A.the scientific references don't advise to increase the hardness of materials 1.3 times more than hardness of abrasive materials because this extra doesn't improve resistance against wear. In another words the only solution to improve the resistance against wear in conditions of wear by abrasion is to apply the following relation: H_M=1.3 H_A.

The Technical Specification of Coating in Traibology Friction Conditions

We have determined the resistance of detonation coating against wear and abrasion, in traibology friction conditions in air and industrial lubricant I–20, by diagram column-bearing, the column is made of quenched–steel CT-45 and the coating was precipitated on bearing surface which is made of steel CT-45, the tests was performed by the laboratory machines of friction. The coating without lubricant consist of the following materials have been tested in traibology friction conditions: Al_2O_3+TiO_2, Al_2O_3 according to variable of bearing force within range 0.5-4 MP in constant rolling speed 1 m/sec for distance 1 km. During the test the friction strength, linear and weight wear, and temperature in the friction area have been determined. Figure 20.1 shows that if bearing strength increase the wear increase too. The strength of oxide layer on friction surfaces depend on portion of fundamental metal hardness and its oxide. If the oxide layer has a high hardness and the fundamental metal is, tender the layer will be breakdown rapidly. Such as what happen to aluminium oxide when the fundamental is pure aluminium then the hardness portion of aluminium oxide and aluminium is 22000:2Mpa and this effects on layer strength and will be the reason for increases the wear of machine parts, which are made of aluminium alloy.

During friction of precipitated oxide coating on steel parts, which have hardness more than aluminium, crush of oxide layer will be done and break down will be noticed just under the effect of over loading. Second stage of experiment was done in friction traibology conditions without lubricant when temperature of surface layers increase to 100-200°C and in the contact points of micro prominences on the surfaces,

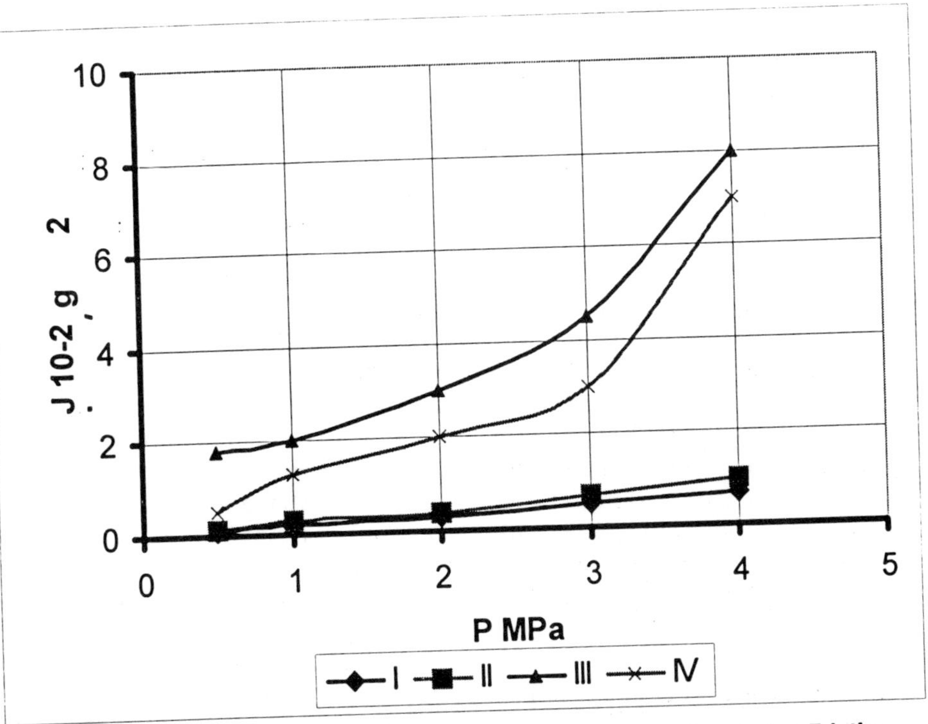

Figure 20.1: Relation Between Wear (J. 10^{-2} g/cm²) and Loading during Friction (1) Coating of Al_2O_3 with quenched steel; (2) Coating Al_2O_3+3 per cent TiO_2; (3) Wear of quenched steel with Al_2O_3; (4) Wear of quenched steel with Al_2O_3+TiO_2.

the temperature may increase whereas in this conditions plastic deformations are happened for thin layers coating and smoothing of grains is happened, and this grains are effected by air and new crystal structure have been created on friction surfaces. If the Al_2O_3 hardness is 2100-2300kg\mm^2 in the normal temperature it will become 600kg\mm^2 in the temperature 90°C.

During the friction of oxide coating with quenched steel, an oxide layer has been created on surface of foundation iron due to this a react between aluminium oxides and iron oxides is happened iron oxides are appeared on the active surface of coating oxide, but with pressure and temperature effect a complex union of spinel was formed on the micro surface prominences. In friction conditions with lubricant the wear strength and friction factor less more than friction without lubricant for all kinds of coating. Elctrocrond coating have the maximum value of friction factor and Al_2O_3 coating have the minimum one which equals 0.05, but other kinds of coating equal 0.09-0.095. It is noticed that maximum wear in friction group of Al_2O_3 coating is 5.5 µm/km under loading 1 MPa and rolling speed 1 m/sec. In this case the temperature in friction area is taken in 1mm of coating surface and it is reached to 40C°. This temperature don't effect on friction operation and in this case the oxidation on active surfaces happens slowly but lubricant in friction area cause formations of secondary crystal structure which decreases friction factor and wear.

Coating Type	*Friction Factor*	*Wear of Coating*	*Temperature*
γ–Al_2O_3	0.05	5.5	40
Al_2O_3+TiO_2	0.09	2.18	55
Eletrocrond	0.095	2.25	60

Coating Resistance Against Wear of Cavitation

Recently different kinds of machines, are used for study the wear occurred by cavitation, some of these machines are magnetic vibrator, ultra sound waves machines, jet current machines which forms bubbles have high dynamics in liquids. The machines which have various nozzles (one of them is Venture pipe), consider more development and provide high rate of formation bubbles. Various diffusion nozzle is used and the samples like rings with thickness 3mm were fixed vertically on path of flow current in order to increase wear strength and decrease time of experiments. Bubbles formation and diffusion them is regulated by changing the pressure between input and output of the liquid which is hydraulic oil AMG-10 the pressure in the input is constant: Pin = 15MPa and the difference between input and output is Dp = 0,93MPa, Velocity of liquid =180m/sec, Pump pressure Pp=0.5Pa. For comparison and in addition to coating experiment,we tested high resistaning materials, and low one for this phenomenon and some of this materials, are cast iron H1-50, bronze, quenched steel 45,chemical treatment steel CT-45 with boron. First stage of wear take important place in studying and this stage provide exact determination of breakdown mechanism for later stage. Use tender and plastic aluminum alloy allow to determine the kinetic of breakdown with the effect of cavitation from exploration of bubbles to strong occurrence of cavitation. Figure 20.2 shows experiment results of some materials in cavitation phenomenon. It is noticed from the figure that the precipitate of detonation coating have high resistance in comparison with other materials except Nickel-Crom steel which a little bit better.

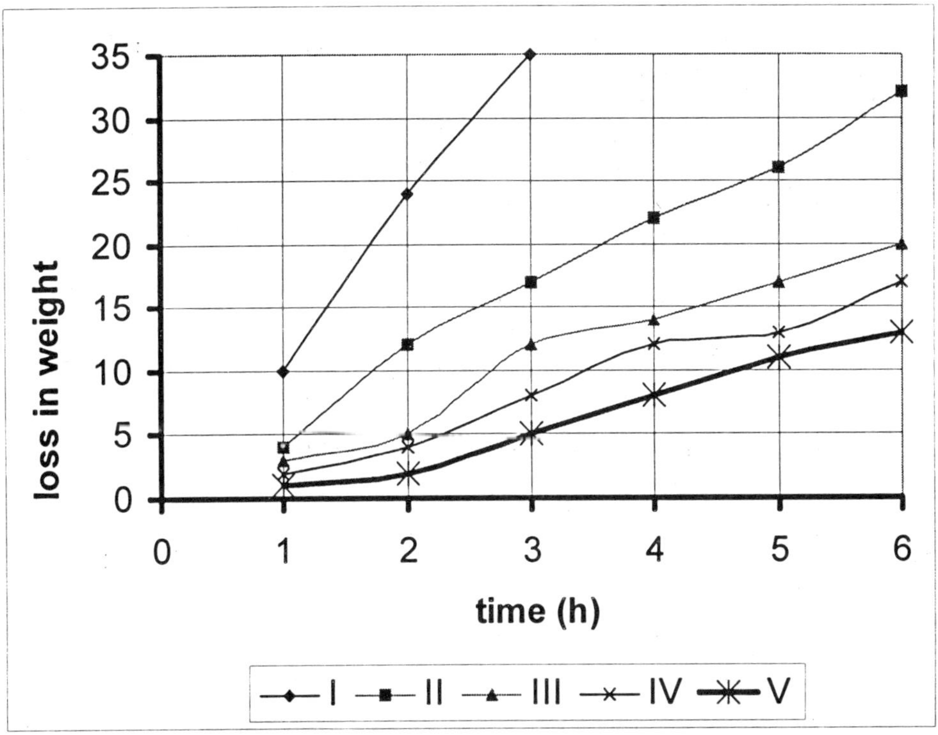

Figure 20.2:The Relation Between Loss in Weight and Time in Cavitation Conditions
(1) High strength cast iron; (2) Bronze; (3) Ct-45 quenched steel;
(4) Ct-45 steel treatment with boron; (5) Steel coated by precipitation with detonation

Analyze of Tested samples, shows that on the active surface of detonation coating which consist of big single grains, which have low porosity, there is no cavitation breakdown in the first hour of experiment. Whereas explosion bubbles don't cause cracks, extractions, and holes. With passing of time a micro deformations is appeared on the coating surface in the holes and micro curves and explosion bubbles places. And this micro deformations with continues of current flow, causes micro cracks, which increases and cause breakdown of coating between molecules and grains and extract grains from the coating and moved it away of cavitation area. Relation between loss in coating weight and time has curve line shape, but aluminium and bronze alloy has straight line shape because a plastic deformation is happened for this materials.

Time rang of deformation for experiment materials has extended to become 5 to 120 minutes, but for detonation coating it was 75 minutes. According to that the Nickel-Crom steel has time range better than the coating one. In accordance with experiment results the detonation coating on CT-45 steel has effected on resistance against cavitation and there are no signs of plastic deformation extraction or breakdown of surface layer. This explain that the detonation coating have big single grains which have low porosity no more than 0.5 per cent and has high cohesion with foundational metal and high resistance against abrasion and corrosion.

Materials and abrasing molecules in the liquids increase wear by cavitation. This case is so sophisticated, and the wear strength in cavitation with abrasion is more than normal cavitation. This explain that in wear by cavitation and with abrasing materials the breakdown doesn't happened due to just bubbles explosion and micro impacts but also from impacts which happened due to abrasing molecules with velocity more than current liquid velocity which causes the cavitation. In addition to that the impacts in cavitation cause thermal,electrical and chemical operations. Then extract of metal molecules and increasing in wear by abrasion are happened simply. According to that, we can specify requirements, which increase, resistance against wear of coating in cavitation phenomenon and abrasive phenomenon.

1. We should choose work system, which provide us a coating with low porosity when we precipitate detonation coating.
2. If there are abrasing particles, the coating should have a high resistance against wear by abrasion.

Field Experiment

The field experiments have been done on samples of submerged axial pump type ЭЦНИ 5. The results show that the resistance of pump against wear and corrosion is 8 times more better service live than the similar. This pump precipitated with detonation coating of the following powders Al_2O_3+3% TiO_2, Al_2O_3+3% Cr_2O_3, Al_2O_3

The experiment results as the following :

Submerged Axial Pump

- Type of pump : ЭЦНИ 5 No. 080
- Pumping flow : 40 m^3/day
- Delivery head : 650 m
- Speed of rotation : 3000 r.p.m
- Material of pump parts is austenite cast iron
- Pumping efficiency : 99 per cent
- Percentage of mechanical gangues : 0.1 g/h
- pH : 6
- Working day : 735

The Results

- Wear in non coated parts 1.62-2,33mm
- Wear in parts 1 to 6 coated by Al_2O_3 is 0.2mm

Figure 20.3: Submerged Pump Diagram

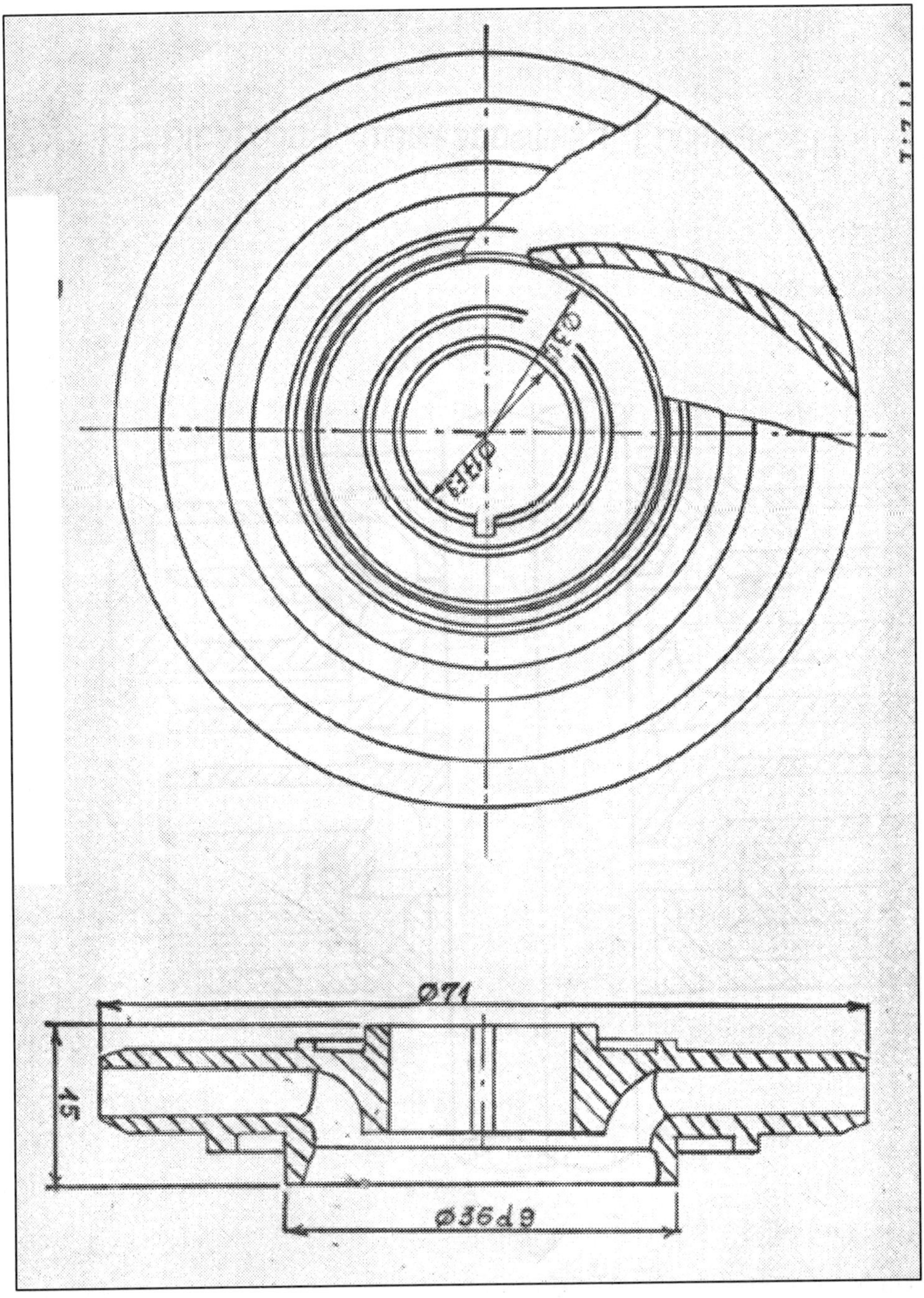

Contd...

Figure 20.3–Contd...

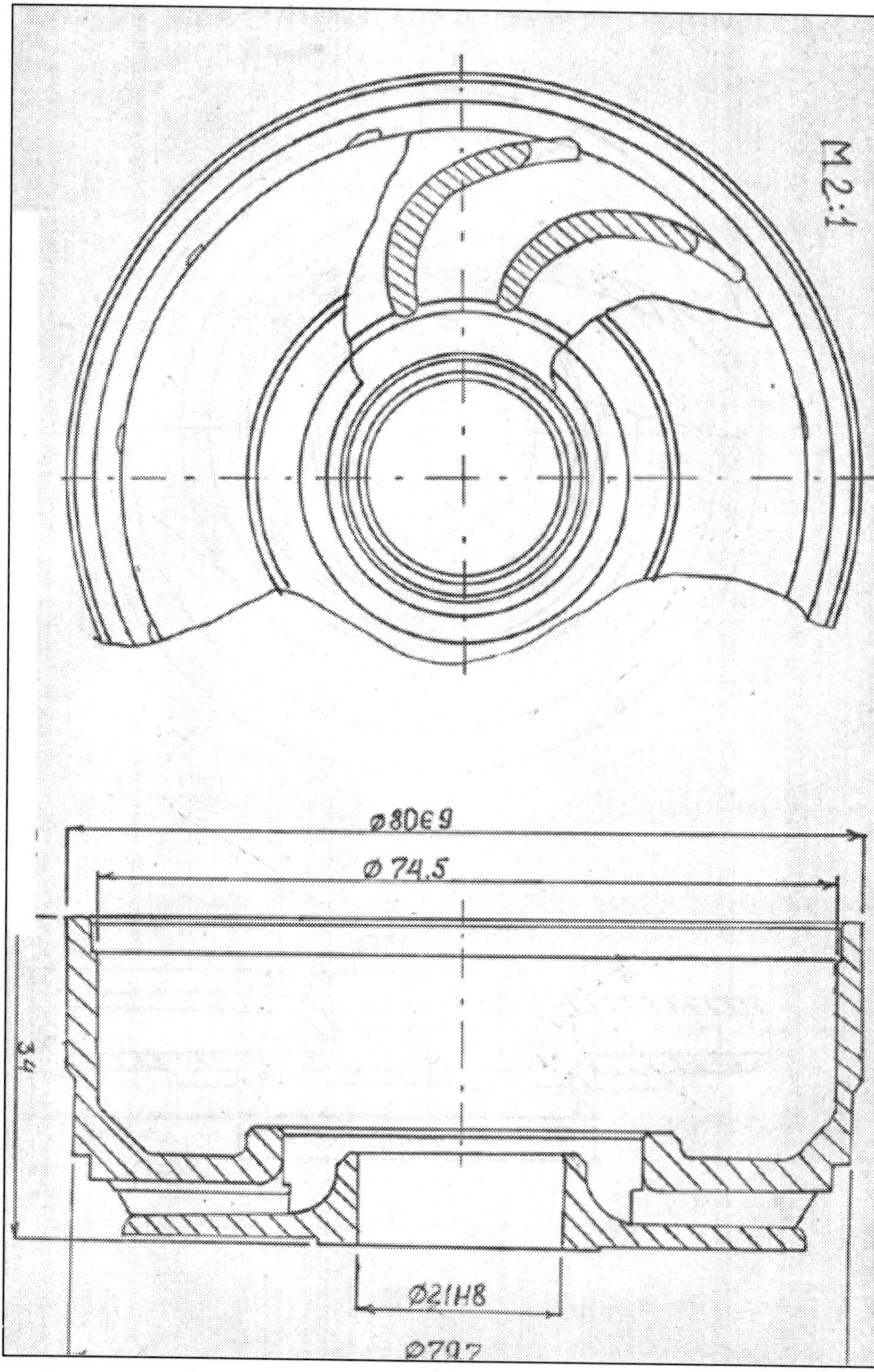

Contd...

Figure 20.3–Contd...

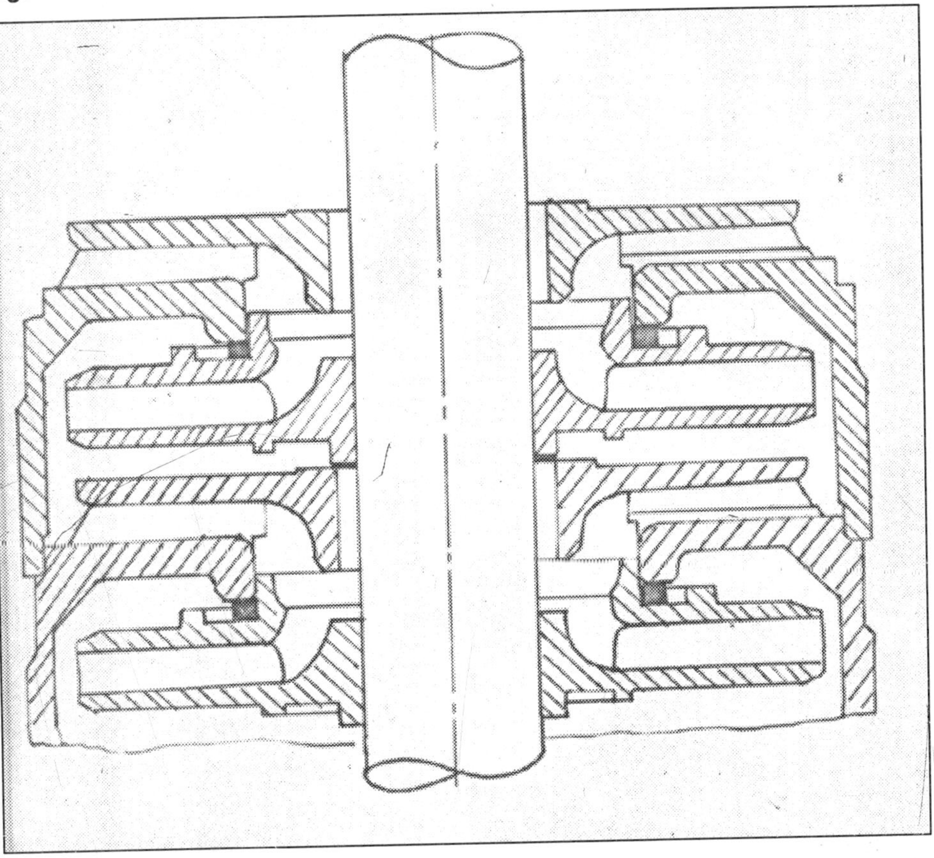

- Wear in parts 7 to 8 coated by Al_2O_3 is 0,02mm
- Wear in parts 9 to 14 coated by Al_2O_3+3% Cr_20_3 is 0,35mm
- Wear in parts–15-is not present
- Wear in parts 16 to 26 coated by Al_2O_3+3 per cent TiO_2 is 0,25-0,45mm

Due to that development of precipitation by detonation coating, and extend practical field to use it in preparation of new machine parts, and restore wearing parts in order to return it back to service, have taken important place. Moreover this methods provide big ability to increase resistance of equipment parts against wear and improve work quality of this parts and reduce using the metals. So it has economic importance and help to better exploitation of industrial wealth.

Results

1. In abrasion conditions it is noticed that the reason of resistance of coating against wear is its hardness and if it increases the resistance against wear will increase too. Maximum resistance is for (α–Al_2O_3) coating which is 4,22 times more than quenched steel.

2. The operations on coating surface have been considered during tribology friction. We conclude that in this conditions,oxidation operations on fricting surface have been done. In case of loading 3mpa without lubricant accompanying operation of abrasion, on the coating surface is happened in addition to oxidation. In this conditions the maximum resistance against wear is for quenched steel with Al_2O_3 coating.
3. In conditions of wear by cavitation surrounding with AMG-10 aircrafts engine lubricant it is noticed that there is no strong wear and with over looking that the oxides are fragile, the high resistance against wear and corrosion and low porosity/not more 2,5 per cent/and high cohesion with foundation provide high resistance against wear by bubbles. The coating resistance against wear by bubbles is more than cast iron, bronze, quenched steel but Nickel Chrom steel is little bit more than coating.
4. The designed coating passed production test of submerged axial pump parts of oil extracting. Moreover the coating increase service life from 5 to 8 times.

References

1. E. Lugscheider, P. Remer, C. Herbst, K. Yushchenko, Y. Borisov, A. Chernishov, P. Vitiaz, A. Verstak, B. Wielage, S. Steinhauser, in: H. Ohmori Ed., Proc. 14th Int. Thermal Spray Conf., Vol. 1, High Temp. Soc. of Japan, Kobe, Japan, 1995, 235–240.
2. K. Azimeh, Optimization of technology detonation coating and application on equipment used for extraction of petroleum Kiev. 1994.
3. K. Niemi, P. Vuoristo, T. Mantyala, E. Lugscheider, J. Knuuttila, H. Jungklaus, Wear characteristics of oxide coatings deposited by plasma spraying, high power plasma spraying and detonation gun spraying, in: Proc. 8th Nat. Thermal Spray Conf., Houston, TX, 11–15 September 1995, ASM International, Metals Park, OH, 1995, pp. 645–650.
4. M. Uma Devi, New phase formation in Al_2O_3-based thermal spray coatings Ceramics International 30 (2004) 555–565.
5. P. Saravanan, V. Selvarajan, D.S. Rao, S.V. Joshi, G. Sundararajan, Influence of process variables on the quality of detonation gun sprayed alumina coatings. Surface and Coatings Technology 123 (2000) 44–54.
6. V. Fervel, B. Normand, C. Coddet, Tribological behavior of plasma sprayed Al_2O_3–based cermet coatings. Wear 230 1999 70–77.
7. Y.N. Wu, F.H. Wang, W.G. Hua, J. Gong, C. Sun, L.S. Wen, Oxidation behavior of thermal barrier coatings obtained by detonation spraying Surface and Coatings Technology 166 (2003) 189–194.

Chapter 21

Corrosion Protection and Finishing of Automobiles and Status of Surface Engineering

Shahid Tufail Sheikh PSO

MSRC, PCSIR Laboratories Complex Ferozepur Road, Lahore-54600, Pakistan

E-mail: shahidsheikh200@hotmail.com, smtp.pcsir@brain.net.pk

ABSTRACT

Finishing of automobiles is an important aspect. There have been considerable reductions of weight in automobiles by the use of composites components replacing heavy metallic components. Fenders previously based on metal have been replaced with plastic and painted with the same colour shade as of the metallic body. This has necessitated that finishing process takes proper steps for proper adhesion of the paints on the plastic fender to avoid chipping off the paint form it. This paper discusses the necessary processes required for finishing of an automobile along with the corrosion protection measures. Automobiles contains a variety of engineering materials, engine main body fuel tanks connecting rods heat radiators and other mechanical parts are made from different types of engineering alloys having varying chemical compositions. Other parts like dashboard, front panel and other are made from composites. The main body made from cold rolled steel having various contours in it due to the different designs is the potential site for corrosion attack. The main body is exposed to the hostile environment through out its life period. An automobile is given a particular finish with a view to counter the hostile environments as they are not limited for plying in a limiting conditions and are taken to different weather conditions in one day thus facing severe stresses and strains. Thus, it is essential that an automobile before rolling out of the assembly line should be properly corrosion resistant and aesthetically pleasant.

Keywords: *Corrosion resistance, Automobile finishing, Alloys, Plastic components, Surface treatments, Surface modification, Pakistan.*

Introduction

Finishing for automobiles is a very specialized job, its main requirement being maximum durability with minimum numbers of coats baked at the fastest possible schedule, high gloss with a range of good eye catching colours. This is important to increase sales appeal. In the past, the car finishing has been based on alkyd-amino resins baking materials and force drying lacquers, which give excellent appearance originally and maintain it on aging. The finishing system for the synthetic baking type may consist of one primer coat and a double finish coat. The two finishing coats are applied one immediately after the other, and both are baked simultaneously. An alternate system is to apply a red iron oxide epoxy primer followed by a gray epoxy primer and to bake the two coats at 200°C for about 35 minutes. The dry film thickness is about 1.5 mils. This coating is wet sanded, washed and dried, and then top-coated with a double (wet-on-wet) coat of alkyd–amino resin enamel. The enamel is baked at 120°C for about 35 minutes. The lacquer system consists of one prime coat followed by several coats of lacquer finish. Number of coats depend on the price range of the car. All the efforts are made to make the metal surface as smooth as possible and free from rough places due to spot wielding and filing. This means a minimum of sanding on the primer, thus saving in labour cost but also makes possible less pigment in the primer resulting in better hold-out of the finish. However, the primer must be hard enough to sand easily, because rubbery primers are slow sanding and tend to show scratch marks from the sand paper. All metal surfaces are given a passivating treatment before application of the primer.

Pre-treatment

Cold roll steel sheets used for making car bodies are covered with grease materials, rust, dirt etc. It is therefore essential that the surface is thoroughly cleaned of from the dirt, grease, corroded products, rust and all other contaminants in order to ensure maximum durability. Thus a pre-treatment sequence is carried out in order to achieve the desired results. Following pre-treatment processes are carried out.

Kerosene Wiping

This is done to remove grease and other oily material from the surface of the car

Wiping with Warm Water at 35-40°C

This involves the removal of dirt, inorganic contaminants and loosely adhering rust from the surface of the car body.

Degreasing

Degreasing is carried by immersing the car bodies into a tank containing degreasing solution. This solution mainly contains phosphates, sulphates and neutral surfactants. These chemicals assist in complete removal of oil, grease and drawing compound from surface of the substrate. Temperature is kept at 45-55°C, immersion time should be 2-3 minutes and total alkali content should range between 18-27.0 ml. The degreasing bath is replenished with alkali solution to maintain the alkali content within the range as due to contamination alkaline contents decrease. During the operation monitoring of alkali contents is carried by titration against 0.1hydtrochloric acid using bromophenol as an indicator.

After degreasing, the car bodies are subjected to intense water showering in order to optimally eliminate alkali, sulphate, phosphate and any other chemical from the surface of the car body.

Water Dipping

The car body after degreasing and extensive showering is submerged in tanks containing water at 35-40°C, in order to completely remove any residual chemical left on the car body, this is followed by another water showering.

Surface Conditioning

The surface conditioning of car body is carried out before actual phosphating, so as to develop fine and tight crystalline grains during the Zinc Phosphating process. The compound utilized for this purposes is alkaline in nature and contains alkaline sulphate and this process is carried out without rinsing with water. The total alkali content is kept with in 1.5-3.0 ml, pH 8.5-9.0, temperature is ambient and time period for this conditioning process is 0.2-1.0 minute. The pH of the surface-conditioning bath is maintained by replenishing as per required after determining the alkali contents by titration against 0.1 HCl.

Phosphating

The phosphating of the steel surface not only provides temporary protection against corrosion but in case of car bodies phosphating is carried out in order to provide better adhesion of the paint to the body as is required for maximum longevity of the car paint.

Zinc Phosphating is carried out for car bodies by immersing the car body in to a tank containing zinc phosphating solution. This process is carried at temperature of 40-50°C. Total alkali content is kept at 20-35 ml and time period for immersion is 1-3 minutes.

After the completion of phosphating process the car bodies are again subjected twice to extensive water showering followed by water dipping subsequently. After final showering the car body is wiped manually to remove any traces of water before dipping into tank holding electrophoretic paint. Removal of moisture is essential in order to prevent any water molecules entrapping under the electrophoretic paint to be applied subsequently on the car body, The entrapment of the moisture may result in the formation of two different potential zones, the active zone being the anodic zone while the less reactive zone is cathode. This may lead to extensive galvanic cell activity, the development of the different zones assist in the passage of electric current between anodes and cathodes by transfer of electron generated at the anode during the formation of soluble ferrous ions. A simplified form of the reaction that occurs between iron, water and oxygen to produce rust is given in the following equations:

$$4Fe \rightarrow 4Fe+$$

$$4Fe_2O + 2O_2 + 8e^- \rightarrow 8OH^-$$

$$4Fe^{++} + 8OH^{-} \rightarrow 4Fe\,(OH)^{2}$$

$$4Fe\,(OH)^{2} + O_2 \rightarrow 2Fe_2O_3.H_2O + 2H_2O$$

$$4Fe + 2H_2O + 3O_2 \rightarrow 2Fe_2O_3.H_2O$$

Corrosion beneath an organic film is an example of galvanic activity. Adsorbed water collects at the anodic areas where rusting occurs and the coating delaminates at the cathode areas. Galvanic corrosion may frequently appear on the surface of the car bodies particularly when they are continuously used in marine atmosphere. The protection mechanism obtained from the paints is either one or combination of the following:

- Preventing corrosive agents from reaching the substrates.
- Reducing the conductivity of any electrolytes that penetrates the paint layer.

Therefore, specialized anti corrosive primers are applied onto the car bodies before they are given finishing layer on them in order to protect them against galvanic and filiform corrosion which makes its appearance on the car body.

- Depressing the substrate potential into the region of immunity;
- Raising the substrate potential into the region of passivation in which the substrate is more resistant to corrosion.

Preventing the access of corrosive agents by the barrier approach is probably the oldest method of protecting metal products. A simple coating across a substrate provides a reasonable barrier that limits the rate of corrosion. All coatings, unless they are very thick, will unfortunately allow more than sufficient water and oxygen to permeate down to the substrate to start the electrochemical process. The rate of transmission is reduced, as the film thickness increases, but a practicable barrier system must function adequately in thin films. This is practically important in shop primers, where the organic coating must not cause excessive toxic fumes during flame cutting and welding, or adversely affect the quality of the weld. Barrier coatings do not therefore simply prevent the access of water and oxygen but, if the pigments are chosen carefully, will also reduce the conductivity of any electrolytes that permeate into the film.

Compounds that perform an active role in preventing a substrate corroding must be very near the substrate, and are therefore incorporated into the primer layer. Subsequent paint layers prevent the ingress of moisture and provide chemical resistance or resistance to continuous immersion in seawater, if needed. The film thickness of the primer must be related to the surface roughness of the abraded/ cleaned substrate.

Electrophoretic Paint

Electrophoretic paints are now used for applying primer on the car bodies. The car body is made either anodic or cathodic and electric current is supplied to thoroughly cover the car body. The major reasons for the application of electric current is to enable the paint to reach each and every corner of the car body and an even layer of same thickness is deposited every where. The epoxy esters are the main binders in the baking primers for car, as these have excellent adhesion combined with toughness and resistance to water and chemicals.

They also have better colour retention and colour stability. The most common baking schedules are 30 minutes at temperatures ranging from 170-175°C. The dry film thickness of the primer is checked carefully and generally it is not less than 1.5 mils, after sanding not more than 2.0 mils. Since these coatings are water base, thus there is less fire hazards but it introduces other problems not associated with the organic solvent type. Being emulsion type they have the advantage of using high polymeric binders and still, maintain good solid contents at applications consistency. High polymers though provide tough, resistant coatings, but their cohesive strength may be greater than their adhesive force which may cause decrease in their adhesion to the steel surface. This problem in case of Epoxy ester based electrophoretic primers is overcome with a slight increase in time for baking. The consistency of the emulsified type primer should be proportional to the molecular weight of the binder. In these types of primers the pigments are dispersed directly in the vehicles to avoid flocculation that occurs in the film forming process of straight emulsion coatings. After the Electro deposition process the car bodies are then dipped in UF dipping Viz: urea formaldehyde dipping for 2-3 minutes, this is followed by extensive water showering, water dipping, and again water showering followed by manual water rinsing in order to make sure that any excess paint is removed followed by baking at 203°C for 30 minutes.

Middle Coat Primer

Middle coat based on ABS (Acryl nitrile Butadiene Styrene) is applied after Electro deposition paint is baked. After that sanding and air blowing of the car body is carried out to remove any loosely adhering paint. Then naphtha is sprayed on the primed car body followed by tack rag wiping (tacked with naphtha). Middle coat ABS primer is then sprayed at air pressure of 4Kg/cm^2; viscosity of the primer is kept in the range of 11-12 sec/#4FC/20°C. Then a flash time of 10 minute is given and middle paint is baked at 170°C for 30 minutes.

Base Coat and Top Coat

Both the synthetic enamel baking type and clear lacquers are used as a finish over the primer. Usually a double coat of enamel is applied and baked simultaneously at schedules ranging form 45 minutes at 130°C or 30 minutes at 150°C depending on the heat sensitivity of the colour pigment. The topcoat in case of white paint is polyester melamine based and the viscosity for application 11seconds NK2 cup. After the application of base coat, an average flash time of 10 minutes is given. After that, top clear coat based on acrylic/Melamine formaldehyde resin is applied by spraying and the application viscosity is 16-20 seconds. Later a flash time of 7-10 minutes is given. Then both the base coat and top clear coat is baked at 175°C for 39 minutes.

The synthetic enamel in case of assorted colour is based on acrylic/melamine formaldehyde resin, the supply viscosity is in the range of 57±3 KU at 25°C solid content at 45± 3 per cent, application viscosity is 11±1 second Nk2 cup, flash time is 7-10 minutes and stoving schedule is 30 minutes at 130°C, hardness of this system after coated with clear lacquer is F~H, cross hatch adhesion should be 100/100, while gloss should be more than 92 per cent @ 60°C sheen gloss head. Also there should not be any coating defect. The top coat should have the hiding ability to cover a dry film thickness of 20 ±5μ and is sprayed manually at 70-80 PSI. Similar conditions apply when the base coat is metallic.

After the completion of spraying process the car bodies are subjected to a series of stringent tests in order to determine the efficacy of painted car body in actual environment.

The following tests are carried out.

1. *Corrosion Resistance on Ed System*: There should not be any blistering or peeling after 960 hours. Rust creep not more than 3 mm from the cross cut.
2. *Stain resistance*: After polishing the body there should not be any discoloration.
3. *Impact Resistance*: There should not be any cracking or peeling with direct impact of 500 grams weight with12 mm point from a height of 200mm.
4. *Flexibility*: This test is carried out over a panel coated under the same conditions as actual finishing of the car body.
5. *Accelerated weathering*: No peeling, cracking rust or blister after polishing or change in colour and gloss more than 80 per cent.
6. *Moisture Resistance*: The surface should remain intact. There should not be any peeling or blister formation after tape test of 240 hours exposure at 49°C and 97 per cent Relative humidity.
7. *Xylene Resistance*: No change in colour is acceptable after 8 rubs with a tag soaked in xylene.

In case of any defect, the remedial measures are carried out in order to ensure the car rolling out of the assembly lines has zero defects and have a pleasant and appealing aesthetic appeal for the potential customers.

Status of Surface Engineering in Pakistan

Pakistan came into existence on August 14, 1947. The areas included in Pakistan were very backward with little or no Industrial establishments except for a few textile units, floor mills and small cottage scale engineering units. During the last 57 years of its existence it had made significant headway in industrialization. Now a large number of industrial units have sprung up through out the country.

Engineering industry is mainly established at Lahore, Karachi, Gujeranwala, Sialkot, Gujarat, Rawalpindi, Taxila beside a number of units are also scattered through out the country. Till the establishing of automobile assembly plants, most of the engineering units were involved in manufacturing consumables and domestic based items. They were involved in manufacturing of surgical goods, electric fans, and sanitary items. Most of these goods were sold locally having indifferent quality and except surgical instruments no worthwhile export was done. Surface Engineering Industry has also made considerable progress producing goods of reproducible quality and are being exported to a number of countries through out the world. Surface Engineering can be divided into the following:

1. Electroplating units
2. Anodizing units
3. Powder coating units
4. Liquid paints spraying units
5. Metallising units
6. Plasma coating units
7. Galvanizing units

Electroplating Units

Electroplating units are involved in a large number of activities, they are carrying out electroplating of Zinc, cadmium, copper, brass, nickel chromium, silver and gold. They can be further subdivided into three categories.

(*i*) Small Scale units.

(*ii*) Medium scale units

(*iii*) Large scale units

Small Scale Units

The units are involved in producing goods of low value like chains for controlling cattle, nuts and bolts, hinges, fasteners, optical frames, door handles buckles and belts. Majority of these goods are consumed locally. They are involved mostly in electroplating of zinc, nickel, copper and cadmium. Beside a small number of them is also involved in electroplating of silver and gold for Jewelry and commemorative medals. They form the local vendors and some of them are even involved in making their own electroplating baths. Since they do not involve any technical help their produce is of inconsistent quality. Besides, their units are not environmentally friendly, thus they are prone to many diseases and adding to the environmental degradation.

Medium Scale Units

The units are better established, they are involved in electroplating of surgical, cultlery, sanitary, domestic fixtures, like door handles, fasteners for engineering industry, nuts and bolts. Some of them are also involved in making automobile parts for car and motor cycle assembly units. They are involved in electroplating of decorative nickel and bright brass, copper, decorative and hard chromium. They are better organized, they utilize the technical know how provided by the electrolytes suppliers, beside they also approach the R&D laboratories established in the public sector for solving their quality and other problems. Most of them utilize imported electrolytes, brightener, levelers, stress relievers etc. Due to the increasing competition some of them are approaching established laboratories to develop local substitutes. Their product quality is also better but still they face lot of problems in matching the quality of the imported ones and are less resourceful. Most of them do not employ good working practices.

Large Scale Units

They are involved in electroplating of wide range of products. Their product ranges involves domestic fixtures like door handles, bolts and nuts, catchers, door locks, both office and domestic steel furniture. Sanitary items, surgical goods, cutlery, motor cycle rims, handles, muffles, shock absorbers covers, car insignias, commemorative medallions, decoration pieces, artificial jewelry and wide range of machinery parts.

They employ trained technical staff and use good quality electrolytes and other electroplating chemicals; hence their quality is better and matches the quality of imported items. Some of them are also involved in selling their goods abroad.They are involved in both decorative and engineering electroplating like bright nickel, zinc, cadmium, machinery parts to withstand wear. Majority of these units have better working environments and are less polluting.

Anodizing Units

They are mostly involved in anodizing of aluminium section for architectural purposes particularly for doors and windows and other purposes in houses and commercial buildings. Their scale of working is also same, medium and large. Small units catering mostly to cheap housing, while medium and large units cater to the needs of big houses, buildings, industry etc. Some of them are also involved in anodizing of aluminium sheets for further coatings for pre-sensitized positive plates. A number of units are also involved in the manufacturing of Aluminum conductors for power supply.

Powder Coating Units

These units are involved in powder coatings for car oil filters, sanitary articles, furniture, refrigerators, air conditioner and electric freezer bodies. They are using imported as well as locally manufactured powder. Most of them use imported guns which have good control units, thus their quality matches the quality of the imported ones.

Wet Paint Spray Units

Wet paint spray units are scattered through out the country. They are involved in providing protection for a very wide range of engineering and consumables goods. Their product range includes motor cycle and motor car assembly, electric fans, steel furniture, automotive components, machinery parts, electric geysers, water and air coolers, LPG gas cylinders and many other consumables and industrial components.

Metallizing Units

These units are involved in spraying metal powders onto the substrates and then subjecting them to baking at high temperature. Their product range includes both domestic and industrial components.

Plasma Coating Units

There is few plasma coating units established in the country catering to the needs of specialized customers. They are involved in repairing machinery and automobile parts. Since this is quite expensive so they are only used for specific purposes particularly where wear resistance is of specific importance especially in textile machinery.

Galvanizing Units

They are mostly involved in the manufacture of galvanized iron pipes and sheets. Most of the pipe manufacturing plants are located in Karachi and Lahore while steel galvanizing plants are scattered through out the country. Some of the pipe manufacturing units also export their products.

Acknowledgements

The author wishes to thank Mr. Syed Asad Mustafa Gilani, Director General, PCSIR Laboratories Complex Lahore, Syed Fayyaz Mahmood, Head, Material Science Research Centre PCSIR Lahore for their permission to make this presentation. Thanks are also due to M/S Berger Paints (Pvt.) Ltd, for providing data and M/S Honda Atlas Cars, for their permission to visit their plant.

Chapter 22

An Overview of the Beneficiation, Environmental Issues and R&D of Industrial Minerals in Indonesia

Husaini

Lecturer, Faculty of Chemical Engineering, University of Achmad Yani
L. Terusan Jenderal Sudirman
PO. Box–148, Cimahi, TLP, Indonesia
Email:husaini_husaini2004@yahoo.com, husaini@tekmira.esdm.go.id

ABSTRACT

Industrial minerals find wide applications in practically all industries, including metallurgical, in one form or another. Some of them find large use as extenders, additives or fillers imparting some special properties to various products. In many cases they form the starting point for basic industries like cement, fertilizer, insulating, refractory and abrasive material and a host of chemicals. Indonesia has a large mineral resources, particularly the industrial mineral including rock and gemstone are scattered throughout the country, while the quality and quantity of the deposit are varied. Some minerals which have been investigated both for laboratory and pilot scales, include manufacturing light calcium carbonate; activation of Ca-bentonite and zeolite, manufacturing of kieserite from dolomite; manufacturing of alum from kaolin and bauxite; up grading feldspar by flotation. Research related to surface engineering includes modification of natural zeolite by polymer coating such as using vinyl pyridine, chitosan, hexa decyl three methyl ammonia (HDTMA) and impregnation of zeolite by tin oxide.

Keywords: *Industrial minerals, Surface engineering, Zeolite, Environment impact, Research and Development, Indonesia.*

Introduction

Although industrial minerals do not possess the glamour of metals like gold, silver, platinum, etc. they are of immense industrial importance. They find wide applications in practically all industries, including metallurgical, in one form or another. Some of them find large use as extenders, additives or fillers imparting some special properties to various products. In many cases they form the starting point for basic industries like cement, fertilizer, insulating, refractory and abrasive material and a host of chemicals. One very significant point about industrial minerals is that based on single mineral, several industries can be set up.

Indonesia has a large mineral resources, particularly the industrial mineral including rock and gemstone and are scattered throughout the country, while the quality and quantity of the deposit are varied. The major industrial mineral products in Indonesia (*e.g.* bentonite, zeolite, kaolin and building construction rocks) are mostly for domestic consumption due to rapid industrial development particularly in Java and some other areas outside Java (Arifin, 2003).

The development of mining industrial mineral resources, in the period of last decade in Indonesia is very positive. In all aspects of activities and development in this sector, a lot of progresses have been achieved. However, the economics of this sector is not isolated from consequences of world economic recession and world political change, such as the low price of some industrial minerals causing to close down some mining companies due to unbalance between production costs and market prices. However, on the other hand the development of increasing production and efficiencies as well as opening new mines on a large scale, utilization diversification etc, can really overcome the consequences of those recessions and show big progress (Wahyu, 2002).

In national policy of industrial mineral resources development such as technology transfer, human resource development, Indonesianization, increased value added, mineral commodity promotion, regional development, domestic mining role, environmental management, etc, much has been successfully achieved. The target of some mineral production in the last decade has been met or even for some commodities, has been exceeded.

The industrial mineral development is very important to support the economic development in providing raw materials for down-stream industries, reducing import and increasing export to enhance government revenues. In the Second Long-term National Development, it was decided that the increased added value of industrial minerals should become the focus of the government achievements in the field of mineral resources development (Ardha, 2002).

Beneficiation, Utilization and Trade

Developments in mining of industrial mineral resources are limited to the potential minerals currently available in Indonesia, such as. bentonite, kaolin, zeolite, feldspar, dolomite, limestone, quartz sand, etc. These are:

Bentonite

The reserves of the bentonite have been estimated to be around 538.07 million tons (Appendix A) spread out through out Indonesia with 29 locations in Java, Sumatra, Kalimantan, and Sulawesi. Only a few deposits of sodium bentonite are available, such as found in Boyolali (Central Java), Pangkalan Brandan (North Sumatra) and Sarolangun (Jambi) (Anonim, 2003).

Mostly, the consumption of bentonite is used for bleaching, drilling and foundry. Most of bentonite deposits in Indonesia are of the calcium bentonite type, which has less swelling index, but has a good absorbing property both being natural as activated. Activation process for bleaching purposes, the bentonite firstly is dried, then crushed, and ground to get powder of bentonite with the particle size of–100 meshes. Bentonite powder is then activated by sulphuric acid at its boiling point of the solution to obtain high adsorption ability, so the activated bentonite is suitable for adsorbing the color of the beta carotene in the crude palm oil (Husaini, 2003; Azis dan Husaini; 2003).

Bentonite production amounted to 225,000 tons in 2001 increased to 270,000.00 tons in 2002 (Appendix B). Most of the bentonite products were natural bentonite in powder form and the rest was acid activated bentonite. The consumption of the bentonite was 196,928.00 tons in 2001 increased to 200,000.00 tons in 2002. Bentonite consumption (outside petroleum industry) in 2003 was 21.000 tonnes (Appendix B) by coconut and palm oil industries (78 per cent of total consumption) and the rest was consumed by soap, cosmetic and metallic casting industries. Consumption of bentonite in the petroleum industry (sodium bentonite) is not available. In 1992 the import of bentonite reached 12.970 tonnes, whereas the exported bentonite of calcium type amounted to 22.140 tonnes. The significant development in Na-bentonite demand that is met from import, is mainly due to the requirement of oil drilling mud in petroleum exploration activities. Similarly, the development of Ca-bentonite demand that is obtained from domestic mining is mostly due to the increase of palm oil (cooking) industries (Arifin, 2003).

Kaolin

Indonesia has large deposits of kaolin of around 597.26 million tonnes (Appendix A) located at over 30 locations. In general, the deposits are classified into filler, ceramic and porcelain grades and have been mined, in particular, in Bangka and Belitung Islands. Other deposits are found in Aceh, South Sumatra, West Sumatra, Bengkulu, West Java, Central Java, Yogyakarta, East Java, South Kalimantan, West Kalimantan, and North Sulawesi (Anonim, 2003).

Beneficiation of kaolin from mine includes washing, scrubbing, screening, and then treated by using various types of hydro cyclone in several steps to produce very fine particle of kaolin, which is separated from impurities such as quartz (Husaini dan Soenara, 2003).

Kaolin production was 337,000 tons in 2001 and decreased to become 312,000.00 tons in 2002. The domestic consumption was 343,520 tons in 2001 and increased to become 350,000.00 tons (Appendix B). Most kaolin production was consumed by the paper industry (40.8 per cent), ceramic and porcelain (30.4 per cent) and paint industries (11.4 per cent). Most kaolin production was consumed by the paper industry (40.4 per cent), ceramic and porcelain (30.4 per cent) and paint industries (11.4 per cent). Coating grade kaolin is still imported. In 2002, imported kaolin reached 120,752 tonnes and 82,851 tonnes was exported (Appendix B) as unprocessed product (Anonim, 2002). Simple quartz separation, washing and desliming are general processes to produce high commercial grades (Husaini, 2003).

Zeolite

Zeolite is volcanic tuff containing large molecular cavities filled with water molecules that are interconnected by pores in some directions. This important property has good purposes for absorption and ion exchange, usually as a catalyst absorbent ion exchange.

Zeolite reserves have been estimated to be around 196.382 million tons (Appendix A) are spread out at lest in 43 locations in Indonesia, such as West Java, Central Java, East Java, Lampung (South Sumatra), West Nusatenggara, and, Sulawesi (Anonim, 2003). In general, the type of zeolite in Indonesia is mordenite and clinoptilolite with high quality, *i.e.* its cation exchange capacity (CEC) is more than 170 meq/100 g. Range of chemical composition of Indonesian zeolite are as follows: SiO_2 : 58.44–76.85 per cent ; Al_2O_3 : 6.92–15.00 per cent ; Na_2O : 0.0–2.80 per cent ; K_2O : 0.05–7.15 per cent ; CaO : 0.20–7.15 per cent ; Fe_2O_3 : 0,42–4,11 per cent ; MgO : 0,0–5,23 per cent.

Zeolite coming from mine is beneficiated by drying followed by crushing and grinding, sieving (classification), and activation (either physical or chemical method). The aim of the classification is to obtain certain of the particle size suitable for its utilization. The activation is aimed to enhance the quality of the zeolite in order to have a higher adsorption ability and cation exchange capacity. Besides, the powder zeolite can be converted into pellet (granule) form by granulation process, which is usually used for slow release of fertilizer (Husaini, 2003).

The exploitation of zeolite was started in mid 1980's. At present, zeolite quarries in West Java are in operation. The total production in 1992 was around 30.000 tons, produced by five companies. It is consumed mostly by agriculture as soil conditioner and slow release fertilizer, shrimp pond, and water and waste- water treatment. Other utilizations are normally applied as an additive to livestock food (cows, cattle, sheep, pigs, poultry etc), adsorbent, filler in rubber industries and plastic goods.

Until now most of zeolite products is produced by small-scale companies. In 2001 zeolite production was 57,500.00 tons and increased to become 60,000.00 tons in 2002. The domestic consumption was 57,474.49 tons in 2001 and increased to become 60,000.00 tons (Appendix B) (Anonim, 2002).

Feldspar

Feldspar is mostly found as a low to medium grade in Indonesia. Whilst, high-grade of feldspar is only found in a limited quantity. Feldspar reserves are estimated about 6.47 billion tons (Appendix A) located in more than 26 locations. The large deposits are found in Central Java, East Java and Aceh (Sumatra). Other deposits are distributed in Sumatra, Java, West Kalimantan, Sulawesi, East NusaTtenggara and Maluku (Anonim, 2003).

At present, many feldspar quarries are in operation, such as in Cipanas (West Java), Banjarnegara (Central Java), Pacitan (Eat Java) and North Sumatra. In 2001 feldspar production was 119,500.00 tons and increased to become 131,200.00 tons in 2002. The domestic consumption was 239,973.84 tons in 2001 and increased to become 250,000.00 tons (Appendix B) (Anonim, 2002). The main consumer of this mineral is ceramic, porcelain and glass industries (around 90 per cent of total domestic consumption) followed by sheet glass, paint, rubber and plastic industries. In 1996 Indonesia was the 6th biggest ceramic and porcelain producer in the world (Ardha, 2002).

Dolomite

Dolomite is a double carbonate of calcium and magnesium. Dolomite has many applications, such as: as raw materials in glass, refractory and steel industries. Other usage is for agricultural industry especially for fertilizer and soil conditioner materials. The technology for improving the quality of dolomite is conducted by calcinations to become magnesium oxide to be used for refractory. Besides, it can be dissolved into magnesium sulphate by sulphuric acid for making kieserite as fertilizer. This technology can be developed in the commercial scale.

In Indonesia, Dolomite reserves are estimated around 1.79 billion tonnes (Appendix A) found in 10 locations (Anonim, 2003). The mining of deposits has been carried out intensively in Tuban and Gresik. Consumption of dolomite is presently used for fertilizer Industries. Simple treatment like crushing and grinding have been applied to beneficiate dolomite (Husaini, 2003).

Right now, the mining of deposits has been carried out intensively in Tuban and Gresik (East Java) and Central Java. This is caused by the high quality of the dolomite. Total production of the Indonesian dolomite increased from 1,460,000.00 ton in 2001 to 1,500,000.00 ton in 2002 (Appendix B) (Anonim, 2002). There are some companies producing dolomite such as P.T. Polowijo Gosari, C.V. Agung Karya, C.V. Dewi, etc. The product is in the form of powder. Dolomite consumption in 2001 amounted to 1,479,166.53 tons (Appendix B), 95 per cent of which is consumed by agriculture and the rest is used in industrial application, in particular for glass industries. To meet the demand of dolomite, about 3,750 tons is imported. Dolomite is presently used for fertilizer industries (Anonim, 2002).

Calcite

Limestone is calcareous sedimentary rock composed of the mineral calcite ($CaCO_3$), which upon calcinations yields lime for commercial use. Limestone is calcined at elevated temperature (above 900°C) to produce quick lime. By reacting the quick lime with water, it can produce lime [$Ca(OH)_2$]. Both the quick lime and lime can be used as a raw material for paint, filler of paper, and other uses such as pH controller in mining industries (Husaini, 2003).

In Indonesia, calcite reserves are estimated to be 34.92 billion tonnes including 16.40 million tonnes of the measured reserves (Appendix A) (Anonim, 2003). The deposits are mainly encountered in West Sumatra (23.27 million tonnes), South Kalimantan (1.00 billion tonnes), West Nusa Tenggara (1.92 billion tonnes), West java (672.82 million tonnes) and South Sulawesi. Indonesia limestones have various compositions ranging from very pure (55 per cent CaO) and low grades of about 40 per cent CaO.

In 2002 calcium carbonate production was 520 million tonnes. Major production originates from small mines producing lime and dead lime. In the same year, domestic consumption of lime stone was 500 million tonnes (Appendix B). Cement industries consumed about 75 per cent of the total and the rest was used by iron steel, paint and paper industries and agricultures. The imported products are in the form of lime or dead lime and light carbonates (Anonim, 2002).

Quartz Sand

Quartz sand reserves have been estimated to be around 3.19 billion tons (Appendix A), scattered over 46 locations through out Indonesia. Most of quartz is found in West Sumatra (72.76 per cent), North Sumatra (11.91 per cent), West Kalimantan (9.23 per cent), and the rest (6.1 per cent) spread out in other locations with the reserve less than 2 per cent (Anonim, 2003). They are found in various qualities, *i.e.* very pure (99.8 per cent SiO_2), medium grade (90 per cent SiO_2) and low grades (85 per cent SiO_2). In general, the chemical composition of quartz in Indonesia has SiO_2 contents 55.30–99.87 per cent, Fe_2O_3 0.01–9.14 per cent. The best quartz is found in East Kalimantan with SiO_2 content between 98.7–99.90 per cent.

Quartz sand is processed by grinding and screening to produce rounded silica sand suitable for gravel pack sand mostly used for drilling of the oil. The other common method to upgrade the natural silica sand to be of commercial grade is washing and desliming (Husaini, 2003).

During the past two decades, quartz sand production has increased steadily. Quartz sand has been intensively and commercially mined at several locations such as in Bangka and Belitung Islands and in Rembang and Tuban (Outside the deposits owned by Cement Industries). In 2001 Silica production was 5,650,000 tons and it increased to 6,000,000.00 tons in 2002. The domestic consumption was 5,649,518.07 tons in 2001 and increased to 6,000,000.00 tons (Appendix B) (Anonim, 2002).

Perlite

Perlite is a glassy mass of volcanic origin, chiefly of rhyolite. The composition may vary from rhyolite to andesite.

Perlite reserves have been estimated to be around 366,6 million tons (Appendix A)., and scattered in North Sumatra, West Sumatra, Jambi, South Sumatra, Lampung, West Java, West Nusa Tenggara, and North Sulawesi (Anonim, 2003). Uses of perlite are based on its properties of acquiring light weight and inert characteristic on expansion. Expansion is carried out in both the vertical and rotary furnace at a temperature of 850–1150 °C (Husaini, 2003).

Pumice

Pumice reserves have been estimated to be around 1.02. billion tons (Appendix A), and scattered in Jambi, West Java, Yogyakarta, West Nusa Tenggara, East Nusa Tenggara and Maluku (Anonim, 2003). In 2002 Pumice production was 500 million tonnes. In the same year, Pumice domestic consumption was 80 million tonnes (Appendix B) and the rest was exported (Anonim, 2002).

Zircon

Geologically, zircon is found in steam sediments and concentrated mainly together with ilmenite, monazite, and other heavy minerals. In Indonesia, this mineral is concentrated together with cassiterite of which its impurities consist of ilmenite, magnesite, monazite, quartz, etc (Husaini, 2003).

The deposits of zircon in Sumatra (Bangka, Belitung, Karimun Kundur) and part of western Kalimantan and Papua (Anonim, 2003). While in Bangka, Belitung and Karimun Kundur area, zircon is exploited by PT. Tambang Timah as by product of tin washing plant, but in a very small capacity and utilized for supplement in ceramic industries. In 2002 Zircon production was 10.5 million tonnes. In the same year, import was 16.07 million tonnes (Appendix B) (Anonim, 2002).

Sulphur

Sulphur is the most widely used industrial mineral and the occurrence is perhaps most widely dispersed among the industrial minerals.

In Indonesia, sulphur reserves are estimated around 1.95 billion tonnes (Appendix A) found several location, such as Aceh, North Sumatra, West Sumatra, Jambi, West Java, Central Java, East Java, North Sumatra and Maluku (Anonim, 2003). The applications of sulphur are used in the form of sulphuric acid, which used in the manufacture of fertilisers. A small quantity of elemental sulphur is used in the manufacture of rubber, detergent, fungicides, and sugar (Husaini, 2003).

In 2002 sulphur production was 341 million tonnes, consumption was 200 million tonnes, export was 114.5 million tonnes and import was 43.8 million tonnes (Appendix B) (Anonim, 2002).

Trass

Ground trass is mixed with lime in a certain composition to be pozolan cement which is suitable for building material cheaper than that of the Portland cement. This kind of plant has been in operation in Lampung, Sumatra, with the capacity of about 75 ton pozoland cement per day (Husaini, 2003).

In Indonesia, trass reserves are estimated to around 6,76 billion tonnes (Appendix A) found in several locations, such as Aceh, North Sumatra, West Sumatra, West Java, South Kalimantan, and East Nusa Tenggara (Anonim, 2003).

Outlook for the Future Development

In Indonesia, future development in beneficiation and utilization is limited because of the lack of the related technologies. However, some of minerals have been tried to develop in order to increase its value added and to reduce the dependency on imports. Some of these are as follows:

Future bentonite utilization will be expanded for drilling mud by modifying calcium bentonite mostly found in Indonesia to be natrium bentonite by applying the cation exchange process. In the mean time, kaolin will be treated using ultra finer grinding techniques in order to comply with the requirement of paper coating. High quality ceramics, China and porcelain need a high quality of kaolin, which is one of the requirement of having ultra fine particles.

The development of zeolite and bentonite beneficiation and utilization is related to the production of a natural catalyst for refining oil to avoid the use of expensive synthetic catalysts. Zeolite is promising minerals, which is particularly used in refrigerator technology. For this long term research is to be carried out in Indonesia.

On the other hand, the development of feldspar beneficiation and utilization is carried to obtain the highest quality to fulfil the demand of the electronic and advanced ceramics. The advanced technology to treat the low natural quality of feldspar currently used is flotation, which needs to be established in Indonesia (Ardha, 2002).

Dolomite and limestone are minerals, which are familiar as carbonate minerals. As at present, their use is limited for most of Indonesian people. Therefore, the utilization of both dolomite and limestone should be developed to diversify their use and to reduce their imports. One of the prospective features of dolomite and limestone in Indonesia is for making refractory bricks and light carbonates, respectively.

Some research activities, which have been conducted for surface engineering, are the studies on chemical modification of Indonesian natural zeolite using organic polymer (vinyl pyridine, chitosan, and hexa decyl three methyl ammonia) at the laboratory scale. Research activities on other minerals, which have been carried out include: upgrading feldspar, kaolin, gypsum, phosphate; activation of bentonite, manufacturing of pozoland cement, alum, and kieserite. Some research activities were conducted at the pilot plant scale like manufacturing of pozoland cement, alum, and gold processing.

Research on impregnation of natural zeolite by tin oxide and coating zeolite by urea is the concern for the immediate future because Indonesia has a lot of tin deposit and natural zeolite. Tin oxide impregnated zeolite has a good chemical and physical properties, including stability on high temperature and varied pH.

Environmental Issues

While mining activities play an important role in Indonesia's economic growth, these activities cause environmental damage. The environmental impacts of mining may be physical, chemical, biological, social and economic. Mining and processing of minerals can have both beneficial and detrimental effects.

The scale of the impact of the industrial mineral industry on the environment is determined by several factors including the location of the minerals and the methods used in mining and processing. In general mining industry can cause environmental problems, namely detrimental effects on the landscapes, water pollution, and air pollution, biological problems including loss of bio-diversity and affecting animals, and also raising problems in the socio-economic-cultural life (Soelistijo, 1999).

Open pit mining may cause more environmental problems than underground mining. In open pit mining, in order to extract products, top soil is often lost. Open pit mining leads to the removal of large quantities of over burden. Then, the processing of minerals produces tailing. Both tailing and overburden disposal can cause the main environmental problems for mining industries. In the industrial mineral exploitation in the monsoon season in tropical areas, the disposal can overflow on to the surrounding including streams and agricultural area. Further more, the acid mine drainage phenomena may almost occur in the exploitation of certain industrial mineral.

To protect the environmental damage due to the mining activities, the government has passed Mining Law in 1967. Article No. 30 of this law stated:

" After completion of mining for minerals in any time, the holder of the relevant mining authorization is obliged to restore the land in such condition so as not to evoke any danger of disease or any other hazards to the people living in the surrounding of the mine ".

Then, in 1982, The Indonesian government introduced the environmental management Act No.4 1982, the first formal national legislation on the environment in Indonesia. Since 1982, environmental impact assessment document is compulsory for every activity, which will have an impact on the environment. The environmental Impact Assessment document deals with the Environmental Management Plan (RKL) and Environmental Monitoring Plan (RPL) specifically designed to prevent and minimize negative impact and promote positive impacts (Wijayanti, 2002). These plans are being implemented by means of a comprehensive set of environmental management and monitoring programmers.

The key objectives of these programmes are to:

- Minimize the impact on water quality of rivers
- Minimize the impact on air quality
- Rehabilitate disturbed areas after mining
- Protect the existing environment
- Adhere rigorously to all requirements of the Department of Mines and Energy

However, in implementing this environmental management, some problems may occur, they are:

- Insufficient number of environmental experts
- Remoteness of activity area
- Lack of budget to inspect the environmental management in the field
- The existence of illegal mining.

Research and Development

R&D on industrial mineral technology constitutes an integral part of mining and mineral resources development, because the results of R&D can be used to develop industrial mining industries and mineral resources management. In the framework of facing the challenge in mining industries and increase of mineral utilization, R&D is being carried out intensively and a bigger role has been given to R&D institutions involved in mining in Indonesian.

In the near future, the world industrial mineral mining is likely to face a variety of challenges such as the tendency of increasing industrial mineral commodity prices in the world market, increased production costs, and the depletion of high industrial mineral grades. In consequence, mining industries will have to struggle harder to recover industrial minerals from low grade and complex ores.

Indonesia as a producer country of industrial minerals should be capable to overcome these challenge by increasing efficiency through mastering in mining and beneficiation technologies and increasing the quality of human resources in managing mining and mineral resources development. Mastering the technology constitutes the key of successfulness in the future. Therefore, the indigenous technology should be developed and mastered by R&D Institutions.

Although the role of R&D on mining and mineral resources development is small, but its activities are significant, starting from giving data and information, raw material testing and study, process technology and feasibility studies. The R& D results have had double effects of both increased industrial mineral value added and employment opportunity.

R and D programme on industrial mineral resources is devoted into four areas, *i.e.* R&D on mining and industrial mineral beneficiation technologies including study on environment feasibility in order to increase industrial mineral added value and the information systems and techno-economic and regional development.

R&D activity on the value-added processing of minerals in Indonesia have been mainly carried out by several institutions, including Mineral and Coal Technology Research and Development Centre (MCTRDC), Ceramic Research Centre, Building Materials Research Centre, Applied Technology Research Board (BPPT), Indonesian Institute of Science (LIPI) and universities. Some results of the research achieved are shown in Table 22.1. High values added minerals are produced through a series of processing steps, which is achieved by carrying out intensive research starting from laboratory scale, bench scale, and pilot plant prior to its production.

Some pilot plants have been built to include beneficiation process of limestone to produce light calcium carbonate mainly for filler of paper, paint, rubber, tooth paste, and pharmaceutical needs; activation of Ca-bentonite using sulphuric acid to get activated bentonite suitable for treating crude palm oil to produce cooking oil; zeolite processing and utilization, particularly for agriculture, industry, water and waste water treatment; utilization of the dolomite for kieserite manufacturing as fertilizer; manufacture of alum from kaolin and bauxite using sulphuric acid; upgrading feldspar by flotation as ceramic raw materials. Some research related to surface engineering include modification of natural zeolite by polymer coating such as using vinyl pyridine, chitosan, hexa decyl three methyl ammonia (HDTMA) and impregnation of zeolite by tin oxide (Husaini dan Soenara, 2003).

Table 22.1: Some Results of the Industrial Mineral Research Achieved in Indonesia

Mineral Commodity (1)	*Products (2)*	*Purpose (3)*	*Status (4)*
Zeolite	• Activated zeolite • Modified zeolite	• Absorbency in drinking water, waste water, fish pond water • Slow release of fertilizer • Feed additive in live stock	Laboratory Pilot Plant scale
Calcium carbonate	• Light-Ca carbonates • Lime	• Paper Industry • Various industry	Pilot Plant Scale
Feldspar	• High grade feldspar	• Ceramic and Porcelain	Pilot Plant Scale
Dolomite	• High grade MgO • Magnesium Sulphate	• Refractory brick industry • Fertilizer and chemical	Pilot Plant Scale
Kaolin	• Alum • High grade kaolin	• Drinking water conditioning • Paper industry	Pilot Plant Scale
Jarosite	• Red Iron concentrate • Potassium sulphate	• Red Iron base paint • Fertilizer	Pilot Plant Scale
Bentonite	• Na-Bentonite • Ca bentonite	• Drilling mud • Palm oil clarifier	Laboratory Pilot Plant Scale
Phosphate	• Multicomponent fertilizer	• Fertilizer	Pilot Plant scale

The competition in international mining industries is increasing requiring improvement in the cost structure essential for the industries to stay competitive. In this relationship, an appropriately structured research and development programmes plays a vital role. In line with this, research is needed to make sure that research and development and broader economic policies are working in the same direction, in particular, on value added processes of industrial mineral for supporting the availability of raw materials in down stream industries.

Government Policy on Minerals Development

Based on the Indonesian government regulation numbers 27 of 1980, minerals are classified into three categories, namely A, B and C. The minerals classified in A category are considered as strategic minerals, and the minerals in B categories are considered as vital minerals, while the minerals in C categories that are dominated by non metallic minerals are considered neither strategic nor vital. This regulation stipulates that the minerals included in first two categories (A and B) are controlled by the Central Government through the Ministry of Energy and Mineral Resources. While, control on the minerals of C category has been decentralized to the Provincial Government who can issue a mining permit to the domestic investors. Foreign investors, who need to develop the C category minerals in Indonesia, may be permitted as joint venture with domestic company. The above regulation, however, is now being reviewed, since the Regulation Number 22 of 1999 and the Regulation Number 25 of 1999 had been released.

Rapid development of construction industries and manufacturing industries in Indonesia would require more mineral as raw materials. Thus, to anticipate the increase of demand of mineral raw material in the future, the Indonesian government has issued such policies, which are expected to be able to promote the growth of investment with the target of increasing production for domestic demand and for export as well. The policies among others are:

- The government regulation was number 37 of 1986 regulated that the central government has decentralized a part of authority concerning the management, supervision and development matters to the Provincial Government. Establishment of the provincial mining office is aimed to create better services, management, supervision and development particularly in the effort of encouraging the small and medium scale of mining activities.
- The Law number 22 of 1999 concerns Regional Government (under Provincial Government), and the law number 25 of 1999 concerns the financial arrangement between Central and Regional Government. These Laws regulate full autonomy to the Regional Government in managing natural resources located in relevant areas.
- In addition, the government is now also reviewing law number 11 of 1967 concerning basic provision on mining. The draft of the proposed Mining Law is currently being discussed intensively in the Indonesian Parliament. The main issues of the draft among others are:
  - (*a*) Control of minerals will be fully decentralized to the Regional Government, except for radioactive mineral and minerals located offshore.
  - (*b*) Simplication on the mining permit for better process and legal assurance.
  - (*c*) Mineral are not classified in A, B and C categories for optimum use.
- Reviewing other laws such as the Law number 18 of 1997 concerning Regional Taxation and Retribution and the Law number 41 of 1999 concerning forestry have been proposed.
- Carrying out improvement of potential mineral inventory. The institution responsible for conducting exploration is Directorate of Mineral Inventory (now is Centre for Geological Resources) under Geological Agency, Ministry of Energy and Mineral Resources.
- Enhancing the quality of minerals products. The product quality must fulfil the industry specification. This must be successfully done by treatment and processing of the raw materials. Mineral and Coal Technology Research and Development Centre (MCTRDC) undertakes the relevant activities of research and development in mining and processing.

In order to successfully implement the policies mention above, several actions have been conducted especially through the management of mining activity efficiencies and productivity. The other actions included are conservation and inventory to gain the added value, marketing and promoting mineral products for present and future need, protection against environmental disruption etc (Soelistijo, 1999).

Conclusions

- Some industrial mineral deposits are abundant in Indonesia with various qualities, so beneficiation of mineral are necessary to be conducted to get added value. To produce a good quality of minerals as a raw material suitable for downstream industry from low to medium quality of minerals will need high cost for processing. As a result, it is seen that most of minerals, which should be processed are uncompetitive to the imported product (example: feldspar).
- Most industrial minerals have not been exploited yet, because of some reasons including very small deposits, location of deposit in remote area, lack of infrastructure, low selling price, limited diversification of utilization, existence of a good mineral substitution (the use of kaolin replaced by light calcium carbonate as coater of paper).
- Research on processing and utilization of some minerals has been successfully conducted starting from laboratory to pilot plant, include processing of zeolite, bentonite, feldspar, kaolin, dolomite, phosphate and trass; but the development of the industrial mineral resources from pilot scale to the commercial scale (such as puzoland cement manufacturing) are facing obstacle because of some factors, such as limited capital and skill, regional policy and regulation as well as taxes, environmental issues.

References

1. Arifin M., Cs, "Study on Mining Data", Research and Development Centre for Mineral and Coal Technology, Agency of Research and Development for Energy and Mineral Resources, 2003.
2. Azis and Husaini, "The Experiment of Calcium Bentonite Upgrading by Using Hydrocyclone for Bleaching of Crude Palm Oil", Indonesia Mineral Bulletin, Vol. 7, No. 18, Research and Development Centre for Mineral and Coal Technology Bandung, 2003.
3. Anonim, "National Mineral Resources Balance", Directorate of Mineral Resources Inventory, 2003.
4. Anonim, "Mining and Energy Year Book, A New Paradigm Towards The Third Millenium", Department of Energy and Mineral Resources, 2000.
5. Anonim, "Statistic Indonesia, modified by Research and Development Centre for Mineral and Coal", Bulletin Statistik Komoditi Mineral Indonesia No. 29 Tahun 2002.
6. Anonim, "Mining Technology in Indonesia", Deparment of Mining and Energy, Directorate General of Mine, Mineral Technology Research and Development Centre, 1995.

7. Husaini, "Mineral Resources and Development in Indonesia", Paper presented in Workshop in Kerman, Iran, 2004.

8. Husaini and Soenara, "Modification of Natural Zeolite from Cikalong, West Java, by Hexa Decyl Trimethyl Ammonia (HDTMA) and Testing its Adsorption ability for Sulphate and Chrom Ions", Indonesia Zeolite Journal, Vol. 2, No. 1, November, 2003.

9. Husaini and Soenara, "The Upgrading of Kaolin from Semin, Yogyakarta, for Ceramic Raw Material", Proceeding of the Second Ceramic Nasional Seminar, Bandung, 2003.

10. Husaini, "Mineral Processing", Paper Presented in Class of Chemical Engineering Student, Achmad Yani University, Cimahi, 2003.

11. Komaruddin Atangsaputra, Cs, " Cooperation in Mining and Mineral Resource Development", Directorate General of Mines, Department of Mines and Energy Republic of Indonesia, 1993.

12. Ngurah Ardha, "Non Metallic Mineral Development in Indonesia", 2002.

13. Purnomo Yusgiantoro, "National Policy of Indonesian Mineral Resources Development", Buletin TekMIRA, No. 28, Year 11, Mei 2003.

14. Retno Wijayanti, "Present Situation of Mining Industry in Indonesia", 2002.

15. Soelistijo U. W., Cs., "Mining and Environmental Technology in Indonesia: Present and Future", "Indonesian Mining Journal", Vol. 5, No. 2, June 1999.

16. Wahyu, B.N., " Indonesian Mining Industry in the Period of Transition Beween 1997- 2001", 2002.

Appendix A
Industiral Mineral Resources in Indonesia (Year 2003)

Sl.No.	Commodity	Total Resources (tons)	Sl.No.	Commodity	Total Resources (tons)
1.	Ametise	86.68	26.	Quartz	3,194,789,600.00
2.	Andesite	55,786,882,946.00	27.	Clay	25,122,067,755.00
3.	Asbestos	20,000.00	28.	Magnecite	25,000.00
4.	Ball Bond Clay	157,421,000.00	29.	Marble	39,694,095,314.00
5.	Barite	677,000.00	30.	Mica	42,300,000.00
6.	Basalt	2,393,900,000.00	31.	Obsidian	70,670,756.00
7.	Pumice	1,027,201,013.00	32.	Oqer	39,926,500.00
8.	Limestone	269,657,888,880.00	33.	Onyx	70,265,000.00
9.	Ornamental Stones	100,996,250.00	34.	Opal	1.67
10.	Batusabak	16,375,000.00	35.	Sand	2,007,534,200.00
11.	Sulphur	1,951,393.00	36.	Quartz sand	16,339,777,251.00
12.	Bentonite	538,068,543.00	37.	Perlite	366,687,000.00
13.	Dasite	2,532,435,080.00	38.	Pirofilite	104,827,100.00
14.	Diatomite	369,564,701.00	39.	Rijang	27,395,500.00
15.	Diorite	946,653,200.00	40.	Serpentine	2,910,000.00
16.	Dolomite	1,793,737,112.00	41.	Gravel sand	3,620,644,931.00
17.	Feldspar	6,466,088,585.00	42.	Talc	3,096,700.00
18.	Phosphate	20,473,135.00	43.	Tanah urug	910,970,000.00
19.	PeridotiteGabbro	8,379,397,000.00	44.	Toseki	250,717,512.00
20.	Gypsum	7,233,326.30	45.	Trass	6,767,415,033.00
21.	Granite/Granodiorite	50,657,956,480.00	46.	Trackite	5,000,000.00
22.	Diamond	100,640.00	47.	Travertine	7,500,000.00
23.	Chalcedony	1,767,352.00	48.	Ultrabasa	386,100,000.00
24.	Calcite	90,200,500.00	49.	Zeolite	196,381,502.00
25.	Kaolin	597,263,856.00			

Appendix B
Industrial Mineral Statistic (1997–2002)
(Tonnes)

		1997	1998	1999	2000	2001	2002
Asbestos	Production						
	Consumption	39,040.39	18,709.97	23,988.65	24,459.72	27,426.18	30,500.00
	Export			197.75		4.01	14.49
	Import	41,845.12	21,079.04	30,250.01	54,891.06	49,885.08	45,148.54
Barite	Production						
	Consumption	75,721.38	60,063.88	43,482.09	42,459.32	42,453.33	42,500.00
	Export	4,030.59	3,161.20	1,370.14	2,708.40	7,511.64	992.77
	Import	79,742.21	62,938.38	44,827.83	45,212.27	72,699.32	52,615.48
Pumice	Production	31,000.00	24,500.00	78,000.00	118,000.00	185,600.00	500,000.00
	Consumption	3,363.23	21,824.62	39,384.71	62,020.82	76,244.93	80,000.00
	Export	28,131.08	3,188.40	39,671.00	58,756.14	110,019.56	411,023.93
	Import	528.94	591.40	894.63	2,692.51	656.24	398.30
Sulphur	Production	850,000.00	50,000.00	100,000.00	267,000.00	202,500.00	341,000.00
	Consumption	1,175,554.25	415,856.15	426,376.33	543,987.85	578,433.71	600,000.00
	Export	82.65	25.22	400.72	4,270.95	976.69	27,776.51
	Import	343,384.86	367,356.36	330,408.54	284,278.71	377,091.28	286,973.89
Bentonite	Production	108,500.00	117,500.00	155,500.00	231,000.00	225,000.00	270,000.00
	Consumption	107,404.40	108,251.15	128,607.52	193,031.14	196,928.23	200,00.00
	Export	20,257.36	18,614.09	41,650.99	63,083.28	62,834.77	114,502.29
	Import	19,337.68	9,288.40	14,784.89	25,004.90	35,513.77	43,882.59

Contd...

Appendix B–Contd...

		1997	1998	1999	2000	2001	2002
Diatomite	Production		500.00	1,900.00	3,500.00	3,900.00	3,300.00
	Consumption	247.75	120.50	128.88	128.47	365.12	400.00
	Export	646.73	2,804.59	4,348.73	6,724.93	7,355.96	6,097.05
	Import	4,231.19	2,476.74	2,610.06	3,404.67	3,907.15	3,186.54
Dolomite	Production	180,000.00	190,000.00	440,000.00	600,000.00	1,460,000.00	1,500,000.00
	Consumption	182,737.07	190,104.65	436,411.67	593,955.17	1,479,166.53	1,500,000.00
	Export	0.5	181.98	2,104.32	1,436.00	105.41	141.95
	Import	1,170.67	129.72	4,411.97	3,742.37	3,301.38	3,778.09
Feldspar	Production	127,000.00	37,500.00	210,300.00	280,500.00	119,500.00	131,200.00
	Consumption	207,682.25	129,818.17	310,248.51	393,751.63	239,973.84	250,000.00
	Export	2.22					21.97
	Import	80,513.94	92,373.61	100,000.00	105,271.35	120,516.61	118,819.35
Phosphate	Production	8,500.00	10,500.00	15,000.00	18,500.00	5,000.00	5,000.00
	Consumption	1,016,803.18	1,057,860.94	1,179,019.95	910,492.70	703,546.42	700,000.00
	Export	1,565.75	174.55	2,280.13	9,061.95	17,263.00	4,907.97
	Import	680,022.41	787,527.87	731,329.40	391,102.30	989,009.85	939,453.00
Gypsum	Production	2,110,000.00	555,500.00	542,000.00	810,000.00	406,500.00	728,000.00
	Consumption	2,623,857.22	1,137,278.84	1,046,440.33	1,353,925.50	1,442,459.25	1,500,000.00
	Export		373.33	3,540.82	1,012.36	36.38	51.55
	Import	513,603.56	562,126.55	507,791.34	546,721.24	1,036,124.34	772,130.39

Contd...

Appendix B–Contd...

		1997	1998	1999	2000	2001	2002
Calcium	Production	750,000.00	375,000.00	205,000.00	375,000.00	481,000.00	520,000.00
Carbonate	Consumption	746,394.36	372,091.34	202,200.13	365,328.14	480,950:74	500,000.00
	Export	1,457.00	2,900.00	4,240.65	10,271.06	142.70	21,822.37
	Import	1,037.66	535.60	260.14	428.12	171.79	476.48
Kaolineite	Production	205,000.00	164,000.00	204,500.00	211,000.00	337,000.00	312,000.00
	Consumption	223,332.64	207,771.14	176,812.99	240,686.57	343,520.63	350,000.00
	Export	126,462.03	90,975.29	128,922.52	114,137.89	88,242.34	82,851.45
	Import	144,963.93	134,909.67	101,231.81	144,107.85	94,988.15	120,752.80
Silica sand	Production	6,850,000.00	1,420,000.00	2,400,000.00	3,650,000.00	5,650,000.00	6,000,000.00
	Consumption	6,828,182.06	1,424,660,06	2,389,547.87	3,649,381.53	5,649,518.07	6,000,000.00
	Export	24,277.45	5,195.77	3,378.81	27,545.99	13,730.92	19,299.98
	Import	16,792.65	12,733.85	6,789.55	37,191.31	67,768.16	41,427.12
Talc	Production		40,000.00	257,000.00	245,500.00	158,000.00	166,200.00
	Consumption	55,335.34	91,149.26	305,243.46	326,768.02	220,642.11	220,000.00
	Export	21.58		19.63	111.29	3,246.79	3,078.42
	Import	88,156.07	50,920.55	48,382.50	81,486.11	65,851.64	56,912.75
Yodium	Production		140.00	170.00	1,800.00	2,500.00	1,500.00
	Consumption	65.59	136.5	156.25	933.09	1,151.07	1,200.00
	Export	33.46	10.00	37.09	940.29	1,427.50	75.41
	Import	170.91	18.2	25.16	34.13	24.80	57.47

Contd...

Appendix B–Contd...

		1997	1998	1999	2000	2001	2002
Zeolite	Production	90,000.00	55,000.00	30,000.00	30,000.00	57,500.00	60,000.00
	Consumption	89,487.20	54,520.78	26,834.59	28;319.91	57,474.49	60,000.00
	Export						
	Import						
Zircon	Production	3,800.00	5,800.00	4,300.00			
	Consumption	4,823.26	5,000.00	7,299.97	9,341.48	10,230.27	10,500.00
	Export	1,000.00	2,001.00			9.01	
	Import	2,065.56	1,169.82	3,009.77	12,184.39	18,030.56	16,075.38